Reaction Mechanisms in Environmental Engineering

Reaction Mechanisms in Environmental Engineering

Analysis and Prediction

James G. Speight
CD & W Inc., Laramie, Wyoming, United States

Butterworth-Heinemann is an imprint of Elsevier
The Boulevard, Langford Lane, Kidlington, Oxford OX5 1GB, United Kingdom
50 Hampshire Street, 5th Floor, Cambridge, MA 02139, United States

Notices
Knowledge and best practice in this field are constantly changing. As new research and experience broaden our understanding, changes in research methods, professional practices, or medical treatment may become necessary.

Practitioners and researchers must always rely on their own experience and knowledge in evaluating and using any information, methods, compounds, or experiments described herein. In using such information or methods they should be mindful of their own safety and the safety of others, including parties for whom they have a professional responsibility.

To the fullest extent of the law, neither the Publisher nor the authors, contributors, or editors, assume any liability for any injury and/or damage to persons or property as a matter of products liability, negligence or otherwise, or from any use or operation of any methods, products, instructions, or ideas contained in the material herein.

Library of Congress Cataloging-in-Publication Data
A catalog record for this book is available from the Library of Congress

British Library Cataloguing-in-Publication Data
A catalogue record for this book is available from the British Library

ISBN: 978-0-12-804422-3

For information on all Butterworth-Heinemann publications visit our website at https://www.elsevier.com//books-and-journals

Publisher: Matthew Dean
Acquisition Editor: Ken McCombs
Editorial Project Manager: Peter Jardim
Production Project Manager: Anitha Sivaraj
Designer: Miles Hitchen

Typeset by TNQ Technologies

Contents

Biography xi
Preface xiii

Part I
Introduction

1. Environmental Chemistry

1.	**Introduction**	3
2.	**Chemical Types**	9
	2.1 Inorganic Chemicals	10
	2.2 Organic Chemicals	11
3.	**Chemicals and the Environment**	18
4.	**Chemistry in the Environment**	20
5.	**Chemical Transformations**	23
	5.1 Chemistry in the Atmosphere	26
	5.2 Chemistry in the Aquasphere	29
	5.3 Chemistry in the Lithosphere	34
6.	**Physical Chemistry in the Environment**	39
	References	40
	Further Reading	41

2. Chemicals in the Environment

1.	**Introduction**	43
2.	**Sources**	48
	2.1 Air Pollution	49
	2.2 Water Pollution	55
	2.3 Land Pollution	61
3.	**Distribution in the Environment**	63
4.	**Chemistry in the Environment**	66
5.	**Industrial Chemicals and Household Chemicals**	67
	5.1 Industrial Chemicals	69
	5.2 Fly Ash, Bottom Ash, and Boiler Slag	70
	5.3 Household Chemicals	76
	References	77

3. Chemical and Physical Properties

1.	**Introduction**	81
2.	**Acids and Bases**	87
	2.1 Inorganic Acids and Bases	90
	2.2 Organic Acids and Bases	92
3.	**Acid–Base Chemistry**	94
	3.1 Reactions of Acids and Alkalis	99
	3.2 Brønsted–Lowry Acids and Bases	100
	3.3 Lewis Acids and Bases	100
4.	**Vapor Pressure and Volatility**	101
	4.1 Vapor Pressure	102
	4.2 Volatility	103
5.	**Water Solubility**	107
	5.1 Effect of Temperature	110
	5.2 Effect of Pressure	110
	5.3 Rate of Dissolution	110
	5.4 Quantification of Solubility	111
	5.5 Solubility of Inorganic Chemicals	111
	5.6 Solubility of Organic Chemicals	112
	References	113
	Further Reading	114

4. Mechanisms of Introduction Into the Environment

1.	**Introduction**	115
2.	**Minerals**	125
3.	**Types of Chemicals**	127
4.	**Release Into the Environment**	138
	4.1 Dispersion	142
	4.2 Dissolution	144
	4.3 Emulsification	145
	4.4 Evaporation	146
	4.5 Leaching	147
	4.6 Sedimentation and Adsorption	148
	4.7 Spreading	150
	4.8 Sublimation	150
5.	**Distribution in the Environment**	151
	References	161

Part II
Transformation Processes

5. Sorption, Dilution, and Dissolution

1.	**Introduction**	165
2.	**Sorption**	171
	2.1 Adsorption	172

2.2 Absorption 178
2.3 Effects of the Various Parameters on Adsorption–Adsorption 179
2.4 Ion Exchange 182
2.5 Desorption 183
2.6 Mechanism 183
3. **Dilution** 186
3.1 Dilution Capacity 187
3.2 Stratification 188
3.3 Mixing Zone 188
3.4 Flushing Time 189
3.5 Sediment Type 189
3.6 Biodilution 190
4. **Solubility** 190
5. **Vapor Pressure** 196
References 198

6. Hydrolysis

1. **Introduction** 203
2. **Nucleophilic Substitution Reactions** 205
2.1 S_N1 and S_N2 Reactions 207
2.2 Reactive Nucleophiles 209
2.3 Bond Strength and Anion Stability 210
2.4 Kinetic Order and Steric Effects 212
2.5 S_E1 and S_E2 Reactions 214
2.6 Effect of Substrate 215
3. **Hydrolysis Reactions** 215
3.1 Acid Hydrolysis 218
3.2 Alkaline Hydrolysis 221
3.3 Reaction Profiles 222
References 226
Further Reading 229

7. Redox Transformations

1. **Introduction** 231
2. **Oxidation Reactions** 239
2.1 Permanganate 241
2.2 Fenton Reagent 244
2.3 Persulfate 248
2.4 Ozone 250
3. **Reduction Reactions** 251
3.1 Zero Valent Metals 252
3.2 Iron Minerals 253
3.3 Polysulfides 253
3.4 Dithionite 253
3.5 Bimetallic Chemicals 254

4. **Photochemical and Photocatalytic Transformations** 255
4.1 Photochemistry 256
4.2 Photochemistry in the Environment 260
4.3 Photocatalysis in the Environment 263
References 265
Further Reading 267

8. Biological Transformations

1. **Introduction** 269
2. **Biotransformation** 271
2.1 Natural Methods 278
2.2 Traditional Methods 278
2.3 Enhanced Methods 279
2.4 Biostimulation and Bioaugmentation 280
2.5 In Situ and Ex Situ Methods 281
3. **Mechanisms and Methods** 282
3.1 Aerobic Biodegradation 283
3.2 Anaerobic Biodegradation 284
3.3 Microbe Selection 284
3.4 Extracellular Electron Transfer 285
3.5 Bioavailability and Transport of Pollutants 286
3.6 Crude Oil Biodegradation 286
3.7 Emerging Technologies 287
4. **Advantages and Disadvantages** 287
5. **The Future** 290
References 300
Further Reading 306

9. Molecular Interactions, Partitioning, and Thermodynamics

1. **Introduction** 307
2. **Molecular Interactions** 313
3. **Partitioning and Partition Coefficients** 318
3.1 Single Chemicals 319
3.2 Mixtures of Chemicals 326
3.3 Partition Coefficients 328
4. **Thermodynamics** 330
4.1 First Law of Thermodynamics 332
4.2 Second Law of Thermodynamics 332
4.3 Free Energy 334
References 335

10. Mechanisms of Transformation

1. **Introduction** 337
2. **Transformation in the Environment** 344

2.1 Chemical Transformation 345
2.2 Physical Transformation 363
2.3 Catalytic Transformation 369
3. Biological Transformation 371
3.1 Biodegradation 372
3.2 Bioremediation 374
4. Transport of Chemicals 375
4.1 Advection 378
4.2 Diffusion 379
4.3 Concentration 381
References 383
Further Reading 384

Part III
Conversion Tables and Glossary

Conversion Tables 387
Glossary 391

Index 431

Biography

Dr. James G. Speight has doctorate degrees in Chemistry, Geological Sciences, and Petroleum Engineering and is the author of more than 75 books in petroleum science, petroleum engineering, biomass and biofuels, and environmental sciences.

Dr. Speight has 50 years of experience in areas associated with (1) the properties, recovery, and refining of reservoir fluids, conventional petroleum, heavy oil, and tar sand bitumen, (2) the properties and refining of natural gas, gaseous fuels, (3) the production and properties of petrochemicals, and (4) the properties and refining of biomass, biofuels, biogas, and the generation of bioenergy. His work has also focused on safety issues, environmental effects, and remediation, as well as reactors associated with the production and use of fuels and biofuels. Although he has always worked in private industry which focused on contract-based work, he has served as Adjunct Professor in the Department of Chemical and Fuels Engineering at the University of Utah and in the Departments of Chemistry and Chemical and Petroleum Engineering at the University of Wyoming. In addition, he was a Visiting Professor in the College of Science, University of Mosul, Iraq, and has also been a Visiting Professor in Chemical Engineering at the following universities: University of Missouri–Columbia, the Technical University of Denmark, and the University of Trinidad and Tobago.

Dr. Speight was elected to the Russian Academy of Sciences in 1996 and awarded the Gold Medal of Honor that same year for outstanding contributions to the field of Petroleum Sciences.

He has also received the Scientists without Borders Medal of Honor of the Russian Academy of Sciences. In 2001, the Academy also awarded Dr. Speight the Einstein Medal for outstanding contributions and service in the field of Geological Sciences.

Preface

Pollution of the environment by any type of chemicals is a global issue, and toxic chemicals are found practically in all ecosystems because at the end of the various chemical life cycles the chemicals have either been recycled for further use or sent for disposal as waste. However, it is the inappropriate mismanagement of such waste (e.g., through haphazard and unregulated burning) that can cause negative impacts on the flora and fauna (including humans) of the environment.

Advanced technologies for the rapid, economical, and effective elimination of industrial and domestic chemical wastes have been developed and employed on a large scale and, in fact, advanced technologies for the control and monitoring of chemical pollutants on regional and global scales continue to be developed and implemented. Satellite-based instruments are able to detect, quantify, and monitor a wide range of chemical pollutants. In addition, an understanding of the fate and consequences of chemicals in the environment has increased dramatically and there are now available the means of predicting many of the environmental, ecological, and biochemical consequences of the inadvertent introduction of organic chemicals into the environment with much greater precision.

Chemicals are an essential component of life, but some chemicals can severely damage the floral (plant life) and faunal (animal life) environment. There is an increase in health problems that can be partially explained using chemicals, and many manmade chemicals are found in the most remote places in the environment. Specific groups of chemicals, such as biocides, pesticides, pharmaceuticals, and cosmetics, are covered by various pieces of legislation. In addition, the challenges posed by endocrine disruptors (i.e., chemicals that interfere with the hormone system causing adverse health effects) are also being addressed. However, in order to successfully manage the environment and protect the flora and fauna from such chemicals, a knowledge of chemical behavior is a decided advantage.

The discharge of chemicals into the environment and the fate of the chemical can take many forms. For example, there can be adsorption, dilution, dissolution, hydrolysis, oxidation–reduction, and biological transformation. In addition, acid–base reactions can cause partitioning of the chemical (especially when the discharge is a mixture), as well as neutralization of acids (by bases) or bases (by acids). In addition, decomposition reactions involve the decomposition or cleavage of one molecule to one or more product molecules and *displacement reactions* involve displacement of one or more cation or anion between two molecules. All of these reactions are subject to the laws of chemistry and the products can take many forms.

It is not surprising, therefore, that there is no one remediation process that can be claimed as the remedy for application to all forms of chemical discharge. Given this conclusion, it must also be concluded that the remediation must meet and defeat the chemistry of the pollution and be able to sever any chemical bonds or physical arrangements that might exist between the chemical and the relevant parts of an ecosystem. This is where the knowledge of the interaction(s) between the discharged chemical and the ecosystem will be invaluable.

Thus, the intent of this book is to focus on the various chemical issues that are at the core of any environmental remediation and explains the chemical and the physical methods by which chemicals reside in the environment. Remediation is a term that is used often as it relates to the cleanup of chemicals from the environment. The properties of the chemicals and the properties of the site offer many variations of interactions. However, there is no one method of remediation that will be sufficient to clean up all sites. Thus, this text relates to an introduction to the various reactions that occur when a chemical is released to the environment and the effects of the reaction products on the environment. Once this is understood, plans can be made to remediate a contaminated site.

The book will serve as an information source to the engineers in presenting details of the various aspects of inorganic and organic chemicals, as they pertain to pollution of the environment. To accomplish this goal, the book focuses on the various aspects of environmental science and engineering and the potential chemical reactions that can occur once a chemical is released into the environment and during environmental remediation.

Dr. James G. Speight
Laramie, Wyoming
April 2018

Part I

Introduction

Chapter 1

Environmental Chemistry

1. INTRODUCTION

Environmental chemistry is the study of the chemical and biochemical phenomena that occur in natural places and is the study of the sources, reactions, transport, effects, and fates of chemical species in the air, water, and soil environments, as well as the effect of human activity and biological activity on these (Tinsley, 2004). More specifically, environmental chemistry is the study of chemical processes occurring in the environment which are impacted by humankind's activities. These impacts may be felt on a local scale, through the presence of urban air pollutants or toxic substances arising from a chemical waste site, or on a global scale, through depletion of stratospheric ozone or global warming.

Furthermore, the study of the environment is an interdisciplinary subject that integrates physical, chemical, and biological sciences; some of the fields of interest are: (1) chemistry, which includes: constitution of environmental matter (air, water, soil, and selected chemicals), materials, and energy balances, (2) biology, which includes: microbiology, botany, zoology, sociology, and biodiversity, and (3) physics, which includes: meteorology, climatology, hydrology, oceanography, and the oceans—atmosphere system. The focus of this book is the category dealing with chemistry, particularly the reaction mechanisms of chemical pollutants with the environment and the various aspects of chemical remediation, as well as the interactions of any chemical used in the remediation process.

Almost any chemical from natural sources or anthropogenic sources can pollute the environment. However, it is the synthetic and other industrial chemicals that are emphasized here. Whether the chemical is present in a small amount or present in the ecosystem in large amount (the effective amount of the chemical depending upon the chemical), the potential for pollution is real (Table 1.1) (Hill, 2010). The presence of a chemical in a large amount causes acute toxicity in the form of having an adverse effect on the flora and fauna of an ecosystem. On the other hand, the presence of a chemical in a small amount can give rise to chronic effects, in which the flora and fauna will suffer when exposed to the long-term exposure to very low concentrations of a substance.

Reaction Mechanisms in Environmental Engineering. https://doi.org/10.1016/B978-0-12-804422-3.00001-8

TABLE 1.1 General Categories of Types of Pollutants

Category	Examples
Organic chemicals	Polychlorinated biphenyls (PCBs), oil, many pesticides
Inorganic chemicals	Salts, nitrate, metals and their salts
Organometallic chemicals	Methylmercury, tributyltin, tetraethyl lead
Acids[a]	Sulfuric, nitric, hydrochloric, acetic
Physical	Eroded soil, trash
Radiological	Radon, radium, uranium
Biological	Microorganisms, pollen

[a]Acids, as well as physical and radioactive pollutants, can be either organic or inorganic— sulfuric acid is inorganic, acetic acid (found in vinegar) is organic. Biological pollutants are mostly organic chemicals, but often contain inorganic chemicals.

Furthermore, pollution by chemicals is a definitive case of habitat destruction (Miller, 1984; Speight, 2017a,b) and involves (predominantly) chemical destruction rather than the more obvious physical destruction. The overriding theme of the definition is the ability (or inability) of the environment to absorb and adapt to changes brought about by human activities. Thus, environmental pollution occurs when the environment is unable to accept, process, and neutralize any harmful byproduct of human activities (such as the gases that contribute to acid rain).

Briefly, in the context of this book, a pollutant is a (1) nonindigenous chemical that is present in an ecosystem or (2) an indigenous chemical that is present in an ecosystem in greater than the natural concentration. Both types of pollutants are the result of human activity and have an overall detrimental effect upon the ecosystem or upon floral (plants) or faunal (animals, including humans) species in an ecosystem. In addition, the term ecosystem represents an assembly of mutually interacting organisms and their environment in which materials are interchanged in a largely cyclical manner and is, essentially, a community of organisms together with their physical environment, which can be viewed as a system of interacting and interdependent relationships. This can also include processes such as the cycling of chemical elements (such as heavy metals) and chemical compounds through the floral and faunal components of the system.

A contaminant is a chemical that is present in nature at a level higher than fixed levels or that would not otherwise be there. This may be due to human activity and bioactivity. The term contaminant is often used interchangeably with pollutant, which is a substance that has a detrimental impact on the surrounding environment. Finally, the word *waste* differs in meaning from the definition of a pollutant, although a waste can be a pollutant too. In the simplest sense, *waste* refers to material such as garbage, trash, construction debris, insofar as waste is typically composed of materials that have reached the end of their useful life. By way of further definition, a chemical waste is a broad term and encompasses many types of materials and can be a solid, liquid, or gaseous chemical (or collection of chemicals) material that displays properties and behavior that are detrimental to an ecosystem. Typically, a chemical waste is classified as hazardous based on four characteristics, which are: (1) ignitability, (2) corrosivity, (3) reactivity, and (4) toxicity. This type of waste (typically classed as hazardous waste) must be categorized as to its identity, constituents, and hazards, so that it may be safely handled and managed.

While a contaminant is sometimes defined as a chemical present in the environment because of human (anthropogenic) activity, but without harmful effects, it is sometimes the case that toxic or harmful effects from contamination only become apparent from a derivative of the original chemical that is caused by transformation after introduction to the environment. Also, the *medium* (such as the air, water, or soil) affected by the pollutant or contaminant is the *receptor*, while a *sink* is a chemical medium or species that retains and interacts with the pollutant.

Environmental chemistry (Schwarzenbach et al., 2003; Manahan, 2010; Baird and Cann, 2012) is a relatively young science but interest in this subject has grown over the last six decades and there is to be increasing interest in the development of environmental topics which are based on chemistry. Thus, one of the first objectives of environmental chemistry is the study of the environment and of the natural chemical processes which occur in the environment or in the various ecosystems. Environmental chemistry is concerned with reactions in the environment and involves a study of the distribution and equilibria (i.e., the reactions, pathways, thermodynamics, and kinetics) between components of an ecosystem.

Environmental chemistry focuses on environmental concerns about materials, energy, as well as production cycles, and demonstrates how fundamental chemical principles and methodologies can protect the floral (plant) and faunal (animal, including human) species within the environment (Speight, 2017a,b). More specifically, the principles of chemistry can be used to develop how global sustainability can be supported and maintained. For this, the environmental chemists and engineers of the future must acquire the scientific and technical knowledge to design products and chemical processes. They must also acquire an increased awareness of the impact of chemicals on the environment, as well as develop an enhanced awareness of the importance of sustainable strategies in chemical research and the industries that produce chemical wastes, as well as the need for the managed disposal and/or destruction of these wastes (Table 1.2) (Chenier, 1992). Such work must include the necessary aspects of chemistry, chemical engineering, microbiology, and hydrology as they can be applied to

TABLE 1.2 Sources and Types of Chemical Waste

Source	Waste Type
Chemical manufacturers	Strong acids and bases
	Spent solvents
	Reactive materials
Vehicle maintenance shops	Heavy-metal paints
	Ignitable materials
	Used lead-acid batteries
	Spent solvents
Printing industry	Heavy-metal solutions
	Waste ink
	Spent solvents
	Spent electroplating wastes
	Ink sludge containing heavy metals
Leather products	Waste toluene and benzene
Paper industry	Paint wastes containing heavy metals
Construction industry	Ignitable paint wastes
	Spent solvents
	Strong acids and bases
Cleaning agents and cosmetics manufacturing	Heavy-metal dusts
	Ignitable materials
	Flammable solvents
	Strong acids and bases
Furniture and wood manufacturing and refinishing	Ignitable materials
	Spent solvents
Metal manufacturing	Paint wastes containing heavy metals
	Strong acids and bases
	Cyanide wastes
	Sludge containing heavy metals

solve environmental problems (Pickering and Owen, 1994; Schwarzenbach et al., 2003; Tinsley, 2004).

This book relates to an introduction to the chemical interactions caused by the planned and unplanned effects of chemicals on the environment. The focus is on developing a fundamental understanding of the nature of these chemical processes, so that human activities can be accurately evaluated. Thus, an important purpose of this book is to aid understanding of the basic distribution and chemical reaction processes which occur in the environment. The book also presents a view of various aspects of the chemistry of the environment and the physical and chemical processes that affect the mechanisms of the chemical reactions that occur in the environment. Thus, throughout the pages of this book, the reader will be presented with explanations of the behavior of inorganic and organic chemicals after release into the environment and the potential hazards than can occur during remediation activities (Speight and Arjoon, 2012). In this way, the book will assist the chemist and the engineer gain an understanding of the behavior of chemicals in the environment.

By way of introduction, a general classification of chemical pollutants is based on the chemical structure of the pollutant and includes (1) pollution by inorganic chemicals and (2) pollution by organic chemicals (Speight, 2017a,b); by way of further introduction, substances of mineral origin (such as ceramics, metals, synthetic plastics, as well as water) are inorganic chemicals, as opposed to those of biological or botanical origin (such as crude oil, coal, wood, as well as food), which are organic chemicals. In addition, minerals that occur in the various ecosystems are the inorganic, crystalline chemicals that are constituents of the various rocks.

Contamination of the environment by any type of chemical is a global issue and toxic chemicals are found in almost all ecosystems (some observers would omit the word "almost") (Tables 1.3 and 1.4) because at the end of the various chemical life cycles the chemicals have either been recycled for further use or sent for disposal as waste. In fact, human life underwent significant changes in the 18th century with the commencement of the Industrial Revolution, when huge amounts of useful energy could be produced with heat engines, greatly expanding transport capabilities, manufacturing, and household appliances. But the associated problem is that there followed an unsustainable path in energy utilization, with >90% of the primary energy sources being nonrenewable (fossil fuels and nuclear fuels), and with the accompanying major disadvantages (from sudden accidents to progressive poisoning).

However, it is the inappropriate mismanagement of such waste (e.g., through haphazard and unregulated burning) that can cause negative impacts on the flora (plant life) and fauna species (animal life, including humans) in the environment (Speight and Lee, 2000; Speight, 2005, 2017a,b, 2011a, 2014; Lee et al., 2014; Speight and Singh, 2014; Speight and Islam, 2016).

TABLE 1.3 The Various Ecosystems Associated With the Environment

Air, the atmosphere, the tightest life-supporting media (we die after a few minutes without). We need fresh air to breathe and air provides the oxidizer (O_2) used in metabolism, and we also make use of air as a cooling medium (temperature conditioning and heat sink). Besides air availability and composition, air temperature and all other meteorological phenomena, ordinary (like rain and wind) and extraordinary (like draughts and storms), have a strong influence on people's way of living.
Water, the hydrosphere, sometimes including solidified water (the cryosphere), mostly used as a solvent and carrier for matter and energy transport inside our body. Water has a deeper role for us because living matter is basically water (an aqueous solution with some macromolecules, enclosed in permeable membranes formed by macromolecules too), and, focusing on the thermodynamic aspects, water is the only substance present in its three phases (solid, liquid, and gas) in our environment. The water cycle is the main controller of matter and energy flows on earth, providing plentiful of distilled water for direct human use and plant growth, and controlling earth's radiation budget.
Land, the lithosphere (or its most external part, the crust, or even the most superficial layer, the soil or earth), so fundamental to land animals like us, which feed mainly from land flora and land fauna. From land we take most of our raw materials and energy, including the food, to act as reducer in our metabolic energy release, and as building matter in our growth.

TABLE 1.4 Environmental Impact on Humans

Air revitalization. Until the 20th century, the only need was getting rid of smoke and odoriferous materials, and the simple solution was ventilation (i.e., allowing a fresh air flow) or going away. With the advent of submarine, aircraft, and spacecraft vehicles, another more basic need arises: procuring fresh air (either from the atmosphere, or by in situ generation), and maintaining an appropriate gas pressure.
Water access and purification. Until the ocean sailing bloom in the 16th century, water supply was based on directly catching water from natural stores (rivers, lakes, or wells), and disposing the water waste through the soil (by natural infiltration), or down the supply river, or far in the same lake. On ocean vessels, seawater was made drinkable by distillation. In the 19th century sewage systems proliferated (it was not until the end of the 20th century that sewage treatment became the rule).
Fertile soils. Soil is a mixture of mineral and organic materials mainly in solid state, but with gases and liquids dispersed and dissolved. On a volume basis, a good quality soil may have 45% minerals, 25% water, 25% air, and 5% organic material (both live and dead). Fertile soils are rich in nutrients necessary for basic plant nutrition, including nitrogen, phosphorus and potassium. Soil depletion occurs when the components which contribute to fertility are removed and not replaced.

While chemicals are recognized as an essential component of life, it is also recognized that some chemicals can severely damage the floral and the faunal life (in an ecosystem) in the environment. Specific groups of chemicals, such as biocides, pesticides, pharmaceuticals, and cosmetics are covered by various legislative acts. In addition, the challenges posed by endocrine disruptors (i.e., chemicals that interfere with the hormone system causing adverse health effects) are also being addressed. In all cases, to successfully manage the environment and protect the flora and fauna from such chemicals, knowledge of chemicals is absolutely necessary.

Thus, the chief reason for studying this subject is not only the effects of chemicals on the environment but also on human health, which may be caused by unforeseen side effects of a chemical substance during its production, transport, use, and disposal processes.

It is only in the case of very rare incidents that pollutants seldom stay at the point of release (Chapter 4). Pollutants move, or are transported, among air, water, soil, and sediments. They often move transboundary: across state and national boundaries traveling with air or water currents; sometimes, through biotransport, in which pollutants are carried in body tissues of migrating animals. After or during transport, a pollutant can be transformed into end products different than the chemical form in which it was initially emitted (Chapter 10). It may be transformed into chemicals that are no longer pollutants as when biological matter is broken down (transformed) by microorganisms and incorporated into normal biological material within these organisms. On the other hand, a pollutant can take years, even decades, to be transformed into harmless products (or into harmful products). Furthermore, the process leading to the final fate can be complex (Chapters 5–10). These effects provide the motivation for the assembly of databases of scientific and engineering knowledge that document the effects of chemicals on the floral and faunal environments.

At this point a general description of the types of chemicals that exist is warranted to enhance the understanding of the remainder of the book.

2. CHEMICAL TYPES

Ideally, scientists and engineers should be able to predict the possible effects of a chemical directly on the environment even before the chemical substance is released and enable a more realistic appraisal of the effects of the chemical. Indeed, a first approximation to predicting a potentially harmful chemical in an ecosystem involves the following criteria as they relate to the flora and fauna: (1) whether or not the chemical is biologically essential or nonessential, (2) the toxicity of the chemical in small amounts or in larger amounts, (3) the potential for the chemical to form stable, inert, and nontoxic compounds in the environment, (4) the persistence of the chemical, either in the original form or in a changed form, in the environment, and (5) the potential mobility of the chemical, in the original form or in a changed form, in the environment and the influence of this mobility on any of the essential biogeochemical cycles.

Thus, like any technical discipline, the chemist and the engineer are faced with understanding many aspects of the behavior of inorganic and organic chemicals in their many forms, as well as the way these chemicals can form from other chemicals and the means by which they react with each other. Chemistry can be used to study the molecular size and the structure of the smallest of ions that exist in the floral and faunal environments to the much large-scale workings of the core of the earth. An understanding of basic chemistry brings with it the realization that chemicals released to the atmosphere by human activities can have serious health effects not only from the original chemicals but also from the changes that occur to these chemicals once released and the behavior of the form of these chemicals during the various forms of remediation (Speight and Arjoon, 2012). And this is where an understanding of chemistry can play an important role in dealing with the various issues of chemicals in the environment.

As an aid to understanding the properties and behavior of chemicals in the environment, it is necessary to understand that chemicals are divided into two major subcategories: (1) inorganic chemicals and (2) organic chemicals, from which various subcategories can be derived.

2.1 Inorganic Chemicals

An inorganic chemical is a chemical that does not contain carbon with the notable exceptions of the carbonate-type chemicals that contain carbon and yet are considered inorganic. These include sodium bicarbonate ($NaHCO_3$, baking soda) and sodium carbonate (Na_2CO_3, washing soda). Inorganic chemicals may contain almost any element in the periodic table—such as nitrogen, sulfur, lead, and arsenic. Many inorganic chemicals are found naturally—salts in the ocean, minerals in the soil, the silicate skeleton made by a diatom, or the calcium carbonate skeleton made by a coral. Inorganic chemicals are also often manufactured by humans. Ammonia (NH_3) used as a nitrogen fertilizer is a major example. Simple inorganic chemicals can be manipulated to make more complicated ones.

Inorganic chemistry focuses on the classification of inorganic compounds based on the properties of the compound(s). With certain exceptions, inorganic chemicals do not contain carbon or its compounds—the carbonate ($-CO_3$) minerals are notable exceptions—and, thus do not contain carbon chemically bound to hydrogen (hydrocarbons) or any of their derivatives that contain elements such as nitrogen, oxygen, sulfur, and metals. On occasion there will be observations by some scientists that no clear line divides organic and inorganic chemistry. This is untrue and such statements should be treated with the utmost caution (even with a high degree of skepticism). However, organometallic compounds that have an organic moiety in the molecule may be considered as hybrid compounds. More specifically, the classification of inorganic chemicals focuses on the position in the periodic table (Fig. 1.1) of the heaviest element (the element with the highest atomic weight) in the compound, partly by grouping compounds based on structural similarities. Also, inorganic compounds generally involve ionic bonds and examples of inorganic compounds include sodium chloride (Na^+Cl^-, written as NaCl) and calcium carbonate ($Ca^{2+}CO_3^{2-}$, written as $CaCO_3$) and pure elements, such as iron (Fe).

2.2 Organic Chemicals

On the other hand, organic compounds are classified according to the presence of functional groups in the molecule (Tables 1.5–1.8). A functional group (Table 1.9) is a molecular moiety that typically dictates the behavior (reactivity) of the organic compound in the environment and the reactivity of that functional group is assumed to be the same in a variety of molecules, within some limits and if steric effects (that arise from the three-dimensional structure of the molecule) do not interfere. Thus, most organic functional groups feature heteroatoms (atoms other than carbon and hydrogen, such as nitrogen, oxygen, and sulfur). Functional groups are a major concept in organic chemistry, both to classify the structure of organic compounds and to predict their physical and chemical properties, especially as these properties are exhibited in the environment (Table 1.10) (Speight, 2017a).

For example, when comparing the properties of ethane (CH_3CH_3) with the properties of propionic acid ($CH_3CH_2CO_2H$), which is a chemical that is formed due to the replacement of a hydrogen atom in the ethane molecule by a

Group→ ↓Period	1	2		3	4	5	6	7	8	9	10	11	12	13	14	15	16	17	18
1	1 H																		2 He
2	3 Li	4 Be												5 B	6 C	7 N	8 O	9 F	10 Ne
3	11 Na	12 Mg												13 Al	14 Si	15 P	16 S	17 Cl	18 Ar
4	19 K	20 Ca		21 Sc	22 Ti	23 V	24 Cr	25 Mn	26 Fe	27 Co	28 Ni	29 Cu	30 Zn	31 Ga	32 Ge	33 As	34 Se	35 Br	36 Kr
5	37 Rb	38 Sr		39 Y	40 Zr	41 Nb	42 Mo	43 Tc	44 Ru	45 Rh	46 Pd	47 Ag	48 Cd	49 In	50 Sn	51 Sb	52 Te	53 I	54 Xe
6	55 Cs	56 Ba	*	71 Lu	72 Hf	73 Ta	74 W	75 Re	76 Os	77 Ir	78 Pt	79 Au	80 Hg	81 Tl	82 Pb	83 Bi	84 Po	85 At	86 Rn
7	87 Fr	88 Ra	**	103 Lr	104 Rf	105 Db	106 Sg	107 Bh	108 Hs	109 Mt	110 Ds	111 Rg	112 Cn	113 Nh	114 Fl	115 Mc	116 Lv	117 Ts	118 Og
			*	57 La	58 Ce	59 Pr	60 Nd	61 Pm	62 Sm	63 Eu	64 Gd	65 Tb	66 Dy	67 Ho	68 Er	69 Tm	70 Yb		
			**	89 Ac	90 Th	91 Pa	92 U	93 Np	94 Pu	95 Am	96 Cm	97 Bk	98 Cf	99 Es	100 Fm	101 Md	102 No		

FIGURE 1.1 Periodic table of the elements showing the groups and periods including the lanthanide elements and the actinide elements.

TABLE 1.5 General Classes of Hydrocarbons

Chemical Class	Group	Formula	Structural Formulae
Alkane	Alkyl	$R(CH_2)_nH$	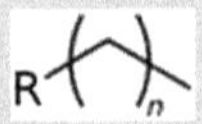
Alkene	Alkenyl	$R_2C{=}CR_2$	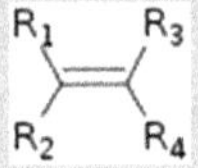
Alkyne	Alkynyl	$R_1C{\equiv}CR_2$	$R_1C{\equiv}CR_2$
Benzene derivative	Phenyl	RC_6H_5	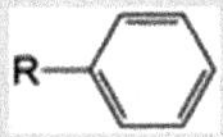

carboxylic acid functional group (CO_2H), the change in properties and behavior is spectacular. Alternatively, the replacement of a methyl group (CH_3) into the ethane molecule by the carboxylic acid function to produce acetic acid (CH_3CO_2H) [or the replacement of a hydrogen in the methane molecule (CH_4) by the carboxylic acid function] produces significant changes in the properties of the product vis-à-vis the original molecule.

Thus, organic chemicals range from very simple compounds such as methane (CH_4) to organic chemicals that contain more than one carbon atom, as many as 10 carbon atoms to chemicals that contain hundreds or more carbon atoms that are linked in carbon–carbon bonds. Those that contain only carbon and hydrogen are called hydrocarbons; a simple example is $H_3C(CH_2)_3CH_3$ (pentane). Organic chemicals commonly contain other elements too, such as oxygen, nitrogen, or sulfur. An organometallic chemical has a carbon atom bonded to a metal as in tetraethyl lead. Some organometallic chemicals are found naturally.

Many organic chemicals are synthetic, that is, produced not by living creatures, but manufactured by human beings. However, the feedstocks from which the chemicals are made come from nature. Commonly synthetic organic chemicals are made from petroleum or natural gas feedstocks, which are referred to as petrochemicals. Coal or wood also sometimes serve as feedstocks for organic chemicals. Plastics are synthetic, organic chemicals and the so-called bioplastics, which humans produce from plant materials, involve some synthetic chemistry. Some commercial organic chemicals are produced

TABLE 1.6 General Classes of Oxygen Compounds

Chemical Class	Group	Formula	Structural Formula
Alcohol	Hydroxyl	ROH	R—O—H
Ketone	Carbonyl	RCOR′	R—C(=O)—R'
Aldehyde	Aldehyde	RCHO	R—C(=O)—H
Acyl halide	Haloformyl	RCOX	R—C(=O)—X
Carbonate	Carbonate ester	ROCOOR	R_1—O—C(=O)—O—R_2
Carboxylate	Carboxylate	$RCOO^-$	R—C(=O)—$O^{\ominus}$
Carboxylic acid	Carboxyl	RCOOH	R—C(=O)—OH
Ester	Ester	RCOOR′	R—C(=O)—OR'
Methoxy	Methoxy	$ROCH_3$	R—O—CH_3

TABLE 1.7 General Classes of Nitrogen Compounds

Class	Group	Formula	Structural Formula
Amide	Carboxamide	$RCONR_2$	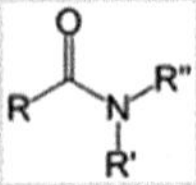
Amines	Primary amine	RNH_2	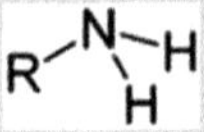
	Secondary amine	R_2NH	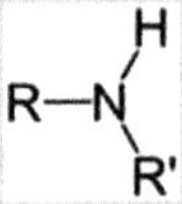
	Tertiary amine	R_3N	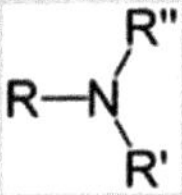
	4 degrees ammonium ion	R_4N^+	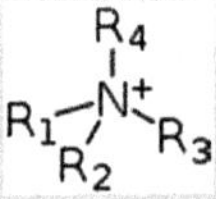
Imine	Primary ketimine	$RC(=NH)R'$	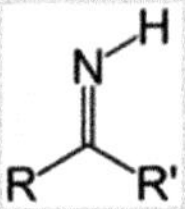
	Secondary ketimine	$RC(=NR)R'$	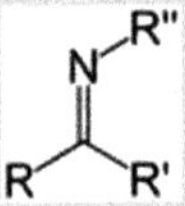
	Primary aldimine	$RC(=NH)H$	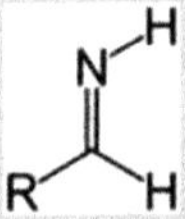
	Secondary aldimine	$RC(=NR')H$	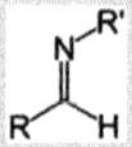

TABLE 1.7 General Classes of Nitrogen Compounds—cont'd

Class	Group	Formula	Structural Formula
Imide	Imide	$(RCO)_2NR'$	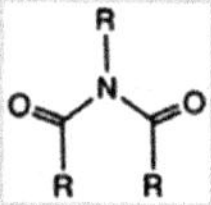
Azide	Azide	RN_3	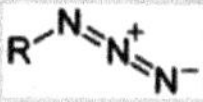
Azo compound	Azo (Di-imide)	RN_2R'	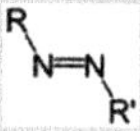
Cyanates	Cyanate	ROCN	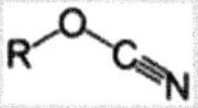
	Isocyanate	RNCO	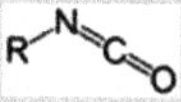
Nitrate	Nitrate	$RONO_2$	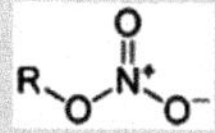
Nitrile	Nitrile	RCN	$RC\equiv N$
	Isonitrile	RNC	$RN^{+}\equiv C$
Nitrite	Nitroso-oxy	RONO	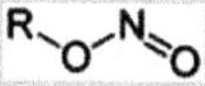
Nitro compound	Nitro	RNO_2	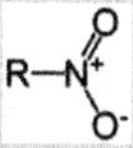

TABLE 1.8 General Classes of Sulfur Compounds

Chemical Class	Group	Formula	Structural Formula
Thiol	Sulfhydryl	RSH	R–S–H
Sulfide (Thioether)	Sulfide	RSR'	R–S–R'
Disulfide	Disulfide	$RSSR'$	R–S–S–R'
Sulfoxide	Sulfinyl	$RSOR'$	R–S(=O)–R'
Sulfone	Sulfonyl	RSO_2R'	R–S(=O)(=O)–R'
Sulfinic acid	Sulfino	RSO_2H	R–S(=O)–OH
Sulfonic acid	Sulfo	RSO_3H	R–S(=O)(=O)–OH
Thiocyanate	Thiocyanate	RSCN	R–S–C≡N
	Isothiocyanate	RNCS	R–N=C=S
Thione	Carbonothioyl	$RCSR'$	R–C(=S)–R'
Thial	Carbonothioyl	RCSH	R–C(=S)–H

TABLE 1.9 Potential Reactions of Organic Functional Groups in the Environment

Functional Group	Interaction
Carboxylic acid, $-COOH$	Ion exchange, complexation
Alcohol, phenol, $-OH$	Hydrogen bonding, complexation
Carbonyl, $>C{=}O$	Reduction–oxidation
Hydrocarbon, $[-CH_2-]_n$	Hydrophobic

TABLE 1.10 General Characteristics of Soil

Property	Characteristic	Comments
Composition	Porous Body	Surface horizons have approximately equal volumes of solids and pores containing water or air
	Weathering	Weatherable minerals decompose and more resistant minerals form
	Transported chemicals	Small particles may be transported by water, wind, or gravity
Physical properties	Texture	Surface area per unit volume and particle size; influence on properties and reactivity
	Structure	Influence on porosity and large and small pore ratio; affects adsorption-absorption
	Consistency	Physical behavior of soil changes as moisture content changes
Chemical properties	Cation Exchange	Ability of soil clay minerals to reversibly adsorb and exchange cations
	pH	A measure of the acidity or alkalinity; can change soil chemistry
Environment	Water	Can change soil chemistry; hydrolysis
	Temperature	Can change soil chemistry; decomposition
	Aeration	Can change soil chemistry; redox reactions

too by cultures of molds or bacteria; such chemicals must then be purified from these cultures by human actions.

A biochemical is an organic chemical synthesized by a living creature. Proteins, fats, and carbohydrates are biochemicals. Sucrose (table sugar) and the acetic acid (CH_3CO_2H) (in vinegar) are examples of simple biochemicals. Many biochemicals can also be made synthetically; not only simple chemicals such as vinegar or the sugars, sucrose and xylose, but also complex ones. If the structure of a chemical made by synthetic means is exactly the same as that found in nature, it is indeed the same chemical—the body treats both in exactly the same way, so there is no biological difference between them. Chemicals synthesized by living creatures can also be extensively manipulated during extraction and purification and still legally be called natural.

3. CHEMICALS AND THE ENVIRONMENT

The use of chemicals for domestic and commercial purposes was initiated in the late 18th century with the onset of the Industrial Revolution and increased phenomenally during the 19th century and has continued during the 20th and 21st centuries. In the early days of the release of chemicals into the environment, either because of the potential benefits of such releases or because of the unmanaged disposal of the chemicals, their negative impacts on human health and safety, as well as on the integrity of terrestrial and marine ecosystems and on air and water quality, became obvious. In fact, the lack of definitive plans to manage the use of chemicals has threatened (and continues to threaten) the sustainability of the environment. Whatever the chemical, there are risks to its use—known and unknown—and some chemicals, including heavy metals, persistent organic pollutants, and polychlorinated biphenyls present risks to the environment that have been known for decades. In addition, there has been the release of chemicals into the environment, many of which are long-lived (environmentally stable), but can transform into byproducts whose behavior, synergies, and impacts on the environment may be even more drastic than the original chemical and the effects cannot always be identified on the basis of property data alone (Jones and De Voogt, 1999; Mackay et al., 2006; Speight, 2017c). For example, property data (without any accompanying knowledge of chemical behavior) does not always define the interaction of the chemical with any of the environmental constituents.

Nevertheless, chemicals are a significant contributor to the human lifestyle (Speight, 2017a,b) and as long as there is sound chemical management across the lifecycle of a chemical—from extraction or production to disposal—it is possible (under current legislative guidelines) and essential to avoid risks to the floral and faunal environments. As always, there are always two sides to the statement: chemicals are a blessing but also can be a curse. There are benefits to the use of chemicals, but they must be treated with a sound basis of knowledge so that harmful impact from exposure of the environment to the inorganic and organic chemicals can be mitigated.

In fact, any organic chemical while being considered to be necessary for life can also cause harm. Understanding chemistry, perhaps not to the extent of the dyed-in-the-wool inorganic chemist or organic chemist, is a part of understanding the use and effect of chemicals. To many nonchemists, chemicals are an unknown danger and (often without justification) the general perception is that all chemicals are dangerous and use of chemicals should be avoided and there is the necessity to treat any chemical with respect and caution. Water too is a chemical that is essential to life but when allowed to envelop and submerge land-based floral and faunal organisms will cause irreparable harm (i.e., death).

Some chemicals can be notoriously hazardous and should always be handled with care, as evidenced by the advisory (warning) statements on the packaging of the various chemicals, which are presented for the sake of safety. The risk faced from exposure to a chemical is based on the perceived (or real) danger of the chemical multiplied by the exposure to the chemical both in terms of the amount of the chemical and the time of exposure. A simple example is the chemical curare (an alkaloid—a nitrogen containing natural product), which is a common name for various plant extracts which are used as arrow-tip poisons (often fatal) that originated in Central America and South America. On the other hand, it has also been used as a muscle relaxant (in extremely small dosages) but not without some risk to the patient. Nevertheless, it has been possible for the medical community to adjust the dosage from a death-dealing quantity (on an arrowhead) to a medicinal quantity under strict supervision (EB, 2015).

Any chemical, and its various derivatives, may have found wide use in many sectors of the modern world including the chemical industry, the fossil fuel industry, agriculture, mining, water purification, and public health. However, not only the dedicated use of chemicals but also the production, storage, transportation, and removal of these substances can pose risks to the environment, if specific handling protocols are not followed. Developing an effective management system for the use of chemicals requires addressing the specific challenges that arise because of the individual chemicals and chemical mixtures because the irregular management of obsolete organic chemicals and chemical mixtures, stockpiles, and waste presents serious threats to the environment. As the use of chemicals and chemical production increase, the management of chemicals, which already has limited resources and capacity, will be further constrained and may fail if not regulated (Speight and Singh, 2014).

Along with the increased use of chemicals, there has been the realization that many widely used organic compounds are more toxic to the environment than was previously suspected. Some are carcinogenic while others may contribute to the destruction of the ozone layer in the upper atmosphere—the ozone layer protects all life from the strong ultraviolet radiation from the sun—while other organic chemicals are concentrated and persist in living tissue, often with an, as yet, unknown effect. Nonetheless, the modern world has adapted to the use of synthetic chemicals, and there are continuing debates that crude oil, the largest source of organic chemicals—while in good supply at the present time—may be in short supply in the next 50 to 100 years, and there will be the need to rely on alternate sources of energy, which are not immune from causing damage to the environment (Speight, 2011a, 2014; Lee et al., 2014; Speight and Islam, 2016).

Thus, in order to develop an effective management system that protects the environment from organic chemicals, there is the need to recognize that the modern world relies on both natural and synthetic chemicals which can be tailored to serve such specific purposes. In fact, the gasification of coal (or, for that matter, the gasification of other carbonaceous materials as biomass) to produce synthesis gas (a mixture of carbon monoxide and hydrogen) is an established process from which a variety of organic chemicals can be synthesized (Davis and Occelli, 2010; Chadeesingh, 2011; Speight, 2011b, 2013; Speight and Singh, 2014).

There are two types of codes related to the use/misuse of chemicals and their subsequent disposal: (1) enforceable codes of conduct and (2) aspirational codes of conduct. An *enforceable code* of conduct deals with the necessary protocols for regulation and enforcement of the code while an *aspirational code* of conduct presents the ideals of performance so that those bound to the code may be reminded of their obligations to perform ethically and responsibly. Nevertheless, there are many observers who are in serious doubt about the practical effectiveness of such codes, which may even prescribe ambiguous (and often unattainable) ideals which can be circumvented if the producers and/or the users of the chemicals wish to do so.

Thus, effective management of chemicals to protect all types of flora and fauna from chemicals should carry with it the reminder that to ensure the proper use of chemistry and chemicals there is the need to develop and adhere to strict codes of conduct that establish guidelines for ethical scientific development and protection of the environment.

4. CHEMISTRY IN THE ENVIRONMENT

The 20th century came into being in much the same manner as the 19th century ended insofar as there was a continuation of the less-than-desirable disposal methods for chemical waste, which included gaseous waste, liquid waste, and solid waste. As the 20th century evolved, the use and disposal of chemicals expanded by several orders of magnitude and this expansion seemed to be unstoppable. In fact, it was not only industrial waste that was disposed of in a manner that was dangerous to the environment but also household chemicals (in considerable quantities when measured on a city-wide basis) that were used to paint, clean, and maintain homes and gardens.

At the time, there was not the realization that many of these products were toxic to the flora and fauna (including humans) of the environment, whether or not they are used or disposed off improperly. However, during the later quarter of the 20th century and by the beginning of the 21st century, there came the realization that chemicals (some in large concentrations, other in small concentrations) were toxic and the unabated disposal of chemicals had to change. This awakening of an (almost global) environmental consciousness led to the legislation in many countries that chemical disposal must be organized and carried out by legislatively sanctioned methods and the unabated and dangerous disposal of chemicals must cease.

The chemicals industry (which, within the context of this book, includes the fossil fuels industry) and its products provide many real and potential benefits, particularly related to improving and sustaining human health and nutrition as well as, on the economic side, financial capital through new opportunities for employment. At the same time that benefits accrue, the production and use of chemicals create risks to the environment at all stages of the production cycle. The generation and intentional and unintentional release of the produced chemicals (and the process byproducts) has contributed to environmental contamination and degradation at multiple levels—local, regional, and global—and in many instances the impact will, more than likely, continue to be felt for generations.

As a result, it is now (some observers would use the word *finally* instead of the word *now*) recognized and legislated that any process waste (including hazardous and nonhazardous wastes) should never be discarded without proper guidance and authority. The effects of these errant, irresponsible, irregular (and often illegal) methods of disposal were being observed in the atmosphere (the occurrence of smog in cities such as London and Manchester in the late 1950s are often cited as examples), the waterways (dead fish floating in rivers, streams, lakes, and oceans) and in landfills (or anywhere), where the waste was dumped on to solid ground (leading to objectionable odors and poisonous runoff material). Many forms of disposal (which are considered to be illegal by modern disposal standards or protocols) continued unchecked and unmonitored during the early part of the 20th century.

The initial moves in the development of an efficient management system is to ensure that there are education programs that prepare professionals to enter the field of environmental technology, as well as prepare individuals to meet the challenges of environmental management in the forthcoming decades (Speight and Singh, 2014). There is no single discipline by which these challenges can be met—young professionals should be skilled in the sciences, the engineering technologies, and the relevant subdisciplines that enable them to cross over from one discipline to another as the occasion demands.

When an event occurs that is detrimental to any floral or faunal ecosystem, allocation of chemical responsibility to the company or persons disposing of the chemicals is often a difficult process. The issues relating to the responsibility for the development and dispersal of chemicals continue to be debated. However, on the basis that the use and disposal of chemicals is a global problem, the responsibility to deal with the problem must fall to policy makers in the various levels of government (local, state, and federal), who are involved in the creation of regulatory laws not only in the United States and the industrialized nations of the world; it is important to create codes of conduct to guide behavior and actions regarding this complex problem. For example, signatory countries to the Kyoto Protocol—an international treaty which extends the 1992 United Nations Framework Convention on Climate Change (UNFCCC) that commits the signatories to reduce greenhouse gas emissions which was adopted in Kyoto, Japan, on December 11, 1997 and entered into force on February 16, 2005—cannot assume that immunity to other forms of pollution is afforded by the Protocol. In addition, governments that fail to create responsible regularity laws must also share some of the blame for the misuse of chemicals. The politicians cannot consider themselves immune from blame when the necessary laws are not passed or especially when such laws are passed but are not policed or enforced.

Briefly, climate change on a global scale has been attributed to increased emissions of carbon dioxide (CO_2), a greenhouse gas, although the actual contribution of anthropogenic effects is not fully understood or known (Speight and Islam, 2016). A global average temperature rise of only 1°C (1.8°F) has been estimated to produce serious climatological implications. Possible consequences include melting of polar ice caps; an increase in sea level; and increases in precipitation and severe weather events like hurricanes, tornadoes, heat waves, floods, and droughts. Other atmospheric effects of air pollution include urban smog and reduced visibility, associated with ozone-forming nitrogen oxides and volatile compound emissions. Sulfur dioxide (SO_2) and nitrogen oxides (NO_x) combine with water in the atmosphere to cause acid rain, which is detrimental to forests and other vegetation, soil, lakes, and aquatic life. Acid rain also causes monuments and buildings to deteriorate.

There are two types of codes related to the use/misuse of chemicals and their subsequent disposal: (1) enforceable codes of conduct and (2) aspirational codes of conduct. An *enforceable code* of conduct deals with the necessary protocols for regulation and enforcement of the code while an *aspirational code* of conduct presents the ideals of performance so that those bound to the code may be reminded of their obligations to perform ethically and responsibly. Nevertheless, there are many observers who are in serious doubt about the practical effectiveness of such codes, which may even prescribe ambiguous (and often unattainable) ideals, which can be circumvented if the producers and/or the users of the chemicals wish to do so.

Typically, the value of a code of conduct is usually evident to the creators and writers of the code. Those who must consider every word and phrase included in the code must also explain the importance of expressing the meaning of the code in an unambiguous, straightforward, understandable, and effective manner. Furthermore, it is also essential to involve the various groups with different interests and perspectives at the time when the code is being formulated to inform the various groups of the issues addressed in the code, as well as to remind all participants and the users of the responsible use of chemicals. In doing so, a code of conduct can be written to be highly effective which should assist the scientists, the engineers, and the public of the issues at hand. From this understanding, the responsibilities and the guidelines for each party to act in a responsible and ethical manner should be provided.

Thus, effective management to protect all types of flora and fauna from harmful chemicals should carry with it the reminder that to ensure the proper use of chemistry and chemicals there is the need to develop and adhere to strict codes of conduct that establish guidelines for ethical scientific development and protection of the environment (Speight and Singh, 2014).

5. CHEMICAL TRANSFORMATIONS

The chemistry of the various ecosystems involves the role of chemical interactions between living organisms and their environment (sometimes referred to as chemical ecology) since the consequences of those interactions on the ethology and evolution of the organisms involved can be adverse (Speight, 2017a,b). Ecosystems are subject to periodic disturbances and are in the process of recovering from some past disturbance. When a perturbation occurs (such as the discharge of a contaminating chemical), the ecosystem responds by moving away from its initial state. The tendency of an ecosystem to remain close to the equilibrium state, despite that disturbance, is termed *ecosystem resistance*. On the other hand, the speed with which an ecosystem returns to its initial state after disturbance is called *ecosystem resilience* (Chapin et al., 2002).

In terms of inorganic chemicals, a chemical transformation is the conversion of a substrate chemical (or reactant chemical) to a chemical product. In more general terms, a chemical transformation is a chemical reaction which is characterized by a chemical change and yields one or more products. These products usually have properties substantially different from the properties of the individual reactants. Reactions often consist of a sequence of individual substeps that can be described by means of chemical equations, which are symbolic representations of the starting materials, intermediate products, and the end products, as well as the reaction parameters (such as temperature, pressure, and time).

Chemical reactions occur at a characteristic rate (the reaction rate) at a given temperature and chemical concentration. Typically, reaction rates increase with increasing temperature because there is more thermal energy available to reach the activation energy necessary for breaking bonds between atoms. The general rule of thumb is that for every 10°C (18°F) increase in temperature the rate of an inorganic chemical reaction is doubled and there is no reason to doubt that this would not be the case for inorganic chemicals discharged into the environment.

Chemicals can enter the environment (air, water, and soil) when they are produced, used, or sent for disposal and their impact on the environment is

TABLE 1.11 Melting Point and Boiling Point Temperatures (at Normal Pressure) for Hydrides of the Elements of Groups 15, 16, and 17 in the Periodic Table That Are Analogous to Water

Melting Points (°C)		
NH_3: −78	H_2O: 0	HF: −83
PH_3: −133	H_2S: −86	HCl: −115
AsH_3: −116	H_2Se: −60	HBr: −89
SbH_3: −88	H_2Te −49	HI: −51
Boiling Points (°C)		
NH_3: −33	H_2O: 100	HF: 20
PH_3: −88	H_2S: −61	HCl −85
AsH_3: −55	H_2Se: −42	HBr −67
SbH_3: −17	H_2Te: −2	HI: −35

determined by the amount of the chemical that is released, the type and concentration of the chemical, and where it is found, as well as through any chemical transformation that occurs after the chemical has entered the environment, whether it is in the atmosphere, the aquasphere, or the lithosphere.

For the most [part], the lithosphere is typically represented by soil and it is the characteristics of the soil (Table 1.11) that influence the behavior of discharged chemicals in the lithosphere. In fact, soil and the sediments in the aquasphere are major sinks for discharged chemicals. Some chemicals (such as herbicides) may be introduced directly to the soil while chemicals such as acid rain (Chapter 2) may reach the soil indirectly. Consequently, the processes by which compounds distribute into and are bound in soil or sediment will influence their environmental distribution.

In terms of soil structure, the soil is conveniently divided into three size groups: (1) sand, (2) silt, and (3) clay minerals from which the texture of soil is defined by the relative amounts of these components. For example, sand constituents (the larger particles in soil) are composed primarily of weathered grains of quartz (SiO_2) and which have a relatively small surface area because of the large particle size and typically do not provide any significant binding sites for organic compounds. Silt constitutes the medium-sized particles that are essentially broken down fragments of such minerals as quartz and feldspar (minerals distinguished by the presence of alumina, Al_2O_3, and silica). Silt, like sand, is relatively inert and would not be involved to any significant extent in the binding of discharged chemicals. Clay minerals are the smallest particles and are composed of layer silicates, and clay surfaces can assume a negative charge that is pH dependent, resulting from the dissociation of the hydroxyl hydrogens and which account for the cation exchange capability. These minerals may absorb considerable quantities of water and swell and shrink as they gain and lose water.

Another constituent of soil that is often ignored but needs consideration is the soil organic matter which is often referred to as *humus* and is derived primarily from the degradation of plant material— lignin, carbohydrates, protein, fats, and waxes. Operationally, the alkali-soluble materials can be subdivided into three fractions: (1) humin, which is the organic material that cannot be extracted from soil by alkaline agents, (2) humic acid, which is the material that precipitates from the alkaline extract on acidification, and (3) fulvic acid, which is the material that remains in solution.

The *atmosphere* (*air*) is essential to life and in addition to air availability and composition, air temperature, and all other meteorological phenomena, ordinary (like rain and wind) and extraordinary (like draughts and storms); it has a strong influence on the existence of the flora and the fauna. The *hydrosphere* (*water*), sometimes including solidified water (the cryosphere), is mostly used as a solvent and carrier for matter and energy transport inside our body. The water cycle is the main controller of matter and energy flows on earth, providing plentiful of distilled water for direct floral and faunal growth. The *lithosphere* (*land*) or the most external part of the earth, the crust, or even the most superficial layer (the soil or earth), is fundamental to land flora and fauna. Cultivated soil, pastures, and forests cannot be further increased on earth, although improvements in the production of food and goods through efficient farming and forestry have been achieved with modern machinery, fertilizers, and pesticides; there has been a corresponding stress in energy consumption and environmental impact (contamination of soils and ground water, acidification, salination, and desertification) through the use of chemicals.

Some chemicals are of particular concern because they can work their way into the food chain and accumulate and/or persist in the environment for prolonged periods, including years. This is in direct contradiction of the general *conventional wisdom* (or unbridled optimism) that assumed that inorganic chemicals would either (1) degrade into harmless byproducts as a result of microbial or chemical reactions, (2) immobilize completely by binding to soil solids, or (3) volatilize to the atmosphere where dilution to harmless levels was assured. This false assurance led to years of agricultural chemical use and chemical waste disposal with no monitoring of atmosphere, groundwater (the aquasphere), or soil (the lithosphere) near the discharge (Barcelona et al., 1990).

As an example, the volatility of an inorganic chemical is of concern predominantly for surface-located chemicals and is affected by (1) temperature of the soil, (2) the water content of the soil, (3) the adsorptive interaction of the chemical and the soil, (4) the concentration of the chemical in the soil, (5) the vapor pressure of the chemical, and (6) the solubility of the chemical in water—water is the predominant liquid in the typical (undried) soil.

5.1 Chemistry in the Atmosphere

The chemistry of the atmosphere has emerged as a central theme in studies of global change. Atmospheric chemistry provides the scientific foundations to understand several phenomena that are part of global change. These phenomena include (1) changes in the ultraviolet rays at the surface of the earth owing to the intrinsically chemical nature of the catalytic loss of stratospheric ozone, (2) changes in the dynamics and radiative structure of the climate system through altered thermal forcing by ozone in the upper troposphere, (3) changes in the concentration of highly oxidizing species in urban as well as remote rural regions, and (4) changes in the acid levels of depositions in a variety of ecosystems. In addition, the chemistry of the atmosphere is often represented by simple equations which fit into the concepts of this chapter, such as the hydrolysis reactions and nucleophilic substitution reactions.

Atmospheric chemistry is a branch of science in which the chemistry of the atmosphere of the earth is the prime focus. Atmospheric chemistry includes studying the interactions between the atmosphere and living organisms. The composition of the atmosphere changes as result of natural processes such as emission from volcanoes, lightning, and bombardment by solar particles from the corona of the sun. It has also been changed by human activity, examples of which include acid rain, ozone depletion, photochemical smog, greenhouse gas emissions, and climate change, although there are other geological factors involved such as the earth being in an interglacial period (Speight and Islam, 2016).

Atmospheric chemistry is an important discipline for understanding air pollution and its impacts. However, atmospheric composition and chemistry are complex, being controlled by the emission and photochemistry. The transport of many trace gases is often represented by convenient chemical equations. Understanding the timescale, as well as the chemical and spatial patterns of perturbations to trace gases, is needed to evaluate possible environmental damage (e.g., stratospheric ozone depletion or climate change) caused by anthropogenic emissions (Finlayson-Pitts, 2010).

Furthermore, chemical reactions that occur rapidly upon mixing of two stable reactants are quite rare in the atmosphere, as very often the energy required to activate the reaction is considerably larger than the available thermal energy. In the troposphere and stratosphere, the available heat content is often not sufficient, and most atmospheric reaction cycles are initiated by solar irradiation. Light absorption creates a transient excited state that can fall apart (photodissociate), change to a new structure (photoisomerize), undergo various reactions (for example, via electron transfer or transfer of hydrogen atoms), or transfer of the excitation energy to other molecules.

Chemicals can be emitted directly into the atmosphere or formed by chemical conversion of precursor species during chemical reactions. In these reactions, highly toxic inorganic chemicals can be converted into less toxic products, but the result of the reactions can also be products having a higher toxicity that the starting chemicals. In order to understand these reactions, it is also necessary to understand the chemical composition of the natural atmosphere; the way gases, liquids, and solids in the atmosphere interact with each other and with the surface of the earth and the associated flora and fauna; and how human activities may be changing the chemical and physical characteristics of the atmosphere (Finlayson-Pitts, 2010; Finlayson and Pitts, 2010).

In addition, the atmosphere is composed of a number of important regions that are defined by thermal structure as a function of altitude (Finlayson-Pitts and Pitts, 2010) that contribute to various physical effects. For example, as altitude increases, air pressure drops off rapidly and, furthermore, the temperature structure of the atmosphere is somewhat complex because of exothermic chemical reactions in the upper atmosphere that are caused by the absorption of high-energy photons (ultraviolet-C radiation—wavelength: 180–280 nm, and ultraviolet-B radiation—wavelength: 260–315 nm) from the incoming solar radiation.

The rupture of molecular bonds produces atomic species and results in recombination reactions that form ozone (O_3). In this process, an oxygen molecule is split (photolyzed) by higher frequency ultraviolet (UV) light into two oxygen atoms after which each oxygen atom then combines with an oxygen molecule to form an ozone molecule:

$$O_2 + h\nu \rightarrow 2O\bullet$$

$$O\bullet + O_2 \rightarrow O_3$$

Other important gases may also be formed, which can filter the incoming ultraviolet rays and act as a protective optical shield against harmful solar radiation. Most of this shielding occurs in the stratosphere, where a great deal of recent interest has been focused because of the depletion of the stratospheric ozone layer by anthropogenic use of various chemicals. As altitude increases from sea level, temperature in the troposphere decreases up to the tropopause, where temperature begins to increase in the stratosphere.

Thus, there are a number of critical environmental issues associated with a changing atmosphere, including photochemical smog, global climate change, toxic air pollutants, acidic deposition, and stratospheric ozone depletion. In fact, photochemical smog has been observed in almost every major city on the planet, and the release of nitrogen oxides (as well as hydrocarbon derivatives) into the air has been found to lead to the photochemical production of ozone, as well as a variety of secondary species including aerosols. Much of this anthropogenic (human) impact on the atmosphere has been associated with the increasing use of fossil fuels as an energy source for heating, transportation, and electric power production. Photochemical smog and the generation of tropospheric ozone are serious environmental problems that have been associated with the use of fossil fuels. In fact, the combustion of fossil fuels (which are in fact, inorganic chemicals) is one of the most common causes of carbon dioxide pollution in the atmosphere. This phenomenon may not be classified as direct pollution but is certainly an indirect form of pollution.

Over the past several decades, it has become clear that aqueous chemical processes occurring in cloud droplets and wet atmospheric particles are an important source of atmospheric particulate matter that is organic in nature. Reactions of water-soluble volatile organic compounds or semivolatile organic compounds in these aqueous media lead to the formation of highly oxidized organic particulate matter (secondary organic aerosol) and key tracer species, such as organosulfate derivatives. These processes are often driven by a combination of anthropogenic and biogenic emissions, and therefore their accurate representation in models is important for effective air quality management (Speight and Singh, 2014). One of the results of these various processes is the formation and deposition of acid rain with the ensuing effects on the environment (Wondyfraw, 2014).

Acid rain is formed when sulfur dioxide and nitrogen oxides react with water vapor and other chemicals in the presence of sunlight to form various acidic compounds in the air. These reactions are a combination of hydrolysis (Chapter 6) and oxidation reactions (Chapter 7). The principle source of acid rain—causing pollutants, sulfur dioxide and nitrogen oxides, are from combustion fossil fuels and their derivatives:

Oxidation reactions:

$$2[C]_{\text{fossil fuel}} + O_2 \rightarrow 2CO$$

$$[C]_{\text{fossil fuel}} + O_2 \rightarrow CO_2$$

$$2[N]_{\text{fossil fuel}} + O_2 \rightarrow 2NO$$

$$[N]_{\text{fossil fuel}} + O_2 \rightarrow NO_2$$

$$[S]_{\text{fossil fuel}} + O_2 \rightarrow SO_2$$

$$2SO_2 + O_2 \rightarrow 2SO_3$$

The oxides of sulfur and nitrogen then react with water in the atmosphere to produce the sulfur-containing acids and the nitrogen-contain acids that constitute the now-infamous acid rain:

Hydrolysis reactions:

$$SO_2 + H_2O \rightarrow H_2SO_3 \text{ (sulfurous acid)}$$

$$SO_3 + H_2O \rightarrow H_2SO_4 \text{ (sulfuric acid)}$$

$$NO + H_2O \rightarrow HNO_2 \text{ (nitrous acid)}$$

$$3NO_2 + 2H_2O \rightarrow HNO_3 \text{ (nitric acid)}$$

Hydrogen sulfide and ammonia are produced from processing sulfur containing and nitrogen containing feedstocks:

$$[S]_{\text{fossil fuel}} + H_2 \rightarrow H_2S + \text{hydrocarbons}$$

$$2[N]_{\text{fossil fuel}} + 3H_2 \rightarrow 2NH_3 + \text{hydrocarbons}$$

Two of the pollutants that are emitted are hydrocarbons (e.g., unburned fuel) and nitric oxide (NO). When these pollutants build up to sufficiently high levels, a chain reaction occurs from their interaction with sunlight in which the NO is converted to nitrogen dioxide (NO_2)—a brown gas that at sufficiently high levels can contribute to urban haze. However, a more serious problem is that nitrogen dioxide (NO_2) can absorb sunlight and break apart to produce oxygen atoms that combine with the oxygen in the air to produce ozone (O_3), a powerful oxidizing agent (Chapter 7), and a toxic gas.

5.2 Chemistry in the Aquasphere

Water is the most abundant molecule on the earth's surface and one of the most important molecules to study in chemistry and plays an important role as a chemical substance. Water has a simple molecular structure—it is composed of one oxygen atom and two hydrogen atoms (H_2O). Each hydrogen atom is bonded to the oxygen by means of a shared pair of electrons—the covalent bond. Oxygen also has two unshared pairs of electrons and, thus, oxygen is an electronegative atom compared with hydrogen. Also, water is a polar molecule because of the uneven distribution of electron density—water has a partial negative charge ($\delta-$) near the oxygen atom due the unshared pairs of electrons, and partial positive charges ($\delta+$) near the hydrogen atoms resulting in the formation of hydrogen bonds. In fact, many other unique properties of water are due to the hydrogen bonds. For example, the unique physical properties, including a high heat of vaporization, strong surface tension, high specific heat, and nearly universal solvent properties of water are also due to hydrogen bonding. The hydrophobic effect or the exclusion of compounds containing carbon and hydrogen (nonpolar compounds) is another unique property of water caused by the hydrogen bonds. The hydrophobic effect is particularly important in the formation of cell membranes.

Thus, the many important functions of water include (1) being a good solvent for dissolving many solids, (2) serving as an excellent coolant both mechanically and biologically, and (3) acting as a reactant in many chemical reactions. For example, the ability of acids to react with bases depends on the tendency of hydrogen ions to combine with hydroxide ions to form water. This tendency is very great, so the reaction is practically *complete*.

$$H^+(aq) + OH^-(aq) \rightarrow H_2O$$

No reaction, however, is really 100% complete; at *equilibrium* (when there is no further net change in amounts of substances) there will be at least a minute concentration of the reactants in the solution. Another way of expressing this is to say that any reaction is at least slightly *reversible*. In pure water, the following reaction will proceed to a very slight extent. Thus:

$$H_2O \rightarrow H^+(aq) + OH^-(aq)$$

Water pollution has become a widespread phenomenon and has been known for centuries, particularly the pollution of rivers and groundwater. By way of example, in ancient time up to the early part of the 20th century, many cities deposited waste into the nearby river or even into the ocean. It is only very recently (because of serious concerns for the condition of the environment) that an understanding of the behavior and fate of toxic chemicals, which are discharged into the aquatic environment because of these activities, has made it essential to control water pollution. In rivers, the basic physical movement of pollutant molecules is the result of advection (Chapters 2 and 10), but superimposed upon this are the effects of dispersion and mixing with tributaries and other discharges.

Some of the chemicals discharged are relatively inert, so their concentration changes only due to advection, dispersion, and mixing. However, many inorganic chemicals are not conservative in their behavior and undergo changes due to chemical processes or biochemical processes, such as oxidation. The dispersability of an inorganic chemical is a measure or indication of the potential of the chemical to spread in the air, water, or on the land. In some spill situations and under appropriate conditions, dispersants (to interfere with the tendency of the chemical to adhere to) adsorb on mineral surfaces. This may be an effective countermeasure for minimizing contamination of the air, shorelines, and/or land-based sites.

In addition, many indications that the chemical materials in the aquasphere (also called, when referring to the sea, the marine aquasphere) are subject to intense chemical transformations and physical recycling processes imply that a total inorganic-carbon approach is not sufficient to resolve the numerous processes occurring.

The effects of a chemical released into the marine environment (or any part of the aquasphere) depend on several factors such as (1) the toxicity of the chemical, (2) the quantity of the chemical, (3) the resulting concentration of the chemical in the water column, (4) the length of time that floral and faunal organisms are exposed to that concentration, and (5) the level of tolerance of the organisms, which varies greatly among different species and during the life cycle of the organism. Even if the concentration of the chemical is below what would be considered as the lethal concentration, a sublethal concentration of an inorganic chemical can still lead to a long-term impact within the aqueous marine environment.

For example, chemically induced stress can reduce the overall ability of an organism to reproduce, grow, feed or otherwise function normally within a few generations. In addition, the characteristics of some inorganic chemicals can result in an accumulation of the chemical within an organism (*bio-accumulation*) (Chapter 8) and the organism may be particularly vulnerable to this problem. Furthermore, subsequent biomagnification may also occur if the inorganic chemical (or a toxic product produced by one or more transformation reactions) can be passed on, following the food chain up to higher flora or fauna.

In terms of the marine environment and a spill of crude oil, complex processes of crude oil transformation start developing almost as soon as the oil contacts the water although the progress, duration, and result of the transformations depend on the properties and composition of the oil itself, parameters of the actual oil spill, and environmental conditions. The major operative processes are (1) physical transport, (2) dissolution, (3) emulsification, (4) oxidation, (5) sedimentation, (6) microbial degradation, (7) aggregation, and (8) self-purification (Chapters 5–8).

In terms of *physical transport*, the distribution of a chemical oil spilled into the aquasphere is subject to transportation, as well as changes due to the presence of oxygen in the water. However, the major changes take place under the combined impact of meteorological and hydrological factors and depend mainly on the power and direction of wind, waves, and currents. A considerable part of the chemical will disperse in the water as ions and droplets that can be transported over large distances away from the place of discharge of the chemical into the water. This is analogous to dilution but should not be construed as being in order because of the high volume of water. Water, like any other solvent, can only tolerate so much in the way of the amount of solute before the properties of the water have an adverse effect on the floral and faunal species in the water. *Oxidation*, a complex process even in water, can also occur as a series of chemical reactions involving the hydroxyl radical (•OH).

As these processes occur, some of the crude oil constituents are adsorbed on any suspended material and deposited on the ocean floor (sedimentation), the rate of which is dependent upon the ocean depth—in deeper areas remote from the shore, sedimentation of oil (except for the heavy fractions) is a slow process. Simultaneously, the process of *biosedimentation* occurs (a reaction that falls under the banner of biodegradation) (Chapter 8)—in this process, plankton and other organisms absorb the emulsified oil—and the crude oil constituents are sent to the bottom of the ocean as sediment with the metabolites of the plankton and other organisms. However, this situation radically changes when the suspended oil reaches the sea bottom—the decomposition rate of the oil on the ocean bottom abruptly ceases—especially under the prevailing anaerobic conditions—and any crude oil constituents accumulated inside the sediments can be preserved for many months and even years. These products can be swept to the edge of the ocean (the beach) by turbulent condition at some later time.

The fate of most of the constituents of crude oil in the marine environment is ultimately defined by their transformation and degradation due to *microbial degradation*. The degree and rates of biodegradation depend upon the structure of the chemicals, and the constituents of the mixture biodegrade faster than aromatic constituents and naphthenic constituents, and with increasing complexity of molecular structure as well as with increasing molecular weight, the rate of microbial decomposition usually decreases. Besides, this rate depends on the physical state of the oil, including the degree of its dispersion, as well as environmental factors such as temperature, availability of oxygen, and the abundance of oil-degrading microorganisms.

In the chemical sense, *aggregation* (especially *particle agglomeration*) refers to formation of assemblages in a suspension and represents a mechanism leading to destabilization of colloidal systems. During this process, particles dispersed in the liquid phase stick to each other, and spontaneously form irregular particle clusters, flocs, or aggregates. This phenomenon (also referred to as *coagulation* or *flocculation*) and such a suspension is also called *unstable*. Particle aggregation can be induced by adding salts or another chemical referred to as coagulant or flocculant.

Particle aggregation is normally an irreversible process and once particle aggregates have formed, they will not easily disrupt. In the course of aggregation, the aggregates will grow in size, and as a consequence they may settle to the bottom of the container, which is referred to as sedimentation. Alternatively, a colloidal gel may form in concentrated suspensions, which changes the rheological properties. The reverse process whereby particle aggregates are disrupted and dispersed as individual particles, referred to as peptization, hardly occurs spontaneously, but may occur under stirring or shear. Colloidal particles may also remain dispersed in liquids for long periods of time (days to years). This phenomenon is referred to as *colloidal stability* and such a suspension is said to be *stable*. Stable suspensions are often obtained at low salt concentrations or by addition of chemicals referred to as *stabilizers* or *stabilizing agents*. Similar aggregation processes occur in other dispersed systems. For example, in an emulsion, they may also be coupled to droplet coalescence, and not only lead to sedimentation but also to creaming. In aerosols, airborne particles may equally aggregate and form larger clusters. In the chemical sense, creaming is the migration of the dispersed phase of an emulsion, under the influence of buoyancy. The particles float upwards or sink, depending on how large they are and how much less dense or more dense they may be than the continuous phase, and also how viscous or how thixotropic (a time-dependent shear thinning property) the continuous phase might be. For as long as the particles remain separated, the process is called *creaming*.

Self-purification is a result of the processes described above in which crude oil in the marine environment rapidly loses its original properties and disintegrates into various fractions. These fractions have different chemical composition and structure and exist in different migrational forms and undergo chemical transformations that slow after reaching thermodynamic equilibrium with the environmental parameters. Eventually, the original and intermediate compounds disappear, and carbon dioxide and water form. This form of self-purification inevitably happens in water ecosystems if the toxic load does not exceed acceptable limits.

While acid–base reactions (Chapter 3) are not the only chemical reactions important in aquatic systems, they do present a valuable starting point for understanding the basic concepts of chemical equilibria in such systems. Carbon dioxide (CO_2), a gaseous inorganic chemical substance of vital importance to a variety of environmental processes, including growth and decomposition of biological systems, climate regulation, and mineral weathering, has acid–base properties that are critical to an understanding of its chemical behavior in the environment. The phenomenon of acid rain is another example of the importance of acid–base equilibria in natural aquatic systems.

Furthermore, the almost unique physical and chemical properties of water as a solvent are of fundamental concern to aquatic chemical processes. For example, in the liquid state water has unusually high boiling point and melting point temperatures compared to its hydride analogs from the periodic table such as ammonia (NH_3), hydrogen fluoride (HF), and hydrogen sulfide (H_2S) (Table 1.12). Hydrogen bonding between water molecules means that there are strong intermolecular forces making it relatively difficult to melt or vaporize.

TABLE 1.12 Properties That are Relevant to the Physical/Chemical Treatment of Soils

Property	Comment
Heterogeneity	The variability of the soil types at a site, such as sandy, clayey, gravels, the presence of sand or clay lenses or fractures
Permeability	The ability to move air and water through the soil, usually measured in cm/s, with a lower limit of 10–4 cm/s for in situ passage of fluids
Clay content	Low permeability, typically 10-6 cm/s; -contaminants are usually associated with small particles, such as silt and clay minerals at <60 microns, and are difficult to separate or react; handling for ex situ treatment, such as drying, screening, feeding, become more difficult
Humus content	The natural organic matter present in soils; -affinity for contaminants, both organic and inorganic, is higher; -the humus or organic matter in soils represents an additional oxidation load when chemical oxidants are considered to destroy or transform organic chemical contaminants

In addition, pure water has a maximum density at 4°C (39°F), higher in temperature than its freezing point (0°C, 32°F), and that ice is substantially less dense than liquid water is important (and fortunate) in several contexts. Thus, as water in a lake is cooled at its surface by loss of heat to the atmosphere, the ice structures formed will float. Furthermore, a dynamically stable water layer near 4°C (39°F) will tend to accumulate at the bottom of the lake, and the overlying, less dense water will be able to continue cooling down to the freezing point. This means that ice will eventually coalesce at the surface, forming an insulating layer that greatly reduces the rate of freezing of the underlying water. This situation is obviously important for plants and animals that inhabit lake waters.

In addition, the presence of salt components means that the temperature of maximum density for seawater is shifted to lower temperatures: in fact, the density of seawater continues to increase right down to the freezing point. The high concentration of electrolytes in seawater assists in breaking up the open, hydrogen-bonded ice-like structure of water near its freezing point. Because the salt components tend to be excluded from the ice formed by freezing seawater, sea ice is relatively fresh and still floats on water. Much of the salt it contains is not truly part of the ice structure but contained in brines that are physically entrained by small pockets and fissures in the ice.

The dielectric constant of water (78.2 at 25°C, 77°F) is high compared to most liquids. Of the common liquids, few have comparable values at this temperature, for example, hydrogen cyanide (HCN, 106.8), hydrogen fluoride (HF, 83.6), and sulfuric acid (H_2SO_4, 101). By contrast, most nonpolar liquids have dielectric constants in the order of 2. The high dielectric constant helps liquid water to solvate ions, making it a good solvent for ionic substances, and arises because of the polar nature of the water molecule and the tetrahedrally coordinated structure in the liquid phase.

In many electrolyte solutions of interest, the presence of ions can alter the nature of the water structure. Ions tend to orient water molecules that are near to them. For example, cations attract the negative oxygen end of the water dipole towards them. This reorientation tends to disrupt the ice-like structure further away. This can be seen by comparing the entropy change on transferring ions from the gas phase to water with a similar species that does not form ions.

5.3 Chemistry in the Lithosphere

The biosphere consists of the parts of earth where life exists and extends from the deepest root systems of trees to the dark environment of ocean trenches to the rain forests and high mountaintops. The earth in terms of spheres: (1) the atmosphere is the layer of air that stretches above the lithosphere, (2) the water on the surface of the earth, in the ground, and in the air constitute the hydrosphere, and (3) the solid surface layer of the earth is the lithosphere (sometimes referred to as the terrestrial biosphere).

Since life exists on the ground, in the air, and in the water, the biosphere overlaps all these spheres. Although the biosphere measures about 20 km (12 miles) from top to bottom, almost all life exists between about 500 m (1640 ft) below the ocean's surface to about 6 km (3.75 miles) above sea level. The addition of oxygen to the biosphere allowed more complex lifeforms to evolve. Millions of different plants and other photosynthetic species developed. Animals, which consume plants (and consume other animals) evolved, as well as bacteria and other organisms evolved to decompose, or break down, dead animals and plants. The biosphere benefits from this food web—the remains of dead plants and animals release nutrients into the soil and into the ocean. These nutrients are reabsorbed by growing plants. This exchange of food and energy makes the biosphere a self-supporting and self-regulating system. The biosphere is sometimes considered to be one large ecosystem—which is, in fact many ecosystems that exists as a complex community of living and nonliving things functioning as a single unit. This delicate balance can be upset by the discharge of toxic inorganic chemicals into the biosphere.

In a general context, the terrestrial biosphere is predominantly the soil that is on the surface of the earth and which can house many different types of inorganic chemicals (Table 1.13) (Fox, 1996). Soil is a matrix of solids including sand, silt, clay, and organic matter particles, as well as aggregates of various sizes formed from them, and pore space, which may be filled with air or water. However, soils are chemically different from the rocks and minerals from which they are formed in that soils contain less of the water-soluble weathering products, calcium, magnesium, sodium, and potassium, and more of the relatively insoluble elements such as iron and aluminum. Older, highly weathered soil typically has a high concentration of aluminum oxide and iron oxide.

TABLE 1.13 The Various Clay Mineral Groups

Group	Layer Type	Layer Charge (x)	Type of Chemical Formula
Kaolinite	1:1	<0.01	$[Si_4]Al_4O_{10}(OH)_8 \cdot nH_2O$ (n = 0 or 4)
Illite	2:1	1.4–2.0	$M_x[Si_{6.8}Al_{1.2}]$ $Al_3Fe.025Mg_{0.7}5O20(OH)_4$
Vermiculite	2:1	1.2–1.8	$M_x[Si_7Al]AlFe.05Mg0.5O_20(OH)_4$
Smectite	2:1	0.5–1.2	$M_x[Si_8]Al_{3.2}Fe_{0.2}Mg_{0.6}O_20(OH)_4$
Chlorite	2:1:1	Variable	$(Al(OH)_{2.55})4[Si_{6.8}Al0_{1.2}]Al_{3.4}Mg_{0.6})$ $20(OH)_4$

An ecosystem is a community of living organisms in conjunction with the nonliving components of their environment (such as air, water, and mineral soil), interacting as a system. Ecosystems are controlled both by (1) external factors and (2) internal factors. External factors such as climate, the parent material that forms the soil, and topography control the overall structure of an ecosystem and the way things work within it, but are not themselves influenced by the ecosystem. Internal factors not only control ecosystem processes but are also controlled by them and are often subject to feedback loops. Other internal factors include disturbance, succession, and the types of species present. Although humans exist and operate within ecosystems, their cumulative effects are large enough to influence external factors like climate. An important aspect of the chemistry of the biosphere is the chemistry that occurs in the soil.

The soil itself is a complex mixture of inorganic chemicals—even before pollution occurs. Eight chemical elements comprise the majority of the mineral matter in soils. Of these eight elements, oxygen, a negatively charged ion (anion) in crystal structures, is the most prevalent on both weight and volume basis. The next most common elements, all positively charged ions, in decreasing order, are silicon, aluminum, iron, magnesium, calcium, sodium, and potassium. Ions of these elements combine in various ratios to form different minerals.

The most chemically active fraction of soils consists of colloidal clays and organic matter. Colloidal particles are so small (<0.0002 mm) that they remain suspended in water and exhibit a very large surface area per unit weight. These materials also generally exhibit net negative charge and high adsorptive capacity. Several different silicate clay minerals exist in soils, but all have a layered structure. It is these clay minerals that act as adsorbents for inorganic pollutants (as well as organic pollutants).

The soil transports and moves water, provides refuge for bacteria and other fauna, and has many different arrangements of weathered rock and minerals. When soil and minerals weather over time, the chemical composition of soil also changes. However, many of the problems of soil chemistry are concerned with environmental sciences such as the accidental or deliberate disposal of chemicals. This can involve several physical-chemical interactions that can cause changes to the soil (1) ion exchange, (2) soil pH, (3) sorption and precipitation, and (4) oxidation-reduction reactions.

Ion exchange (Chapter 5) involves the movement of cations (positively charged elements such as calcium, magnesium, and sodium) and anions (negatively charged elements such as chloride, and compounds like nitrate) through the soils. More specifically, *cation exchange* is the interchanging between a cation in the solution of water around the soil particle and another cation that is on the surface of the clay mineral.

A cation is a positively charged ion and most inorganic chemical contaminants are cations, such as: ammonium (NH_4^+), calcium (Ca^{2+}), copper (Cu^{2+}), magnesium (Mg^{2+}), manganese (Mn^{2+}), potassium (K^{+}), and zinc (Zn^{2+}). These cations are in the soil solution and are in dynamic equilibrium with the cations adsorbed on the surface of clay and organic matter. It is the cation exchange interaction of ions with clay minerals in the soil that gave soil the ability to adsorb and exchange cations with those in soil solution (water in soil pore space). The adsorption capacity of clay minerals can cause a mixture of inorganic pollutants to physically transform when one or more constituents of the mixture adsorbs on to a clay mineral. In addition, the total amount of positive charges that the clay can absorb (the *cation exchange capacity*, CEC) impacts the rate of movement of the pollutants through the soil and the physical and chemical changes that can occur to the pollutants. For example, soil (i.e., clay) with a low cation exchange capacity is much less likely to retain pollutants and, furthermore, the soil (i.e., the clay) is less able to retain inorganic chemicals with the potential that the chemicals are released into the groundwater.

The acidity or alkalinity of the soil (measured as the pH) is a commonly measured soil chemical property and is also one of the more informative properties since the data imply certain characteristics that might be associated with a soil. For example, the pH of *soil* is a measure of the acidity or alkalinity: values 0 to 7 indicate acidity while values from 7 to 14 indicate alkalinity. The pH is one of the most important properties involved in plant growth, as well as understanding how rapidly reactions occur in the soil. For example, the element iron becomes less available to plants at higher alkalinity ($pH > 7$) which creates iron deficiency problems. Crops usually have optimal growth at pH values between 5.5 and 8, but the value depends on the crop. The pH of soil also affects the ability of organisms to feed on contaminants. Thus, any chemical spillage on to the soil can seriously affect the overall chemistry of the soil and the ability of flora and fauna to survive.

The other principal variables affecting life in soil environment include moisture, temperature, and aeration, and the balance of these factors controls the abundance and activities of the floral and faunal inhabitants in the soil, which in turn have a marked influence on the critical processes of soil aggregation retention of pollutants.

Soil water is usually derived from rainfall or some other form of overland flow. The amount of water that enters the soil is a function of soil structure which in turn is a function of texture and higher-order structure, that is, aggregation (e.g., well-aggregated, porous soils allow greater infiltration). Larger pores created by root channels and animal burrows (ant tunnels and worm holes), as well as other types of macropores, also greatly facilitate movement of water into and through the soil profile. This is why these soil animals are usually regarded as highly beneficial in soil and why some scientists often use the abundance of earthworms as an indicator of a healthy soil, though it should be remembered that highly fertile, productive soils do not always contain large numbers of earthworms and other soil faunal species.

Sorption (Chapter 5) is the process in which one substance takes up or holds another. In the current context, soil that has a high sorption capacity can hold a high level of environmental contaminants by sorption of the contaminants on to the soil particles, which have the ability to capture different nutrients and ions. *Precipitation* occurs during chemical reactions when a nutrient or chemical in the soil solution (water around the soil particles) transforms into a solid. The presence or absence of oxygen determines the manner by which soil can react chemically. *Oxidation* is the loss of electrons, and *reduction* is the gain of electrons at the soil surface.

Although soil is an inorganic matrix of metals and minerals, the main reason why the soil becomes contaminated is due to the presence of anthropogenic waste. This waste is typically (1) chemicals in one form or another that are not indigenous to the soil or (2) chemicals that are indigenous and are deposited into the soil in amounts that exceed the natural abundance. The typical actions that lead to pollution of the soil by inorganic chemicals are (1) industrial activities, (2) agricultural activities, (3) waste disposal, and (4) acid rain.

Industrial activities have been the biggest contributor to the soil pollution since the time of the *Industrial Revolution*, especially since the amount of mining and manufacturing increased almost logarithmically in the 160 years immediately following 1800. Most industries are dependent on minerals extracted from the earth and the byproducts have not been sent for disposal in a manner that can be considered safe. Thus, the industrial waste lingers in the soil surface for a long time and makes it unsuitable for use. At the same time, physical changes and chemical changes to the inorganic pollutant also occur for the better or for worse.

Agricultural activities have caused a rise in the use of inorganic chemicals since technology provided the means of producing inorganic pesticides and inorganic fertilizers. They are full of chemicals that are not produced in nature and cannot be broken down by it. These chemicals seep into the ground after they mix with water and slowly change the character of the soil by adsorption and by changing the acidity/alkalinity of the water in the soil. Other chemicals damage the composition of the soil and make it easier to erode by water and air.

Waste disposal has also been a major cause for concern while many types of waste are sure to cause contamination of the environment. Most types of waste contain chemicals which pass into the environment and cause pollution. Finally, acid rain is caused when gaseous pollutants present in the air mix up with the moisture and fall back to the earth as acidified rain. The polluted (acidified) water has the ability to dissolve away some of the soil constituents and, if more drastic, change the structure of the soil. This latter action could release pollutants that were adsorbed (and may have even remained adsorbed), and once released the pollutants can find their way into groundwater.

6. PHYSICAL CHEMISTRY IN THE ENVIRONMENT

Inorganic chemical reactions and organic chemical reactions are certainly an important aspect of environmental chemistry but there is another aspect of chemistry that must not be ignored, and this is physical chemistry.

When a chemical is released into the environment, it becomes distributed among the four major environmental compartments: (1) air, (2) water, (3) soil, and (4) flora and fauna, that is, living organisms. Each of the first three categories can be further subdivided in floral (plant) environments and faunal (animal, including human) environments. The fraction of the chemical that will move into each compartment is governed by the physicochemical properties of that chemical.

In addition, the distribution of chemicals in the environment is governed by physical processes such as (1) sedimentation, (2) adsorption, and (3) volatilization and the chemical can then be degraded by chemical and/or biological processes. Chemical processes generally occur in water or the atmosphere and follow one of four reactions: oxidation, reduction, hydrolysis, and photolysis. Biological mechanisms in soil and living organisms utilize oxidation, reduction, hydrolysis, and conjugation to degrade chemicals. The process of degradation will largely be governed by the compartment (water, soil, atmosphere, biota) in which the chemical is distributed, and this distribution is governed by the physical processes already mentioned (i.e., sedimentation, adsorption, and volatilization).

The impact of the changes in the chemical state of organic chemicals on the environment is only partially elucidated, but will be significant in many cases. Atmospheric abundance of radioactive gases could lead to substantial drift in the climate of the earth, including changes in temperature and precipitation, and in the frequency of occurrence of extreme events (such as hurricanes). Ecosystem damage and problems related to human health also result from regional and global pollution, such as acid rain (also called *acidic precipitation*), which can vary between mildly acidic and strongly acidic on the pH scale (Fig. 1.2) and can suppress life and, together with enhanced ozone levels, can lead to forest damage. In fact, when assessing the impact of chemicals on the environment, the most critical characteristics are: (1) the type of chemical discharged, which depends on the type of industries and processes

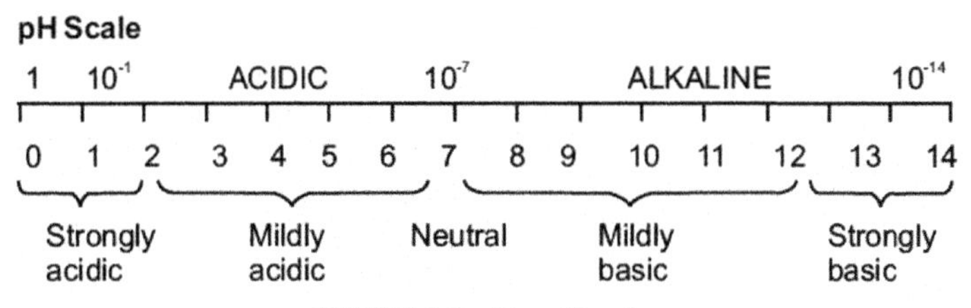

FIGURE 1.2 The pH scale.

used, and (2) the amount and concentration of the chemical. Solid wastes (containing chemicals) and/or gaseous emissions generated from industrial sources also contribute to the amount and concentration of chemicals in the environment.

REFERENCES

Baird, C., Cann, M., 2012. Environmental Chemistry, fifth ed. W. H. Freeman, Macmillan Learning, New York.

Barcelona, M., Wehrmann, A., Keeley, J.F., Pettyjohn, J., 1990. Contamination of Ground Water. Noyes Data Corp., Park Ridge, New Jersey.

Chadeesingh, R., 2011. The Fischer-Tropsch process. In: Speight, J.G. (Ed.), The Biofuels Handbook. The Royal Society of Chemistry, London, United Kingdom, pp. 476–517 (Part 3, Chapter 5).

Chapin, F.S., Matson, P.A., Mooney, H.A., 2002. Principles of Terrestrial Ecosystem Ecology. Springer, New York.

Chenier, P.J., 1992. Survey of Industrial Chemistry, second ed. VCH Publishers Inc., New York.

Davis, B.H., Occelli, M.L., 2010. Advances in Fischer-Tropsch Synthesis, Catalysts, and Catalysis. CRC Press, Taylor & Francis Group, Boca Raton, Florida.

EB, 2015. Curare. Encyclopedia Britannica, Chicago, Illinois.

Finlayson-Pitts, B.J., Pitts, J.N., 2010. Chemistry of the Upper and Lower Atmosphere: Theory, Experiments, and Applications. Academic Press, San Diego, California.

Finlayson-Pitts, B.J., 2010. Atmospheric chemistry. Proceedings of the National Academy of Sciences 107 (15), 6566–6567.

Fox, R.D., 1996. Physical/chemical treatment of organically contaminated soils and sediments. Journal of the Air & Waste Management Association 46 (5), 391–413.

Hill, M.K., 2010. Understanding Environmental Pollution, third ed. Cambridge University Press, Cambridge, United Kingdom.

Jones, K.C., De Voogt, P., 1999. Persistent organic pollutants (POPs): state of the science. Environmental Pollution 100, 209–221.

Lee, S., Speight, J.G., Loyalka, S., 2014. Handbook of Alternative Fuel Technologies, second ed. CRC Press, Taylor & Francis Group, Boca Raton, Florida.

Mackay, D., Shiu, W., Ma, K., Lee, S., 2006. Handbook of Physical-Chemical Properties and Environmental Fate for Organic Chemicals, second ed. CRC Press, Taylor & Francis Group, Boca Raton, Florida.

Manahan, S.E., 2010. Environmental Chemistry, ninth ed. CRC Press, Taylor & Francis Group, Boca Raton, Florida.

Miller, D.R., 1984. Chemicals in the environment. In: Sheehan, P.J., Miller, D.R., Butler, G.C., Bourdeau, P. (Eds.), Effects of Pollutants at the Ecosystem Level, pp. 7–14 (Chapter 2).

Pickering, K.T., Owen, L.A., 1994. Global Environmental Issues. Routledge Publishers, New York.

Schwarzenbach, R.P., Gschwend, P.M., Imboden, D.M., 2003. Environmental Organic Chemistry, second ed. John Wiley & Sons Inc., Hoboken, New Jersey.

Speight, J.G., 2005. Environmental Analysis and Technology for the Refining Industry. John Wiley & Sons Inc., Hoboken, New Jersey.

Speight, J.G., 2011a. An Introduction to Petroleum Technology, Economics, and Politics. Scrivener Publishing, Beverly, Massachusetts.

Speight, J.G. (Ed.), 2011b. The Biofuels Handbook. Royal Society of Chemistry, London, United Kingdom.

Speight, J.G., 2013. The Chemistry and Technology of Coal, third ed. CRC Press, Taylor and Francis Group, Boca Raton, Florida.

Speight, J.G., 2014. The Chemistry and Technology of Petroleum, fifth ed. CRC Press, Taylor and Francis Group, Boca Raton, Florida.

Speight, J.G., 2017a. Environmental Organic Chemistry for Engineers. Butterworth-Heinemann, Elsevier, Oxford, United Kingdom.

Speight, J.G., 2017b. Environmental Inorganic Chemistry for Engineers. Butterworth-Heinemann, Elsevier, Oxford, United Kingdom.

Speight, J.G. (Ed.), 2017c. Lange's Handbook of Chemistry, seventeenth ed. McGraw-Hill Education, New York.

Speight, J.G., Arjoon, K.K., 2012. Bioremediation of Petroleum and Petroleum Products. Scrivener Publishing, Beverly, Massachusetts.

Speight, J.G., Islam, M.R., 2016. Peak Energy – Myth or Reality. Scrivener Publishing, Beverly, Massachusetts (Chapter 8).

Speight, J.G., Lee, S., 2000. Environmental Technology Handbook, second ed. Taylor & Francis, New York (Also, CRC Press, Taylor and Francis Group, Boca Raton, Florida).

Speight, J.G., Singh, K., 2014. Environmental Management of Energy from Biofuels and Biofeedstocks. Scrivener Publishing, Beverly, Massachusetts.

Tinsley, I.J., 2004. Chemical Concepts in Pollutant Behavior, second ed. John Wiley & Sons Inc., Hoboken, New Jersey.

Wondyfraw, M., 2014. Mechanisms and effects of acid rain on environment. Journal of Earth Science & Climatic Change 5 (6), 204–206.

FURTHER READING

Johnson, R.W., Gordon, G.E. (Eds.), 1987. The Chemistry of Acid Rain: Sources and Atmospheric Processes. Symposium Series No. 349. American Chemical Society, Washington, D.C.

Chapter 2

Chemicals in the Environment

1. INTRODUCTION

The increasing awareness among the public about the chemicals in the environment, their disposition, and the fate of these chemicals have created the need to find reliable mechanisms to assess the environmental behavior and effects of new chemicals. Whether a chemical will pose a hazard to the environment or constitute a benign insert into the environment will depend on the concentration levels it will reach in various ecosystems and whether or not those concentrations are toxic to the floral and faunal species within the ecosystem. It is, therefore, important to determine expected environmental distribution patterns of chemicals and the reactions of chemical with the environment to identify which areas will be of primary environmental concern (Table 2.1) (Tinsley, 2004).

In order to assess the behavior of a chemical in the environment, typical physical and chemical properties play a role but a property that is not often considered in the partitioning of a chemical in an ecosystem is a function of the chemical structure, the physical structure, and the properties of the chemical (Chapters 3 and 10). This leads to a determination of the way in which the chemicals (or the mixture of chemicals) are distributed among the different environmental phases. These phases may include air, water, organic matter, mineral solids, and even organisms.

However, in the absence of contradictory evidence it is best to assume that chemicals released into the environment do, without many exceptions, cause pollution which is the contamination of the physical and biological components of the earth system (atmosphere, aquasphere, and geosphere) to such an extent that the normal environmental processes, as well as the flora and fauna, are adversely affected. Thus, chemicals in the environment, in the context of this chapter and this book, is the introduction of chemicals or their byproducts such as radiant energy, noise, heat, or light into the environment that cause harm or discomfort to the floral and faunal species.

Pollution by chemicals can take the form of overloading an ecosystem by (1) returning or disposing of naturally occurring chemicals to the ecosystem in amounts that exceed the natural abundance of the chemicals or (2) anthropogenic chemicals that are released into the ecosystem (Chapter 1). In both

Reaction Mechanisms in Environmental Engineering. https://doi.org/10.1016/B978-0-12-804422-3.00002-X

TABLE 2.1 Summary of the General Mechanisms That Occur After the Discharge of a Chemical Into the Environment

Environmental Process	Definition
Adsorption/hetero-aggregation′	The association of the discharged chemical with other solid surfaces
Desorption	Detachment of the chemical from other surfaces
Agglomeration	Reversible coagulation of primary particles to form clusters
Aggregation (homoaggregation)	Irreversible fusing of primary particles to form larger particles
Sedimentation	A chemical in suspension settles out of (say) a water phase
Dissolution	A chemical dissolves (release of individual ions or molecules) in water
Precipitation	The process of dissolved species forming a solid phase which separates from the liquid phase
Oxidation (chemical)	The chemical is oxidized by the loss of electrons
Reduction (chemical)	The chemical is reduced by the uptake of electrons
Photocatalytic degradation	Chemical change induced by light, which includes excitation of photocatalytic chemicals (absorption of a photon causing generation of free radical species) and photolysis of the chemical or components of a mixture
Biologically mediated processes	A chemical undergoes a transformation due to the presence of living organisms, such as biological oxidation and biological degradation and interactions with bio-macromolecules

cases, pollution occurs when a (natural or manufactured) chemical is released to an ecosystem at a rate faster than the ecosystem can accommodate, by any of the usual processes such as dispersion, breakdown, recycling, or storage in some benign form.

Thus, pollution occurs when (1) the natural environment is incapable of decomposing the unnaturally generated chemicals (*anthropogenic pollutants*), and (2) when there is a lack of knowledge on the means by which these chemicals can be treated for disposal. This leaves the environment (or any ecosystem) subject to any negative impacts on crucial environmental chemistry.

Although pollution has been known to exist for a very long time (at least since people started using fire thousands of years ago), it had seen the growth of truly global proportions only since the onset and expansion of the Industrial Revolution during the 19th century. In fact, the Industrial Revolution brought with it technological progress such as discovery of crude oil (and uses for crude oil products). The technological progress facilitated the manufacture of chemicals—with the accompanying chemical waste—that became one of the main causes of serious deterioration of natural resources and pollution of the environment. At the same time, of course, development of natural sciences and the various applied sciences led to the better understanding of negative effects produced by pollution on the environment.

Pollution of the environment is caused by the release of chemical waste that has detrimental effects on the environment (Miller, 1984; Speight, 2017a,b). Environmental pollution is often divided into pollution of (1) the air, (2) the water, and (3) the land or soil. The range of chemicals is broad and often refers to hundreds of thousands of several different types of pollutants, including toxic as well as high concentrations of normally innocuous compounds (Miller, 1984; Speight, 2017a,b). While the blame for much of the pollution is caused by the chemicals industry, domestic sources of pollution include automobile exhaust and municipal waste. The chemicals in such wastes can react with tissues in the body and change the structure and function of the organ, cause abnormal growth and development of the individual, or bind with the genetic material of cells and cause cancer (Goyer, 1996; Goyer and Clarkson, 1996; Gallo, 2001).

However, if the discharged chemical is indigenous to the ecosystem (i.e., the chemical is a naturally occurring compound) it can be (should be) classified as a pollutant when it is released into the system in amounts that are more than the natural concentration of the chemical in the ecosystem and by this increased concentration the chemical can cause harm to (or is destructive to) the flora and/or the fauna of the ecosystem. More specifically, pollutants are elements or compounds found in water supplies and may be natural in the geology or caused by activities of man through mining, industry, or agriculture. It is common to have trace amounts of many contaminants in water supplies, but amounts of these pollutants that are above the maximum contaminant levels may cause a variety of damaging effects to an ecosystem, depending upon the contaminant and level of exposure.

Persistent pollutants, like any chemical pollutant, can enter an ecosystem through the gas phase, the liquid phase, or solid phase and can resist degradation and are mobile over considerable distances (especially in the gas phase or through transportation in river systems) before being redeposited in a location that is remote to the location of their introduction (into the ecosystem). Furthermore, persistent pollutants can be present as vapors in the atmosphere or bound to (adsorbed on) the surface of soil or mineral particles and have variable solubility in water.

Many persistent pollutants currently (or were in the past) arise from the extensive use of agrochemicals (agricultural chemicals). Although some persistent pollutants arise naturally, for example, from various natural pathways, most are products of human industry and tend to have higher concentrations and are eliminated more slowly. If not removed, persistent pollutants will bioaccumulate and have significant impact on the flora and fauna of the environment. The most frequently used measure of the potential for bioaccumulations and persistence of any compound in the environment is a result of the physicochemical properties (such as partition coefficients and reaction rate constants).

However, there is the possibility that through the judicious use of resources and the application of the principles of environmental science, environmental engineering, and environmental analysis (disciplines involved in the study of the environment, as well as determining the *purity* of the environment), a state can be reached where pollution is minimal and does not pose a threat to the existing floral and faunal species. Such a program must, of necessity, involve not only well-appointed suites of analytical tests but also subsequent studies that cover the effects of changes to the environmental conditions in the ecosystem on the flora and fauna of that system. These suites of studies should include the relevant aspects of chemistry, chemical engineering, biochemistry, microbiology, hydrology, and climatology, as they can be applied to solve environmental problems.

The potential for pollution of chemicals starts during the production stage when not only the desired products are produced but there is also the potential (often the reality in many processes) to produce various byproducts. These products must be converted to useful products or discarded, to the detriment of the process and (in the past) to the detriment of the environment. The typical chemical synthesis process involves combining one or more feedstocks in one or more unit process operations. Typically, commodity chemicals tend to be synthesized in a continuous reactor (one unit process) while specialty chemicals are usually produced in one or more batch reactors (perhaps involving two or more unit processes). The yield of the chemical and the efficiency of the chemical process will determine the type and quantity of the byproducts.

Many of the process reactions (1) take place at high temperatures, (2) include one or two additional reaction components, and (3) involve the use of catalysts that may be based on a discharged chemical. Thus, many specialty chemicals may require a series of two or three reaction steps, each involving a different reactor system and each capable of producing one or more byproducts. Once the reaction is complete, the desired product must be separated from the byproducts, often by a separate unit operation in which any one (or more) of several separation techniques such as settling, distillation, or refrigeration may be used. In addition, to produce the saleable item, the final product may be further processed using process techniques such as spray drying or pelletizing. Frequently, byproducts are also sold and their value can influence the production efficiency and economics of the process.

Despite numerous safety protocols that have been adopted by industry and which are in place, as well as the care taken to avoid environmental incidents that are harmful to the environment, every industry suffers accidents that lead to contamination of the environment by chemicals. It is therefore often helpful to be knowledgeable of the nature (the chemical and physical properties) of the chemical contaminants and the products of chemical transformations (when the ecosystem parameters interact with the chemicals) in order to understand not only the nature of the immediate and continuing chemical contamination but also of the chemical changes to the contaminants from which cleanup methods can be chosen (Speight and Lee, 2000). By way of clarification, a pollutant is a chemical that is not indigenous to the ecosystem.

In the past, the existence and source of such information was generally unknown and, if known, was not always used to determine any potential environmental issues (contamination and cleanup). When the existence and sources of the relevant information are known, decisions must be made for environmental scientists and engineers to make an informed, and often quick, decision on the next steps, even if it is decided later not to use the information for a site, on the basis that not two sites are exactly alike.

However, on the basis that *it is better to know than to not know*, knowledge of the relevant data gives investigators and analysts the ability to assess whether a chemical discharge into the environment should be addressed and efforts made to halt the discharge of the chemical or whether the environment can take care of itself through biodegradation of the chemical. This is especially true for scientists and engineers involved in (1) site cleanup operations, (2) assessment of ecological risk, (3) assessment of ecological damage, and (4) the steps necessary to protect the environment. Modern databases relating to the properties of chemicals have been created, and there can be no reasons (or excuses) for not knowing or understanding the fundamental aspects of the behavior of chemicals that are discharged to the environment.

Furthermore, the capacity of the environment to absorb the effluents and other impacts of process technologies is not unlimited, as some would have us believe. The environment should be considered as an extremely limited resource, and discharge of chemicals into it should be subject to severe constraints. Indeed, the declining quality of raw materials, dictates that more material must be processed to provide the needed chemical products. Moreover, the growing magnitude of the products and effluents from industrial processes has moved above the line where the environment has the capability to absorb such process effluents without disruption.

As a commencement to this process of data examination and data use, this chapter introduces the terminology of environmental technology as it pertains to the sources and types of pollutants. Briefly, a *contaminant*, which is not usually classified as a pollutant unless it has some detrimental effect, can cause deviation from the normal composition of an environment. A *receptor* is an object (animal, vegetable, or mineral) or a locale that is affected by the pollutant. A *chemical waste* is any solid, liquid, or gaseous waste material that, if improperly managed or disposed of, may pose substantial hazards to human health and the environment. At any stage of the management process, a chemical waste may be designated by law as a *hazardous waste*. Improper disposal of these waste streams, such as solvents, in the past has created hazards to human health and the need for very expensive cleanup operations. Correct handling of these chemicals, as well as dispensing with many of the myths related to chemical processing can mitigate some of the problems that will occur, especially problems related to the flammability of liquids that will occur when incorrect handling is practiced. Chemical waste is also defined and classified into various subgroups.

2. SOURCES

The sources of discharged chemicals can be categorized as either (1) point sources or (2) nonpoint sources. Point sources are discrete discharges of chemicals that are usually identifiable and measurable, such as industrial or municipal effluent outfalls, chemical or petroleum spills and dumps, smokestacks, and other stationary atmospheric discharges. Nonpoint sources are more diffuse inputs over large areas with no identifiable single point of entry such as agrochemical (pesticide and fertilizer) runoff, mobile sources emissions (automobiles), atmospheric deposition, desorption, or leaching from very large areas (contaminated sediments or mine tailings), and groundwater inflow. Nonpoint sources often include multiple smaller point sources, such as septic tanks or automobiles that are impractical to consider on an individual basis. Thus, the identification and characterization of a source is relative to the environmental compartment or system being considered. For example, there may be many chemical sources to a river; each must be considered when assessing the hazards of the chemicals to aquatic life in the river or to humans who might drink the water or consume the fish and shellfish. However, these chemical sources can be well mixed in the river (dilution, see Chapter 5), resulting in a rather homogeneous and large point source to a downstream lake or estuary.

Thus, environmental pollutants arise from a variety of sources that are constituent parts of the pollution process. The actual causative agents of environmental pollution can occur in *gaseous*, *liquid*, or *solid* form. Most important, pollutants (or any pollutants for that matter) are transboundary insofar as they (the pollutants) do not recognize boundaries. In addition, many pollutants cannot be degraded by living organisms and therefore stay in the ecosphere for many years, even decades. pollutants also destroy biota and the habitat and, because of this, it is emphasized that pollutants can present a serious long-term global problem that affects more or less every country and, therefore, can only be solved by a coordinated set of actions and the necessary commitment of nations of the earth (with no exceptions) to international environmental agreements.

The contamination of the environment by chemicals (Miller, 1984; Speight, 2017a,b) can occur from several process sources (Speight, 2017a,b) and the effects of the chemical on the environment are determined by the chemical and physical properties of the chemical (Chapters 3 and 10): for example, most toxic pollutants originate from human-made sources, such as mobile sources (such as cars, trucks, buses) and stationary sources (such as factories, refineries, power plants), as well as indoor sources (such as building materials and activities such as cleaning).

2.1 Air Pollution

Chemicals can be emitted directly into the atmosphere or formed by chemical conversion of precursor species through chemical reactions. In these reactions, highly toxic chemicals can be converted into less toxic products, but the result of the reactions can also be products having a higher toxicity that the starting chemicals. In order to understand these reactions, it is also necessary to understand the chemical composition of the natural atmosphere, the way gases, liquids, and solids in the atmosphere interact with each other and with the earth's surface and associated biota, and how human activities may be changing the chemical and physical characteristics of the atmosphere (Table 2.2).

Thus, air pollution occurs in many forms but can generally be thought of as gaseous and particulate contaminants that are present in the atmosphere of the earth. Air pollution can further be classified into two sections: (1) visible air pollution and (2) invisible air pollution. The former is customarily referred to as *smog* while the latter remains unnamed. Furthermore, there are a number of critical environmental issues associated with a changing atmosphere, including photochemical smog, global climate change, toxic air pollutants, acidic deposition, and stratospheric ozone depletion.

Air pollution is caused by harmful gases and particulate matter (PM) in the air, which originate from various sources (Miller, 1984; Speight and Lee, 2000; Speight, 2017a,b). In fact, the sources of air pollution can impact many areas even though the vast majority of air pollution is created elsewhere, even outside of national boundaries. There are four main types of air pollution sources: (1) mobile sources, such as cars, buses, planes, trucks, and trains, (2) stationary sources, such as power plants, oil refineries, industrial facilities, and factories, (3) area sources, such as agricultural areas, cities, and wood burning fireplaces, and (4) natural sources, such as wind-blown dust, wildfires, and volcanoes. Mobile sources account for more than half of all the air pollution in the United States and the primary mobile source of air pollution is the automobile, according to the United States Environmental Protection Agency. Stationary sources, like power plants, emit large amounts of pollution from a single location; these are also known as point sources of pollution.

TABLE 2.2 Atmospheric Pollutants and Their Effects

Atmospheric Dust
Atmospheric dust consists of a mixture of solid and liquid particles suspended in the atmosphere varying in composition, source, and size. Atmospheric dust particles can be removed out of the atmosphere by dry and wet deposition and fall back on soil, vegetation, or watercourses. Atmospheric dust particles can be classified according to their diameter (measured in micrometers or μm. 1000 μm equivalent to 1 mm) ranging from 0.005 to 100 μm. Sources of dust particles can be natural (from volcanic eruption, sea aerosols, spores, pollen, and soil erosion) or manmade (vehicular traffic, industrial emissions and combustion processes).
Benzene
Benzene is classified as a polycyclic aromatic hydrocarbon (PAH). It is a liquid substance, but at high temperatures it has a rapid volatilization process passing from a liquid phase into a gas phase. Benzene is either natural or manmade and can, for example, be generated by volcanic eruptions.
Acid Deposits
The atmosphere contains acid-reaction substances that deposit on the earth's surface and contaminate it: they are the so-called "acid deposits." The substances that make these deposits acid are generally nitric acid and sulfuric acid that form by the reaction of water and nitrogen oxides and sulfur oxides (SO_x) contained in polluted air. Nitrogen oxides are produced by the combustion of fossil energy sources rich in sulfur—especially coal and lignite—and by volcanic eruptions. Sulfur oxides can have a natural origin (lightning, fires, bacterial decomposition of organic materials, biological processes of the oceans), or an anthropogenic origin, deriving from the combustion of fossil energy sources.
Ozone
Ozone (O_3) is a gas found in high levels in the stratosphere, in a region also known as the ozone layer, in the order of 50,000 and 130,000 ft above the surface where it plays an important role screening the sun's ultraviolet radiations which are harmful for living organisms. Ozone in the lower atmosphere ozone pollution refers to high ozone concentration in the troposphere, the only atmospheric layer which can support life, and which should not be mistaken for the ozone hole. Tropospheric ozone is formed by the interaction of sun radiations and primary pollutants, especially nitrogen dioxide.
Radioactive Pollution
The sudden explosion that occurred in April 1986 in the Chernobyl plant, in the former Soviet Union, brought the whole world face to face with the tragic consequences of the nuclear pollution of the air, related in particular to the international dimension of this risk of pollution. Damage caused by nuclear pollution is not limited to a specific area, but it can affect large regions, even very far from its source.

TABLE 2.2 Atmospheric Pollutants and Their Effects—cont'd

Photochemical Pollution
Photochemical smog is a typical form of pollution of all the main urban and industrial areas of the world. It occurs in or near areas with a high traffic density, in the presence of specific climatic conditions (no wind or weak winds, high temperatures, etc.) that cause the concentration of polluting gases to increase and prevent them from dispersing. In these areas, the concentrations of some gases (tropospheric ozone, carbon monoxide, particulate, VOC, nitrogen oxides, etc.) very often exceed the threshold values, above which they are risks for human health, farming, and natural vegetation.

In the current context, *toxic air pollutants* are a class of chemicals which may potentially cause health problems in a significant way. Persistent toxic pollutants, such as mercury, are of particular concern because of their global mobility and ability to accumulate in the food chain. *Primary air pollutants* are those pollutants that are emitted directly into the air from pollution sources. On the other hand, *secondary air pollutants* are formed when primary pollutants undergo chemical changes in the atmosphere.

Fossil fuels which are sources of pollutants include coal, natural gas and crude oil, as well as the more familiar fuels refined from crude oil refining, which include gasoline, diesel fuel, and fuel oil (Speight, 2014, 2017c). The burning, or combustion, of fossil fuels is a major source of pollutants which contribute to acid rain, which is a general term used to describe several ways that acids fall out of the atmosphere (Speight, 2013, 2014, 2017a,b; Wondyfraw, 2014).

The major air pollutants are: (1) carbon monoxide, CO, (2) nitrogen oxides, NO_x, such as nitric oxide, NO, and nitrogen dioxide, NO_2 (3) ozone, O_3, (4) sulfur oxides, such as dioxide (SO_2), as well as sulfur trioxide (SO_3), and (5) PM.

Carbon monoxide (CO) is a colorless and odorless gas that is generated during the combustion (mostly in automobiles) of the typical hydrocarbon fuels derived from crude oil. Carbon monoxide is released when engines burn fossil fuels and emissions of this gas are higher when engines are not tuned correctly and when fuel there undergoes incomplete combustion—complete combustion of hydrocarbon fuels results in the production of carbon dioxide (CO_2) and water (H_2O). Industrial and domestic furnaces and heaters can also emit high concentrations of carbon monoxide if they are not properly maintained.

Carbon monoxide has a 4-month lifetime in the atmosphere, where it reacts with the hydroxyl radical to form carbon dioxide. The sequence of reactions that transform carbon monoxide to carbon dioxide and regenerate the hydroxyl radical are:

$$CO + HO^{\bullet} \rightarrow CO_2 + H^{\bullet}$$

$$H^{\bullet} + O_2 + M \rightarrow HOO^{\bullet} + M$$

$$HOO^{\bullet} + HOO^{\bullet} \rightarrow H_2O_2 + O_2$$

$$H_2O_2 + \text{energy} \rightarrow 2HO^{\bullet}$$

Carbon monoxide, generated by the incomplete combustion of hydrocarbons, displaces and prevents oxygen from binding to hemoglobin (also spelled: haemoglobin) in the blood and causes asphyxiation.

Complete combustion of heptane (a fuel constituent):

$$C_7H_{16} + 11O_2 \rightarrow 7CO_2 + 8H_2O$$

Incomplete combustion of heptane to produce carbon monoxide, water, and carbon:

$$C_7H_{16} + 6O_2 \rightarrow 4CO + 8H_2O + 3C$$

Also, carbon monoxide binds with metallic pollutants and causes them to be more mobile in air and water. These pollutants are emitted from large stationary sources such as fossil fuel—fired power plants, smelters, industrial boilers, petroleum refineries, and manufacturing facilities, as well as from area and mobile sources. They are corrosive to various materials which cause damage to cultural resources, can cause injury to ecosystems and organisms, aggravate respiratory diseases, and reduce visibility.

Nitrogen dioxide (NO_2) is a reddish-brown pungent gas that comes from the burning of fossil fuels that is produced when a nitrogen-containing fuel is combusted or when nitrogen in the air reacts with oxygen at very high temperatures, such as in certain types of industrial furnaces. Nitrogen dioxide can also react in the atmosphere to form ozone, acid rain, and particles.

Ozone (O_3) is formed naturally when oxygen molecules are photochemically dissociated into oxygen atoms that can then react with a second oxygen molecule to make ozone. It is an example of a secondary pollutant and is formed when nitrogen oxides (NO_x) and volatile compounds (VOCs) are mixed and warmed by sunlight. The presence of nitrogen oxides (NO, NO_2) leads to higher than normal background levels of ozone through several well-understood photochemical reactions.

This gas is a desirable substance in the stratosphere, but it is a major environmental hazard at ground level (the troposphere) where it contributes to the formation of smog. This harmful ozone in the lower atmosphere should not be confused with the protective layer of ozone in the upper atmosphere (stratosphere), which screens out harmful ultraviolet rays. The presence of nitrogen oxides (nitric oxide, NO, and nitrogen dioxide, NO_2)

leads to higher than normal background levels of ozone through several well-understood photochemical reactions. Nitrogen oxides (including nitrogen dioxide) arise from the burning of gasoline of various fossil fuel-derived flues.

Sulfur dioxide (SO_2) is a pungent corrosive gas that arises predominantly from the burning of coal or crude oil in power plants, as well as from factories that produce chemicals, paper, or fuel. Nitrogen oxides and sulfur dioxide can combine with water to form acids, which not only irritate the lungs but also contribute to the long-term destruction of the environment due to the generation of acid rain. Thus, *acid rain* is formed when sulfur dioxide and nitrogen oxides react with water vapor and other chemicals in the presence of sunlight to form various acidic compounds in the air. The principle source of acid rain–causing pollutants, sulfur dioxide and nitrogen oxides, are from fossil fuel combustion and from the combustion of fossil fuel–derived fuels (Speight and Lee, 2000):

$$2[C]_{\text{fossil fuel}} + O_2 \rightarrow 2CO$$

$$[C]_{\text{fossil fuel}} + O_2 \rightarrow CO_2$$

$$2[N]_{\text{fossil fuel}} + O_2 \rightarrow 2NO$$

$$[N]_{\text{fossil fuel}} + O_2 \rightarrow NO_2$$

$$[S]_{\text{fossil fuel}} + O_2 \rightarrow SO_2$$

$$2SO_2 + O_2 \rightarrow 2SO_3$$

$$SO_2 + H_2O \rightarrow H_2SO_3 \text{ (sulfurous acid; contributor to acid rain)}$$

$$SO_3 + H_2O \rightarrow H_2SO_4 \text{ (sulfuric acid; contributor to acid rain)}$$

$$NO + H_2O \rightarrow HNO_2 \text{ (nitrous acid; contributor to acid rain)}$$

$$3NO_2 + 2H_2O \rightarrow HNO_3 \text{ (nitric acid; contributor to acid rain)}$$

Rainfall that has a pH value of less than 5.6 is considered as acid rain. A more precise term for acid rain is acid deposition, which has two parts: (1) wet deposition and (2) dry deposition. *Wet deposition* refers to acidic rain, fog, and snow. Dry deposition refers to acidic gases and particles.

PM— the term particulates is also used— refers to the mixture of solid particles that are suspended in the air– to remain in the air, particles usually must be less than 0.1 mm wide and can be as small as 0.00005 mm. PM can either be carbonaceous matter (such as soot) or noncarbon matter (such as mineral matter). More generally, PM can be divided into two types: (1) coarse particles and (2) fine particles. Coarse particles are formed from sources such as road dust, sea spray, and construction dust. Fine particles are formed when fuel is burned in automobiles and power plants and includes sulfate ($-SO_4^{2-}$) derivatives and nitrate ($-NO_3^-$) derivatives.

A major health hazard is from PM that is smaller than 2.5 microns (μm) in diameter and contains lead, beryllium, mercury, cadmium, and chromium. These particles are removed from the atmosphere by settling out over time or by precipitation events. Mercury is the only metal that exists as a gas and is therefore the only metal that exists in a steady-state concentration in air. All other metals are emitted to the atmosphere from natural or anthropogenic sources, and then removed by settling or precipitation events. However, gaseous substances remain in the atmosphere for long residence times and would continue to build up to much higher levels than observed today if it were not for the gas phase reactions that convert these substances to water-soluble species.

Chlorine is another atmospheric pollutant that, because of its structure, can form a reactive species (a free radical):

$$Cl_2 \rightarrow 2Cl\bullet$$

However, the most common atmospheric form of chlorine, however, is as chloride (Cl^-) which is water-soluble and washes out of the atmosphere quickly. chloride derivatives, therefore, have very short lifetimes and do not always constitute a threat to the environment.

Chlorofluorocarbons (CFCs) and some volatile compounds (VOCs) contain covalently bonded chlorine atoms, and being relatively inert substances, will remain in the atmosphere for many years. By remaining in the atmosphere for such prolonged periods, these molecules migrate across the troposphere/stratosphere boundary where they become exposed to much higher levels of ultraviolet radiation than experienced at ground level. Ultraviolet radiation can cause covalent bonds to separate into individual atoms containing unpaired electrons (free radicals, such as the chlorine free radical, Cl•) that can react further to create environmental pollutants.

It is often stated that the production of energy from renewable sources (such as biomass) may mitigate many pollution problems. However, on the downside, biomass as proposed for use as a fuel (or fuel source) can also be a source of environmental pollution and knowledge of the components of biomass feedstock is important for process control and for handling coproducts and wastes resulting from energy and fuel utilization of biomass. In fact, many elements listed in the periodic table of elements (Fig. 2.1) may be present in biomass. In fact, majority of the alkali metals (lithium, Li, sodium, Na, potassium, K, and rubidium, Rb) will substantially modify soil if applied as a fertilizer. Only magnesium (Mg), calcium (Ca), and strontium (Sr) of the alkali earth metals; manganese (Mn), copper (Cu), and zinc (Zn) of the period 4 transition metals; and molybdenum (Mo) and cadmium (Cd) of the period 5 transition metals may be permitted (in certain regions) to exceed regulatory limits if used as a fertilizer. The heavy elements occur in concentrations too low to cause concern except for selenium (Se).

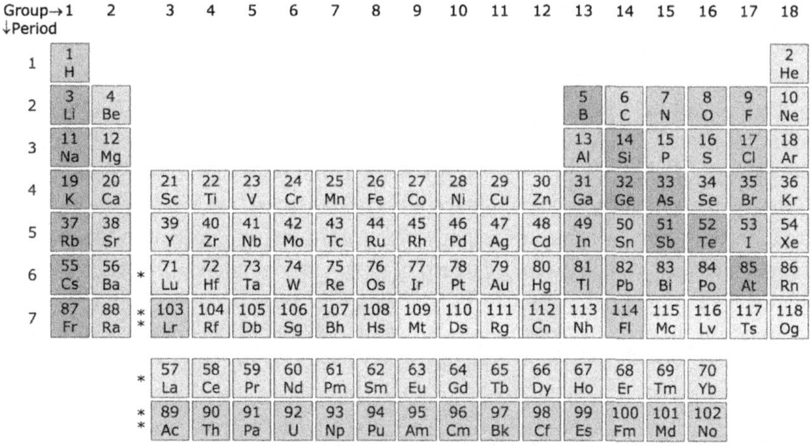

Group→ ↓Period	1	2		3	4	5	6	7	8	9	10	11	12	13	14	15	16	17	18
1	1 H																		2 He
2	3 Li	4 Be												5 B	6 C	7 N	8 O	9 F	10 Ne
3	11 Na	12 Mg												13 Al	14 Si	15 P	16 S	17 Cl	18 Ar
4	19 K	20 Ca		21 Sc	22 Ti	23 V	24 Cr	25 Mn	26 Fe	27 Co	28 Ni	29 Cu	30 Zn	31 Ga	32 Ge	33 As	34 Se	35 Br	36 Kr
5	37 Rb	38 Sr		39 Y	40 Zr	41 Nb	42 Mo	43 Tc	44 Ru	45 Rh	46 Pd	47 Ag	48 Cd	49 In	50 Sn	51 Sb	52 Te	53 I	54 Xe
6	55 Cs	56 Ba	*	71 Lu	72 Hf	73 Ta	74 W	75 Re	76 Os	77 Ir	78 Pt	79 Au	80 Hg	81 Tl	82 Pb	83 Bi	84 Po	85 At	86 Rn
7	87 Fr	88 Ra	**	103 Lr	104 Rf	105 Db	106 Sg	107 Bh	108 Hs	109 Mt	110 Ds	111 Rg	112 Cn	113 Nh	114 Fl	115 Mc	116 Lv	117 Ts	118 Og

*	57 La	58 Ce	59 Pr	60 Nd	61 Pm	62 Sm	63 Eu	64 Gd	65 Tb	66 Dy	67 Ho	68 Er	69 Tm	70 Yb
**	89 Ac	90 Th	91 Pa	92 U	93 Np	94 Pu	95 Am	96 Cm	97 Bk	98 Cf	99 Es	100 Fm	101 Md	102 No

FIGURE 2.1 The periodic table of the elements showing the groups and periods including the lanthanide elements and the actinide elements.

Also, wood ash is a carrier of many elements (such as barium, calcium, lithium, magnesium, potassium, rubidium, and strontium) and some transition elements (copper, cadmium manganese, molybdenum, silver, and zinc). In contrast, ash of herbaceous plant material in addition to potassium contains varying amounts of cadmium, calcium, lithium, magnesium, molybdenum, selenium, and sodium. Water leaching results in significant losses for anionic chlorine and sulfur as well as for most of the alkali metals, thus making resulting ash from such treated feedstock less attractive as potassium-containing fertilizers although fuel properties are enhanced for thermal conversion (Thy et al., 2013).

2.2 Water Pollution

Water pollution is the contamination of the various bodies of water (e.g., ponds, lakes, rivers, oceans, as well as aquifers and groundwater) which occurs when (in the context of this book) chemicals are directly or indirectly discharged into the water bodies without adequate pretreatment to remove harmful or hazardous chemicals (Miller, 1984; Speight, 2017a,b). Water pollution affects the entire floral and faunal communities living in these bodies of water. In almost all cases the effect is damaging not only to individual floral and/or faunal species and population, but also to any natural biological community.

Water pollution has become a widespread phenomenon and has been known for centuries, particularly the pollution of rivers and groundwater. By way of example, in ancient time up to the early part of the 20th century, many cities deposited waste into the nearby river or even into the ocean. It is only very recently (because of serious concerns for the condition of the environment) that an understanding of the behavior and fate of chemicals, which are discharged to the aquatic environment as a result of these activities, is essential to the control of water pollution. In rivers the basic physical movement of pollutant molecules is the result of advection but superimposed upon this are the effects of dispersion and mixing with tributaries and other discharges. Some of the chemicals discharged are relatively inert, so their concentration changes only due to advection, dispersion, and mixing. However, many substances are not conservative in their behavior and undergo changes due to chemical or biochemical processes, such as oxidation.

In addition, there are many indications that the chemicals in the aquasphere (also called, when referring to the sea, the marine aquasphere) are subject to intense chemical transformations and physical recycling processes implying that a total organic-carbon approach is not sufficient to resolve the numerous processes occurring. The transport of anthropogenically produced or distributed organic compounds such as petroleum hydrocarbons and halogenated hydrocarbons, including the polychlorobiphenyl derivatives, the DDT family, and the Freon derivatives and the chemistry of these organic chemicals in water is not fully understood.

Typically, water is referred to as polluted when it is impaired by anthropogenic contaminants after which it either does not support a human use (for example, as drinking water) or undergoes a marked shift in its ability to support its constituent floral and faunal biotic communities. It must also be recognized that natural phenomena such as volcanic activity, algae bloom activity, storms, and earthquakes also cause major changes in water quality and the ability of the water to support any ecological communities therein.

The specific contaminants leading to pollution in groundwater include, amongst other effects, a wide range of chemicals (Miller, 1984; Speight, 2017a,b), as well as physical changes such as elevated temperature and discoloration due to dissolution of the chemicals. In fact, chemicals have been reported to constitute by far the greatest proportion of chemical contaminants in drinking water (Fawell, 1993). These chemicals are present in greatest quantity as a consequence of natural processes, but several important contaminants are present because of human activities.

While many of the chemicals (such as, for example, calcium, sodium, iron, and manganese) that are regulated may be indigenous to the water body or to the region—from which these chemicals could be transported into the water from mineral deposit—it is the concentration of these chemicals in the water that is often the key in determining what is a natural component of water and what is a contaminant. Indeed, a high concentration (higher than the indigenous concentration) of even a naturally occurring chemical can have a negative impact on aquatic flora and fauna. In addition, some chemicals undergo chemical change (Speight, 2017a,b) over long periods of time in groundwater reservoirs or in aquifers where the chemicals contact clay minerals that can serve as adsorbents of the chemicals (Nutting, 1943; Swineford, 1960; Velde, 1992; Velde and Meunier, 2008; Bergaya et al., 2011).

Generally, groundwater pollution is much more difficult to abate than surface pollution because groundwater can move great distances through unseen aquifers. On the beneficial side, nonporous aquifers such as clay minerals can partially purify water of bacteria or various contaminants by simple filtration, association with minerals (adsorption and absorption), dilution, and, in some cases, chemical reactions through biological activity. However, on the adverse side, the pollutants may be transformed from water contaminants to soil contaminants. Furthermore, groundwater that moves through (natural) open fractures and is not filtered can be transported as easily as surface water. As part of the transportation process, there is a variety of secondary effects that are not due to the presence of the original pollutant but from a chemical derivative that has occurred as part of a chemical transformation process (Speight, 2017a,b).

Heavy metals are naturally occurring elements in soil environments and have high atomic weights and densities, at least five times greater than the density of water (Fergusson, 1990). The processes of weathering of parent materials at levels that are regarded as *trace amounts* (<1000 mg/kg) and rarely toxic (Pierzynski et al., 2000; Kabata-Pendias and Pendias, 2001). Due to the disturbance and acceleration of nature's slowly occurring geochemical cycle of metals by man, most soils of rural and urban environments may accumulate one or more of the heavy metals above defined background values high enough to cause risks to human health, plants, animals, ecosystems, or other media (D'Amore et al., 2005). The heavy metals essentially become contaminants in the soil environments because (1) their rates of generation via anthropogenic processes and cycles are more rapid relative to natural processes and ones, (2) they become transferred from mines to random environmental locations where higher potentials of direct exposure occur, (3) the concentrations of the metals in discarded products are relatively high compared to those in the receiving environment, and (4) the chemical form (species) in which a metal is found in the receiving environmental system may render it more bioavailable (D'Amore et al., 2005).

With the assumption that density of the metal and toxicity are interrelated, heavy metals also include metalloids, such as arsenic, that are able to induce toxicity at low level of exposure. In recent years, there has been an increasing ecological and global public health concern associated with environmental contamination by these metals. Reported sources of heavy metals in the environment include natural geological sources; industrial, agricultural, pharmaceutical, domestic effluents; and atmospheric sources (He et al., 2005). Environmental pollution is very prominent in point source areas such as mining, foundries, and smelters, and areas of other metal-based industrial operations (Fergusson, 1990; Shallari et al., 1998; Herawati et al., 2000; Bradl, 2002; He et al., 2005).

Heavy metals are also considered as trace elements because of their presence in trace concentrations (ppb range to less than 10 ppm) in various environmental matrices (Kabata-Pendias and Pendia, 2001). Their bioavailability is influenced by physical factors such as temperature, phase association, adsorption, and sequestration. It is also affected by chemical factors that influence speciation at thermodynamic equilibrium, complexation kinetics, lipid solubility, and octanol/water partition coefficients. Biological factors, such as species characteristics, trophic interactions, and biochemical/physiological adaptation, also play an important role (Hamelink et al., 1994).

Water pollution by heavy metals can also occur through metal corrosion, atmospheric deposition, soil erosion of metal ions and leaching of heavy metals, sediment resuspension, and metal evaporation from water resources to soil and groundwater. Natural phenomena such as weathering and volcanic eruptions have also been reported to significantly contribute to heavy metal pollution (Nriagu, 1989; Fergusson, 1990: Shallari et al., 1998; Bradl, 2002; He et al., 2005). Industrial sources include metal processing in refineries, coal burning in power plants, petroleum combustion, nuclear power stations and high-tension lines, plastics, textiles, microelectronics, wood preservation and paper processing plants (Pacyna, 1996; Arruti et al., 2010).

Water pollution by silt or sediments involves the disruption of sediments that flow from rivers into the sea. This reduces the formation of beaches, increases coastal erosion (the natural destruction of cliffs by the sea), and reduces the flow of nutrients from rivers into seas (potentially reducing coastal fish stocks). Increased sediments can also present an environmental problem such as during construction work, when soil, rock, and other fine powders sometimes enter nearby rivers in large quantities, causing them to become turbid (muddy or silted).

The effects of a chemical released into the marine environment (or any part of the aquasphere) depends on several factors such as (1) the toxicity of the chemical, (2) the quantity of the chemical, (3) the resulting concentration of the chemical in the water column, (4) the length of time that floral and faunal organisms are exposed to that concentration, and (5) the level of tolerance of the organisms, which varies greatly among different species and during the life cycle of the organism. Even if the concentration of the chemical is below what would be considered as the lethal concentration, a sublethal concentration of a chemical can still lead to a long-term impact within the aqueous marine environment. For example, chemically induced stress can reduce the overall ability of an organism to reproduce, grow, feed, or otherwise function normally within a few generations. In addition, the characteristics of some chemicals can result in an accumulation of the chemical within an organism (*bioaccumulation*) and the organism may be particularly vulnerable to this problem. Furthermore, subsequent biomagnification may also occur if the chemical (or a toxic product produced by one or more transformation reactions) can be passed on, following the food chain up, to higher flora or fauna.

Considering the marine environment and a spill of crude oil, complex processes of crude oil transformation start developing almost as soon as the oil contacts the water although the progress, duration, and result of the transformations depend on the properties and composition of the oil itself, parameters of the actual oil spill, and environmental conditions. The major operative processes are (1) physical transport, (2) dissolution, (3) emulsification, (4) oxidation, (5) sedimentation, (6) microbial degradation, (7) aggregation, and (8) self-purification.

In terms of *physical transport*, the distribution of oil spilled on the sea surface occurs under the influence of gravitation forces and is controlled by the viscosity of the crude oil, as well as the surface tension of water. In addition, during the first several days after the spill, a part of the oil is lost as evaporation oil (into the gaseous phase) and any water-soluble constituents disappear into the sea. The portion of the crude oil that remains is the more viscous fraction. Further changes take place under the combined impact of meteorological and hydrological factors and depend mainly on the power and direction of wind, waves, and currents. A considerable part of oil disperses in the water as fine droplets that can be transported over large distances away from the place of the spill.

Crude oil is not particularly *soluble* in water although some constituents may be water-soluble to a certain degree, especially low molecular—weight aliphatic and aromatic hydrocarbons. Polar compounds formed as a result of oxidation of some oil fractions in the marine environment also dissolve in seawater. Compared to evaporation process, the dissolution of cured oil constituents in water is a slow process. However, the *emulsification* of crude oil constituents in the marine environment does occur but depends predominantly on the presence of functional groups in the oil, which can increase with time due to oxidation. Emulsions usually appear when heavy oil is spilled into the ocean because of the higher proportion of polar constituents compared to conventional (lighter) crude oil (Speight, 2014). The rate of emulsification process can be decreased by use of emulsifiers—surface-active chemicals with strong hydrophilic properties used to eliminate oil spills—which help to stabilize oil emulsions and promote dispersing oil to form microscopic (invisible) droplets that accelerate the decomposition of the crude oil constituents in the water.

Oxidation is a complex process that can ultimately result in the destruction of the crude oil constituents. The final products of oxidation (such as hydroperoxide derivatives, phenol derivatives, carboxylic acid derivatives, ketone derivatives, and aldehyde derivatives) usually have increased water solubility. This can result in the apparent disappearance of the crude oil from the surface of the water. What is actually happening is the incorporation of functional groups into the oil constituents which results in a change in density with an increase in the ability of the constituents to become miscible (or emulsify) and sink to various depths of the ocean as these changes intensify. These chemical changes also result in an increase in the viscosity of the crude oil which promotes the formation of solid oil aggregates. The reactions of photooxidation and photolysis, in particular, also initiate transformation of the more complex (polar) constituents in the crude oil (Speight, 2014).

As these processes occur, some of the crude oil constituents are adsorbed on any suspended material and get deposited on the ocean floor (sedimentation), the rate of which is dependent upon the ocean depth—in deeper areas remote from the shore, sedimentation of oil (except for the heavy fractions) is a slow process. Simultaneously, the process of *biosedimentation* occurs —in this process, plankton and other organisms absorb the emulsified oil—and the crude oil constituents are sent to the bottom of the ocean as sediment with the metabolites of the plankton and other organisms. However, this situation radically changes when the suspended oil reaches the sea bottom—the decomposition rate of the oil on the ocean bottom abruptly ceases—especially, under the prevailing anaerobic conditions—and any crude oil constituents accumulated inside the sediments can be preserved for many months and even years. These products can be swept to the edge of the ocean (the beach) by turbulent condition at some later time.

The fate of most of the constituents of crude oil in the marine environment is ultimately defined by their transformation and degradation due to *microbial degradation*. The degree and rates of biodegradation depend, first of all, upon the structure of the crude oil constituents—alkanes biodegrade faster than aromatic constituents and naphthenic constituents—and with increasing complexity of molecular structure as well as with increasing molecular weight, the rate of microbial decomposition usually decreases. Besides, this rate depends on the physical state of the oil, including the degree of its dispersion, as well as environmental factors such as temperature, availability of oxygen, and the abundance of oil-degrading microorganisms.

Aggregation occurs when crude oil forms lumps or tar balls after the evaporation and dissolution of its relatively low-boiling fractions, emulsification of oil residuals, and chemical and microbial transformation. The chemical composition of oil aggregates is changeable but typically includes asphaltene constituents (up to 50%) and other highmolecular–weight constituents of the oil (Speight, 2014). These tar balls have an uneven shape and vary from 1 mm to 10 cm in size (sometimes reaching up to 50 cm) and complete their life cycle by slowly degrading in the water column, on the shore (if they are washed there by currents), or on the sea bottom (if they lose their floating ability).

Self-purification is a result of the processes previously described above in which crude oil in the marine environment rapidly loses its original properties and disintegrates into various fractions. These fractions have different chemical composition and structure and exist in different migrational forms and they undergo chemical transformations that slow after reaching thermodynamic equilibrium with the environmental parameters. Eventually, the original and intermediate compounds disappear, and carbon dioxide and water form. This form of self-purification inevitably happens in water ecosystems if the toxic load does not exceed acceptable limits.

As an example of a chemical transformation that can occur in a water system, the chemistry of methyl iodide (which is thermodynamically unstable in seawater) is known and its chemical fate is kinetically controlled. The equations showing the fate of methyl iodide are as follows:

$$CH_3I + Cl^- = CH_3Cl + I^-$$

$$CH_3I + Br^- = CH_3Br + I^-$$

$$CH_3Br + Cl^- = CH_3Cl + Br^-$$

$$CH_3X + H_2O = CH_3OH + X^-$$

In this equation, $X = Cl^-, Br^-, I^-$.

Chloride ion was theoretically predicted to be the most kinetically reactive species, with water second, and other anions of lesser importance. This suggested that methyl iodide in seawater would react predominantly via a nucleophilic substitution reaction with chloride ion to yield methyl chloride. Methyl iodide and the methyl chloride produced would also react with water, although more slowly, to yield methanol and halide ions. According to these experiments, substantial amounts of methyl chloride should be formed in seawater. Methyl chloride has a long half-life for decomposition by known reactions in seawater. Hence, its presence could be a useful label for some surface-derived water masses. Methyl chloride is in fact found in the atmosphere, where compared to methyl iodide, it is less stable to photodegradation reactions.

2.3 Land Pollution

Soils are formed by the decomposition of rock and matter over many years and the properties of soil vary from place to place with differences in bedrock composition, climate, and other factors. Certain chemical elements occur naturally in soils as components of minerals, yet may be toxic at some concentrations, while other potentially harmful substances can end up in soils through human activities. Moreover, the properties of soil are affected by past land use, current activities on the site, and nearness to pollution sources. A prime exposure pathway, either directly or indirectly, for soilborne chemical contaminants is via transport in the pore-water solution though the structured and chemically reactive medium of the various soils, which are home to a variety of floral and faunal species. In predicting pollutant transport, it is important to distinguish between whether the cause for pollution lies in the soil itself, or in the receiving water contained in the soil.

For the most part, soil pollution is also caused by means other than the direct addition of xenobiotic (manmade) chemicals such as agricultural runoff water, industrial waste material, acidic material, and (not forgetting or ignoring) radionuclides (radioactive substances and any radioactive fallout). Thus, soil pollution comprises the pollution of soils with materials, mostly chemicals that are out of place or are present at concentrations higher than normal, which may have adverse effects on humans or other organisms. Thus, soil pollution (soil contamination) as part of land degradation is caused by the presence of anthropogenic chemicals (humanmade or xenobiotic chemicals) or other alterations in the natural environment. This form of pollution is typically caused by industrial activity, agricultural chemicals, or improper disposal of waste chemical products, and pollution can typically be correlated with the degree of industrialization and intensity of chemical usage.

The term *heavy metals* includes transition metals such as cadmium, mercury, and lead (Fig. 2.1) and occur naturally in the soil environment as a result of pedogenesis (the process of soil formation) at levels that are regarded as *trace* amounts (<1000 mg/kg) (Pierzynski et al., 2000; Khan et al., 2008). Heavy metals essentially become contaminants in the soil environments because (1) their rates of generation via anthropogenic cycles are more rapid relative to natural ones, (2) they become transferred from mines to random environmental locations where higher potentials of direct exposure occur, (3) the concentrations of the metals in discarded products are relatively high compared to those in the receiving environment, and (4) the chemical form (species) in which a metal is found in the receiving environmental system may render it more bioavailable (D-Amore et al., 2005).

A simple mass balance of the heavy metals in the soil can be expressed as follows (Sposito and Page, 1984; Alloway, 1995; Lombi and Gerzabek, 1998):

$$M_{\text{total}} = (M_p + M_a + M_f + M_{ag} + M_{ow} + M_{ip}) - (M_{cr} + M_l)$$

In this equation, M is the heavy metal, p is the parent material, a is the atmospheric deposition, f is the fertilizer sources, ag are the agrochemical sources, ow are the waste sources, ip are other pollutants, cr is crop removal, and l is the losses by leaching and volatilization.

Heavy metals in the soil from anthropogenic sources tend to be more mobile, hence bioavailable than pedogenesis or lithogenesis (the formation of sedimentary rocks) (Kuo et al., 1983; Kaasalainen Yli-Halla, 2003). Metal-bearing solids at contaminated sites can originate from a wide variety of anthropogenic sources in the form of metal mine tailings, disposal of high metal wastes in improperly protected landfills, land application of fertilizer, animal manure, biosolids (i.e., sewage sludge), compost, pesticides, coal combustion residues, petrochemicals, and atmospheric deposition (Pierzynski et al., 2000; Basta et al., 2005; Khan et al., 2008; Zhang et al., 2010; Wuana and Okieimen, 2011).

A typical chemical process generates products and wastes from raw materials such as substrates and excess reagents. If most of the reagents and the solvents can be recycled, the mass flow looks quite different and the prevention of waste (to be disposed into the environment) can be achieved. Furthermore, there has been the suggestion that an efficiency factor (E-factor, i.e., the mass efficiency in terms of mass of the reactants) of a chemical process can be used to assess the transformation of a chemical in the environment:

E-factor = (mass of original waste)/(mass of transformed product)

Alternatively:

E-factor = (mass of raw feedstock − mass of product)/mass of product

Typically, E-factor used in the manufacturing processes for bulk and fine chemicals gives reliable data but in those processes the conditions are controlled, and any examination of amounts and properties can be achieved relatively conventionally (Chapter 3). The examination of chemical properties and chemical transformation in an ecosystem is much more difficult and the E-factor may, more than likely, not be as meaningful. In addition, any such E-factor and any related factors do not account for any type of toxicity of the chemical waste. Efforts (to determine an E-factor) are at a very preliminary stage and the parameters for the calculation of this factor needs much more consideration and development before being applied to chemical wastes in the environment.

3. DISTRIBUTION IN THE ENVIRONMENT

The concentration of chemicals in an ecosystem is controlled primarily by (1) the amount of chemicals present at the source, (2) the rate of release of the chemical(s) from the source, and various ecosystem-related factors, including aqueous adsorption (or desorption), precipitation, and diffusion. To accurately predict chemical transport through the subsurface, it is essential that the important geochemical processes affecting chemical transport be identified. Adsorption/desorption and dissolution/precipitation are usually the most important processes affecting interaction of the pollutant(s) with soils. In fact, dissolution/precipitation is more likely to be the key process where a chemical exists in a nonequilibrium state, such as at a point source, an area where high concentration of the chemical(s) or where high pH or oxidation–reduction (redox) gradients exist.

An adsorption/desorption process (Chapters 5 and 10) will likely be the key process controlling migration of the chemical(s) in areas where the naturally present constituents are already in equilibrium and only the anthropogenic chemicals are in a nonequilibrium state, such as in areas far from the point source. Diffusion is a dominant transport mechanism when advection is insignificant and is usually a negligible transport mechanism when water is being advected in response to various forces.

Thus, the distribution of chemical within the environment is subset to (1) the properties of the chemical or the properties and interactions of the chemicals in a mixture, (2) the properties of an environment, and (3) the potential of the chemical for transformation (Table 2.3) (Speight and Lee, 2000).

By way of recall, a chemical transformation is the conversion of a substrate (or reactant) to a product and involves a (or is)chemical reaction which is characterized by a chemical change, and yields one or more products, which usually have properties substantially different from the properties of the individual reactants. Reactions often consist of a sequence of individual substeps that can be described by means of chemical equations, which symbolically present the starting materials, end products, and sometimes intermediate products and reaction conditions.

Chemicals can enter the environment (air, water, and soil) when they are produced, used, or disposed and the impact on the environment is determined by the amount of the chemical that is released, the type and concentration of the chemical, and where it is found, as well as through any chemical transformation that occurs after the chemical has entered the environment, whether it is in the atmosphere, the aquasphere, or the terrestrial biosphere.

However, the entry of chemicals into the environment and the distribution of these chemicals within the environment is often complex and there have been many occasions when a significant amount of a chemical (or a mixture of chemicals) has entered an ecosystem and the effects of contamination are well defined. Generally, the assumption is that the chemical (or a mixture of chemicals) does not rapidly diffuse away but remains in the immediate vicinity at a noticeably high concentration or perhaps moves, but in such a way that concentration levels of the chemical remain high as it moves. Such cases would normally occur when large quantities of a substance were being stored, transported, or otherwise handled in concentrated form. Thus, due to leakages, spills, improper disposal, and accidents during transport, many chemicals have become subsurface contaminants that threaten various ecosystems.

Some chemicals do not degrade but accumulate in the bodies of human beings and animals. Other chemicals are harmful in that they give rise to problems such as allergies. It is best to assume that any chemical, in a small dose or in a large dose, has the potential to be toxic to flora and fauna. Therefore, background knowledge of the sources, chemistry, and potential risks of toxic heavy metals in contaminated soils is necessary for the selection of appropriate remedial options. Remediation of soil contaminated by heavy metals is necessary to reduce the associated risks, make the land resource available for agricultural production, enhance food security, and scale down land tenure problems. Immobilization, soil washing, and phytoremediation are frequently listed among the best available technologies for cleaning up heavy metal—contaminated soils. These technologies are recommended for field applicability and commercialization in the developing countries also, where agriculture, urbanization, and industrialization are leaving a legacy of environmental degradation.

TABLE 2.3 Processes by Which Chemicals Are Introduced and Dispersed Into the Environment

Dissolution
The equilibrium solubility (amount of dissolved material) of a chemical will influence the environmental fate and toxicity. Although solubility is material-dependent, the solubility of a solid material is not an inherent property as such but also depends on the media composition (e.g., ionic strength, ligands, pH, and temperature). Upon dissolution, the dissolved ions or molecules may form dissolved complexes with, for example, anions or organic matter (complexation) in the media or the ions may form a solid phase and sediment out (precipitation).
Agglomeration and Aggregation
Occurs as a result of attractive forces between particles, causing them to cluster together which can occur during production, storage, and use and after emission to the environment—independent of whether the nanoparticles are in solution, powder form, or in the gas phase. Aggregates are clusters of particles held together by strong chemical bonds or electrostatic interactions and are (generally) an irreversible process. Agglomerated particles are held together by weaker forces that do not exclude reversible processes. The direction of such processes will depend on the conditions of the surrounding media. The processes of aggregation and agglomeration significantly influence the fate and behavior of chemicals in the environment, with a dependency of particle properties (e.g., size, chemical composition, surface charge) as well as environmental conditions (e.g., mixing rates, pH, and NOM).
Sedimentation
Related to aggregation/agglomeration insofar as the velocity of the settling of particles in water depends on both the viscosity and density of water as well as particle radius and density. The implication is that larger agglomerates and aggregates will settle more quickly compared to more dispersed particles. Thus, agglomeration is the rate limiting factor for sedimentation of chemicals in the aquatic environment.
Interaction With Solid Surfaces (adsorption)
Is of importance for the transport and fate of chemicals in the environment, such as adsorption on to soil particles and clay minerals. Retention can be strongly linked to aggregation and agglomeration as larger particle sizes are more likely to adhere to or in the micropores of the soil. Soils with a higher fraction of natural colloids, e.g., soils with high content of clay, will generally show the highest retention of chemicals. Changes in pH affect the sorption process.
Biodegradation
A biological process that involves the decomposition of a chemical (especially an organic chemical) by microorganisms. Considered not to be relevant for many inorganic chemicals but more relevant for the carbon-based chemicals.

Continued

TABLE 2.3 Processes by Which Chemicals Are Introduced and Dispersed Into the Environment—cont'd

Biomodification
A biologically mediated transformation process which includes intraorganism processes that occur after uptake of the chemical or processes that are " indirectly" mediated by an organism, for example, by release exudates that bind to the chemical and change the properties. Biomodifications may also include changes in solubility and the agglomeration state as a result of organism uptake.

To effectively monitor changes in the environmental behavior of chemicals that are of most concern, it is extremely important to understand how these chemicals typically behave in natural systems and in specific ecosystems. Equally important is an understanding of how these chemicals might respond to specific best management practices. Some of the discharged chemicals only become problems at high concentrations that impair the beneficial uses of these ecosystems. An effective monitoring program explicitly considers how these chemicals may change as they move from a source into the groundwater, surface water, or into the soil. This includes an understanding of how a specific chemical may be introduced or mobilized within an ecosystem, how the chemical moves through an ecosystem, and the transformations that may occur during this process.

4. CHEMISTRY IN THE ENVIRONMENT

Chemical reactions occur at a characteristic rate (the reaction rate) at a given temperature and chemical concentration. Typically, reaction rates increase with increasing temperature because there is more thermal energy available to reach the activation energy necessary for breaking bonds between atoms. The general rule of thumb (see above) is that for every 10°C (18°F) increase in temperature the rate of an organic chemical reaction is doubled and there is no reason to doubt that this would not be the case for organic chemicals discharged into the environment. However, the composition of the chemicals is varied and there are very few indications of how a chemical (or a mixture of chemicals) will behave once the chemical (or the mixture) is discharged into the environment either as a single chemical or as a mixture. It is at this stage that a knowledge of chemical properties can bring some knowledge of predictability about chemical behavior.

Some chemicals can be harmful if released to the environment even when there is not an immediate, visible impact. Some chemicals are of concern as they can work their way into the food chain and accumulate and/or persist in the environment for prolonged periods, even for years; this is in direct contradiction of the earlier *conventional wisdom* (or unbridled optimism) that assumed that chemicals would either (1) degrade into harmless byproducts as a result of microbial or chemical reactions, (2) immobilize completely by binding to soil solids, or (3) volatilize to the atmosphere where dilution to harmless levels was assured.

This false assurance led to years of agricultural chemical use and chemical waste disposal with no monitoring of atmosphere, or groundwater (the aquasphere), or soil (the terrestrial biosphere) in the vicinity of discharge.

Long-term trends of chemical species in the environment are determined by emissions from anthropogenic and natural sources, as well as by transport of the chemical, physical, and chemical processes that affect the behavior of the chemical, and its deposition. while continually increasing emissions of such trace species as carbon dioxide (CO_2), nitrous oxide (N_2O), and methane (CH_4) that can arise from transformation occurring during the life cycle of chemicals are predicted to raise global temperatures via the *greenhouse effect*, growing emissions of sulfur dioxide (SO_2), which forms sulfate ($-SO_4$) aerosol through oxidation most likely will have a cooling effect by reflecting solar radiation back to space. However, these postulates do not take into account the fact that the earth is in an interglacial period during which time there will be an overall rise in climatic temperature as the natural order of climatic variation. Therefore, the extent of the anthropogenic contributions to temperature rise (climate change) cannot be accurately assessed (Speight and Islam, 2016). Complicating matters is the fact that the chemical reactions are sensitive to climatic conditions, being functions of temperature, water vapor, as well as a variety of other physical parameters.

5. INDUSTRIAL CHEMICALS AND HOUSEHOLD CHEMICALS

Before delving into the realm of chemicals introduced into the environment, it is necessary for any investigator to recognize that there are chemicals that exist naturally in the environment, and they must be considered before accurate assessment can be made of the behavior and effects of chemicals on the environment.

Naturally occurring organic chemicals are often grouped under the umbrella name natural organic matter (NOM) is an inherently complex mixture of polyfunctional organic molecules. Because of their universality and chemical reversibility, oxidation/reductions (redox) reactions of NOM have an especially interesting and important role in geochemistry. Variabilities in NOM composition and chemistry make studies of its redox chemistry particularly challenging, and details of NOM-mediated redox reactions are only partially understood. This is in large part due to the analytical difficulties associated with NOM characterization and the wide range of reagents and experimental systems used to study NOM redox reactions.

The natural-occurring inorganic chemicals are, in fact, the minerals that occur in the crust of the earth. A mineral is a naturally occurring chemical, usually in crystalline form and abiogenic in origin (not produced by life processes). A mineral has one specific chemical composition whereas a rock is typically an aggregate of two or more different minerals.

In fact, clay minerals as well-known adsorbents (Nutting, 1943; Swineford, 1960; Velde, 1992; Velde and Meunier, 2008; Bergaya et al., 2011). These minerals form flat hexagonal sheets and are common products of weathering processes (including weathering of feldspar minerals—a group of minerals distinguished by the presence of alumina, Al_2O_3, and silica, SiO_2, in their chemistry) and low-temperature hydrothermal alteration products. Clay minerals are very common in soils and in fine-grained sedimentary rocks such as shale, mudstone, and siltstone. Clay minerals are usually (but not necessarily) ultrafine-grained (normally considered to be less than 2 μm (2 microns, 2×10^{-6}) in size on standard particle size classifications) and so may require special analytical techniques for their identification and study.

Like any chemical, a mineral is distinguished by its various chemical and physical properties. Differences in chemical composition and crystal structure distinguish the various minerals; characters which were determined by the geological environment of the mineral during the formation period. Changes in the temperature, pressure, or bulk composition of a rock mass cause changes in its minerals. As with any type of mineral, there may be variation in physical properties or minor amounts of impurities.

Minerals are classified by key chemical constituents. Silicon and oxygen make up 75% w/w of the crust of the earth's, which gives rise to the overall predominance of silicate minerals, which are, in turn, subdivided into six subclasses by the degree of polymerization in the chemical structure. All silicate minerals have a base unit of a $[SiO_4]^{4-}$ silica tetrahedron—that is, a silicon cation coordinated by four oxygen anions, which gives the shape of a tetrahedron. These tetrahedra can be polymerized to give the subclasses: orthosilicate minerals (no polymerization, thus single tetrahedra), disilicates (two tetrahedra bonded together), cyclosilicates (rings of tetrahedra), inosilicates (chains of tetrahedra), phyllosilicates (sheets of tetrahedra), and tectosilicates (three-dimensional network of tetrahedra). Other important mineral groups include the native (unchanged) elements, sulfide minerals, oxide minerals, halide minerals, carbonate minerals, sulfate minerals, and phosphate minerals. It is the various types of mineral that are responsible for the retention (adsorption) of chemical contaminants that are spilled into an ecosystem (Chapter 5).

When dealing with chemicals that have been released (advertently or inadvertently, dispensing upon the circumstances), there are several types of chemical transformation reactions that can occur in the environment. These reactions can be grouped into four major categories: (1) oxidation–reduction reactions, also known as redox reactions, (2) carbon–carbon bond formation,

(3) carbon–heteroatom bond formation in which a carbon atom of one molecule forms a bond with the nitrogen atom or oxygen atom or sulfur atom of another molecule, (4), carbon-–carbon bond cleavage, (5) carbon–heteroatom bond cleavage, and (6) organic–inorganic interactions.

Redox reactions would include the hydrogenation of olefin derivatives and acetylene derivatives, the loss of hydrogen through aromatization reactions, the oxidation or reduction of alcohols, aldehydes and ketones, and the oxidative cleavage of olefins. Examples of chemical transformations involving bond formation are polymerization or condensation reactions, esterification or amide ($-CONH_2$) formation, and cyclization (ring formation) reactions. Several types of bond cleavage reactions which might affect the fate or longevity of organic chemicals discharged into the environment are the formation of amino acids from peptides and proteins, and the hydrolysis of esters and amides to form carboxylic acids, as well as other forms of chemical degradation. Organic–inorganic interactions include the formation organometallic compounds and organomineral phase interactions.

There are at least two types of organometallic complexes that are found in the environment: (1) a compound that contains covalently bound metals such as metalloenzymes, as well as anthropogenic alkyl metal compounds, generally written as RM or R^-M^+, where R is the alkyl (organic) group and M is the metal, and (2) the more abundant chelate-type complexes such as metal humate derivatives, where the humate derivatives are formed from humic acid derivatives (produced as a collection of organic acids by the biodegradation of dead organic matter). Some algal products form complexes with metals and there is always the potential for metal detoxification or making the metals otherwise available to the phytoplankton cells as micronutrients. Organomineral phase interactions involve the adsorption of highly surface active dissolved organic matter to ocean PM. The mechanisms by which this takes place include ion-exchange (such materials such as calcium carbonate, $CaCO_3$), interlayering of organic compounds in clay minerals, formation of clathrates, hydrogen bonding, and van der Waals interactions. By way of explanation, van der Waals' forces are the residual attractive or repulsive forces between molecules or atomic (functional) groups that do not arise from a covalent bond or electrostatic interaction of ions or of ionic groups with one another or with neutral molecules (Table 2.4). The resulting van der Waals' forces can be attractive or repulsive.

5.1 Industrial Chemicals

The late 19th century was the time of the commencement of an explosion in both the quantity of production and the variety of chemicals that were manufactured. Large chemical industries also took shape in Germany and later in the United States. Polymers and plastics, especially polyethylene, polypropylene, polyvinyl chloride, polyethylene terephthalate, polystyrene, and

TABLE 2.4 Different Types of Bond Arrangements

Covalent bond: a bond in which one or more electrons (often a pair of electrons) are drawn into the space between the two atomic nuclei. These bonds exist between two particular identifiable atoms and have a direction in space, allowing them to be shown as single connecting lines between atoms in drawings, or modeled as sticks between spheres in models.
Ionic bond: Occurs between ionized functional groups such as carboxylic acids and amines.
Hydrogen bond: Occurs between alcohol derivatives, carboxylic acid derivatives, amide derivatives, amine derivatives, and phenol derivatives. Hydrogen bonding involves the interaction of the partially positive hydrogen on one molecule and the partially negative heteroatom on another molecule. Hydrogen bonding is also possible with elements other than nitrogen or oxygen and can occur intermolecularly or intramolecularly.
Dipole–dipole interaction: Possible between molecules having polarizable bonds, in particular, the carbonyl group (C=O) which has a dipole moment and molecules can align themselves such that their dipole moments are parallel and in opposite directions. Ketones and aldehydes are capable of interacting through dipole–dipole interactions.
Van der Waals interaction: Weak intermolecular bonds between regions of different molecules bearing transient positive and negative charges which are caused by the movement of electrons. Alkanes, alkenes, alkynes, and aromatic rings interact through van der Waals interactions.
Intermolecular bond: Occurs between different molecules and can take the form of ionic bonding, hydrogen bonding, dipole–dipole interactions, and van der Waals interactions.
Intramolecular bond: Occurs within a molecule and can take the form of ionic bonding, hydrogen bonding, dipole–dipole interactions, and van der Waals interactions.

polycarbonate comprise about 80% of the industry's output worldwide. These materials are often converted to fluoropolymer tubing products and used by the industry to transport highly corrosive materials. Chemicals are used in a lot of different consumer goods, but they are also used in a lot of different other sectors, including agriculture manufacturing, construction, and service industries. Major industrial customers include rubber and plastic products, textiles, apparel, petroleum refining, pulp and paper, and primary metals.

5.2 Fly Ash, Bottom Ash, and Boiler Slag

Coal ash is the general term that refers to whatever waste is leftover after coal is combusted, usually in a coal-fired power plant and typically contains arsenic, mercury, lead, and many other heavy metals. Coal ash includes a number of byproducts produced from burning coal, including: (1) fly ash, which is a fine, powdery material composed mostly of silica made from the burning of finely ground coal in a boiler, (2) bottom ash, which is a coarse, angular ash particle that is too large to be carried up into the smoke stacks so it forms in the bottom of the coal furnace, and (3) boiler slag, which is molten bottom ash from slag-type furnaces and from cyclone-type furnaces that turns into pellets that have a smooth glassy appearance after it is cooled with water (Table 2.5).

Coal ash contains contaminants such as arsenic (As), cadmium (Cd), and mercury (Hg) which, without proper management, can pollute waterways, ground water, drinking water, and the air. The related environmental regulations address the risks from coal ash disposal, such as: leaking of contaminants into ground water, blowing of contaminants into the air as dust, and the catastrophic failure of coal ash surface impoundments. Thus, fly ash, bottom ash, and boiler slag are presented here in a separate section because of the composition of these types of waste products, none of which can be classified as a single chemical.

Fly ash (also known as *pulverized fuel ash*) is one of the coal combustion products and is composed of the fine particles that are driven out of the boiler with the flue gases. Ash that falls in the bottom of the boiler is called *bottom ash*. Fly ash, together with bottom ash removed from the bottom of the boiler, is more commonly known as coal ash. *Bottom ash* and *boiler slag* are the coarse, granular, incombustible byproducts that are collected from the bottom of furnaces that burn coal for the generation of steam, the production of electric power, or both. *Boiler slag* is the melted form of coal ash that can be found both in the filters of exhaust stacks and the boiler at the bottom of the stack. Most of these coal byproducts are produced at coal-fired electric utility generating stations, although considerable bottom ash and/or boiler slag are also produced from many smaller industrial or institutional coal-fired boilers and from coal-burning independent power production facilities. The type of byproduct (i.e., bottom ash or boiler slag) produced depends on the type of furnace used to burn the coal.

5.2.1 Fly Ash

The most voluminous and well-known constituent is fly ash, which makes up more than half of the coal leftovers. Fly ash particles are the lightest kind of coal ash and pass upward from the combustor into the exhaust stacks of the power plant. Filters within the stacks capture about 99% w/w of the ash, attracting it with opposing electrical charges and the captured fly ash is recyclable. The fine particles bind together and solidify, especially when mixed with water, making them an ideal ingredient in concrete and wallboard. The recycling process also renders the toxic materials within fly ash safe for use.

In modern coal-fired power plants, fly ash is generally captured by electrostatic precipitators or other particle filtration equipment before the flue gases reach the chimneys. Depending upon the source and makeup of the coal being burned, the components of fly ash vary considerably, but all fly ash

TABLE 2.5 Composition of Selected Bottom Ash and Boiler Slag Samples (% w/w)

Ash Type:	Bottom Ash					Boiler Slag		
Coal Type:	Bituminous			Subbituminous	Lignite	Bituminous		Lignite
SiO_2	53.6	45.9	47.1	45.4	70.0	48.9	53.6	40.5
Al_2O_3	28.3	25.1	28.3	19.3	15.9	21.9	22.7	13.8
Fe_sO_3	5.8	14.3	10.7	9.7	2.0	14.3	10.3	14.2
CaO	0.4	1.4	0.4	15.3	6.0	1.4	1.4	22.4
MgO	4.2	5.2	5.2	3.1	1.9	5.2	5.2	5.6
Na_2O	1.0	0.7	0.8	1.0	0.6	0.7	1.2	1.7
K_2O	0.3	0.2	0.2	–	0.1	0.1	0.1	1.1

includes substantial amounts of silica (silicon dioxide, SiO_2) (both amorphous and crystalline), aluminum oxide (alumina, Al_2O_3) and calcium oxide (CaO), the main mineral compounds in coal-bearing rock strata. The constituents depend upon the specific coal-bed makeup but may include one or more of the following elements or substances found in trace concentrations (up to hundreds ppm): arsenic, beryllium, boron, cadmium, chromium, hexavalent chromium, cobalt, lead, manganese, mercury, molybdenum, selenium, strontium, thallium, and vanadium (Speight, 2013).

In the past, fly ash was generally released into the atmosphere but air pollution control standards now require that it be captured prior to release by fitting pollution control equipment. In the United States, fly ash is generally stored at coal power plants or placed in landfills. About 43% w/w of the fly ash is recycled, often used as a pozzolan to produce hydraulic cement or hydraulic plaster and a replacement or partial replacement for Portland cement in concrete production. Pozzolans ensure the setting of concrete and plaster and provide concrete with more protection from wet conditions and chemical attack. In the case that fly ash or bottom ash is not produced from coal, for example, when solid waste is used to produce electricity in an incinerator, this kind of ash may contain higher levels of contaminants than coal ash. In that case the ash produced is often classified as hazardous waste.

Fly ash material solidifies while suspended in the exhaust gases and is collected by electrostatic precipitators or filter bags. Since the particles solidify rapidly while suspended in the exhaust gases, fly ash particles are generally spherical in shape and range in size from 0.5 to 300 μm. The major consequence of the rapid cooling is that few minerals have time to crystallize, and that mainly amorphous, quenched glass remains. Nevertheless, some refractory phases in the pulverized coal do not melt (entirely), and remain crystalline. In consequence, fly ash is a heterogeneous material. SiO_2, Al_2O_3, Fe_2O_3 and occasionally CaO are the main chemical components present in fly ash. The mineralogy of fly ash is very diverse.

The main phases encountered are a glass phase, together with quartz, mullite, and the iron oxides: hematite, magnetite, and/or maghemite. Other phases often identified are cristobalite, anhydrite, free lime, periclase, calcite, sylvite, halite, portlandite, rutile, and anatase. The calcium-bearing minerals anorthite, gehlenite, akermanite, and various calcium silicates and calcium aluminates identical to those found in Portland cement can be identified in calcium-rich fly ash. The mercury content can reach 1 ppm and the concentrations of other trace elements vary as well according to the kind of coal combusted to form it. In fact, in the case of bituminous coal, with the notable exception of boron, trace element concentrations are generally similar to trace element concentrations in unpolluted soils.

Two classes of fly ash are generally recognized: (1) Class F fly ash and (2) Class C fly ash (ASTM C618). The chief difference between these two classes is the amount of calcium, silica, alumina, and iron in the ash. The chemical properties of the fly ash are largely influenced by the chemical content of the coal burned (i.e., anthracite, bituminous coal, and lignite).

Class F fly ash is produced during the burning of harder, older anthracite and bituminous coal typically produces Class F fly ash. This fly ash is pozzolanic in nature, and contains less than 7% lime (calcium oxide, CaO). Possessing pozzolanic properties, the glassy silica and alumina of Class F fly ash require a cementing agent, such as Portland cement, quicklime, or hydrated lime, mixed with water to react and produce cementitious compounds. Alternatively, adding a chemical activator such as sodium silicate (water glass) to a Class F ash can form a geopolymer.

On the other hand, *Class C fly ash* is produced from the burning of younger lignite or subbituminous coal, and in addition to having pozzolanic properties also has some self-cementing properties. In the presence of water, Class C fly ash hardens and gets stronger over time. Class C fly ash generally contains more than 20% lime (calcium oxide, CaO). Unlike Class F, self-cementing Class C fly ash does not require an activator. Alkali and sulfate (SO_4) contents are generally higher in Class C fly ash.

Fly ash contains trace concentrations of heavy metals and other substances that are known to be detrimental to health in sufficient quantities. Potentially toxic trace elements in coal include arsenic, beryllium, cadmium, barium, chromium, copper, lead, mercury, molybdenum, nickel, radium, selenium, thorium, uranium, vanadium, and zinc. Approximately 10% w/w of the mass of coal burned in the United States consists of unburnable mineral material that becomes ash, so the concentration of most trace elements in coal ash is approximately 10 times the concentration in the original coal.

Crystalline silica (SiO_2) and lime along with toxic chemicals represent exposure risks to human health and the environment. Exposure to fly ash through skin contact, inhalation of fine particulate dust, and ingestion through drinking water may well present health risks. Fly ash contains crystalline silica which is known to cause lung disease, in particular, silicosis. Also, lime (CaO) reacts with water (H_2O) to form calcium hydroxide [$Ca(OH)_2$], giving fly ash a pH on the order of 10–12, which is a medium to strong base.

5.2.2 Bottom Ash

By definition, bottom ash is part of the noncombustible residue of combustion in a furnace or incinerator. In an industrial context, it usually refers to coal combustion and comprises traces of combustibles embedded in forming clinkers and sticking to hot side walls of a coal-burning furnace during its operation. The clinkers fall by themselves into the bottom hopper of a coal-burning furnace and are cooled. In the United Kingdom it is known as furnace bottom ash, to distinguish it from incinerator bottom ash, the noncombustible elements remaining after incineration.

The most common type of coal-burning furnace in the electric utility industry is the dry, bottom pulverized coal boiler. When pulverized coal is burned in a dry, bottom boiler, approximately 80% w/w of the unburned material or ash is entrained in the flue gas and is captured and recovered as fly ash. The remaining 20% w/w of the ash is dry bottom ash, a dark gray, granular, porous, predominantly sand size minus 12.7 mm (½ in.) material that is collected in a water-filled hopper at the bottom of the furnace. When a sufficient amount of bottom ash drops into the hopper, it is removed by means of high-pressure water jets and conveyed by sluiceways either to a disposal pond or to a decant basin for dewatering, crushing, and stockpiling for disposal or use.

Bottom ash is the coarser component of coal ash, comprising about 10% w/w of the waste. Rather than passing into the exhaust stacks, the ash settles to the bottom of the power plant boiler (hence the name bottom ash). Bottom ash is not quite as useful as fly ash, although power plant owners have continued to develop options for beneficial use, such as structural fill and road-base material. However, the bottom ash typically contains toxic constituents and the ash, when recycled, can be the source of heavy metals leaking into the groundwater.

Bottom ash can be extracted, cooled, and conveyed using dry ash technology from various companies. When left dry, the ash can be used to make concrete and other useful materials. There are also several environmental benefits. For example, bottom ash may be used as raw alternative material, replacing earth or sand or aggregates, for example, in road construction and in cement kilns (clinker production). A noticeable other use is as growing medium in horticulture (usually after sieving). An example of the use of bottom ash is in the production of concrete blocks for use in building construction.

Due to the salt content and, in some cases, the low pH of bottom ash (and boiler slag), the material can exhibit corrosive properties. Corrosivity indicator tests normally used to evaluate bottom ash or boiler slag are pH, electrical resistivity, soluble chloride content, and soluble sulfate content. Materials are judged to be noncorrosive if the pH exceeds 5.5, the electrical resistivity is greater than 1500 Ω-centimeters, the soluble chloride content is less than 200 parts per million (ppm), or the soluble sulfate content is less than 1000 parts per million (ppm).

5.2.3 Boiler Slag

Boiler slag is a byproduct produced from a wet-bottom boiler, which is a special type of boiler designed to keep bottom ash in a molten state before it is removed. These types of boilers (slag-tap and cyclone boilers) are much more compact than pulverized coal boilers used by most large utility generating stations and can burn a wide range of fuels and generate a higher proportion of bottom ash than fly ash (50%–80% w/w bottom ash compared to 15%–20% w/w bottom ash for pulverized coal boilers). With wet-bottom boilers, the molten ash is withdrawn from the boiler and allowed to flow into quenching water. The rapid cooling of the slag causes it to immediately crystallize into a black, dense, fine-grained glassy mass that fractures into angular particles, which can be crushed and screened to the appropriate sizes for several uses.

There are two types of wet-bottom boilers: (1) the slag-tap boiler and (2) the cyclone boiler. The slag-tap boiler burns pulverized coal and the cyclone boiler burns crushed coal. In each type, the bottom ash is kept in a molten state and tapped off as a liquid. Both boiler types have a solid base with an orifice that can be opened to permit the molten ash that has collected at the base to flow into the ash hopper below. The ash hopper in wet-bottom furnaces contains quenching water. When the molten slag comes in contact with the quenching water, it fractures instantly, crystallizes, and forms pellets. The resulting boiler slag, often referred to as *black beauty*, is a coarse, hard, black, angular, glassy material.

When pulverized coal is burned in a slag-tap furnace, as much as 50% of the ash is retained in the furnace as boiler slag. In a cyclone furnace, which burns crushed coal, some 70%–80% w/w of the ash is retained as boiler slag, with only 20%–30% w/w leaving the furnace in the form of fly ash. Wet-bottom boiler slag is a term that describes the molten condition of the ash as it is drawn from the bottom of the slag-tap or cyclone furnaces. At intervals, high-pressure water jets wash the boiler slag from the hopper pit into a sluiceway which then conveys it to a collection basin for dewatering, possible crushing or screening, and either disposal or reuse.

Since boiler slag is angular, dense and hard, it is often used as a wear-resistant component in surface coatings of asphalt in road paving. Finer-sized boiler slag can be used as blasting grit and is commonly used for coating roofing shingles. Other uses include raw material for the manufacture of cement and in colder climates, it is spread onto icy roads for traction control. Because there are so many uses and such a limited supply, most of the boiler slag produced in the United States is used and even some is imported from other countries.

5.3 Household Chemicals

Household chemicals are nonfood chemicals that are commonly found and used in and around the typical household and are designed particularly to assist cleaning, pest control, and for general hygiene purposes. Food additives generally do not fall under this category, unless they have a use other than for human consumption. Additives in general (e.g., stabilizers and coloring found in washing powder and dishwasher detergents) make the classification of household chemicals more complex, especially in terms of health and ecological effects —some of these chemicals are irritants or potent allergens. Together with noncompostable household waste, the chemicals found in private household commodities pose a serious ecological problem. In addition, to having slightly adverse effects to seriously toxic effects when swallowed, chemical agents around may contain flammable or corrosive substances.

While useful, many of household chemicals can have adverse effect on people or the environment, either in manufacture or by the way people use them or improperly dispose of them. Many different types of toxic chemicals are found in waterways in North America.

A toxic chemical is any substance that is capable of harming a person if ingested, inhaled, or absorbed through the skin. Toxic substances vary widely in the types of harm they may cause and the conditions under which they become harmful. The effects of the toxic substances vary widely, too. Acute reactions are sudden ones such as vomiting, dizziness, or even death. Chronic reactions occur over longer periods and include symptoms such as decline in mental alertness, change in behavior, cancer, and mutations that can harm unborn children of exposed parents. Because toxins can cause both acute and chronic reactions, they are a broader category than poisons, which produce acute reactions only. For this reason, the words toxin and poison are not interchangeable.

Many consumers do not realize that household chemicals can be toxic. Most of the dangerous substances in the home are found in cleaners, solvents, pesticides, and products used for automotive care.

REFERENCES

Alloway, B.J., 1995. Heavy Metals in Soils. Blackie Academic and Professional Publishers, London, United Kingdom.

Arruti, A., Fernández-Olmo, I., Irabien, A., 2010. Evaluation of the contribution of local sources to trace metals levels in urban $PM_{2.5}$ and PM_{10} in the Cantabria region (Northern Spain). Journal of Environmental Monitoring 12 (7), 1451–1458.

ASTM C618. Standard Specification for Coal Fly Ash and Raw or Calcined Natural Pozzolan for Use in Concrete. Annual Book of Standards. ASTM International, West Conshohocken, Pennsylvania.

Basta, N.T., Ryan, J.A., Chaney, R.L., 2005. Trace element chemistry in residual-treated soil: key concepts and metal bioavailability. Journal of Environmental Quality 34 (1), 49–63.

Bergaya, F., Jaber, M., Lambert, J.F., 2011. Clays and Clay Minerals. John Wiley & Sons Inc., Hoboken, New Jersey.

Bradl, H. (Ed.), 2002. Heavy Metals in the Environment: Origin, Interaction and Remediation, vol. 6. Academic Press, New York.

D'Amore, J.J., Al-Abed, S.R., Scheckel, K.G., Ryan, J.A., 2005. Methods for speciation of metals in soils: a review. Journal of Environmental Quality 34 (5), 1707–1745.

Fawell, J.K., 1993. The impact of inorganic chemicals on water quality and health. Annali dell'Istituto Superiore di Sanità 29 (2), 293–303. https://www.ncbi.nlm.nih.gov/pubmed/8279720.

Fergusson, J.E. (Ed.), 1990. The Heavy Elements: Chemistry, Environmental Impact and Health Effects. Pergamon Press, Oxford, United Kingdom.

Gallo, M., 2001. History and scope of toxicology. In: Klaasen, C.D. (Ed.), Casarett and Doull's Toxicology: The Basic Science of Poisons, sixth ed. McGraw-Hill, New York.

Goyer, R.A., 1996. Toxic effects of metals. In: Klaasen, C.D. (Ed.), Cassarett and Doull's Toxicology: The Basic Science of Poisons. McGraw-Hill, New York, pp. 811–867.

Goyer, R.A., Clarkson, T.W., 1996. Toxic effects of metals. In: Klaasen, C.D. (Ed.), Casarett and Doull's Toxicology: The Basic Science of Poisons, fifth ed. McGraw-Hill, New York.

Hamelink, J.L., Landrum, P.F., Harold, B.L., William, B.H. (Eds.), 1994. Bioavailability: Physical, Chemical, and Biological Interactions. CRC Press, Taylor & Francis Group, Boca Raton, Florida.

He, Z.L., Yang, X.E., Stoffella, P.J., 2005. Trace elements in agroecosystems and impacts on the environment. Journal of Trace Elements in Medicine & Biology 19 (2–3), 125–140.

Herawati, N., Suzuki, S., Hayashi, K., Rivai, I.F., Koyoma, H., 2000. Cadmium, copper and zinc levels in rice and soil of Japan, Indonesia and China by soil type. Bulletin of Environmental Contamination and Toxicology 64, 33–39.

Kaasalainen, M., Yli-Halla, M., 2003. Use of sequential extraction to assess metal partitioning in soils. Environmental Pollution 126 (2), 225–233.

Kabata-Pendias, A., Pendias, H., 2001. Trace Metals in Soils and Plants, second ed. CRC Press, Taylor & Francis Group, Boca Raton, Florida.

Khan, S., Cao, Q., Zheng, Y.M., Huang, Y.Z., Zhu, Y.G., 2008. Health risks of heavy metals in contaminated soils and food crops irrigated with wastewater in Beijing, China. Environmental Pollution 152 (3), 686–692.

Kuo, S., Heilman, P.E., Baker, A.S., 1983. Distribution and forms of copper, zinc, cadmium, iron, and manganese in soils near a copper smelter. Soil Science 135 (2), 101–109.

Lombi, E., Gerzabek, M.H., 1998. Determination of mobile heavy metal fraction in soil: results of a Pot experiment with sewage sludge. Communications in Soil Science and Plant Analysis 29 (17–18), 2545–2556.

Miller, D.R., 1984. Chemicals in the environment. In: Sheehan, P.J., Miller, D.R., Butler, G.C., Bourdeau, P. (Eds.), Effects of Pollutants at the Ecosystem Level, pp. 7–14 (Chapter 2).

Nriagu, J.O., 1989. A global assessment of natural sources of atmospheric trace metals. Nature 338, 47–49.

Nutting, P.G., 1943. Adsorbent Clays: Their Distribution, Properties Production, and Uses. Bulletin 928-C. United States Geological Survey, Reston, Virginia.

Pacyna, J.M., 1996. Monitoring and assessment of metal contaminants in the air. In: Chang, L.W., Magos, L., Suzuli, T. (Eds.), Toxicology of Metals. CRC Press, Taylor & Francis Group, Boca Raton, Florida, pp. 9–28.

Pierzynski, G.M., Sims, J.T., Vance, G.F., 2000. Soils and Environmental Quality, second ed. CRC Press, Taylor & Francis Group, Boca Raton, Florida.

Shallari, S., Schwartz, C., Hasko, A., Morel, J.L., 1998. Heavy metals in soils and plants of serpentine and industrial sites of Albania. The Science of the Total Environment 192 (09), 133–142.

Speight, J.G., 2013. The Chemistry and Technology of Coal, third ed. CRC Press, Taylor & Francis Group, Boca Raton, Florida.

Speight, J.G., 2014. The Chemistry and Technology of Petroleum, fifth ed. CRC Press, Taylor & Francis Group, Boca Raton, Florida.

Speight, J.G., 2017a. Environmental Organic Chemistry for Engineers. Butterworth-Heinemann, Elsevier, Cambridge, Massachusetts.

Speight, J.G., 2017b. Environmental Inorganic Chemistry for Engineers. Butterworth-Heinemann, Elsevier, Cambridge, Massachusetts.

Speight, J.G., 2017c. Handbook of Petroleum Refining. CRC Press, Taylor & Francis Group, Boca Raton, Florida.

Speight, J.G., Islam, M.R., 2016. Peak Energy – Myth or Reality. Scrivener Publishing, Salem, Massachusetts (Chapter 8).

Speight, J.G., Lee, S., 2000. Environmental Technology Handbook, second ed. Taylor & Francis, New York (Also, CRC Press, Taylor and Francis Group, Boca Raton, Florida).

Sposito, G., Page, A.L., 1984. Cycling of metal ions in the soil environment. In: Sigel, H. (Ed.), Metal Ions in Biological Systems, Circulation of Metals in the Environment, vol. 18. Marcel Dekker, Inc., New York, pp. 287–332.

Swineford, A. (Ed.), 1960. Clays and Clay Minerals, vol. 5. Elsevier, Amsterdam, Netherlands.

Thy, P., Yu, C., Jenkins, B.M., Lesher, C.E., 2013. Composition and environmental impact of biomass feedstock. Energy & Fuels 27 (7), 3969–3987.

Tinsley, I.J., 2004. Chemical Concepts in Pollutant Behavior, second ed. John Wiley & Sons Inc., Hoboken, New Jersey.

Velde, B.B., 1992. Introduction to Clay Minerals: Chemistry, Origin, Uses and Environmental Significance. Chapman & Hall, London, United Kingdom.

Velde, B.B., Meunier, A., 2008. The Origin of Clay Minerals in Soils and Weathered Rocks. Springer-Verlag, Berlin, Germany.

Wondyfraw, M., 2014. Mechanisms and effects of acid rain on environment. Journal of Earth Science & Climatic Change 5 (6), 204–206.

Wuana, R., Okieimen, F.E., 2011. Heavy metals in contaminated soils: a review of sources, chemistry, risks and best available strategies for remediation. ISRN Ecology 2011, 402647. https://doi.org/10.5402/2011/402647, 2011. https://www.hindawi.com/journals/isrn/2011/402647/.

Zhang, M.K., Liu, Z.Y., Wang, H., 2010. Use of single extraction methods to predict bioavailability of heavy metals in polluted soils to rice. Communications in Soil Science and Plant Analysis 41 (7), 820–831.

Chapter 3

Chemical and Physical Properties

1. INTRODUCTION

Chemicals are the basic building blocks that make up all living and nonliving species that exist on the earth. Many chemicals occur naturally in the environment and are harvested for use while some are manufactured synthetically and used everyday in products from medicines to computers to fabrics and fuels. There is also a class of chemicals that is not manufactured deliberately but is the byproduct of chemical processes and is often sent to the relevant part of the manufacturing plant for discharge or, in some cases, recycling.

All chemicals have distinct chemical and physical properties and may undergo physical or chemical changes. Chemical properties, such as flammability and acidity, and chemical changes, such as rusting, involve production of matter that differs from that present beforehand. Physical properties, such as hardness and boiling point, and physical changes, such as the melting point or freezing point, do not involve a change in the composition of matter (Tables 3.1–3.3).

However, chemicals (no matter what the source or the properties) can enter the air (the atmosphere), water (the aquasphere), and soil (the lithosphere, sometimes called the terrestrial biosphere) when they are produced, used, or disposed. The impact of chemicals on the environment is determined by (1) the amount of the chemical that is released, (2) the type and concentration of the chemical, and (3) the locale into which the chemical is released. Some chemicals can be harmful if released to the environment even when there is not an immediate, visible impact (Speight and Lee, 2000; Sánchez-Bayo et al., 2011; Speight, 2017a,b). Some chemicals are of concern as they can work their way into the food chain and accumulate and/or persist in the environment for many years. The effect of the released chemical is dependent upon the chemical properties and/or the physical properties of the chemical.

The chemical and physical properties of a chemical are characteristics of the chemical that become evident when the chemical undergoes a chemical

Reaction Mechanisms in Environmental Engineering. https://doi.org/10.1016/B978-0-12-804422-3.00003-1

TABLE 3.1 Differentiation of Chemical and Physical Properties

Chemical Properties	Physical Properties
Reaction with acids	Melting point
Reaction with alkalis (bases)	Boiling point
Reaction with oxygen (oxidation)	Vapor pressure
Reaction with oxygen (combustion)	Color
Ability to act as an oxidizing agent	State (gas, liquid, solid)
Ability to act as a reducing agent	Density
Reaction with another chemical	Electrical conductivity
Decomposition into another chemical	Solubility
Can cause corrosion	Adsorption to a surface
Contains functional groups	Hardness

reaction or chemical change, such as a redox reaction (Chapter 7) or a physical association such as adsorption (Chapter 5). These properties cannot be observed or determined by simply viewing (only under safe conditions) or by touching the chemical (not recommended because of safety and health concerns). The properties of the chemical, including the presence of inorganic polyatomic ions (Table 3.4) that are likely to interact with minerals when discharged into the environment, must be determined by application of the series of standard test methods (see for example, ASTM, 2017 and the test methods cited therein).

Furthermore, while a chemical property is only revealed by the behavior of a substance in a chemical reaction, a physical property may be observed and measured without changing the composition of a sample. Once determined, the relevant chemical and physical properties of a chemical can be used to predict whether the chemical can participate in a chemical reaction or in a physical association (such as adsorption) and may be used to classify compounds and find applications for them.

Briefly, and by way of definition, a chemical property is any property of a chemical that becomes evident during, or after, a chemical reaction and is a quality of the chemical that can be established only by changing the chemical identity. When a chemical undergoes a chemical reaction, the properties will change drastically, resulting in one or more chemical changes.

TABLE 3.2 Various Physical and Chemical Properties of Chemicals

Property	Rationale for Inclusion
Flammability	Associated with flammability hazard
Corrosivity	Associated with ability to gradually destroy materials by chemical reactions
Oxidizing ability	Associated with ability to give off oxidizing substances or oxidize combustible materials, increasing fire or explosion hazards
Melting and boiling point	Impacts environmental fate and transport, as well as potential bioavailability
Vapor pressure	Impacts environmental fate and transport, as well as potential bioavailability
Acidity (pK_a)	Determines ionization state in the environment as well as in biological compartments
Aqueous solubility	Reflects ability to partition into aquatic environment
Octanol-water partition coefficient ($\log P$)	Important determinant of human/mammalian oral and skin bioavailability; relevance to acute and chronic aquatic toxicity (narcosis) and directly related to bioconcentration
Henry's law constant ($\log P_{w/g}$)	Relevance to environmental partitioning and transport as well as human/mammalian alveolar absorption
Molecular electronic dipole moments, μ, and dipole polarizabilities, α	Important in determining the energy, geometry, and intermolecular forces of molecules, often related to biological activity
Biodegradation	Indicator of persistence, and persistence is tied to ecotoxicity
Bioconcentration factor (BCF)	Bioconcentration enhances the hazard potential of lipophilic chemicals; provides a comparative basis for assessing the potential for a chemical to have effects that resonate through the food chain.

TABLE 3.3 Examples of Important Common Physical Properties of Gases, Liquids, and Solids

Gases
Density
Critical temperature, critical pressure
Solubility in water
Solubility in organic solvents
Odor threshold
Color
Diffusion coefficient
Liquids
Vapor pressure-temperature relationship
Density, specific gravity
Viscosity
Miscibility with water
Miscibility with organic solvents
Odor
Color
Coefficient of thermal expansion
Interfacial tension
Solids
Melting point
Density
Odor
Solubility in water
Solubility in organic solvents
Coefficient of thermal expansion
Hardness/flexibility
Particle size distribution
Physical form (powder, granules, pellets, lumps)
Porosity

Reactions may proceed in the forward direction and processed to completion, as well as in the reverse direction until they reach equilibrium.

$$A + B \rightarrow C + D$$

$$C + D \rightarrow A + B$$

Thus,

$$A + B \leftrightarrow C + D$$

Reactions that proceed in the forward direction to approach equilibrium are often described as spontaneous, requiring no input of free energy to go

TABLE 3.4 Common Polyatomic Ions

Name: Cation Anion	Formula
Ammonium ion	NH_4^+
Hydronium ion	H_3O^+
Acetate ion	$C_2H_3O_2^-$
Arsenate ion	AsO_4^{3-}
Carbonate ion	CO_3^{2-}
Hypochlorite ion	ClO^-
Chlorite ion	ClO_2^-
Chlorate ion	ClO_3^-
Perchlorate ion	ClO_4^-
Chromate ion	CrO_4^{2-}
Dichromate ion	$Cr_2O_7^{2-}$
Cyanide ion	CN^-
Hydroxide ion	OH^-
Nitrite ion	NO_2^-
Nitrate ion	NO_3^-
Oxalate ion	$C_2O_4^{2-}$
Permanganate ion	MnO_4^-
Phosphate ion	PO_4^{3-}
Sulfite ion	SO_3^{2-}
Sulfate ion	SO_4^{2-}
Thiocyanate ion	SCN^-
Thiosulfate ion	$S_2O_3^{2-}$

forward. Nonspontaneous reactions require input of free energy to go forward (for example, application of heat for the reaction to proceed). In organic chemical synthesis, different chemical reactions are used in combinations during chemical synthesis in order to obtain a desired product. Also, in organic chemistry, a consecutive series of chemical reactions (where the product of one reaction is the reactant of the next reaction) are often catalyzed by a variety of catalysts which increase the rates of biochemical reactions, so that syntheses and decompositions impossible under ordinary conditions can occur at the temperatures, pressures, and reactant concentrations present within a reactor and, by inference, within the environment.

$$A + B \rightarrow C$$

$$C \rightarrow D + E$$

This simplified equation illustrates the potential complexity of organic chemical reaction, and such complexity must be anticipated when an organic chemical is transformed in an environmental ecosystem.

Chemical properties and, hence, chemical reactions can be used for building the classification of chemicals and are also useful for establishing the identity of an unknown substance or to separate or purify it from other substances. Thus, chemical properties can be contrasted with physical properties, which can be discerned without changing the structure of the chemical.

On the other hand, a physical property (often referred to as an *observable*) is any property that is measurable and the value of which can be used to describe the state of a physical system. The changes in the physical properties of a chemical can be used to describe its changes between momentary (intermediate or transitory) states. Physical properties are often characterized as *intensive properties* and *extensive properties*.

All properties of matter are either physical or chemical and physical properties are either intensive or extensive. Extensive properties, such as mass and volume, depend on the amount of matter being measured. An extensive property shows an additive relationship. Intensive properties, such as density and color, do not depend on the amount of the substance present. An intensive property does not depend on the size or extent of the system, nor on the amount of matter in the object. These classifications are in general only valid in cases when smaller subdivisions of the sample do not interact in some physical or chemical process when combined. Properties may also be classified with respect to the directionality of their nature. For example, isotropic properties do not change with the direction of observation, and anisotropic properties do have spatial variance.

It may be difficult to determine whether a given property is a material property or not. For example, the color of a chemical can be seen and measured; however, what one perceives as color is really an interpretation of the reflective properties of a surface and the light used to illuminate it. In this sense, many ostensibly physical properties are called supervenient. A supervenient property is one which is actual but is secondary to some underlying reality. This is similar to the way in which objects are supervenient on atomic structure. Physical properties are contrasted with chemical properties which determine the way a chemical will behave in a chemical reaction.

Both chemical and physical properties are relevant to the manner in which a chemical will behave when discharged (advertently or inadvertently into an ecosystem. It is necessary to know specific properties about a chemical to predict whether it will be a danger to the environment, such as (1) the amount of the chemical—the threshold quantity of the chemical—that can be released before it is a danger to the ecosystem, (2) if the chemical will persist in the ecosystem or if it will be transformed into an equally or more seriously ecosystem-toxic material, and (3) if the chemical find its way to target flora and faunal species by having some kind of affinity for these life forms. Thus, knowledge of the chemical and physical properties of chemicals is and aid to understanding the role of chemicals as environmental contaminants, as well as an aid to help better understand processes that affect their movement and fate in ecosystems (Lyman et al., 1982, 1990; Andrews et al., 1996; Boethling and Mackay, 2000; Mackay et al., 2006; Manahan, 2010; Schwarzenbach, et al., 2003; Speight, 2017a,b).

The potential for a discharged chemical to distribute among environmental ecosystems is determined by its chemical properties and physical chemical properties and it is necessary to understand these properties in the context of the character (and properties of the ecosystem. For example, environmental conditions are often determined by the ambient temperature and the predominant solvent of any consequence is water, which is not an effective solvent for many categories of chemicals (especially nonpolar organic chemicals) organic compounds.

In fact, there are many categories of chemicals. A chemical category is a group of chemicals whose physicochemical and human health and/or ecotoxicological properties and/or environmental fate properties are likely to be similar or follow a regular pattern, usually as a result of structural similarity. The similarities may be based on the following: (1) a common functional group (e.g., aldehyde, epoxide, ester, specific metal ion), (2) common constituents or chemical classes and similar carbon range numbers, (3) an incremental and constant change across the category (e.g., a chain-length category), (4) the likelihood of common precursors and/or breakdown products, via physical or biological processes, which result in structurally similar chemicals (e.g., the metabolic pathway approach of examining related chemicals such as acid/ester/salt). However, for convenience, only two categories will be considered briefly at this point: (1) acids, and (2) bases.

2. ACIDS AND BASES

In simple terms, acids are chemical compounds charged with highly reactive hydrogen ions, as indicated by the dissociation constant (Table 3.5). Acids are used in experiments demonstrating acid–base reactions but must be compatible with any other acids added to the mixture (Table 3.6). The uncontrolled mixing of acids can have a serious adverse effect on any process in which the mixture is involved. Acids can be highly toxic and dangerous and can be classified as stronger or weaker depending on the degree of reactivity. Thus, an acid is a molecule or ion capable of donating a proton or hydrogen ion H^+), or,

TABLE 3.5 Examples of Inorganic Acids, Inorganic Bases, and Inorganic Salts

Inorganic Acids

- Carbonic acid (H_2CO_3): a weak inorganic acid.
- Hydrochloric acid (HCl): a highly corrosive, strong inorganic acid with many uses.
- Hydrofluoric acid (HF): a weak inorganic acid that is highly reactive with silicate, glass, metals, and semimetals.
- Nitric acid (HNO_3): a highly corrosive and toxic strong inorganic acid.
- Phosphoric acid (H_3PO_4): not considered to be a strong inorganic acid. It is found in solid form as a mineral and has many industrial uses.
- Sulfuric acid (H_2SO_4): a highly corrosive inorganic acid. It is soluble in water and widely used.

Inorganic Bases

- Ammonium hydroxide (ammonia water, NH_4OH): a solution of ammonia in water.
- Calcium hydroxide [lime water, $Ca(OH)_2$]: a weak base with many industrial uses.
- Magnesium hydroxide [$Mg(OH)_2$]: referred to as brucite when found in its solid mineral form.
- Sodium bicarbonate (baking soda, $NaHCO_3$): a mild alkali.
- Sodium hydroxide (caustic soda, NaOH): a strong inorganic base. It is widely used in industrial and laboratory environments.

Inorganic Salts

- Calcium chloride ($CaCl_2$): many industrial uses.
- Potassium dichromate ($K_2Cr_2O_7$): commonly used as an oxidizing agent.
- Sodium chloride (NaCl): common table salt used in the food industry.

TABLE 3.6 Acid Compatibility

Acid	Properties	Incompatibility
Nitric Acid	Very reactive	All organic acids
		Chlorosulfonic acid
		Hydrofluoric acid
		Perchloric acid
		Sulfuric acid
Hydrofluoric acid	Acute toxin	All organic acids
		Chlorosulfonic acid
		Nitric acid
		Perchloric acid
		Sulfuric acid
All organic acids	Can be toxic	All inorganic (mineral) acids

alternatively, capable of forming a covalent bond with an electron pair (a Lewis acid) (Chapter 1).

More specifically, an acid is a substance that donates hydrogen ions. Because of this, when an acid is dissolved in water, the balance between hydrogen ions and hydroxide ions is shifted. Now there are more hydrogen ions than hydroxide ions in the solution. This kind of solution is acidic. On the other hand, a base is a substance that accepts hydrogen ions. When a base is dissolved in water, the balance between hydrogen ions and hydroxide ions shifts the opposite way. As the base "soaks up" hydrogen ions, the result is a solution with more hydroxide ions than hydrogen ions. This kind of solution is alkaline.

Acidity and alkalinity are measured with a logarithmic scale (the pH scale) (Fig. 3.1). Each one-unit change in the pH scale corresponds to a tenfold change in hydrogen ion concentration. The pH scale is theoretically open-ended, but most pH values are in the range from 0 to 14. To be more precise, pH is the negative logarithm of the hydrogen ion concentration:

$$pH = -\log [H^+]$$

The square brackets around the H^+ automatically mean "concentration" to a chemist. In the equation, for each one-unit change in pH, the hydrogen ion concentration changes ten-fold. Pure water has a neutral pH of 7. pH values lower than 7 are acidic, and pH values higher than 7 are alkaline (basic).

On the other hand, bases are chemical compounds with hydroxide, a reactive ion with one hydrogen ion and one oxygen ion. When hydroxide mixes with the hydrogen in an acid, the two react to produce water (H_2O). The remainder of the chemicals in the acids and bases combine to form salts. Like acids, bases can be considered stronger or weaker depending on their reactivity, which is indicated by the dissociation constant (Table 3.7).

Thus, bases are chemicals which, in aqueous solution, (1) react with acids to form salts, (2) promote certain chemical reactions, (3) accept protons from any proton donor, and/or (4) contain completely or partially displaceable hydroxide (OH^-) ions. Examples of bases are the hydroxides of the alkali metals (lithium, sodium, potassium) and the alkaline earth metals, such as calcium (Ca), magnesium (Mg), and barium (Ba).

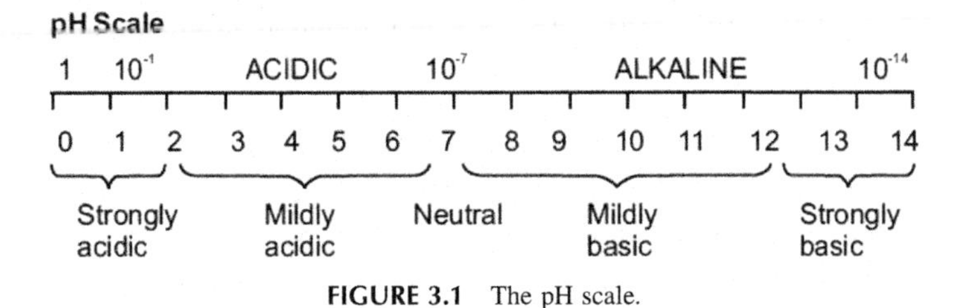

FIGURE 3.1 The pH scale.

TABLE 3.7 Examples of Acid Dissociation Constant K_a

Acid	Formula	K_{a1}	K_{a2}	K_{a3}
Arsenic	H_3AsO_4	5.8×10^{-3}	1.1×10^{-7}	3.2×10^{-12}
Carbonic	H_2CO_3	4.45×10^{-7}	4.69×10^{-11}	
Hydrogen Cyanide	HCN	6.2×10^{-10}		
Hydrofluoric	HF	6.8×10^{-4}		
Hydrogen Peroxide	H_2O_2	2.2×10^{-12}		
Hydrogen Sulfide	H_2S	9.6×10^{-8}	1.3×10^{-14}	
Hydrochloric	HCl	Strong		
Hypochlorous	HOCl	3.0×10^{-8}		
Iodic	HIO_3	1.7×10^{-1}		
Nitric	HNO_3	Strong		
Nitrous	HNO_2	7.1×10^{-4}		
Perchloric	$HClO_4$	Strong		
Periodic	H_5IO_6	2×10^{-2}	5×10^{-9}	
Phosphoric	H_3PO_4	7.11×10^{-3}	6.32×10^{-8}	4.5×10^{-13}
Phosphorous	H_3PO_3	3×10^{-2}	1.62×10^{-7}	
Sulfamic	H_2NSO_3H	1.03×10^{-1}		
Sulfuric	H_2SO_4	Strong	1.02×10^{-2}	
Sulfurous	H_2SO_3	1.23×10^{-2}	6.6×10^{-8}	
Thiosulfuric	$H_2S_2O_3$	0.3	2.5×10^{-2}	

pK_{a1}, pK_{a2}, and pK_{a3} are the constants for the release of one proton at a time.

2.1 Inorganic Acids and Bases

An inorganic acid (also called a mineral acid) is an acid derived from one or more inorganic compounds. All inorganic acids form hydrogen ions and the conjugate base ions when dissolved in water. Commonly used inorganic acids are sulfuric acid (H_2SO_4), hydrochloric acid (HCl), and nitric acid (HNO_3). Inorganic acids range from superacids (such as perchloric acid, $HClO_4$) to very weak acids (such as boric acid, H_3BO_3). Inorganic acids tend to be very soluble in water and insoluble in organic solvents.

Inorganic acids are used in many sectors of the chemical industry as feedstocks for the synthesis of other chemicals, both organic and inorganic. Large quantities of these acids—especially sulfuric acid, nitric acid, and hydrochloric acid—are manufactured for commercial use in large plants. Inorganic acids are also used directly for their corrosive properties. For example, a dilute solution of hydrochloric acid is used for removing the deposits from the inside of boilers, with precautions taken to prevent the corrosion of the boiler by the acid. This process is known as descaling.

Although HF can be named hydrogen fluoride, it is given a different name for emphasis that it is an acid—a substance that dissociates into hydrogen ions (H^+) and anions in water. A quick way to identify acids is to see if there is an H (denoting hydrogen) in front of the molecular formula of the compound. To name acids, the prefix *hydro-* is placed in front of the nonmetal and modified to end with *-ic*. The state of acids is aqueous (*aq*) because acids are found in water. Some common binary acids include:

$$HF(g) \text{ (hydrogen fluoride)} \rightarrow HF(aq) \text{ (hydrofluoric acid)}$$

$$HBr(g) \text{ (hydrogen bromide)} \rightarrow HBr(aq) \text{ (hydrobromic acid)}$$

$$HCl(g) \text{ (hydrogen chloride)} \rightarrow HCl(aq) \text{ (hydrochloric acid)}$$

$$H_2S(g) \text{ (hydrogen sulfide)} \rightarrow H_2S(aq) \text{ (hydrosulfuric acid)}$$

The term *inorganic base* represents a large class of inorganic compounds with the ability to react with acids, that is, neutralize acids to form salts. An inorganic base causes an indicator to take on characteristic colors and usually refers to water-soluble hydroxides, such as sodium hydroxide (NaOH), potassium hydroxide (KOH), or ammonium hydroxide (NH_4OH). The term also includes weak bases, such as water-soluble carbonate derivatives ($-CO_3^{2-}$) or bicarbonate derivatives ($-HCO_3^-$).

Some chemicals can act either as an acid and as a base—an example is water which may either donate a hydrogen ion (to form the hydroxyl ion, OH^-) or accept a hydrogen ion to form the hydroxonium ion (H_3O^+), also called the hydronium ion. This property makes water an amphoteric solvent. Briefly, an amphoteric compound is a molecule or ion that can act both as an acid as well as a base, and this property can influence the behavior of chemicals in the environment.

Metal oxides which react with both acids as well as bases to produce salt and water are amphoteric oxides, and include lead oxide (PbO) and zinc oxide (ZnO), among many others such as the oxides of aluminum (Al_2O_3) and copper (CuO). Other examples of amphoteric compounds are oxides and hydroxides of elements that lie on the border between the metallic and nonmetallic elements in the periodic table (Fig. 3.2). For example, aluminum hydroxide [$Al(OH)_3$] is insoluble at neutral pH (pH = 7.0) but can accept protons in an acid solution to produce $[Al(H_2O)_6]^{3+}$ or accept a hydroxide ion (OH^-) in a basic solution to produce $[Al(OH)_{4+}]^-$ ions. Consequently, aluminum oxide is soluble in acid and in base, but not in neutral water. Other examples of amphoteric oxides are beryllium oxide (BeO), gallium oxide

Group→ ↓Period	1	2	3	4	5	6	7	8	9	10	11	12	13	14	15	16	17	18
1	1 H																	2 He
2	3 Li	4 Be											5 B	6 C	7 N	8 O	9 F	10 Ne
3	11 Na	12 Mg											13 Al	14 Si	15 P	16 S	17 Cl	18 Ar
4	19 K	20 Ca	21 Sc	22 Ti	23 V	24 Cr	25 Mn	26 Fe	27 Co	28 Ni	29 Cu	30 Zn	31 Ga	32 Ge	33 As	34 Se	35 Br	36 Kr
5	37 Rb	38 Sr	39 Y	40 Zr	41 Nb	42 Mo	43 Tc	44 Ru	45 Rh	46 Pd	47 Ag	48 Cd	49 In	50 Sn	51 Sb	52 Te	53 I	54 Xe
6	55 Cs	56 Ba *	71 Lu	72 Hf	73 Ta	74 W	75 Re	76 Os	77 Ir	78 Pt	79 Au	80 Hg	81 Tl	82 Pb	83 Bi	84 Po	85 At	86 Rn
7	87 Fr	88 Ra **	103 Lr	104 Rf	105 Db	106 Sg	107 Bh	108 Hs	109 Mt	110 Ds	111 Rg	112 Cn	113 Nh	114 Fl	115 Mc	116 Lv	117 Ts	118 Og

*	57 La	58 Ce	59 Pr	60 Nd	61 Pm	62 Sm	63 Eu	64 Gd	65 Tb	66 Dy	67 Ho	68 Er	69 Tm	70 Yb
**	89 Ac	90 Th	91 Pa	92 U	93 Np	94 Pu	95 Am	96 Cm	97 Bk	98 Cf	99 Es	100 Fm	101 Md	102 No

FIGURE 3.2 The periodic table of the elements showing the groups and periods including the Lanthanide elements and the Actinide elements.

(Ga_2O_3), and antimony oxide (Sb_2O_3). Increasing the oxidation state of a metal increases the acidity of its oxide by withdrawing electron density from the oxygen atoms. For example, antimony pentoxide (Sb_2O_5) is acidic but antimony trioxide (Sb_2O_3) is amphoteric.

One other phenomenon that deserves consideration here is the phenomenon known as *solvent leveling*, which is an effect that occurs when a strong acid is placed in a solvent such as (but not limited to) water. Because strong acids donate their protons to the solvent, the strongest possible acid that can exist is the conjugate acid of the solvent. In aqueous solution, this is the hydroxonium ion (H_3O^+). This means that the strength of acids such as hydrochloric acid (HCl) and hydrobromic acid (HBr) cannot be differentiated in water as they both are dissociated 100% to the hydroxonium ion.

2.2 Organic Acids and Bases

The structural unit containing an alkyl group bonded to a carbonyl group is known as an acyl group. A family of functional groups, known as carboxylic acid derivatives, contains the acyl group bonded to different substituents. Examples of acyl groups are.

RC(=O)OR — esters

RC(=O)NR$_2$ — amides

RC(=O)X — acid halides (X = Cl, Br)

RC(=O)OC(=O)R — anhydrides

The conjunction of a carbonyl (acyl) group and a hydroxyl group forms a functional group known as carboxylic acid group. Carboxylic acids all have in common what is known as a carboxyl group, designated by the symbol –COOH. This consists of a carbon atom with a double bond to an oxygen atom, and a single bond to another oxygen atom that is, in turn, wedded to a hydrogen. All carboxylic acids can be generally symbolized by RCOOH, with R as the standard designation of any hydrocarbon. The hydrogen of a carboxyl acid group can be removed (to form a negatively charged carboxylate ion $-CO_2^-$), and, thus, molecules containing the carboxyl acid group have acidic properties.

Esters have an alkoxy (–OR) fragment attached to the acyl group; amides have attached amino groups ($-NR_2$); acyl halides (RCOCl) have an attached halogen, typically a chlorine atom; and anhydrides have an attached carboxyl group. Each type of acid derivative has a set of characteristic reactions that qualifies it as a unique functional group, but all acid derivatives can be readily converted to a carboxylic acid under appropriate reaction conditions. Many simple esters are responsible for the pleasant odors of fruits and flowers. Methyl butanoate, for example, is present in pineapples. Urea, the major organic constituent of urine and a widely used fertilizer, is a double amide of carbonic acid (carbonic acid: H_2CO_3 or HO–CO–OH; urea: H_2NCONH_2). Acyl chlorides and anhydrides are the most reactive carboxylic acid derivatives and are useful chemical reagents, although they are not important functional groups in natural substances.

Vinegar is an example of 5% v/v solution of acetic acid (CH_3CO_2H) in water, and its sharp acidic taste is due to the carboxylic acid (acetic acid) present. Lactic acid provides much of the sour taste of pickles and sauerkraut and is produced by contracting muscles. Lactic acid is also generated by the human body when a person overexerts; the muscles generate lactic acid, resulting in a feeling of fatigue until the body converts the acid to water and carbon dioxide. Another example of a carboxylic acid is butyric acid, responsible in part for the smells of rancid butter and human sweat. Citric acid is a major flavor component of citrus fruits, such as lemons, grapefruits, and oranges.

Illustration of the carboxyl group and selected carboxylic acids:

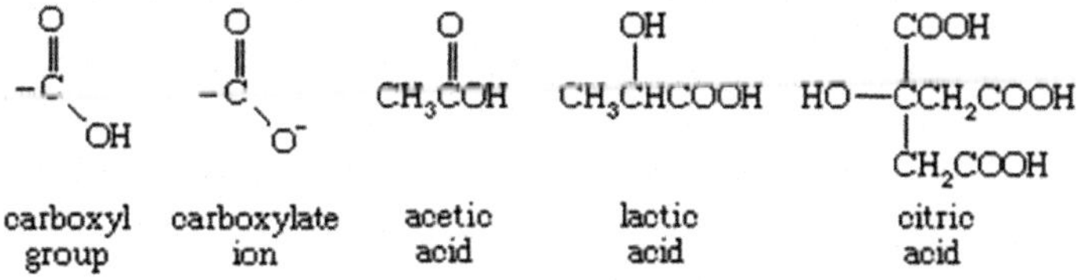

$-C(=O)OH$	$-C(=O)O^-$	$CH_3C(=O)OH$	$CH_3CH(OH)COOH$	$HO-C(COOH)(CH_2COOH)CH_2COOH$
carboxyl group	carboxylate ion	acetic acid	lactic acid	citric acid

Ibuprofen, an effective analgesic and antiinflammatory agent, contains a carboxyl group.

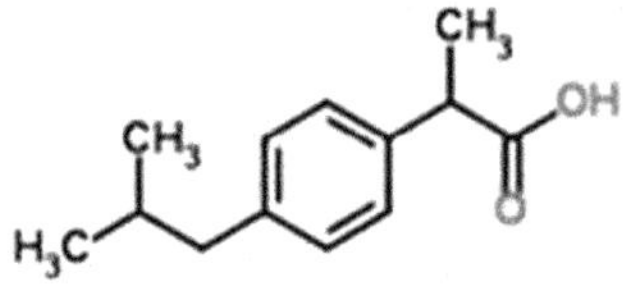

When a carboxylic acid reacts with an alcohol, it forms an ester:

$$R^1CO_2H + R^2OH \rightarrow R^1CO_2R^2 + H_2O$$

An ester has a structure that is similar to the structure described for a carboxylic acid, but with prominent differences. In addition to the bonds (one double, one single) with the oxygen atoms, the carbon atom is also attached to a hydrocarbon, which originates from the carboxylic acid. Furthermore, the single-bonded oxygen atom is attached not to hydrogen, but to a second hydrocarbon moiety from the alcohol. One well-known (and well-used) ester is acetylsalicylic acid—better known as Aspirin:

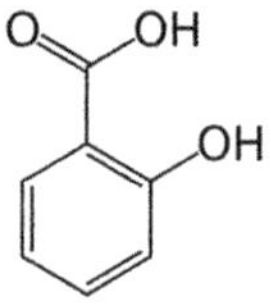

Salicylic acid

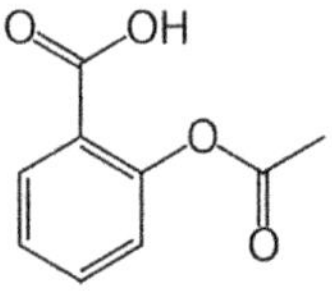

Acetyl salicylic acid (Aspirin).

Esters, which are a key factor in the aroma of various types of fruit, are often noted for their pleasant smell.

3. ACID–BASE CHEMISTRY

When dissolved in an aqueous solution, certain ions were released into the solution. An Arrhenius acid is a compound that increases the concentration of H^+ ions that are present when added to water. These H^+ ions form the hydronium ion (H_3O^+) when they combine with water molecules. This process is represented in a chemical equation by adding H_2O to the reactants side.

$$HCl(aq) \rightarrow H^+(aq) + Cl^-(aq) \tag{3.1}$$

In this reaction, hydrochloric acid (HCl) dissociates completely into hydrogen (H^+) and chlorine (Cl^-) ions when dissolved in water, thereby releasing H^+ ions into solution. Formation of the hydronium ion equation:

$$HCl(aq) + H_2O(l) \rightarrow H_3O^+(aq) + Cl^-(aq) \tag{3.2}$$

The Arrhenius theory, which is the simplest and least general description of acids and bases, includes acids such as $HClO_4$ and HBr and bases such as NaOH or $Mg(OH)_2$. For example, the complete dissociation of HBr gas into water generates free $H_3O + H_3O +$ ions.

$$HBr(g) + H_2O(l) \rightarrow H_3O^+(aq) + Br^-(aq) \tag{3.3}$$

This theory successfully describes how acids and bases react with each other to make water and salt. However, it does not explain why some substances that do not contain hydroxide ions, such as F^- and NO_2^-, can make basic solutions in water. The Brønsted–Lowry definition of acids and bases addresses this problem.

An Arrhenius base is a compound that increases the concentration of OH^- ions that are present when added to water. The dissociation is represented by the following equation:

$$NaOH(aq) \rightarrow Na^+(aq) + OH^-(aq)$$

In this reaction, sodium hydroxide (NaOH) disassociates into sodium (Na^+) and hydroxide (OH^-) ions when dissolved in water, thereby releasing OH^- ions into solution.

Polyatomic ions are often useful in the context of acid–base chemistry or in the formation of salts. A polyatomic ion can often be considered as the conjugate acid/base of a neutral molecule. For example, the conjugate base of sulfuric acid (H_2SO_4) is the polyatomic hydrogen sulfate anion (HSO_4^-). Although there may be an element with positive charge, such as the hydrogen ion (like H^+), it is not joined with another element with an ionic bond. This occurs because if the atoms formed an ionic bond, then it would have already become a compound, thus not needing to gain or lose any electrons. Polyatomic anions have negative charges while polyatomic cations have positive charges. To correctly specify how many oxygen atoms are in the ion, prefixes and suffixes are used.

A class of *inorganic compounds* that has recently emerged falls under the category of *ionic liquids* that are salts in the liquid state or salts with melting points lower than 100°C (212°F). A typical liquid is predominantly electrically neutral while ionic liquids are composed predominantly of ions and short-lived ion pairs. These substances are variously called liquid electrolytes, ionic melts, ionic fluids, fused salts, liquid salts, or ionic glasses. Ionic liquids are powerful solvents and electrically conducting fluids (electrolytes). Any salt that melts without decomposing or vaporizing usually yields an ionic liquid. Conversely, when an ionic liquid is cooled, it often forms an ionic solid which may be either crystalline or glass-like. Examples include compounds based on the 1-ethyl-3-methylimidazolium (EMIM) cation and include: $(C_2H_5)\ (CH_3)C_3H_3N^+\ N(CN)^-$ that melts at −21°C (−6°F) and 1-butyl-3,5-dimethylpyridinium bromide which becomes a glass below −24°C (−11°F).

Acid dissociation constants are most often associated with weak acids, or acids that do not completely dissociate in solution because strong acids are presumed to ionize completely in solution and therefore their K_a values are large (Table 3.3). The larger the value of pK_a, the smaller the extent of dissociation. A strong acid is almost completely dissociated in aqueous solution to the extent that the concentration of the undissociated acid in negligible and becomes undetectable.

Acids are prepared by the interaction of acid oxides with water (for oxyacids):

$$SO_3 + H_2O \rightarrow H_2SO_4$$

$$P_2O_5 + 3H_2O \rightarrow 2H_3PO_4$$

Another method of acid preparation is by the interaction of hydrogen with nonmetals and following dissolution product in water (for acids without oxygen in the molecule):

$$H_2 + Cl_2 \rightarrow 2HCl$$

$$H_2 + S \rightarrow H_2S$$

The exchange reaction between a salt and an acid also produces a new acid:

$$Ba(NO_3)_2 + H_2SO_4 \rightarrow BaSO_4 + 2HNO_3$$

This method also includes the displacement of a weak or slightly soluble acid from a salt by means of a stronger acid:

$$Na_2SiO_3 + 2HCl \rightarrow H_2SiO_3 + 2NaCl$$

$$2NaCl + H_2SO_4(conc.) \rightarrow Na_2SO_4 + 2HCl$$

The chemical properties of acids include (1) interaction with bases, that is, neutralization, (2) Interaction with basic oxides, (3) interaction with metals, and (4) interaction with salts, that is,

$$H_2SO_4 + 2KOH \rightarrow K_2SO_4 + 2H_2O$$

$$2HNO3 + Ca(OH)_2 \rightarrow Ca(NO_3)_2 + 2H_2O$$

$$CuO + 2HNO_3 \rightarrow Cu(NO_3)_2 + H_2O$$

$$Zn + 2HCl \rightarrow ZnCl_2 + H_2$$

$$2Al + 6HCl \rightarrow 2AlCl_3 + 3H_2$$

$$BaCl_2 + H_2SO_4 \rightarrow BaSO_4 + 2HCl$$

$$K_2CO_3 + 2HCl \rightarrow 2KCl + H_2O + CO_2$$

Bases are also complex chemicals in which an atom of a metal is bonded with one or several hydroxyls groups which, when dissociating in water, form a metal cation (or the ammonium cation, NH_4^+) and a hydroxide anion (OH^-).

Bases are prepared by the interaction of active metals (alkaline and alkaline earth metals) with water:

$$2Na + 2H_2O \rightarrow 2NaOH + H_2$$

$$Ca + 2H_2O \rightarrow Ca(OH)_2 + H_2$$

Another method of preparation involves the interaction of the oxides of active metals with water:

$$BaO + H_2O \rightarrow Ba(OH)_2$$

The electrolysis of an aqueous solution of a salt can also produce a base:

$$2NaCl + 2H_2O \rightarrow 2NaOH + H_2 + Cl_2$$

The chemical reactions of bases, like the acids, are variable and include (1) Interaction with acid oxides, (2) interaction with acids—neutralization, and (3) exchange reactions with salts.

$$KOH + CO_2 \rightarrow KHCO_3$$

$$2KOH + CO_2 \rightarrow K_2CO_3 + H_2O$$

$$NaOH + HNO_3 \rightarrow NaNO_3 + H_2O$$

$$Ba(OH)_2 + K_2SO_4 \rightarrow 2KOH + BaSO_4$$

Acids, such as hydrochloric acid, and bases, such as potassium hydroxide, that have a great tendency to dissociate in water are completely ionized in solution and are called strong acids or strong bases. On the other hand, acids, such as acetic acid, and bases, such as ammonia, that are reluctant to dissociate in water are only partially ionized in solution; they are called weak acids or weak bases. Strong acids in solution produce a high concentration of hydrogen ions, and strong bases in solution produce a high concentration of hydroxide ions and a correspondingly low concentration of hydrogen ions. The hydrogen ion concentration is often expressed in terms of its negative logarithm, or pH. Strong acids and strong bases make very good electrolytes insofar as solutions of strong acids and strong bases readily conduct electricity. Conversely, weak acids and weak bases make poor electrolytes.

The chemical interaction between an acid and a base (*acid–base reaction*) varies depending on the species involved. All inorganic acids elevate the hydrogen concentration in an aqueous solution. All inorganic bases elevate the hydroxide concentration in an aqueous solution. Inorganic salts are neutral, ionically bound molecules and do not affect the concentration of hydrogen in an aqueous solution. Inorganic compounds, in a broader sense, consist of an ionic component and an anionic component. A very large number of compounds occur naturally while others may be synthesized. In all cases, charge neutrality of the compound is key to the structure and properties of the compound.

Historically, the theory of acid–base chemistry was thought to be related to the composition of the compound, specifically that it should contain oxygen; this was the theory of Antoine Lavoisier but was disproved some years later by Sir Humphry Davy when he proved nonoxygen containing species, such as hydrogen sulfide and the hydrohalic acids (such as hydrochloric acid) existed. This work was extended by Liebig, who proposed that acids are substances that contain hydrogen that can be displaced by a metal species before being formulated into the currently accepted *Arrhenius definition* of an acid as a substance that dissociates in water to produce hydrogen cations (H^+), while a base dissociates in water to form hydroxide anions (OH^-). It should be appreciated that hydrogen ions (H^+) do not exist as separate species in a solution but rather that the hydronium cation (H_3O^+) is actually formed.

This forms the basis of the statement that the reaction of an acid and a base will produce a salt and water in a neutralization reaction, as the hydrogen cations form a compound (water) with the anion from the base (i.e., hydroxide anion) and the cation from the base, often a metal, will react with the anion from the acid to form a salt. For example, the reaction of potassium hydroxide and nitric acid:

$$\underset{\text{Acid}}{HNO_3} + \underset{\text{base}}{2KOH} \rightarrow \underset{\text{salt}}{K_2NO_3} + \underset{\text{water}}{2H_2O}$$

Note that the Arrhenius definition only holds for aqueous systems and that the dissolution of acids in other solvents would not necessarily be acidic. Similarly, molten metal hydroxides are not basic. The issue of the solvent used was resolved with the generalization offered by the *solvent system theory* in which it was noted that in several solvents, there is a presence of solvonium cations and solvate anions in equilibrium with nondissociated solvent molecules. The simplest example is water, which exists with some hydronium cations and hydroxide anions. The addition of another substance (such as a solute) to the solvent causes a shift in this equilibrium to form either more solvonium cations or solvate anions, depending on its nature. An acid increases the solvonium (in the case of water, this is H_3O^+) cations present, while a base increases the concentration of solvent anions. For example, water exists as the equilibrium:

$$2H_2O \leftrightarrow H_3O^+ + OH^-$$

Addition of the acid (HNO_3) will form more hydronium cations:

$$2H_2O + HNO_3 \leftrightarrow H_3O^+ + NO_3^- + 2OH^-$$

While the presence of the base (KOH) will increase the presence of hydroxide anions

$$2H_2O + KOH \rightarrow H_3O^+ + K^+ + OH^-$$

The strength of this approach, in terms of solvent rather than only water, is demonstrated by consideration of aprotic solvents, such as liquid dinitrogen tetroxide (N_2O_4):

$$\underset{\text{Base}}{AgNO_3} + \underset{\text{acid}}{NOCl} \sim \underset{\text{salt}}{AgCl} + \underset{\text{solvent}}{N_2O_4}$$

In an interaction between the basic/acidic species and the solvent, the nature of a substance can be altered by changing the solvent, for example, perchloric acid, which is a strong acid in water, is a weak acid in ethanoic acid yet a weak base in fluorosulfonic acid. It is important to appreciate the changing nature of compounds with regard to their environment, which can have consequences for reaction and materials choice. The pH scale (Fig. 3.1) is a measure of acidity and alkalinity of an acid or base or a solution of either.

3.1 Reactions of Acids and Alkalis

This is a particular acid—base reaction, where the base is also an alkali species, meaning it is a basic, ionic salt of either an alkali or alkaline earth metal. In this instance, the reaction produces a metal salt and water in a neutralization reaction. The acid either contains hydrogen cations or causes them to be produced in solution, while the alkali is a soluble base either containing hydroxide ions or causing them to be produced in solution. For example, nitric acid and potassium hydroxide:

$$HNO_3 \rightarrow H^+(aq) + NO_3^-(aq)$$

$$NaOH \rightarrow Na^+(aq) + OH^-(aq)$$

As a result of the neutralization of either the acid or base, by the other species, into water, such reactions have a number of applications including antacid formulations, where excess hydrochloric acid in the stomach is neutralized, most commonly, by species including sodium carbonate and aluminum hydroxide. Neutralization can also be used for the regulation of pH in agriculture via the application of fertilizers, or the abatement of acid formation by the removal of sulfur from fuel before combustion to avoid the formation of sulfur dioxide, and, subsequently, acid rain (Chapter 2).

3.2 Brønsted–Lowry Acids and Bases

Working independently, Johannes Nicolaus Brønsted and Martin Lowry developed a definition, underpinned by the concept of base protonation and acid deprotonation, where acids have the ability to *donate* hydrogen cations (H^+) to bases, hence the bases *accept* these H^+ ions. This definition removes the reliance on a full system description, that is, the solvent or salt formed, but rather focuses on the acid and base themselves. The species involved in the transfer of a proton are known as *conjugates* (i.e., a pair), with a conjugate acid and conjugate base produced.

By this definition, acids donate protons and bases accept protons and an acid–base reaction is, therefore, the transfer of a hydrogen cation from an acid to a base. Thus, the acid and base react, not to give a salt and solvent, but rather a new (conjugate) acid, the base with accepted H^+, and new (conjugate) base, the acid with H^+ removed, consequently eliminating the idea of neutralization. For example, the addition of hydrogen cations (H^+) to hydroxide anions (OH^-), which are basic, produces water (H_2O); hence water is the conjugate acid of hydroxide ions:

$$H^+ + OH^- \rightarrow H_2O$$

As mentioned above, the solvent is no longer considered. However, the solvent system can coexist within the Brønsted–Lowry definition, which can also provide an explanation for the low concentrations of hydronium and hydroxide ions produced by the dissociation of water:

$$2H_2O \leftrightarrow H_3O^+ + OH^-$$

In this case, water, which is amphoteric, acts as both acid and base; as one molecule of water donates a hydrogen (H^+) ion, forming the conjugate base, the hydroxide ion, with the second molecule of water accepts the H^+ ion, forming the conjugate acid, the hydroxonium ion. In general, for acid–base reactions, the Brønsted–Lowry definition is:

$$AH + B^- = BH + A^-$$

AH is the acid, B is the base, BH^+ is conjugate acid of B, and $A^-$is the conjugate base of the acid AH.

3.3 Lewis Acids and Bases

In contrast to the Brønsted –Lowry definition of acids and bases, Lewis defined the species in terms of electron transfer rather than hydrogen cations. Hence, a *Lewis base* is a species that donates an electron pair to a *Lewis acid,* which accepts the donated electron pair. This concept is most easily demonstrated by considering the established aqueous acid–base reaction:

$$HCl + NaOH^- + H_2O + NaCl$$

In this case, the acid combines with the base, rather than exchanging atoms with it; the acid is H^+ and the base OH^-, which has an unshared electron pair that it is able to donate to forming a bond with the electron-deficient proton. Hence, the acid–base reaction is not the transfer of H^+ but the donation of an electron pair from OH^-, forming a covalent bond to produce water. The electronic interactions involved are the donation of electrons from the highest occupied molecular orbital of the subsequently known base to the lowest unoccupied molecular orbital of the second species, then known as the acid.

The strength of a Lewis acid and base interaction is determined by the concentrations of all solution species, for a simple system:

$$A + B \rightarrow AB$$

The stability constant is given by:

$$K = [AB]/[A][B]$$

A high value of K indicates a strong interaction as [AB] is large compared to the product of [A] [B] and a low value of K indicates a weak interaction. The interaction is typically governed by the relative strengths of the acid and/or the base.

Hard acids and bases are usually small species that are difficult to polarize, for example, hard acids: H^+, Na^+; hard bases: OH^-, F^-, NH_3. In contrast, *soft acids and bases* are usually large species that are easily polarized, for example, soft acids: Ag^+, Cd^{2+}, Cu^+; soft bases: H^-, 1^-, CO. This is an important distinction since hard acids tend to bind to hard bases, as they both exhibit high ionic character while soft acids tend to bind to soft bases, which both exhibit significant covalent character.

4. VAPOR PRESSURE AND VOLATILITY

The range of chemicals produced by the chemicals industry and the petroleum refining industry is so vast that summarizing the properties and/or the toxicity or general hazards of crude oil in general or even for a specific crude oil is a difficult task. However, petroleum and some petroleum products, because of the hydrocarbon content, are at least theoretically biodegradable but large-scale spills can overwhelm the ability of the ecosystem to break the oil down (Speight, 2005, 2014, 2015).

The toxicological implications from petroleum occur primarily from exposure to or biological metabolism of aromatic structures. These implications change as a chemical spill ages or is weathered. It is in such instances that the vapor pressure and the volatility of the chemical (or chemical constituents of the spill) can play an important role in increasing or diminishing the effect of the spill on the environment.

4.1 Vapor Pressure

The vapor pressure (or equilibrium vapor pressure) of a chemical is the pressure exerted by a vapor in thermodynamic equilibrium with the condensed phase (liquid or solid) at a given temperature in a closed system (Boethling and Mackay, 2000; Mackay et al., 2006). The equilibrium vapor pressure is an indication of the evaporation rate of the liquid and relates to the tendency of particles to escape from the liquid (or a solid). A substance with a high vapor pressure at normal temperature is often referred to as *volatile*. As the temperature of a liquid increases, the kinetic energy of its molecules also increases which leads to an increase in the number of molecules passing from the liquid phase to the vapor phase, thereby increasing the vapor pressure.

By way of definition, evaporation is a kinetic process and involves (1) the escaping tendency of a discharged chemical, which is the distribution between the solid or liquid and vapor phases, (2) diffusion away from a surface and the rate of evaporation is determined predominantly by the rate of diffusion through the thin stagnant layer, usually referred to as the boundary layer, of air at the surface, and (3) dispersion, which is caused by atmospheric currents that move the chemical away from the site of evaporation. Overall, evaporation is an endothermic process and temperature will obviously affect the rate of evaporation.

Vapor pressure controls the volatility of a chemical from soil, and along with its water solubility, determines evaporation from water (Boethling and Mackay, 2000). Predicting the volatility of a chemical in soil systems is important for estimating chemical partitioning in the subsurface between sorbed phase, dissolved phase and gas phase. Vapor pressure is an important parameter in estimating vapor transport of an organic chemical in the air. It is often useful to determine the vapor pressure and other properties of a chemical by assuming that it is a liquid or supercooled liquid at a temperature less than the melting point (Mackay et al., 2006). At very low environmental concentrations such as in liquid solutions or on aerosol particles, pure chemical behavior relates to the liquid rather than the solid state.

The vapor pressure of any substance increases nonlinearly with temperature according to the Clausius–Clapeyron equation, which relates pressure, P, the enthalpy of vaporization, ΔH_{vap}, and the temperature, T:

$$P = A \exp(-\Delta H_{vap}/RT)$$

In this equation, R (=8.3145 J mol^{-1} K^{-1}) and A are the gas constant and unknown constant, respectively. If P_1 and P_2 are the pressures at two temperatures T_1 and T_2, the equation has the form:

$$\ln P_1/P_2 = \Delta H_{vap}/\mathrm{R}(1/\mathrm{T}_1 - 1/\mathrm{T}_2)$$

The Clausius–Clapeyron equation allows an estimate of the vapor pressure at another temperature, if the vapor pressure is known at some temperature, and if the enthalpy of vaporization is known.

The atmospheric pressure boiling point (the normal boiling point) of a liquid is the temperature at which the vapor pressure equals the ambient atmospheric pressure. With any incremental increase in that temperature, the vapor pressure becomes sufficient to overcome atmospheric pressure and lift the liquid to form vapor bubbles inside the bulk of the substance. Bubble formation deeper in the liquid requires a higher pressure, and therefore higher temperature, because the fluid pressure increases above the atmospheric pressure as the depth increases.

4.2 Volatility

The volatility of organic compounds or the individual chemical constituents in a mixture is an important loss mechanism in the overall material balance. The key environmental factors affecting volatilization are the reaeration constant (surface transfer rate of dissolved oxygen per mixed depth of the water body), wind speed, and the mixed depth of the water. Furthermore, when evaluating the volatilization of complex mixtures versus single chemicals, only physicochemical properties can be affected differently by mixtures. Thus, only by altering either aqueous solubility or vapor pressure of a chemical by interactions with other chemicals can volatilization rates be altered. Also, if the composition of a given mixture is known as well as the solubility of each constituent (from knowledge of the physiochemical properties of each constituent) (Table 3.8), it is possible to anticipate that rate of change for volatilization due to chemical interactions.

TABLE 3.8 General Guide for the Solubility of Inorganic Compounds

Soluble	Insoluble
Group I NH_4^+ compounds	Carbonates (except Group I, NH_4^+, and uranium compounds)
Nitrates	Sulfites (except Group I and NH_4^+ compounds)
Acetates	Phosphates (except Group I [except for Li^+] and NH_4^+compounds)
Chlorides Chlorates and perchlorates Bromides and iodides (except Ag^+, Pb^{2+}, Cu^+, Hg^{2+})	Hydroxides and oxides (except Group I, NH_4^+, Ba^{2+}, Sr^{2+}, Tl^+)
Sulfates (except Ag^+, Pb^{2+}, Ba^{2+}, Sr^{2+}, Ca^{2+})	Sulfides (except Group I, Group II, and NH_4^+ compounds)

Chemicals, whether they are produced by the chemical industry or pharmaceutical industry or petroleum industry, can be considered as environmentally transportable materials, the character of which is determined by several chemical and physical properties (i.e., solubility, vapor pressure, and propensity to bind with soil and organic particles). These properties are the basis of measures of leachability and volatility of individual hydrocarbons. Thus, the transport or organic chemicals either as individual chemicals or as mixtures (such as the various crude oil—derived products) can be considered by equivalent carbon number to be grouped into 13 different fractions. The analytical fractions are then set to match these transport fractions, using specific *n*-alkane derivatives to mark the analytical results for aliphatic compounds and selected aromatic compounds to delineate hydrocarbons containing benzene rings.

Although chemicals grouped by transport fraction generally have similar toxicological properties, this is not always the case. For example, benzene is a carcinogen but many alkyl-substituted benzenes do not fall under this classification. However, it is more appropriate to group benzene with compounds that have similar environmental transport properties than to group it with other carcinogens, such as benzo(a)pyrene, that have very different environmental transport properties. Nevertheless, consultation of any reference work that lists the properties of chemicals will show the properties and hazardous nature of the types of chemicals that are found in petroleum. In addition, petroleum is used to make petroleum products, which can contaminate the environment.

The range of chemicals produced by the organic chemicals industry and petroleum refining industry is so vast that summarizing the properties and/or the toxicity or general hazard of crude oil in general or even for a specific crude oil is a difficult task. However, petroleum and some petroleum products, because of the hydrocarbon content, are at least theoretically biodegradable but large-scale spills can overwhelm the ability of the ecosystem to break the oil down. The toxicological implications from petroleum occur primarily from exposure to or biological metabolism of aromatic structures. These implications change as an oil spill ages or is weathered.

4.2.1 Low-Boiling Chemicals

Typically, low-boiling chemicals do not include inorganic chemicals and the term is usually used to indicate organic chemicals. These chemicals vary from simple hydrocarbons of low-to-medium molecular weight to higher molecular—weight organic compounds containing sulfur, oxygen, and nitrogen, as well as compounds containing metallic constituents, particularly vanadium, nickel, iron, and copper and contain one or more functional groups that dictate the behavior of the chemical. However, the behavior of an organic chemical on the basis of functional groups depends upon (1) the type of functional group, (2) the number of functional groups, (3) the position of the functional groups within the molecule, and (4) the ecosystem into which the chemical is discharged.

Thus, the volatility of a chemical is of concern predominantly for surface-located chemicals and is affected by (1) temperature of the soil, (2) the water content of the soil, (3) the adsorptive interaction of the chemical and the soil, (4) the concentration of the chemical in the soil, (5) the vapor pressure of the chemical, and (6) the solubility of the chemical in water, which is the predominant liquid in the soil.

Many of the gaseous and liquid chemicals that are low-boiling materials (including crude oil and crude oil–derived products) fall into the class of chemicals which have one or more of the following characteristics as considered to be hazardous by the Environmental Protection Agency: (1) ignitability, (2) flammability, (3) corrosivity, and (4) reactivity. In summary, many of the specific chemicals in crude oil and crude oil–derived products are hazardous because of their chemical reactivity, fire hazard, toxicity, and other properties. In fact, a simple definition of a hazardous chemical (or hazardous waste) is that it is a chemical substance (or chemical waste) that has been inadvertently released, discarded, abandoned, neglected, or designated as a waste material and has the potential to be detrimental to the environment. Alternatively, a hazardous chemical may be a chemical that may interact with other (chemical) substances to give a product that is hazardous to the environment. Low-boiling organic chemicals (whether crude oil-derived or derived from another source) fit very well into this definition. Examples of some commonly encountered volatile hydrocarbons are:

Aliphatic Hydrocarbons	Aromatic Hydrocarbons
Pentane derivatives	Benzene
Hexane derivatives	Toluene
Heptane derivatives	Ethylbenzene
Octane derivatives	Xylene isomers and derivatives
Nonane derivatives	Naphthalene derivatives
Decane derivatives	Phenanthrene derivatives
	Anthracene derivatives
	Acenaphthylene derivatives

For example, a liquid that has a flash point of less than 60°C (140°F) is considered *ignitable*. Some examples are: benzene, hexane, heptane, benzene, pentane, petroleum ether (low boiling), toluene, and the xylene isomers. An organic chemical is classified as *flammable* if it has the ability to burn or ignite, causing fire or combustion. The degree of difficulty required to cause the combustion of a substance is quantified through standard test methods (Speight, 2014, 2015). The data from such test methods are used in regulations that govern the storage and handling of highly flammable substances inside and outside of structures and in surface and air transportation.

An aqueous solution that has a pH of less than or equal to 2, or greater than or equal to 12.5, is considered *corrosive*. Most organic chemicals, crude oils, and crude oil–derived products are not corrosive but many of the chemicals used in refineries are corrosive—corrosive materials include the inorganic chemicals such as sodium hydroxide, as well as other acids or bases. The term *reactivity* applies to chemicals that react violently with air or water and, as a result, are considered to be hazardous chemicals. Reactive organic chemicals include chemicals capable of detonation (trinitrotoluene, TNT) when subjected to an initiating source.

Gas condensate also falls within the volatile organic compound category and condensate release can be equated to the release of volatile constituents but are often named as such because of the specific constituents of the condensate, often with some reference to the gas condensate that is produced by certain crude oil wells and natural gas wells. However, the condensate is often restricted to the low-boiling alkane derivative as well as benzene, toluene, ethyl benzene, and the xylene isomers (BTEX).

4.2.2 High-Boiling and Nonvolatile Chemicals

In almost all cases of contamination by organic chemicals, attention must be directed to the presence of semivolatile hydrocarbon derivatives and nonvolatile hydrocarbon derivatives.

Among the polynuclear aromatic hydrocarbons, the toxicity of many hydrocarbon liquids (especially crude oils and crude oil–derived products) is a function of its di- and triaromatic hydrocarbon content. Like the single aromatic ring variations, including benzene, toluene, and the xylene isomers, all are relatively volatile compounds with varying degrees of water solubility. However, in the higher boiling hydrocarbons liquids (particular products designed as fuel oil), the two-ring condensed aromatic hydrocarbons, naphthalene, and the various homologs are less acutely toxic than benzene but are more prevalent for a longer period after a spill or discharge. The toxicity of different crude oils and refined products (such as naphtha and naphtha-derived solvents and fuels) depends not only on the total concentration of hydrocarbons but also on the hydrocarbon composition in the water-soluble fraction (WSF) as well as on the (1) the degree of water solubility, (2) the concentrations of the individual component, and (3) the toxicity of the components either individually or collectively. The WSFs prepared from different crude oils will vary in terms of these three parameters. Water-soluble fractions of the refined products (such as naphtha and naphtha-derived solvents, and fuels such as No. 2 fuel oil and bunker C oil) are more toxic to the floral and faunal inhabitants of many ecosystems than the typical water-soluble fraction of crude oil. Organic chemical either having a higher number of condensed rings or with methyl substituents on the rinds are typically more toxic than the less substituted derivatives but tend to be less water-soluble and thus less plentiful in the water-soluble fraction. There are also indications that pure naphthalene (a constituent of moth balls that are, by definition, toxic to moths) and alkyl naphthalene derivatives are from three-to-ten times more toxic to test animals than are benzene and alkylbenzene derivatives. In addition, and because of the low water-solubility of tricyclic and polycyclic (polynuclear) aromatic hydrocarbons (that is, those aromatic hydrocarbons heavier than naphthalene), these compounds are generally present at very low concentrations in the water-soluble fraction of oil. Therefore, the results of this study and others conclude that the soluble aromatics of crude oil (such as benzene, toluene, ethylbenzene, xylene isomers, and naphthalene derivatives) produce the majority of toxic effects of crude oil in the environment.

The higher molecular–weight aromatic structures (with four to five condensed aromatic rings), which are the more persistent in the environment, have the potential for chronic toxicological effects. Since these compounds are nonvolatile and are relatively insoluble in water, their main routes of exposure are through ingestion and epidermal contact. Some of the compounds in this classification are considered possible human carcinogens; these include benzo(a)pyrene, benzo(e)pyrene, benzo(a)anthracene, benzo(b, j, and k)fluorene, benzo(ghi)perylene, chrysene, dibenzo(ah)anthracene, and pyrene.

5. WATER SOLUBILITY

Water solubility is one of the most important properties for evaluating the fate and direct measure of a chemical in an ecosystem since a chemical with high water solubility will partition readily and rapidly into the aqueous phase and will often remain in solution and be available for degradation. A chemical that is sparingly soluble in water will often dissolve slowly into solution and partition more readily into other phases including air, solids, and the surface of solid particles including soil (Boethling and Mackay, 2000). Water solubility is used to determine the maximum theoretical concentration of a chemical in the soil pore water and is also important for estimating the air–water partitioning coefficient (K_{AW}), used to determine the partitioning behavior of an organic chemical in the soil.

Solubility is one of the characteristic properties of a chemical that is commonly used (1) to describe the chemical, (2) to indicate the polarity of the chemical, (3) to assist in distinguishing the chemical from other chemicals, and (4) as a guide to applications for which the chemical is suited. According to IUPAC (Union of Pure and Applied Chemistry), solubility is the analytical composition of a saturated solution expressed as a proportion of a designated solute in a designated solvent. Solubility may be stated in various units of concentration such as molarity, molality, mole fraction, mole ratio, mass(solute) per volume(solvent) and other units. Solubility occurs under dynamic equilibrium, which means that solubility results from the simultaneous and opposing processes of dissolution and phase joining (such as the precipitation of a solid).

Solubility is a property of interest in many aspects of science, including but not limited to: environmental predictions, biochemistry, pharmacy, drug-design, agrochemical design, and protein ligand binding. Aqueous solubility is of fundamental interest owing to the vital biological and transportation functions played by water. In addition, to this clear scientific interest in water solubility and solvent effects; accurate predictions of solubility are important industrially. The ability to accurately predict a molecule's solubility represents potentially large financial savings in many chemical product development processes.

Thus, solubility is of fundamental importance in many scientific disciplines and practical applications, ranging from ore processing and nuclear reprocessing to the use of medicines, and the transport of pollutants. More pertinent to the present text and in the context of environmental technology, the solubility of a chemical in water dictates the degree of mobility of the chemical in water and the potential for the pollution of the aquasphere (water systems). Simply, the solubility of a chemical in water is a measure of the amount of chemical substance that can dissolve in water at a specific temperature. The unit of solubility is generally in mg/L (milligrams per liter) or ppm (parts per million).

Also, the solubility of a chemical is useful when separating mixtures in the laboratory in a chemical process, as well as determining which constituents of a mixture will dissolve in the aquasphere and which will remain in the air or in the soil. For example, a mixture of salt (sodium chloride, NaCl) and silica (SiO_2) may be separated by treating the mixture with water (in which the salt will dissolve) and filtering off the undissolved silica. In fact, the synthesis of chemicals in the laboratory and by an industrial process make use of the relative solubilities of the desired chemical product, as well as any unreacted starting materials, byproducts, and side products to achieve separation.

Water solubility is one of the most important properties affecting bioavailability and environmental fate of chemicals. Chemicals that arewater-soluble bioconcentrate poorly in aquatic species (Mackay, 1982). Furthermore, chemicals that exhibit a high solubility in water tend to degrade more readily by processes such as photolysis or hydrolysis. There is an aquatic ecotoxicological test that can be used to determine the solubility of a chemical in aqueous solution, that is, truly dissolved in the test medium. For poorly soluble substances (substances with low water solubility <100 mg/L), the test methods are usually conducted only up to the maximum dissolved concentration under the test conditions.

At the molecular level, solubility is controlled by the energy balance of intermolecular forces between solute–solute, solvent–solvent, and solute–solvent molecules. Intermolecular forces vary in strength from very weak induced dipole to the much stronger dipole–dipole forces (such as hydrogen bonding). Most organic molecules are relatively nonpolar and are usually soluble in organic solvents (e.g., diethyl ether, dichloromethane, chloroform, organic chemicals ether, hexanes, etc.) but not in polar solvents such as water. However, some organic molecules are more polar and therefore solubility in water is generally indicative of a high ratio of polar group(s) to the nonpolar hydrocarbon chain, that is, a low molecular–weight compound containing a hydroxyl group (–OH) or an amino group ($-NH_2$) or a carboxylic acid group ($-CO_2H$ group) or a higher molecular–weight compound containing several polar groups. The presence of an acidic carboxylic acid group or a basic amino group in a water-soluble compound can be detected by the low or high pH, respectively, of the solution.

The water solubility of a chemical reflects (1) the physical and chemical properties of the chemical (the solute) and water (the solvent) as well as (2) the temperature, (3) the pressure, and (4) the pH of the solution. The extent of the solubility of a chemical in a specific solvent is measured as the saturation concentration, where adding more solute does not increase the concentration of the solution and begins to precipitate the excess amount of solute. The extent of solubility ranges widely, from infinitely soluble (without limit, fully miscible) to poorly soluble (limited solubility). The term insoluble may be (correctly or erroneously) applied to poorly or very poorly soluble chemicals. A common threshold to describe something as insoluble is less than 0.1 g per 100 mL of solvent. Under certain conditions, the equilibrium solubility can be exceeded to give a so-called supersaturated solution, which is metastable, which is a dynamic system other than the state of least energy of the system. Solubility is not to be confused with the ability of water (solvent) to dissolve a chemical (solute) because the solution might also occur because of a chemical reaction. However, it must be noted that running water that is extracting a chemical from the soil may never reach the saturation concentration or a condition of equilibrium solubility.

Crucially, solubility applies to all areas of chemistry, geochemistry, inorganic chemistry, physical chemistry, organic chemistry, and biochemistry. In all cases solubility will depend on the physical conditions (temperature, pressure and concentration) with the enthalpy and entropy directly relating to the solvents and solutes concerned.

Also, solubilization is a mass-transfer process related to the movement of a chemical between a solid phase and the aqueous phase. Mixing often enhances solubilization by enhancing mass transfer. Additionally, chemicals can be added that enhance solubilization. For example, detergents may be used to increase the solubility of liquid-phase chlorinated solvents so that they can be extracted more readily from groundwater. Detergent and solvent-enhanced solubilization are major remediation processes that require the introduction and mixing of chemicals for groundwater remediation.

Precipitation reactions (the precipitate is indicated by "s" following the chemical) are of importance when stabilization of a chemical is desired, such as by its removal from the water phase and formation of a solid phase that does not contaminate or move with groundwater. For example, formation of the precipitate chromium hydroxide [$Cr(OH)_3(s)$] removes chromium from water. The low-solubility product of many sulfide chemicals suggests that they would be good candidates for removal from groundwater. Sulfides for this purpose might be transformed to form metal sulfate derivatives by reduction under anaerobic (oxygen deficient) conditions.

Precipitation of a chemical in an ecosystem, while often beneficial, can give rise to serious problems. An example is the interference with the flow of liquid and/or solution caused by the precipitate [such as calcium carbonate, $CaCO_3(s)$] in a channel that is being used during a groundwater remediation project. The formation of the precipitate (often referred to as *clogging*) may be undesirable because it can result in rerouting the direction of groundwater flow leading to migration of contaminated water into previously uncontaminated regions and/or delivery of added chemicals to regions that are uncontaminated.

5.1 Effect of Temperature

The solubility of a given solute in a given solvent typically depends on temperature. Depending on the nature of the solute the solubility may increase or decrease with temperature. For most solids and liquids, their solubility increases with temperature. In liquid water at high temperatures (e.g., that approaching the critical temperature), the solubility of ionic solutes tends to decrease due to the change of properties and structure of liquid water; the lower dielectric constant results in a less polar solvent.

5.2 Effect of Pressure

For condensed phases (solids and liquids), the pressure dependence of solubility is typically weak and usually neglected in practice. The pressure dependence of solubility does occasionally have practical significance. For example, precipitation fouling of remediation sites by calcium sulfate ($CaSO_4$)—the solubility of which decreases with decreasing pressure—can result in decreased productivity with time.

Furthermore, and very important in the context of environmental chemistry, the solubility of carbon dioxide in seawater is also affected by temperature. The decrease of solubility of carbon dioxide in seawater when temperature increases is also an important retroaction factor (positive feedback) exacerbating past and future estimates of the climate change phenomenon as observed from ice cores (Speight and Islam, 2016).

5.3 Rate of Dissolution

The rate of dissolution (rate of solubilization) (in kg/s) is related to the solubility product and the surface area of the material. The speed at which a solid can dissolve in a solvent (water) may depend on its crystallinity or lack thereof in the case of amorphous solids and the surface area (crystallite size). Furthermore, the rate of dissolution can vary by orders of magnitude between different systems. Typically, a very low dissolution rate is associated with low solubility and a chemical with a high solubility will exhibit a high dissolution rate.

Importantly relatively low-solubility compounds are found to be soluble in more extreme environments resulting in geochemical and geological observations of the activity of hydrothermal fluids in the earth's crust. These are often the source of high-quality economic mineral deposits and precious or semiprecious gems. In the same way compounds with low solubility will dissolve over extended time (geological time) resulting in significant effects such as extensive cave systems or Karstic land surfaces, which are areas of limestone terrane characterized by sinks, ravines, and underground streams.

5.4 Quantification of Solubility

Solubility is commonly expressed as a concentration; for example, as g of solute per kg of solvent, g per 100 mL solvent. The maximum equilibrium amount of solute that can dissolve in a given amount of solvent is the solubility of that solute in that solvent under the specified conditions. The advantage of expressing solubility in this manner is its simplicity, while the disadvantage is that it can strongly depend on the presence of other species in the solvent (for example, the common ion effect).

A solubility constant is used to describe saturated solutions of ionic compounds of relatively low solubility and is a special case of an equilibrium constant. Thus, the solubility constant describes the balance between dissolved ions from the salt and undissolved salt. The solubility constant is also applicable to determining the potential for the precipitation of a solid from the liquid, which is the reverse of the dissolution. As with other equilibrium constants, the temperature can affect the numerical value of solubility constant. However, the value of the solubility constant is generally independent of the presence of other species in the solvent.

5.5 Solubility of Inorganic Chemicals

Many inorganic chemicals (such as salts) are ionic in character and dissolve readily in water because of the attraction between positive and negative charges. For example, the positive ions of a salt such as silver chloride (Ag^+) attract the partially negative oxygen atoms in the water molecule. Likewise, the negative ions of the silver chloride (Cl^-) attract the partially positive hydrogen atoms.

$$AgCl(s) \rightleftharpoons Ag^+(aq) + Cl^-(aq)$$

However, there is a limit to how much salt can be dissolved in a given volume of water which is given by the solubility product, K_{sp}. This value depends on the type of salt (for example, silver chloride compared to sodium chloride), temperature, and any common ion effect.

Thus, it is possible to calculate the amount of silver chloride that will dissolve in 1 L of water. Thus:

$K_{sp} = [Ag^+] \times [Cl^-]$ (definition of solubility product)
$K_{sp} = 1.8 \times 10^{-10}$ (from a table of solubility products)
$[Ag^+] = [Cl^-]$, in the absence of other silver or chloride salts
$[Ag^+]^2 = 1.8 \times 10^{-10}$
$[Ag^+] = 1.34 \times 10^{-5}$

Thus, 1 L of water can dissolve 1.34×10^{-5} mol of AgCl(s) at room temperature. In contrast, sodium chloride (NaCl) has a higher K_{sp} than silver chloride and is, therefore, more soluble than silver chloride.

5.6 Solubility of Organic Chemicals

In the chemical sciences, it is generally understood (it is a useful rule of thumb) that in terms of predicting solubility that *like dissolves like* which indicates that a chemical will dissolve best in a solvent that has a chemical character similar to the chemical. In simple terms a simple ionic chemical (with positive and negative ions) such as sodium chloride (Na^+Cl^-) is easily soluble in a water (with some separation of positive $[\delta^+]$ and negative $[\delta^-]$ charges in the covalent molecule). This principle is also the usual guide to solubility with organic systems. For example, the highly polar acetic acid (CH_3CO_2H) will dissolve readily in water, whereas the less polar hexane ($CH_3CH_2CH_2$ $CH_2CH_2CH_3$) has no (or at best, marginal) solubility in water.

The solubility of organic chemicals is an important property when assessing toxicity since the water solubility of an organic chemical determines the routes of exposure that are possible. In fact, an easy way of assessment of the water solubility of organic chemicals is that the solubility of an organic chemical is approximately inversely proportional to molecular weight—lower molecular–weight hydrocarbon derivatives (excluding the presence of polar functional groups) are typically more soluble in water than the higher molecular–weight compounds. Thus, lower molecular–weight hydrocarbons (specifically, the C_4 to C_8 alkanes, including the aromatic compounds) are relatively soluble, up to approximately 2000 ppm, while the higher molecular–weight hydrocarbon derivatives are much less soluble in water. Typically, the most soluble components are also the most toxic but whether this toxicity is due to the chemical and structural aspects of the lower molecular weight derivatives or whether it is noticed more frequently because of the relative ease of transportation is a subject to debate.

Finally, each functional group has a particular set of chemical properties that allow it to be identified. Some of these properties can be demonstrated by observing solubility behavior, while others can be seen in chemical reactions that are accompanied by color changes, precipitate formation, or other visible effects. The identification and characterization of the organic chemicals is an important aspect of environmental organic chemistry. Although it is often possible to establish the structure of chemicals through spectroscopic test methods, the spectroscopic data should be supplemented with additional information such as (1) the physical state and (2) the relevant properties such as melting point, boiling point, solubility, odor, elemental analysis, and confirmatory tests for the presence of functional groups. The later test methods usually involve simple laboratory tests for solubility from which conclusions can be drawn which, when combined with other test data, can provide valuable information about the nature of the sample, but also whether or not this can be applied to complex mixtures. Nevertheless, the solubility of an organic compound in water, dilute acid, or dilute base can provide useful information about the presence or absence of certain functional groups. For example, (1) solubility in water, (2) solubility in sodium hydroxide, and (3) solubility in hydrochloric acid.

Most organic compounds are insoluble in water, except for low molecular–weight amines and oxygen-containing compounds. Low molecular–weight compounds are generally limited to those with fewer than five carbon atoms. Carboxylic acids (RCO_2H) with fewer than five carbon atoms in the molecule are soluble in water and form solutions that give an acidic response ($pH < 7$) when tested with litmus paper. Amines (RNH_2) with fewer than five carbon atoms in the molecule are also soluble in water, and their amine solution gives a basic response ($pH > 7$) when tested with litmus paper. Ketones, aldehydes, and alcohols with fewer than five carbon atoms are soluble in water and form neutral solutions ($pH = 7$).

In addition, solubility in sodium hydroxide solution (usually 6M NaOH) is a positive identification test for acids and acid derivatives. A carboxylic acid that is insoluble in pure water will be soluble in base due to the formation of the sodium salt of the acid as the acid is neutralized by the base. Also, solubility in hydrochloric acid (usually 6 M HCl) is a positive identification test for bases. Amines that are insoluble in pure water will be soluble in hydrochloric acid due to the formation of an ammonium chloride–type salt (RNH_3^+ Cl^-).

REFERENCES

Andrews, J.E., Brimblecombe, P., Jickells, T.D., Liss, P.S., 1996. An Introduction to Environmental Chemistry. Blackwell Science Publications, Oxford, United Kingdom.

ASTM, 2017. Annual Book of Standards. ASTM International, West Conshohocken, Pennsylvania.

Boethling, R.S., Mackay, D., 2000. Handbook of Property Estimation Methods. Lewis Publishers, Boca Raton, Florida.

Lyman, W.J., Reehl, W.F., Rosenblatt, D.H. (Eds.), 1982. Handbook of Chemical Property Estimation Methods. McGraw-Hill, New York.

Lyman, W.J., Reehl, W.F., Rosenblatt, D.H., 1990. Handbook of Chemical Property Estimation Methods. American Chemical Society, Washington, DC.

Mackay, D., 1982. Correlation of bioconcentration factors. Environmental Science and Technology 16, 274–278.

Mackay, D., Shiu, W.-Y., Ma, K.-C., Lee, S., 2006. Handbook of Physical-chemical Properties and Environmental Fate of Organic Chemicals, second ed. CRC Press, Taylor & Francis Group, Boca Raton, Florida.

Manahan, S.E., 2010. Environmental Chemistry, ninth ed. CRC Press, Taylor & Francis Group, Boca Raton, Florida.

Sánchez-Bayo, F., Van den Brink, P.J., Mann, R.M., 2011. Ecological Impacts of Toxic Chemicals. Bentham Science Publishers, Sharjah, United Arab Emirates.

Schwarzenbach, R.P., Gschwend, P.M., Imboden, D.M., 2003. Environmental Organic Chemistry, second ed. John Wiley & Sons Inc., Hoboken, New Jersey.

Speight, J.G., 2005. Environmental Analysis and Technology for the Refining Industry. John Wiley & Sons Inc., Hoboken, New Jersey.

Speight, J.G., 2014. The Chemistry and Technology of Petroleum, fifth ed. CRC Press, Taylor & Francis Group, Boca Raton, Florida.

Speight, J.G., 2015. Handbook of Petroleum Products Analysis, second ed. John Wiley & Sons Inc., Hoboken, New Jersey.

Speight, J.G., 2017a. Environmental Organic Chemistry for Engineers. Butterworth-Heinemann, Elsevier, Cambridge, Massachusetts.

Speight, J.G., 2017b. Environmental Inorganic Chemistry for Engineers. Butterworth-Heinemann, Elsevier, Cambridge, Massachusetts.

Speight, J.G., Islam, M.R., 2016. Peak Energy – Myth or Reality. Scrivener Publishing, Salem, Massachusetts (Chapter 8).

Speight, J.G., Lee, S., 2000. Environmental Technology Handbook, second ed. Taylor & Francis, New York (Also, CRC Press, Taylor and Francis Group, Boca Raton, Florida).

FURTHER READING

Nutting, P.G., 1943. Adsorbent Clays: Their Distribution, Properties Production, and Uses. Bulletin 928-C. United States Geological Survey, Reston, Virginia.

Chapter 4

Mechanisms of Introduction Into the Environment

1. INTRODUCTION

Environmental pollution is a major global issue insofar as pollution of the atmosphere (the air), the aquasphere (the water), and the lithosphere (the land) are all the recipients of discharged chemical pollutants. Natural and human-induced chemicals can be found in all areas of the environment. For example, as groundwater flows through the ground, metals such as iron and manganese are dissolved and may later be found in high concentrations in the water. Industrial discharges, urban activities, agriculture, groundwater pumpage, and disposal of waste all can affect groundwater quality. Contaminants can be human-induced, as toxic chemical spills and leakage from waste-disposal sites also can introduce chemicals into the water. Or, a well might have been dug on a land that was once used as a garbage dump or a chemical dump site.

In addition, the structural characteristics that enable a chemical to persist in the environment can also help it to resist metabolic breakdown in people or wildlife. For example, synthetic chemicals that contain halogen atoms (particularly fluorine, chlorine, or bromine) are often resistant to degradation in the environment or within organisms.

Thus, there is urgency to control all forms of pollution to preserve the environment. Pollution may be defined as an undesirable change in the physical, chemical, and biological characteristics of air, water, and soil, which affects the flora and fauna of the various ecosystems. A pollutant is, in the current context, a chemical which adversely interferes with an ecosystem. Generally, most pollutants are introduced in the environment by waste and/or accidental discharge, or else they are byproducts or residues from a production process.

The term *chemical contamination*, in the context of this book, is used to indicate those situations where chemical contaminants (often referred to as IOCs) are either present where they should not be or are at higher concentrations than they would have occurred if the chemicals are indigenous to an ecosystem (Miller, 1984). Chemical contaminants can be found in mass-produced products. For many of these substances accumulation into the various subdivisions of

Reaction Mechanisms in Environmental Engineering. https://doi.org/10.1016/B978-0-12-804422-3.00004-3

the environment (Fig. 4.1) can cause environmental problems, although some chemicals do not damage the environment, and for many chemicals the consequences to the environment and the fate of the chemicals are often unknown. But generally, the fate of a chemical in the environment is subject to any one or more of several processes:

Chemical transport processes

- Runoff
- Erosion
- Wind
- Leaching
- Movement in streams or in groundwater

Chemical fate processes

- Transport
- Transformation/Degradation
- Sorption
- Volatilization
- Biological processes

Transformation and degradation processes

- Biological transformations due to microorganism
  - aerobic
  - anaerobic processes

Understanding the chemical and physical nature of these processes (from the bonding interactions involved) is necessary for identifying the most appropriate remediation process. Furthermore, chemicals (Table 4.1), especially the water-soluble contaminants, are often transported by water across the land and other impermeable surfaces. With little prior treatment, many of these

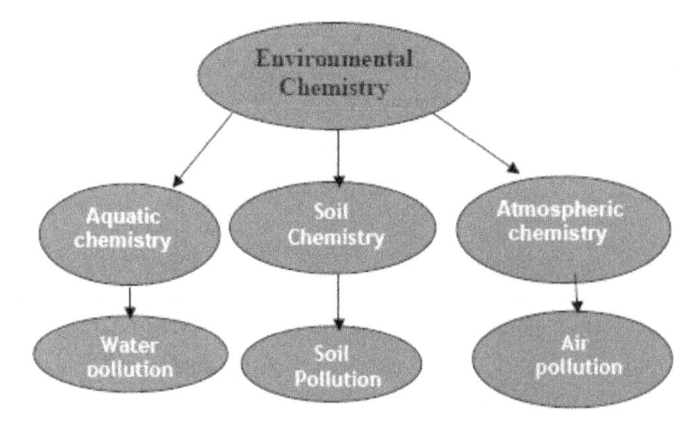

FIGURE 4.1 The subdivisions of environmental chemistry.

TABLE 4.1 Examples of Common Pollutants

Pollutant	Sources of Origin	Directly Polluted Medium	Biological Actions
SO_2	Volcanoes, industry, transports	Air	Expectoration, spasms, respiratory difficulties, bronchitis
NO_2	Volcanoes, industry, transports	Air	Metha-hemoglobin that restrains the transport of oxygen to the tissues
CO	Transports	Air	Carboxyhemoglobin that generates dizziness, asphyxia due to oxygen deficiency
NH_3	Industry, agriculture	Air, soil	Irritations of the nasal mucous
H_2S	Industry, anaerobic fermentations with sulfo-bacteria	Soil, water	Troubles of the nervous system functions and of the sanguine circulation
HCl	Industry, transports	Air	Respiratory diseases, cancerous effects
HF	Industry	Air	Bleedings, loss of the visual acuity, vomit, cerebral diseases
NH_4^+, NO_3^-, NO_2^-	Farms, factories that produce nitrogen	Soil, water	Respiratory diseases, cancer
Pb	Heavy industry, transports	Air	Intoxications, hemoglobin alteration, disturbances of the liver and kidney functions
Hg	Industry	Soil, water	Caught, insomnia, hallucinations
Zn	Industry	Air, water	Corrosive action on the tissues; muscular, cardiovascular, and nervous system diseases

contaminants may eventually be discharged into rivers, lakes, oceans, and other waterways.

Thus, contaminants are elements or compounds found in water supplies and may be natural in the geology or caused by activities of man through mining, industry, or agriculture. It is common to have trace amounts of many contaminants in water supplies, such as remnants of fertilizers including derivatives of nitrogen (N), phosphorus (P), and potassium (K). Increases in these simple chemicals in waterways are nearly always a result of land use activities such as fertilizer runoff or direct discharge from industrial operations. Contaminants also include metals and metal particles. These can be found in storm-water runoff from urban development and will accumulate in drainage systems or low-lying areas of land, such as marshy areas and wetlands. Many of these contaminants eventually are the result of discharge into waterways with little prior treatment to remove chemicals. Some contaminants, such as mercury, may bioaccumulate in animal tissues and occur in fish and eventually end up as part of the human diet.

Chemicals that accumulate in living organisms so that their concentrations in floral and faunal tissues continue to increase are called bioaccumulative or, in aquatic organisms, the process of bioaccumulation is sometimes called bioconcentration. The bioconcentration factor (BCF) is an expression of the extent to which the concentration of a chemical in a fish is higher than the concentration in the surrounding water. Very low concentrations of a bioaccumulative substance in water can result in markedly higher concentrations in the tissue of fish at higher levels of the aquatic food chain, as well as in people or wildlife eating those fish. Concentrations of airborne bioaccumulative chemicals will also be magnified in air-breathing organisms.

The potential for a chemical to bioaccumulate can be predicted by examining whether the chemical preferentially dissolves in a solvent as opposed to water. If the concentration of a chemical in the solvent is more than 1000 times higher than its concentration in water when added to a mixture of solvent and water, the chemical is likely to bioaccumulate in organisms. If that concentration gradient is >5000, the chemical is most likely to bioaccumulate. Many bioaccumulative chemicals are fat-soluble so that they tend to reside primarily in fat deposits or in the fatty substances in blood. This explains why fat-soluble bioaccumulative chemicals are often found at elevated levels in fat-rich breast milk. But bioaccumulative substances may also be deposited elsewhere, including bone, muscle, or the brain.

Both the concentration of these chemicals and how they enter a waterway vary greatly. As with some contaminants, natural global cycling has always been a primary contributor to the presence of chemical elements in air, water, and soil (or sediments). This process can involve transfer of a contaminant between the atmosphere, the hydrosphere, and the lithosphere, where it may eventually be transported and deposited onto surface water and soil. Major anthropogenic sources of contaminants in the environment have been (1) mining operations, (2) industrial processes, (3) combustion of fossil fuels, especially charcoal, (4) production of cement, as well as (5) incineration of municipal, chemical, and medical wastes. Alternatively, industries such as forest processing, meat processing and dairy processing, and waste water treatment may discharge waste water that can potentially contain chemical contaminants: examples are bleach (hypochlorite derivative), curing agents (which include nitrate derivatives—NO_3, and nitrite derivatives—NO_2), and certain metals like mercury, copper, chrome, zinc, iron, arsenic, and lead. Prior treatment of these chemicals before discharge is now strictly regulated and controlled via the resource consenting process, and will vary depending on the type, quantity, and potential environmental reactivity of the chemical to be discharged.

In the case of the discharge of any chemical into the environment is proposed, it is necessary to set standards for acceptable concentrations in air, water, soil, and flora as well as fauna—if there are any such acceptable concentrations. Monitoring of these concentrations of the chemicals in an ecosystem and any resultant biological effects must be undertaken to ensure that the standards as set in any regulation are realistic and provide protection of the environment from any adverse effects. Furthermore, considerable attention continues to be focused on regulation of the use of all chemicals and a primary aspect involves the prediction of the behavior and effects of a chemical from the properties of that chemical (Chapter 3). Also, this concept that the molecular characteristics of the chemical (such as the three-dimensional structure and the presence of functional groups) govern the physical and chemical properties of the compound, which in turn influence the effects of the chemical on the ecosystem, as well as the transformation and distribution of the chemical in the environment. This suggests that the transformation and distribution in the environment as well as any effects on the floral and faunal species can be predicted from the physical properties and/or the chemical properties of the contaminant. However, the prediction of biological effects may also involve a complex set of chemical-floral and/or chemical-faunal interactions.

Industrial chemical manufacturers use and generate large quantities of chemicals. In the past, especially during the 19th century and the first half of the 20th century, when environmental legislation was not in place or, if in place, was not enacted, the chemical industry disposed of chemicals to all types of environmental ecosystems including air through both fugitive emissions (emissions of gas or vapor from pressurized equipment due to leaks and other unintended or irregular releases of gases) and direct emissions (emissions from sources that are owned or controlled by the reporting entity), water (direct discharge and runoff), and land (Table 4.2). However, the types of pollutants a single facility will release depend on (1) the type of process, (2) the feedstocks to the process, (3) the equipment used in the process, such as the reactor, and (4) the equipment and process maintenance practices, which can vary over short periods of time (such as from hour to hour) and can also vary with the part of the process that is underway. For example, for batch reactions in a closed vessel, the chemicals are more likely to be emitted at the beginning and end of a reaction step (that are associated with reactor or

TABLE 4.2 Types of Releases From Industrial Processes

RELEASE is an onsite discharge of a toxic chemical to the environment and include (1) emissions to the air, (2) discharges to bodies of water, (3) releases at the facility to land, as well as (4) the contained disposal into underground injection wells.

Releases to Air (Point and Fugitive Air Emissions):
Include all air emissions from industry activity; point emissions occur through confined air streams as found in stacks, ducts, or pipes while fugitive emissions include losses from equipment leaks, or evaporative losses from impoundments, spills, or leaks.
Releases to Water (Surface Water Discharges):
Include any releases going directly to streams, rivers, lakes, oceans, or other bodies of water; any estimates for storm water runoff and nonpoint losses must also be included.
Releases to Land:
Include disposal of toxic chemicals in waste to onsite landfills, land treated or incorporation into soil, surface impoundments, spills, leaks, or waste piles; these activities must occur within the facility's boundaries for inclusion in this category.
Underground Injection:
Include a contained release of a fluid into a subsurface well for the purpose of waste disposal.

TRANSFER is a transfer of toxic chemicals in wastes to a facility that is geographically or physically separate from the facility reporting under the toxic release inventory. The quantities reported represent a movement of the chemical away from the reporting facility and, except for off-site transfers for disposal, these quantities of chemicals do not necessarily represent entry of the chemicals into the environment.

Transfers to Publicly Owned Treatment Works:
Include waste waters transferred through pipes or sewers to a publicly owned treatment works (POTW); treatment and chemical removal depend on the nature of the chemical and the treatment methods employed; chemicals that are not treated or destroyed by the publicly owned treatments works are generally released to surface waters or land filled within the sludge.
Transfers to Recycling:
Includes chemicals that are sent offsite for the purposes of regenerating or recovering valuable materials; once these chemicals have been recycled, they may be returned to the originating facility or sold commercially.
Transfers to Energy Recovery:
Include wastes combusted off-site in industrial furnaces for energy recovery; treatment of a chemical by incineration is not considered to be energy recovery.
Transfers to Treatment:
Include wastes moved offsite for either neutralization, incineration, biological destruction, or physical separation; in some cases, the chemicals are not destroyed but are prepared for further waste management.
Transfers to Disposal:
Include wastes taken to another facility for disposal generally as a release to land or as an injection underground.

treatment vessel loading and product transfer operations) than during the reaction. Fluid-bed reactors which are used in continuous process operations may emit chemicals at any part of the process.

The chemical pollutants that are most likely to present ecological risks are those that are (1) highly reactive and likely to react with any part of the ecosystem and undergo transformation, (2) highly bioaccumulative, building up to high levels in floral and faunal tissues even when concentrations in the ecosystem remain relatively low, and (3) highly toxic, which can cause adverse effects to the ecosystem itself and to the floral and faunal members of the ecosystem at comparatively low doses. In addition, atmosphere–water interactions that control the input and outgassing of persistent pollutants in aquatic systems are critically important in determining the life cycle and residence time of a chemical in an ecosystem, as well as the extent of any contamination and the resulting adverse effects.

Briefly, in the environment, many chemicals are degraded by sunlight, destroyed through reactions with other environmental substances, or metabolized by naturally occurring bacteria. Some chemicals, however, have features than enable them to resist environmental degradation. They are classified as *persistent* and can accumulate in soil and aquatic environments. Those that can evaporate into air (volatilize) or dissolve in water can migrate considerable distances from where they are released. Floral and faunal species are more likely to be exposed to a chemical if it does not easily degrade or is dispersed widely in the environment.

Metals, such as lead, mercury, and arsenic, are always persistent, since they are basic elements and cannot be further broken down and destroyed in the environment. Although this discussion will focus on synthetic chemicals, the potential health effects of exposure to metals should not be overlooked. For example, lead contamination of air, soil, or drinking water can ultimately result in significant exposures in fetuses, infants, and children, resulting in impaired brain development.

Although the effects of various types of chemical pollutants are usually evaluated independently, many ecosystems are subject to multiple pollutants, and their fate and impacts are intertwined. For example, the effects of nutrient deposition in an ecosystem can alter the methods by which the contaminants are assimilated, bioaccumulated, as well as the methods by which the floral and faunal organisms in the ecosystem are affected.

Of all the chemical pollutants released into the environment by anthropogenic activity, persistent pollutants (PPs) are among the most dangerous chemicals and often require extreme measures for removal (Chapters 1 and 10). PPs are chemical substances that persist in the environment, bioaccumulate through the food chains of the floral and faunal species, and pose a risk of eventually causing adverse effects to human health and the environment. Moreover, persistent pollutants can be transported across international boundaries far from their sources, even to regions where they have never been used or produced. Consequently, PPs can pose a serious threat not only to the environment but also to human health on a global scale.

Releases into the environment of chemicals that persist in an ecosystem (rather than undergo some form of biodegradation) lead to an exposure level that is not only subject to the length of time the chemical remains in circulation (in the environment) but also on the number of times that the chemical is recirculated before it is ultimately removed from the ecosystem. In addition, a chemical that is not a PP may transform into a PP in a manner that is not only dependent upon the chemical and physical properties of the pollutant but also upon the influence of the ecosystem.

Typically, PPs are highly toxic and long-lasting (hence the name *persistent*) and cause a wide range of adverse effects to environmental flora and fauna, including disease and birth defects in humans and animals—some of the severe human health impacts from PPs include (1) the onset of cancer, (2) damage to the central nervous system, (3) damage to the peripheral nervous system, (4) damage to the reproductive system, and (5) disruption of the immune system. Moreover, PPs do not respect international borders and the serious environmental and human health hazards created by these chemicals affect not only developing countries, where systems and technology for monitoring, tracking, and disposing of them can be weak or nonexistent but also developed countries. As long as the chemical can be transported by air, water, and land, no region or country is immune from the effects of these chemicals.

Generally, PPs such as borate, silicate, and sulfur derivatives are minerals that are mined from the earth and ground into a fine powder. Some of these chemicals function as poisons and others function by physically interfering with the pest; pesticides used in earlier times included highly toxic compounds such as arsenic, copper, lead, and tin salts. On the other hand, modern pesticides are relatively low in toxicity and have a low environmental impact. Other contaminants include industrial chemicals or the unwanted byproducts of industrial processes that have been used and subjected to unmanaged disposal for decades prior to the inception of the various regulations and often without due regard for the environment. But all nations have, more recently, found to share several significant characteristics that need consideration before disposal is planned. These characteristics include: (1) persistence in the environment insofar as these chemicals resist degradation in air, water, and sediments, (2) bioaccumulation insofar as these chemicals accumulate in floral and faunal tissues at concentrations higher than those in the surrounding environment, and (3) long-range transport insofar as these chemicals can travel great distances from the source of release through air, water, and the internal organs of migratory animals. Any of these characteristics can result in the contamination not only of local ecosystems but also of ecosystems that are significant distances (even up to thousands of miles) away from the source of the chemicals.

Briefly and by way of explanation, bioaccumulation is a process by which persistent environmental pollution leads to the uptake and accumulation of one or more contaminants, by organisms in an ecosystem. The amount of a pollutant available for exposure depends on its persistence and the potential for its bioaccumulation. Any chemical can be capable of bioaccumulation if the chemical has a degradation half-life in excess of 30 days or if the chemical has a bioconcentration factor (BCF) greater than 1000 or if the log K_{ow} (the octanol–water partition coefficient of the chemical; Chapter 9) is greater than 4.2:

$$\text{BCF} = \text{Concentration in biota/Concentration in ecosystem}$$

The octanol–water partition coefficient (K_{ow}) is the ratio of the concentration of a chemical in the octanol phase relative to the concentration of the chemical in the aqueous phase of a two-phase octanol/water system:

$$K_{ow} = \text{Concentration in octanol phase/Concentration in aqueous phase}$$

The BCF indicates the degree to which a chemical may accumulate in biota (flora and fauna). However, measurement of bioconcentration is typically made on faunal species and is distinct from food-chain transport, bioaccumulation, or biomagnification. The BCF is a constant of proportionality between the chemical concentration in flora or fauna in an ecosystem. It is possible to estimate the BCF for many chemicals from their octanol–water partition coefficients (K_{ow}):

$$\log \text{BCF} = \text{m}\log K_{ow} + \text{b}$$

In terms of actual numbers, for many lipophilic chemicals, the BCF can be calculated using the regression equation:

$$\log \text{BCF} = -2.3 + 0.76 \times (\log K_{ow})$$

Furthermore, empirical relationships between the K_{ow} and the BCF can be developed on a chemical-by-chemical basis.

On this note, it is worth defining the source of chemical contaminant as (1) a point source or (2) a nonpoint source. The point source of pollution is a single identifiable source of pollution which may have a negligible extent, distinguishing it from other pollution source geometries. On the other hand, nonpoint source (NPS) pollution generally results from land runoff, precipitation, atmospheric deposition, drainage, seepage, or hydrologic modification. Thus, NPS pollution, unlike pollution from industrial and sewage treatment plants, originates from many diffuse sources and is often caused by rainfall or snowmelt moving over and through the ground. As the runoff water moves, it picks up and transports humanmade pollutants, as well as natural pollutants away from the site and the pollutants are ultimately deposited into water systems such as groundwater, wetlands, lakes, rivers, and coastal waters.

Indeed, an analysis of the amount of waste formed in processes for the manufacture of a range of fine chemicals and chemical intermediates has revealed that in some processes the amount of waste generated is in excess of the amount of the desired product and such overproduction of waste is not exceptional in the chemical industry. As a means of measuring the amount of waste vis-à-vis the number of products and the amount of each product, the E-factor (environmental factor) (kilograms of waste per kilogram of product) was introduced as an indication of the environmental footprint of the manufacturing process in various segments of the chemical industry (Chapters 9 and 10). This factor is derived from the chemicals and fine chemical industries as a measure of the efficiency of the manufacturing process: Thus, simply:

$$E = \text{kilograms of waste/kilogram of product}$$

The factor can be conveniently calculated from knowledge of the number of tons of raw materials purchased and the number of tons of product sold, the calculation being for a particular product or a production site or even a whole company. A higher E-factor means more waste and, consequently, a larger environmental footprint—thus, since mass cannot be created resulting in a negative E-factor, the ideal E-factor for any process is zero.

However, in the context of environmental protection and to be all inclusive, the E-factor is the total mass of raw materials plus ancillary process requirements minus the total mass of product, all divided by the total mass of product. Thus, the E-factor should represent the *actual amount* of waste produced in the process, defined as everything but the desired product and takes the chemical yield into account and includes reagents, solvent losses, process aids, and (in principle) even the fuel necessary for the process. Water has been generally excluded from the calculation of the E-factor since the inclusion of all process water could lead to exceptionally high E-factors in many cases. Thus, the exclusion of water from the calculation of the E-factor allowed meaningful comparisons of the technical factors (E-factors excluding water use) to be made for the various processes. Recent thinking, in this modern environmentally conscious era, has led to the conclusion that there is no reason for the water requirement or the water product to be omitted (water being a reactant or reaction product), since the disposal of process water is an environmental issue. Moreover, use of the E-factor has been widely adopted by many chemical industries and pharmaceutical industries, in particular. Thus, a major aspect of process development recognized by process chemists and process engineers is the need for determining an E-factor—whether or not it is called by that name (i.e., the E-factor) but *chemicals in compared to chemicals out* has become a major yardstick in many of the chemical industries.

It is clear that the E-factor increases substantially when comparing bulk chemicals to fine chemicals. This is partly a reflection of the increasing complexity of the products, necessitating not only processes which use multistep syntheses, but is also a result of the widespread use of stoichiometric amounts of the reagents, that is, the required amounts of the reagents (some observers would advocate the stoichiometric amounts of the reagents plus 10%) to accomplish conversion of the starting chemical to the product(s)

(Chapters 2 and 10). A reduction in the number of steps of a process for the synthesis of chemicals will, in most cases (but not always), lead to a reduction in the amounts of reagents and solvents used and hence a reduction in the amount of waste generated. This has led to the introduction of the concepts of step economy and function-oriented synthesis (FOS) of some chemicals. The main issue behind the concept of FOS is that the structure of an active compound can be reduced to simpler structures for ease of synthesis while not exerting an adverse influence on the flora and fauna of an ecosystem. This approach can provide practical access to new (designed) structures with novel activities while, hopefully, at the same time allowing for a relatively straightforward synthesis.

As noted above, knowledge of the stoichiometric equation allows the process chemist or process engineer to predict the theoretical minimum amount of waste that can be expected (Chapters 2 and 3). This led to the concept of *atom economy* or *atom utilization* to quickly assess the environmental acceptability of alternatives to a particular product before any experiment is performed. It is a theoretical number, that is, it assumes a chemical yield of 100% and exactly stoichiometric amounts and disregards substances which do not appear in the stoichiometric equation. In short, the key to minimizing waste is precision or *selectivity* in synthesis which is a measure of how efficiently a synthesis is performed. The standard definition of selectivity is the yield of product divided by the amount of substrate converted, expressed as a percentage.

Chemists distinguish between different categories of selectivity in two ways: (1) chemoselectivity, which relates to competition between different functional groups, and (2) regioselectivity, which is the selective formation of one regioisomer. However, one category of selectivity was, traditionally, largely ignored by many chemists: the *atom selectivity* or *atom utilization* or *atom economy* and the virtual complete disregard of this important parameter by chemists and engineers has been a major cause of the generation of large amounts of waste during the manufacture or chemicals. Quantification of the waste generated in the manufacture of chemicals, for example, by way of E-factors, served to illustrate the omissions related to the production of chemical waste and focussed the attention of the companies that manufacture chemicals on the need for a paradigm shift from a concept of process efficiency, which was exclusively based on chemical yield, to a need that more focused on (or a more conscious process-related attitude to) the elimination of waste chemicals and maximization of the utilization of the raw materials used as process feedstocks.

2. MINERALS

Minerals (of which more than 3000 individual minerals are known) are substances formed naturally in the earth. Minerals have a definite chemical composition and structure. Some are rare and precious such as gold (which occurs as the element) and diamond (a crystalline form of carbon), while others are more ordinary and ubiquitous, such as quartz (SiO_2).

A mineral is a naturally occurring solid that possesses an orderly internal structure and a definite chemical composition. The term *rock* is less specific, referring to any solid mass of mineral or mineral-like material. Common rocks are often made up of crystals of several kinds of minerals. There are some substances (*mineraloids*) which have the appearance of a mineral but lack any definite internal structure. The essential characteristics of a mineral are: (1) it must occur naturally, (2) it must be a solid, (3) it must possess an orderly internal structure, that is, its atoms must be arranged in a definite pattern, and (4) it must have a definite chemical composition that may vary within specified limits. This can be confusing when coal, petroleum, and natural gas are often listed as part of the mineral resources of many states and countries.

Minerals are chemical compounds, sometimes specified by crystalline structure, as well as by composition, which are found in rocks (or pulverized rocks or sand). On the other hand, rocks consist of one or more minerals, and fall into three main types, depending on their origin and previous processing history: (1) igneous rocks, which are rocks that have solidified directly from a molten state, such as volcanic lava, (2) sedimentary rocks, which are rocks that have been remanufactured from previously existing rocks, usually from the products of chemical weathering or mechanical erosion, without melting, and (3) metamorphic rocks, which are rocks that have resulted from processing, by heat and pressure (but not melting), of previously existing sedimentary or igneous rocks.

On the other hand, sand is a naturally occurring granular material composed of finely divided rock and mineral particles that is defined by size, being finer than gravel and coarser than silt. The term *sand* can also refer to a textural class of soil or soil type; that is, a soil containing more than 85% sand-sized particles by mass. The composition of sand varies, depending on the local rock sources and conditions, but the most common constituent of sand in inland continental settings and nontropical coastal settings is silica (silicon dioxide or SiO_2), usually in the form of quartz. The second most common type of sand is calcium carbonate ($CaCO_3$), for example, the mineral aragonite, which has mostly been created, over the past half billion years, by various forms of life, such as coral and shellfish. Aragonite is the primary form of sand, apparent in areas where reefs have dominated the ecosystem for millions of years.

The most common minerals are the silica-based (SiO_2-based) minerals because of the abundance of silica in the crust of the earth. There is a great variety of different minerals most of which contain silica but many of which contain other elements (Tables 4.3–4.5). In the crust of the earth, approximately 20 of the minerals are common and fewer than 10 minerals account for over 90% of the crust by mass. Minerals are classified in many ways including properties such as: (1) hardness, (2) optical properties, and (3) crystal structure. Nonsilicate minerals constitute less than 10% of the earth's crust and the

TABLE 4.3 Elements in the Crust of the Earth

Element Name	Symbol	Percentage by Weight of the Earth's Crust
Oxygen	O	47
Silicon	Si	28
Aluminum	Al	8
Iron	Fe	5
Calcium	Ca	3.5
Sodium	Na	3
Potassium	K	2.5
Magnesium	Mg	2
All other elements		1

TABLE 4.4 Mineral Names and Chemical Composition

Mineral Name	Chemical Formula	Useful Element
Galena	PbS	Lead
Pyrite	FeS_2	Sulfur (pyrite is not used as an ore of iron)
Chalcopyrite	$CuFeS_2$	
Chalcocite	Cu_2S	
Bauxite	Al_2O_3	
Magnetite	Fe_3O_4	
Haematite	Fe_2O_3	
Rutile	TiO_2	

most common nonsilicate minerals are the carbonate minerals ($-CO_3$), the oxide minerals ($-O$), and the sulfide ($-S$) minerals. There are also naturally occurring phosphate minerals ($-PO_4$) and salts. There are some elements which occur in pure form, including gold, silver, copper, bismuth, arsenic, lead, and tellurium. Carbon is found in both graphite and diamond form.

3. TYPES OF CHEMICALS

As described above, chemicals can enter the air, water, and soil when they are produced, used, or disposed. The impact of these chemicals on the

TABLE 4.5 Common Elements and Ores

Aluminum	The most abundant metal element in earth's crust. Aluminum originates as an oxide called alumina. Bauxite ore is the main source of aluminum and must be imported from Jamaica, Guinea, Brazil, Guyana, etc. Used in transportation (automobiles), packaging, building/construction, electrical, machinery, and other uses.
Antimony	A native element, antimony metal is extracted from stibnite ore and other minerals. Used as a hardening alloy for lead, especially storage batteries and cable sheaths; also used in bearing metal, type metal, solder, collapsible tubes and foil, sheet and pipes, and semiconductor technology. Antimony is used as a flame retardant, in fireworks, and antimony salts are used in the rubber, chemical, and textile industries, as well as medicine and glassmaking.
Barium	A heavy metal contained in barite. Used as a heavy additive in oil well drilling; in the paper and rubber industries as a filler or extender in cloth, ink, and plastics products; in radiography ("barium milkshake"); as a deoxidizer for copper; a sparkplug in alloys; and in making expensive white pigments.
Bauxite	Rock composed of hydrated aluminum oxides. In the US, it is primarily converted to alumina. See "aluminum."
Beryllium	Used in the nuclear industry and to make light, very strong alloys used in the aircraft industry. Beryllium salts are used in fluorescent lamps, in X-ray tubes, and as a deoxidizer in bronze metallurgy. Beryl is the gem stones emerald and aquamarine. It is used in computers, telecommunication products, aerospace and defense applications, appliances, and automotive and consumer electronics. Also used in medical equipment.
Chromite	The United States consumes about 6% of world chromite ore production in various forms of imported materials, such as chromite ore, chromite chemicals, chromium ferroalloys, chromium metal, and stainless steel. Used as an alloy and in stainless and heat resisting steel products. Used in chemical and metallurgical industries (chrome fixtures, etc.) Superalloys require chromium. It is produced in South Africa, Kazakhstan, and India.

TABLE 4.5 Common Elements and Ores—cont'd

Clay	Used in floor and wall tile as an absorbent, in sanitation, mud drilling, foundry sand bond, iron pelletizing, brick, light-weight aggregate, and cement. It is produced in 40 states. Ball clay is used in floor and wall tile. Bentonite is used for drilling mud, pet waste absorbent, iron ore pelletizing, and foundry sand bond. Kaolin is used for paper coating and filling, refractory products, fiberglass, paint, rubber, and catalyst manufacture. Common clay is used in brick, light aggregate, and cement.
Cobalt	Used primarily in superalloys for aircraft gas turbine engines, in cemented carbides for cutting tools and wear-resistant applications, chemicals (paint dryers, catalysts, magnetic coatings) and permanent magnets. The United States has cobalt resources in Minnesota, Alaska, California, Idaho, Missouri, Montana, and Oregon. Cobalt production comes principally from Congo, China, Canada, Russia, Australia, and Zambia.
Copper	Used in building construction, electric and electronic products (cables and wires, switches, plumbing, heating); transportation equipment; roofing; chemical and pharmaceutical machinery; and alloys (brass, bronze, and beryllium alloyed with copper are particularly vibration resistant); alloy castings; electroplated protective coatings and undercoats for nickel, chromium, zinc, etc. More recently copper is being used in medical equipment due to its antimicrobial properties. The United States has mines in Arizona, Utah, New Mexico, Nevada, and Montana. Leading producers are Chile, Peru, China, United States, and Australia.
Feldspar	A rock-forming mineral; industrially important in glass and ceramic industries; patter and enamelware; soaps; bond for abrasive wheels; cements; insulating compositions; fertilizer; tarred roofing materials; and as a sizing, or filler, in textiles and paper. In pottery and glass, feldspar functions as a flux. End-uses for feldspar in the United States include glass (70%) and pottery and other uses (30%).
Fluorite (fluorspar)	Used in production of hydrofluoric acid, which is used in pottery; ceramics; optical, electroplating, and plastics industries; in the metallurgical treatment of bauxite; as a flux in open hearth steel furnaces and in metal smelting; in carbon

Continued

TABLE 4.5 Common Elements and Ores—cont'd

	electrodes; emery wheels; electric arc welders; toothpaste; and paint pigment. It is a key ingredient in the processing of aluminum and uranium.
Gallium	Gallium is used in integrated circuits, light-emitting diodes (LEDs), photodetectors and solar cells. It has a new use in chemotherapy for some types of cancer. Integrated circuits are used in defense applications, high performance computers, and telecommunications. Optoelectronic devices were used in areas such as aerospace, consumer goods, industrial equipment, medical equipment, and telecommunications. Leading sources are Germany, the United Kingdom, China, and Canada.
Gold	Used in jewelry and arts; dentistry and medicine; in medallions and coins; in ingots as a store of value; for scientific and electronic instruments; as an electrolyte in the electroplating industry. Mined in Alaska and several western states. Leading producers are China, Australia, United States, Russia, and Canada.
Gypsum	Processed and used as prefabricated wallboard or an industrial or building plaster; used in cement manufacturing; agriculture and other uses.
Halite (sodium chloride–salt)	Used in human and animal diet, food seasoning and food preservation; used to prepare sodium hydroxide, soda ash, caustic soda, hydrochloric acid, chlorine, metallic sodium; used in ceramic glazes; metallurgy, curing of hides; mineral waters; soap manufacturing; home water softeners; highway deicing; photography; in scientific equipment for optical parts. Single crystals used for spectroscopy, ultraviolet, and infrared transmission.
Indium	Indium tin oxide is used for electrical conductivity purposes in flat panel devices—most commonly in liquid crystal displays (LCDs). It is also used in solders, alloys, compounds, electrical components, semiconductors, and research. Indium ore is not recovered from ores in the US. China is the leading producer. It is also produced in Canada, Japan, and Belgium.
Iron Ore	Used to manufacture steels of various types. Powdered iron: used in metallurgy products; magnets; high-frequency cores; auto parts; catalyst. Radioactive iron (iron 59): in medicine; tracer element in biochemical and metallurgical

TABLE 4.5 Common Elements and Ores—cont'd

	research. Iron blue: in paints, printing inks, plastics, cosmetics, paper dyeing. Black iron oxide: as pigment; in polishing compounds; metallurgy; medicine; magnetic inks. Most US production is from Michigan and Minnesota. China, Australia, Brazil, and Russia are the major producers.
Lead	Used in lead–acid batteries, gasoline additives (now being eliminated) and tanks, and solders, seals, or bearing; used in electrical and electronic applications; TV tubes and glass, construction, communications, and protective coatings; in ballast or weights; ceramics or crystal glass; X-ray and gamma radiation shielding; soundproofing material in construction industry; and ammunition. Industrial type batteries are used as a source of uninterruptible power equipment for computer and telecommunication networks and mobile power. United States mines lead mainly in Missouri, but also in Alaska and Idaho.
Lithium	Compounds are used in ceramics and glass; batteries; lubricating greases; air treatment; in primary aluminum production; in the manufacture of lubricants and greases; rocket propellants; vitamin A synthesis; silver solder; batteries; medicine. Lithium–ion batteries have become a substitute for nickel–cadmium batteries in handheld/portable electronic devices. There is one brine operation in Nevada. Australia, Chile, and China are major producers.
Manganese	Ore is essential to iron and steel production. Also used in the making of manganese ferroalloys. Construction, machinery and transportation end uses account for most US consumption of manganese. Manganese ore has not been produced in the US since 1970. Major producers are South Africa, Australia, China, Gabon, and Brazil.
Mica	Micas commonly occur as flakes, scales, or shreds. Ground mica is used in paints, as joint cement, as a dusting agent, in oil well-drilling muds; and in plastics, roofing, rubber and welding rods. Sheet mica is fabricated into parts for electronic and electronic equipment. China and Russia are leading producers.
Molybdenum	Used in alloy steels to make automotive parts, construction equipment, gas transmission pipes; stainless steels; tool steels; cast irons; super alloys; and chemicals and lubricants. As a pure metal, molybdenum is used because of its high melting

Continued

TABLE 4.5 Common Elements and Ores—cont'd

	temperatures (4730F) as filament supports in light bulbs, metalworking dies and furnace parts. Major producers are China, the United States, Chile, and Peru.
Nickel	Vital as an alloy to stainless steel; plays key role in the chemical and aerospace industries. End uses were transportation, fabricated metal products, electrical equipment, petroleum and chemical industries, household appliances, and industrial machinery. Major producers are the Philippines, Indonesia, Russia, Australia, and Canada.
Perlite	Expanded perlite is used in building construction products like roof insulation boards; as fillers, for horticulture aggregate and filter aids. It is produced in New Mexico and other western states and is processed in over 20 states. Leading producers are the US, Greece, and Turkey.
Platinum Group Metals (PGM)	Includes platinum, palladium, rhodium, iridium, osmium, and ruthenium. Commonly occur together in nature and are among the scarcest of the metallic elements. Platinum is used principally in catalysts for the control of automobile and industrial plant emissions; in jewelry; in catalysts to produce acids, chemicals, and pharmaceuticals. PGMs are used in bushings for making glass fibers, in fiber-reinforced plastic, and other advanced materials, in electrical contacts, in capacitors, in conductive and resistive films used in electronic circuits; in dental alloys used for making crowns and bridges. South Africa, Russia, the US, and Canada are major producers.
Phosphate rock	Used to produce phosphoric acid for ammoniated phosphate fertilizers, feed additives for livestock, elemental phosphorus, and a variety of phosphate chemicals for industrial and home consumers. US production occurs in Florida, North Carolina, Idaho, and Utah.
Potash	A carbonate of potassium; used as a fertilizer, in medicine, in the chemical industry, and to produce decorative color effects on brass, bronze, and nickel. The leading producers are Canada, Russia, and Belarus.
Pyrite	Used in the manufacture of sulfur, sulfuric acid, and sulfur dioxide; pellets of pressed pyrite dust are used to recover iron, gold, copper, cobalt, nickel; used to make inexpensive jewelry.
Quartz (silica)	As a crystal, quartz is used as a semiprecious gem stone. Crystalline varieties include amethyst,

TABLE 4.5 Common Elements and Ores—cont'd

	citrine, rose quartz, smoky quartz, etc. Cryptocrystalline forms include agate, jasper, onyx, etc. Because of its piezoelectric properties, quartz is used for pressure gauges, oscillators, resonators, and wave stabilizers; because of its ability to rotate the plane of polarization of light and its transparency in ultraviolet rays, it is used in heat-ray lamps, prisms, and spectrographic lenses. Also used in manufacturing glass, paints, abrasives, refractory materials, and precision instruments.
Rare Earth Elements (lanthanum, cerium, praseodymium, neodymium, promethium, samarium, europium, gadolinium, terbium, dysprosium, holmium, erbium, thulium ytterbium, and lutetium)	Used mainly in petroleum fluid cracking catalysts, metallurgical additives and alloys, glass polishing and ceramics, permanent magnets and phosphors. It is estimated that 40 pounds of rare earths are used in a hybrid car for rechargeable battery, permanent magnet motor and the regenerative braking system. The US now has one rare earth (bastnasite) mine in California. More than 85% of global production is in China.
Silica	Aluminum and aluminum alloy producers and the chemical industry are major users of silicon metal. Silica is also used in manufacture of computer chips, glass, and refractory materials; ceramics; abrasives; water filtration; component of hydraulic cements; filler in cosmetics, pharmaceutical, paper, insecticides; anticaking agent in foods; flatting agent in paints; thermal insulator; and photovoltaic cells. China is the leading producer.
Silver	Used in coins and medals, electrical and electronic devices, industrial applications, jewelry, silverware, and photography. The physical properties of silver include ductility, electronic conductivity, malleability, and reflectivity. Used in lining vats and other equipment for chemical reaction vessels, water distillation, etc.; a catalyst in manufacture of ethylene; mirrors; silver plating; table cutlery; dental, medical, and scientific equipment; bearing metal; magnet windings; brazing alloys, solder. Also used in catalytic converters, cell phone covers, electronics, circuit boards, bandages for wound care and batteries. Silver is produced in the US at over 30 bases and precious metal mines primarily in Alaska and Nevada. The leading global producers include Mexico, China, Peru, Chile, Australia, Bolivia, and the United States.

Continued

TABLE 4.5 Common Elements and Ores—cont'd

Sodium Carbonate (soda ash or Trona)	Used in glass container manufacture; in fiberglass and specialty glass; also used in production of flat glass; in liquid detergents; in medicine; as a food additive; photography; cleaning and boiler compounds; pH control of water. Most US production comes from Wyoming.
Sulfur	Used in the manufacture of sulfuric acid, fertilizers, petroleum refining; and metal mining. Elemental sulfur and byproduct sulfuric acid were produced in over 100 operations in 26 states and the Virgin Islands.
Tantalum	A refractory metal with unique electrical, chemical, and physical properties used to produce electronic components, tantalum capacitors (in auto electronics, pagers, personal computers, and portable telephones); for high-purity tantalum metals in products ranging from weapon systems to superconductors; high-speed tools; catalyst; sutures and body implants; electronic circuitry; thin-film components. Used in optical glass and electroplating devices. Leading producers are Mozambique, Brazil, and Congo.
Titanium	Titanium mineral concentrates are used primarily by titanium dioxide pigment producers. A small amount is used in welding rod coatings and for manufacturing carbides, chemicals, and metals. It is produced in Florida and Virginia. Leading producing countries are South Africa, Australia, Canada, and China. The US was 77% reliant in 2012.
	Titanium and titanium dioxide are used in aerospace applications (in jet engines, airframes, and space and missile applications). It is also used in armor, chemical processing, marine, medical, power generation, sporting goods, and other nonaerospace applications. Titanium sponge metal was produced in three operations in Nevada and Utah.
Tungsten	More than half of the tungsten consumed in the United States was used in cemented carbide parts for cutting and wear-resistant materials, primarily in the construction, metalworking, mining, and oil- and gas-drilling industries. The remaining tungsten was consumed to make tungsten heavy alloys for applications requiring high density; electrodes, filaments, wires, and other components for electrical, electronic, heating, lighting, and

TABLE 4.5 Common Elements and Ores—cont'd

	welding applications; steels, superalloys, and wear-resistant alloys; and chemicals for various applications. China is by far the leading producer. Russia, Canada, Austria, and Bolivia also produce tungsten.
Uranium	Nearly 20% of America's electricity is produced using uranium in nuclear generation. It is also used for nuclear medicine, atomic dating, powering nuclear submarines, and other uses in the defense system of the United States.
Vanadium	Metallurgical use, primarily as an alloying agent for iron and steel, accounted for about 93% of the domestic vanadium consumption. Of the other uses for vanadium, the major nonmetallurgical use was in catalysts for the production of maleic anhydride and sulfuric acid. China, South Africa, and Russia are largest producers.
Zeolites	Used in animal feed, cat litter, cement, aquaculture (fish hatcheries, for removing ammonia from the water); water softener and purification; in catalysts; odor control; and for removing radioactive ions from nuclear plant effluent.
Zinc	Of the total zinc consumed in the US, about 55% was used in galvanizing, 21% in zinc-based alloys, 16% in brass and bronze, and 8% in other uses. Zinc compounds and dust were used principally by the agriculture, chemical, paint, and rubber industries. Major coproducts of zinc mining and smelting, in order of decreasing tonnage, were lead, sulfuric acid, cadmium, silver, gold and germanium. Zinc is used as protective coating on steel, as die casting, as an alloying metal with copper to make brass and as chemical compounds in rubber and paints; used as sheet zinc for galvanizing iron; electroplating; metal spraying; automotive parts; electrical fuses; anodes; dry cell batteries; nutrition; chemicals; roof gutter; engravers' plates; cable wrappings; organ pipes and pennies. Zinc oxide is used in medicine, paints, vulcanizing rubber, sun block. Zinc dust used for primers, paints, precipitation of noble metals; removal of impurities from solution in zinc electrowinning. US production is in three states and 13 mines. Leading producers are China, Australia, Peru and the United States.

The U.S. Geological Survey, Facts About Minerals (National Mining Association); Mineral Information Institute; the Energy Information Administration.

environment is determined by the amount of the chemical that is released, the type and concentration of the chemical, and where it is found. Some chemicals can be harmful if released to the environment even when there is not an immediate, visible impact. On the other hand, some chemicals are of concern as they can work their way into the food chain and accumulate and/or persist in the environment for many years.

Chemicals, in fact all chemicals, that enter the environment should be categorized and ranked using hazard assessment criteria. This would not only ensure that truly pressing environmental issues are identified and prioritized but would also maximize the use of limited resources. In the case of soluble chemicals, surrogate data such as persistence and bioaccumulation have been used, in combination with toxicity, for the purpose of hazard categorization. However, for insoluble or sparingly soluble chemicals such as metals and metal compounds, persistence and bioaccumulation are neither appropriate nor useful. Unfortunately, this is not always recognized by regulators or even by scientists.

The use of persistent, bioaccumulative, and toxic (PBT) criteria for chemicals was developed to address the hazards posed by synthetic chemicals. In fact, the criteria and test methods to evaluate persistence (i.e., the lack of degradability of a chemical) and bioaccumulation (the dispersion of a chemical through knowledge of the water–octanol partition coefficient) were developed to be used in combination with toxicity in order to reduce the importance given to the use of toxicity data alone. These test methods were based on an understanding of the chemistry of chemicals of concern at the time and of the biological interactions that the chemicals would have with the surrounding biota. Specifically, it was realized that if some chemicals exerted high intrinsic toxicity under standardized laboratory test conditions but did not persist or bioaccumulate, the environmental hazard of such chemicals would be lower.

As mentioned above, persistence is measured by determining the lack of degradability of a substance from a form that is biologically available and active to a form that is less available. This applies to many chemicals—metals and metal compounds tend to be in forms that are not bioavailable. Only under specific conditions metals or metal compounds will transform into a bioavailable form. Thus, rather than persistence, the key criterion for classifying metals and metal compounds should be their capacity to transform into bioavailable form(s). Furthermore, although bioavailability is a necessary precursor to toxicity, it does not inevitably lead to toxicity. Although metals and metal compounds stay in the environment for long periods of time, the risk they may pose generally decreases over time. For example, metals introduced into the aquatic environment are subject to removal/immobilization processes (e.g., precipitation, complexation, and absorption).

Similarly, the use of bioaccumulation has significant limitations for predicting hazard for metals and metal compounds. Generally, either BCFs or bioaccumulation factors (BAFs) are used for this purpose. BCF is the ratio of the concentration of a substance in an organism, following direct uptake from the surrounding environment (water), to the concentration of the same substance in the surrounding environment. BAF considers uptake from food as well. In contrast to compounds, uptake of metals is not based on lipid partitioning. Further, organisms have internal mechanisms (homeostasis) that allow them to regulate (bioregulate) the uptake of essential metals and control the presence of other metals. Thus, if the concentration of an essential metal in the surrounding environment is low and the organism requires more, it will actively accumulate that metal. This will result in an elevated BCF (or BAF) value which, while of concern in the case of substances, is not an appropriate measure in the case of metals.

The primary determining factor of hazard for metals and metal compounds is therefore toxicity, which requires consideration of dose (indeed, the fundamental tenet of toxicology is *the dose makes the poison*). Historically, it has been the practice to measure the toxicity of soluble metal salts, or indeed the toxicity of the free metal ion. However, in different media, metal ions compete with different types or forms of matter (e.g., fish gills, suspended solids, soil particulate material) to reduce the total amount of metals present in bioavailable form. Toxicity of the bioavailable fraction (i.e., as determined through transformation processes) is the most appropriate and technically defensible method for categorizing and ranking the hazard of metals and metal compounds.

The relative proportion of hazardous constituents present in any collection of chemicals (crude oil—derived products included) is variable and rarely consistent because of site differences. Therefore, the extent of the contamination will vary from one site to another and, in addition, the farther a contaminant progresses from low molecular weight to high molecular weight the greater the occurrence of polynuclear aromatic hydrocarbons, complex ring systems (not necessity aromatic ring systems), as well as an increase in the composition of the semivolatile chemicals or the nonvolatile chemicals. These latter chemical constituents (many of which are not so immediately toxic as the volatiles) can result in long-term/chronic impacts to the flora and fauna of the environment. Thus, any complex mixture of chemicals should be analyzed for the semivolatile compounds which may pose the greatest long-term risk to the environment.

Bioconcentration is also the process by which a chemical concentration in an organism exceeds that in the surrounding environment as a result of exposure of the organism to the chemical. Bioconcentration can be measured and assessed using: (1) K_{ow}, (2BCF, (3) BAF, and (4) the biota—sediment accumulation factor, BSAF. Each of these factors can be calculated using either empirical data or measurements, as well as from mathematical models.

The bioconcentration factor can be predicted from the octanol-water partition coefficient:

$$\log(\text{bioconcentration factor}) = m\log K_{ow} + b$$

$$K_{ow} = (\text{concentration in octanol})/(\text{concentration in water}) = C_O/C_W \text{ at equilibrium}$$

Heavy metals are common chemical pollutants. The most common heavy metals found at contaminated sites, in order of abundance are Pb, Cr, As, Zn, Cd, Cu, and Hg. These metals are important since they are capable of decreasing crop production due to the risk of bioaccumulation and biomagnification in the food chain. There is also the risk of superficial and groundwater contamination. Knowledge of the basic chemistry, environmental, and associated health effects of these heavy metals is necessary in understanding their speciation, bioavailability, and remedial options. The fate and transport of a heavy metal in soil depends significantly on the chemical form and speciation of the metal. Once in the soil, heavy metals are adsorbed by initial fast reactions (minutes, hours), followed by slow adsorption reactions (days, years) and are, therefore, redistributed into different chemical forms with varying bioavailability, mobility, and toxicity (Shiowatana et al., 2001). This distribution is believed to be controlled by reactions of heavy metals in soils such as (1) mineral precipitation and dissolution, (2) ion exchange, adsorption, and desorption, (3) aqueous complexation, (4) biological immobilization and mobilization, and (5) plant uptake (Levy et al., 1992). The toxicity of metals varies greatly with pH, water hardness, dissolved oxygen levels, salinity, temperature, and other parameters. Physiological impacts occur at small concentrations.

The specific type of metal contamination found in soil is directly related to the operation that occurred at the site. The range of contaminant concentrations and the physical and chemical forms of contaminants will also depend on activities and disposal patterns for contaminated wastes on the site. Other factors that may influence the form, concentration, and distribution of metal contaminants include soil and ground-water chemistry and local transport mechanisms.

4. RELEASE INTO THE ENVIRONMENT

For the purposes of this text, it is assumed that any chemicals released into the environment are hazardous. It is not only safe but necessary to assume that any chemicals (except chemicals that are indigenous to the ecosystem in which they exist but in quantities that do not exceed the indigenous amounts) have the potential to be hazardous to the environment and human health (Table 4.6).

Contamination by chemicals is a global issue and there is no single company that should shoulder all of the blame. Past laws and regulations (or the lack thereof) allowed unmanaged disposal and discharge of chemicals into

TABLE 4.6 Chemical Safety Guidelines to Be Followed When Working With Chemicals

1. Assume that any unfamiliar chemical is hazardous and treat it as such.
2. Know all the hazards of the chemicals with which you work.
3. Never underestimate the potential hazard of any chemical or combination of chemicals. Consider any mixture or reaction product to be at least as hazardous as its most hazardous component.
4. Never use any substance that is not properly labeled.
5. Date all chemicals when they are received and again when they are opened.
6. Follow all chemical safety instructions, such as those listed in Material Safety Data Sheets or on chemical container labels, precisely.
7. Minimize human exposure to any chemical regardless of the hazard rating of the chemical and avoid repeated exposure.
8. Use personal protective equipment (PPE), as appropriate for that chemical.
9. Always have a colleague present and do not work alone when working with hazardous chemicals.

the environment. These companies were not breaking the law; it was a matter of there being insufficient laws (the fault of various level of government) enacted to protect the environment. Moreover, the inappropriate management of such chemical waste has resulted in negative impacts on the environment. Thus, chemicals (with varying levels of toxicity) are found practically in all ecosystems on earth along with the adverse effects of these chemicals (1) on biodiversity, (2) on agricultural production, and/or (3) on water resources. At the end of the life cycles of the various chemicals, the chemicals are recycled or sent for disposal as part of waste.

Briefly, environmental pollution by chemicals can be classified using the following criteria according to the origin of the factors, into: (1) natural pollution resulting from volcanoes, hurricanes, earth quakes, sand storms; and (2) artificial pollution resulting from human activities: industry, agriculture, and domestic activities (Table 4.7). A second method of general classifications can be made according to the type of the pollutants, into: (1) physical pollution, such as radiation, (2) chemical pollution, such as by combustion products (carbon monoxide, carbon dioxide, and nitrogen oxides; sulfur compounds; nitrates; phosphates; and heavy metals. The third and final method of pollution is a subdivision according to the polluted medium, into (1) air pollution resulting from gases, powders from factories, vehicle emissions, and odors from agricultural activities, (2) water pollution, discharges of

TABLE 4.7 Examples of Pollutants

Pollutant	Formula	Sources	Polluted Medium
Sulfur dioxide	SO_2	Volcanoes	Air
		Industry	Air
		Transports	Air
Nitrogen dioxide	NO_2	Volcanoes	Air
		Industry	Air
		Transports	Air
Carbon monoxide	CO	Transports	Air
Ammonia	NH_3	Industry	Air, soil
		Agriculture	Air, soil
Hydrogen sulfide	H_2S	Industry	Soil, water
		anaerobic fermentation	Soil, water
Hydrogen chloride	HCl	Industry	Air
		Transports	Air
Hydrogen fluoride	HF	Industry	Air
Ammonium salts	NH_4^+	Farms, factories	Soil, water
Nitrate salts	NO_3^-	Farms, factories	Soil, water
Nitrite salts	NO_2^-	Farms, factories	Soil, water
Lead salts[a]	Pb	Heavy industry	Air
		Transports	Air
Mercury salts[a]	Hg	Industry	Soil, water
Zinc salts[a]	Zn	Industry	Air, water

[a]*May also appear as the metal.*

industrial residues such as metals and salts, and (3) soil pollution, resulting from nonecological tourism, waste grounds, car cemeteries, foams, and insecticides.

The potential for emissions from the manufacture and use of chemicals is high but, because of economic necessity, the potential for emissions is reduced by recovery of the chemicals. In some cases, the manufacturing process is operated as a closed system that allows little or no emissions to escape to the environment (air, water, or land). Emission sources from chemical processes include heaters and boilers; valves, flanges, pumps, and compressors; storage and transfer of products and intermediates; waste water handling; and emergency vents. Regular maintenance of the process equipment reduces the potential for these emissions. However, the emissions that do reach the atmosphere from the chemical industry are generally gaseous emissions that are controlled by a sequence of gas cleaning operations, including an adsorption process or an absorption process. In addition, emission of particulate matter, which could also lead to an environmental issue, since the particulate materials emitted are usually extremely small (typically a collection of particulates in the micron range) also requires (and is subject to) efficient treatment for removal (Mokhatab et al., 2006; Speight, 2007, 2014, 2017; Kidnay et al., 2011; Bahadori, 2014).

As already stated (Chapter 1), of all the pollutants released into the environment by human activity, PPs among the most dangerous to environmental flora and fauna are (1) pesticides, (2) various industrial chemicals, or (3) the unwanted byproducts of industrial processes that have been used for decades but have more recently been found to share several disturbing characteristics. These characteristics include: (1) persistence, which means that the chemicals resist degradation in air, water, and sediments, (2) bioaccumulation, which means that the chemicals accumulate in living tissues at concentrations higher than those in the surrounding environment, and (3) long-range transport, which means that the chemicals can travel a considerable distance from the source of release through air, water, and migratory animals, often contaminating areas miles away from any known source. On the environmental side, persistent pollutants are highly toxic and long-lasting, and cause an array of adverse effects on flora and fauna.

Thus, many chemicals that have toxic, carcinogenic, mutagenic, or teratogenic (causing developmental malformations) effects on environmental flora and fauna are designated either as (1) *acutely hazardous waste* or as (2) *toxic waste* by the United States Environmental Protection Agency (https://www.epa.gov/hw). Substances found to be fatal to humans in low doses, in the absence of data on human toxicity, have been shown to have an oral LD_{50} toxicity (lethal dose at 50% concentration) of less than 2 mg per liter or a dermal LD_{50} of less than 200 mg per kilogram. These substances capable of causing or significantly contributing to an increase in serious, irreversible, or incapacitating reversible illness are designated as *acute hazardous waste* (https://www.epa.gov/hw). Materials containing any of the toxic constituents so listed are to be considered hazardous waste, unless, after considering the following factors it can reasonably be concluded (by the United States Department of Environmental Health and Safety) that (1) the waste is not capable of posing a substantial present or potential hazard to public health or (2) the waste is not capable of posing a substantial present or potential hazard to the environment when improperly treated, stored, transported, or disposed of, or otherwise managed.

However, despite the nature of the environmental regulations and the precautions taken by the chemical industry, the accidental release of nonhazardous chemicals and hazardous chemicals into the environment has occurred and, without being unduly pessimistic, will continue to occur (by all industries—not wishing to select any single industry as the only industry that suffers accidental release of chemicals into the environment). It is a situation that, to paraphrase *chaos theory*: *no matter how well the preparation, the unexpected is always inevitable*. It is, at this point that the environmental scientist and engineer must identity (through careful analysis) the nature of the chemicals and their potential effects on the ecosystem(s). Thus, the predominance of one chemical or any particular class of chemicals may offer the environmental scientist or engineer an opportunity for predictability of behavior of the chemical(s) after consideration of the chemical and physical properties of the chemical (Chapter 3) and the effect of the chemical of the floral and faunal species in an ecosystem.

Thus, when a spill of chemicals occurs the primary processes determining the fate of chemicals are: (1) dispersion, (2) dissolution, (3) emulsification, (4) evaporation, (5) leaching, (6) sedimentation, (7) spreading, and (8) wind. These processes are influenced by the physical and properties of the chemicals (especially if the chemicals are constituents of a mixture), spill characteristics, environmental conditions, and chemical and physical properties of the spilled material after it has undergone any form of chemical transformation.

4.1 Dispersion

For the purposes of this text, the term *dispersion* encompasses all phenomena which give rise to the proliferation of chemicals through the manmade and natural environment. Thus, the disposal of a chemical into the water column and subsequent transportation of the chemical is referred to as dispersion. This is often a result of water surface turbulence, but also may result from the application of chemical agents (dispersants). The dispersed chemicals may remain in the water column or coalesce with other chemicals (the same chemical or different chemicals) and gain enough energy to be adsorbed by minerals, soil, or sediment. In the context of the geosphere, dispersion is a process that occurs in soil that is particularly vulnerable to erosion by water. In a soil layer where clay minerals are saturated with sodium ions (*sodic soil*), the soil can break down very easily into fine particles and wash away. This can lead to a variety of soil and water quality problems, including: (1) large losses of soil losses by gully erosion and tunnel erosion, (2) structural degradation of the soil, clogging, and sealing where dispersed particles settle, and (3) turbidity in water due to suspended soil particles which also cause transportation of nutrients from the land. Dispersive soil is more common in older landscapes where leaching and illuviation processes have had more time to show an effect.

Briefly, gully erosion involves creation of a gully created by running water, eroding sharply into soil and typically occurs on a hillside. On the other hand, tunnel erosion occurs in some soils where channels and tunnels develop beneath the surface. It is insidious as it is not readily noticeable (apart from small sediment fans at the tunnel discharge point) until the surface itself is undermined. Illuvium is a material displaced across a soil profile, from one layer to another, by the action of rainwater.

The dispersion of chemicals released into the environment has been the focus of the realization that the dispersion behavior of chemicals can be markedly different when different chemicals are considered. Accidents that involve chemicals give rise to a new class of problems in dispersion prediction for the following reasons: (1) the material is, in almost all cases, stored as a solid but may also emit a gas after a spill and exposure to the air, (2) the modes of release can vary widely and geometry of the source can take many forms and the initial momentum of the spill may be significant, (3) in some cases, a chemical transformation also takes place as a result of reaction with water vapor in the ambient atmosphere.

Emissions of many chemicals of concern occur into the air initially, from where they are dispersed into other media. Many chemicals emitted into the air, for instance, from combustion processes, tend to become associated with particulate matter. Removal from the air occurs through a range of complex processes involving photodegradation and particle sedimentation and/or precipitation (known respectively as *dry deposition* and *wet deposition*). A discharged chemical may undergo several cycles of physical and/or chemical transformation (Chapters 5–8) which can also make chemicals more accessible to photochemical degradation or biodegradation.

In addition, the physical properties of the chemical may result in one or more interactions with the surrounding ecosystem, especially if the chemical is reactive and has the potential to react quickly with the air, water, or soil. This reactivity will influence the dispersability of the chemical and, moreover, if the release occurs over a short time-scale, compared to the steady-state releases characteristic of many chemical release problems, this can give rise to the complication of predicting dispersion for time-varying releases. There is also the uncertainty of individual predictions resulting from variability about the behavior of a mixture. Also, the dispersing chemical, which is typically denser that air, may form a low-level cloud that is sensitive to the effects of either natural or manmade obstructions in the surrounding topography.

Wind transport (Aeolian transport—relocation by wind) can also occur and is particularly relevant when dust from solids (even coke dust which contains deposited chemicals) and catalyst dust are also considered. Dust becomes airborne when winds traversing arid land with little vegetation cover pick up small particles such as catalyst dust, coke dust, and other debris and send them skyward after which the movement of pollutants in the atmosphere is caused by transport, dispersion, and deposition. Dispersion results from local turbulence, that is, motions that last less than the time used to average the transport. Deposition processes, including precipitation, scavenging, and sedimentation, cause downward movement of pollutants in the atmosphere, which ultimately move the pollutants back to the ground surface but nor necessary in the locales from which the pollutants originated.

4.2 Dissolution

Dissolution is the process whereby gases, liquids, or solids dissolve into or are dissolved into a liquid or other solvent by which these original states (gas, liquid, or solid) become a solute (dissolved component), forming a solution of the gas, liquid, or solid in the original solvent. Solid solutions are the result of dissolution of one solid into another (such as metal alloys) and occur where the formation is governed and described by the relevant phase diagram. In the case of a crystalline solid dissolving in a liquid, the crystalline structure must be disintegrated such that the separate atoms, ions, or molecules are released. For liquids and gases, the molecules must be able to form noncovalent intermolecular interactions with those of the solvent for a solution to form. Dissolution is of fundamental importance in all chemical processes, especially the solubility of chemicals in water which can be enhanced by the occurrence of acid rain.

Generally, most chemicals are polar and are usually soluble in water but insoluble in solvents (such as diethyl ether, dichloromethane, chloroform, and hexane). The solubility (dissolution) characteristics of chemicals in water are complex and very much dependent upon the structure and properties of the chemical(s). The chemical may be a gas, a liquid, or a solid (the solute) that dissolves in the water—the solubility depends on temperature, pressure, and the pH (acidity or alkalinity of the water) as well.

Furthermore, the extent of solubility of the chemical can range from infinitely soluble (without limit) to poorly soluble (such as some lead salts)—the term *insoluble* is often applied to poorly or very poorly soluble compounds and (in the world of chemistry) a common threshold to describe a chemical as insoluble is a solubility less than 0.1 g per 100 mL of water. However, solubility of a chemical in water should not be confused with the ability to of water to dissolve the chemical because apparent solubility of the chemical might also occur because of a chemical reaction (*reactive solubility*).

Solubility of a chemical in water applies not only to environmental chemistry but to areas of chemistry, such as biochemistry, geochemistry, chemistry, and physical chemistry. In all cases, the solubility of the chemical depends on the physical conditions (temperature, pressure, and concentration of the chemical) and the enthalpy and entropy directly relating to the water and the chemicals concerned. Water is, by far, the most common solvent in chemistry, especially for those chemicals that readily form ions. This is a crucial factor in which acidity and alkalinity play a role as in much of the area known as environmental chemistry.

In addition, the term *dissolved chemicals* (sometimes referred to as *total dissolved solids*) is used as a broad classification for chemicals of varied origin and composition within aquatic systems (the aquasphere) that can exert toxicity effects on the aquasphere. The source of the dissolved chemical in freshwater and marine systems depends on the body of water. When water contacts highly ionizable chemicals, these components can drain into rivers and lakes as a dissolved chemical. Whatever the source of the dissolved chemical, it is also extremely important in the transport of metals in aquatic systems—certain metals can form extremely strong metallic complexes with already-dissolved substances which enhance the solubility of the metal in aqueous systems while also reducing the bioavailability of the metal.

In terms of chemicals that are not naturally occurring, knowledge of the structure and properties can be used to examine relationships between the solubility properties of the chemical and its structure, and vice versa. In fact, structure dictates function which means that by knowing the structure of a chemical it may be (but not always) possible to predict the properties of the chemical such as its solubility, acidity or basicity, stability, and reactivity. In the context of environmental distribution of a chemical, predicting the solubility of a chemical is a useful component of knowledge.

Dissolved compounds are generally associated with the various characteristics of drinking water (such as taste, smell, color, and odor) and the quality of drinking water, as perceived by the senses, largely determines the acceptability of water. For health-related contaminants—a contaminant that is unsafe for one is unsafe for all—the general characteristics (taste, smell, and odor) are subject to social, economic, and cultural considerations.

On the other hand, the presence of undissolved suspended solids in water gives rise to turbidity. Suspended solids may consist of clay, silt, airborne particulates, colloidal particles, plankton, and other microscopic organisms. The presence of particulate matter in water may not only be offensive to the general senses but can, in a more serious aspect, protect bacteria and viruses from the action of disinfectants. In addition, the adsorptive capacity of some suspended particulates can lead to entrapment of undesirable compounds present in the water and in this way, turbidity can bear an indirect relationship to the health aspects of water quality.

4.3 Emulsification

An emulsion is a mixture of two or more liquids that are usually immiscible. Examples include crude oil and water which can form an *oil-in-water emulsion*, wherein the oil is the dispersed phase, and water is the dispersion medium or a *water-in-oil emulsion*, wherein the water is the dispersed phase, and the oil is the dispersion medium. The physical structure of an emulsion is based on the droplet size—most emulsions contain droplets with a mean diameter of more than around 1 μm (1 micron, 1×10^{-6} m); however, miniemulsions and nanoemulsions can be formed with droplet sizes in the 100–500 nm range (nanometer, 1×10^{-9} m) range, and with proper formulation, highly stable microemulsions can be prepared having droplets as small as a few nanometers.

The stability of an emulsion refers to the ability of the emulsion to resist change in form and properties over time. There are three types of instability in emulsions: (1) flocculation, (2) creaming, and (3) coalescence. Flocculation occurs when there is an attractive force between the droplets, so they form a flocculant mass (flocs) in the fluid through precipitation or aggregation of suspended particles. Creaming occurs when the droplets rise to the top of the emulsion under the influence of buoyancy. Coalescence occurs when droplets collide and combine to form a larger droplet, so the average droplet size increases over time. Use of a surface-active agent (surfactant) can increase the stability of an emulsion so that the size of the droplets does not change significantly with time and the emulsion is then defined as a *stable emulsion*.

This demulsification process can be slowed with the help of emulsifiers which are surface-active substances with strong hydrophilic properties that are used to eliminate the prolonged effects of spills in which emulsions are formed. Emulsifiers help to stabilize emulsions and promote dispersing the dispersed phase oil to form microscopic (invisible) droplets, which accelerates the decomposition of the pollutant in the water.

The formation of an emulsion based on water and chemicals is less likely. Nevertheless, the properties of chemicals as a function of the solubility in water is important because of the potential use of the chemicals to break, for example, an oil-in-water emulsion that forms after a crude oil spill into the aquasphere and the potential to aid in the clean-up process.

4.4 Evaporation

Evaporation is the process by which water changes from a liquid to a gas or vapor and is the primary pathway that water moves from the liquid state back into the water cycle as atmospheric water vapor. Evaporation is the opposite of condensation and sublimation is the phenomenon that occurs when a solid becomes a gas without the conversion of the solid to a liquid and thence the liquid to a gas.

Evaporation:

$$\text{Liquid} \rightarrow \text{gas}$$

Sublimation:

$$\text{Solid} \rightarrow \text{gas}$$

Both phenomena are important parts of the environmental chemical cycle since both processes involve disappearance of a contaminant in the water or on the land, *but* the contaminant does appear in the atmosphere. Of the two processes, the evaporation is the most common process since chemicals that go through the sublimation processes are not as obvious or as common as chemicals that evaporate. In the field of environmental chemistry, evaporation is more typical when water evaporates from an aqueous solution of the chemical rather than evaporation (volatilization) of the chemical.

Many factors affect the evaporation process. For example, if the air is already saturated with other chemicals (or the humidity is high when the water content is high) there is typically little chance of a liquid evaporating as quickly as when the air is not saturated (or humidity is low). In addition, the air pressure also affects the evaporation process since, under high air pressure, a chemical is much more difficult to evaporate. Temperature also affects the ability of a chemical to evaporate.

4.5 Leaching

Leaching is a natural process by which water-soluble substances (such as water-soluble chemicals or hydrophilic chemicals) are washed out from soil or waste disposal areas (such as a landfill). These leached out chemicals (leachates) can cause pollution of surface waters (ponds, lakes, rivers, and the sea) and subsurface groundwater aquifers).

In the environment, leaching is the process by which chemicals or radionuclides are released from the solid phase into the water phase under the influence of mineral dissolution, desorption, and complexation processes as affected by pH, redox chemistry, and biological activity. The process itself is universal, as any material exposed to contact with water will leach components from its surface or its interior depending on the porosity of the material. In terms of the effect of pH (acidity or alkalinity), the leachability of a chemical (i.e., the ability of the chemical to be leached) from the source can be enhanced by the occurrence of acid rain which can enhance the solubility of the chemical in the (now-acidified) water.

Thus, leaching is the process by which (in the current context) contaminants are released from the solid phase into the water phase under the influence of dissolution, desorption, or complexation processes as affected by acidity or alkalinity (the pH value). The process itself is universal, as any material exposed to contact with water will leach components from its surface or its interior depending on the porosity of the material under consideration. Leaching often occurs naturally with soil contaminants such as chemicals with the result that the chemicals end up in potable waters.

Many chemicals occur in a mixture of different components in a solid (such as the metallic deposits on petroleum coke or metals in spent catalysts). In order to separate the desired solute constituent or remove an undesirable solute component from the solid phase, the solid is brought into contact with a liquid during which time the solid and liquid are in contact and the solute or solutes can diffuse from the solid into the solvent, resulting in separation of the components originally in the solid (*leaching, liquid–solid leaching*). In addition, leaching may also be referred to as *extraction* because the chemical is being extracted from the solid or the process may also be referred to as *washing* because a chemical is removed from a solid with water.

Leaching is affected by (1) the texture of the solid that holds the chemical to be leached, (2) structure of the chemical to be leached, and (3) water content of the solid which holds the chemical. If the solid holding the chemical is soil, then in terms of soil texture, for example, the proportions of sand, silt, and clay affect the movement of water through the soil. Coarse-textured soil contains more sand particles which have large pores and the soil is highly permeable which allow the water to move rapidly through the pore system. On the downside, chemicals (such as phosphate pesticides) carried by water through coarse-textured soil are more likely to reach and contaminate groundwater. On the other hand, clay-textured soils have low permeability and tend to retain more water and adsorb more chemicals from the water. This slows the downward movement of chemicals, helps increase the chance of degradation and adsorption to soil particles, and reduces the chance of groundwater contamination.

In terms of soil structure, loosely packed soil particles allow speedy movement of water through the soil while tightly compacted soil holds water back and does not allow the water to move freely through it. Plant roots penetrate soil, creating excellent water channels when they die and rot away. These openings and channels may permit relatively rapid water movement even through clay-containing soil. On the other hand, the amount of water already in the soil has a direct bearing on whether rain or irrigation results in the recharging of groundwater and possible leaching of chemicals into an aquifer. Soluble chemicals are more likely to reach groundwater when the water content of the soil approaches or is at saturation. Saturation is typical in the spring when rain and snowmelt occur, but when soil is dry the added water just fills the pores in the soil near the soil surface, making it unlikely that the water will reach the groundwater supply.

Leaching processes introduce chemicals into the water phase by solubility and entrainment. In addition, the leaching processes that move chemicals from soil can have a variety of potential scenarios—the spill of chemicals on to soil may cause partitioning in water that has been in contact with the contamination.

4.6 Sedimentation and Adsorption

In the current context, sedimentation is the tendency for chemicals in a suspension in water to settle out of the water and come to rest against a barrier, which is typically the basement of a waterway. In geology, sedimentation is often used as the opposite of erosion, that is, the terminal end of sediment transport. Settling is the falling of suspended particles through the liquid, whereas sedimentation is the termination of the settling process. Sedimentation may pertain to objects of various sizes, ranging from large rocks in flowing water to suspensions of dust which produce significant sedimentation. The term is typically used in geology to describe the deposition of sediment which results in the formation of sedimentary rock. Settling is due to the motion of the particles through the fluid in response to the forces acting on them, which in an ecosystem can be gravity. Sedimentation can be generally classified into three different types that are all applicable to chemicals. Type 1 sedimentation is characterized by particles that settle discretely at a constant settling velocity and typically these particles settle as individual particles and do not flocculate or stick to other during settling. Type 2 sedimentation is characterized by particles that flocculate during sedimentation and because of this the particle size is constantly changing and therefore their settling velocity is changing. Type 3 sedimentation (also known as zone sedimentation) involves particles that are at a high concentration (for example, greater than 1000 mg/L) such that the particles tend to settle as a mass and a distinct clear zone and sludge zone are present. Zone settling occurs in active sludge sedimentation and sedimentation of sludge thickeners.

In terms of the sedimentation of chemicals, some of the chemicals may be adsorbed on the suspended material (especially if the suspended material is clay or another highly adsorptive mineral) and deposited to the bottom. This mainly happens in shallow waters where particulate matter is abundant and water is subjected to intense mixing, usually through turbulence. Simultaneously, the process of biosedimentation can also occur when plankton and other organisms absorb the chemical. The suspended forms of the chemicals undergo intense chemical and biological (microbial) decomposition in the water. However, this situation radically changes when the suspended chemical reaches the lake bed, river bed, or sea bed and the decomposition rate of the chemical(s) that settles on the bottom abruptly drops. The oxidation processes slow down, especially under anaerobic conditions in the bottom environment and the chemical(s) accumulated inside the sediments can be preserved for many months and even years.

Although related to sedimentation, adsorption (a surface phenomenon) has some differences insofar as this process is the adhesion of an chemical from a gas, liquid, or dissolved solid to a surface—it might be considered as a sedimentation process that is encouraged or enhanced by the properties of the chemical and the adsorbing properties of the surface to which the chemical is adsorbed. The (adsorption) process creates a film of the *adsorbate* on the surface of the *adsorbent*. This process differs from absorption in which a fluid (the *absorbate*) is dissolved by or permeates a liquid or solid (the *absorbent*), respectively. Adsorption is a surface-based process while absorption involves the whole volume of the material. The term *sorption* encompasses both processes, while *desorption* is the reverse of it.

4.7 Spreading

Spreading is a form of dispersion in which the movement of chemical contaminants through the subsurface is complex and difficult to predict since different types of chemicals react differently with soils, sediments, and other geologic materials and commonly travel along different flow paths and at different velocity. Chemicals released from a source area (typically on the surface) infiltrate (through solubility in water) into the subsurface and migrate downward by gravity as an aqueous solution through the vadose zone (the zone that extends from the top of the ground surface to the water table). When low-permeability soil units are encountered, the contaminants in aqueous solution can also spread laterally along the permeability contrast.

Most chemical contaminants are introduced into the subsurface by percolation through soil strata and the interactions between the soil and the chemical are important for assessing the fate and transport of the contaminant in the subsurface, especially in the groundwater system. Contaminants that are highly soluble, such as salts and any potentially ionic derivatives, can move readily from surface soil to saturated materials below the water table as often occurs during and after rainfall events. Those contaminants that may not be highly soluble may have considerably longer residence times in the surface strata (the soil zone). Some chemicals adsorb readily onto soil particles and slowly dissolve during precipitation events, resulting in migration into the water table. Once below the water table, chemical contaminants are also subject to dispersion (mechanical mixing with uncontaminated water) and diffusion (dilution by concentration gradients). Many contaminants begin to spread immediately after entry into the environment.

4.8 Sublimation

Sublimation is the transition of a substance directly from the solid phase to the gas phase without passing through the intermediate liquid phase (Table 4.8, Fig. 4.2). Sublimation is an endothermic phase transition that occurs at temperatures and pressures below the triple point of a chemical in the phase diagram. The reverse process of sublimation is the process of *deposition* in which some chemicals pass directly from the gas phase to the solid phase, again without passing though the intermediate liquid phase.

The term *sublimation* refers to a *physical* change of state and is not used to describe transformation of a solid to gas in a chemical reaction. For example, the dissociation on heating of solid ammonium chloride (NH_4Cl) into

TABLE 4.8 Phase Transformations From Gas to Liquid to Solid and the Reverse

	To:	Gas	Liquid	Solid
From:	Gas	N/A	Condensation	Deposition
	Liquid	Evaporation[a]	N/A	Freezing
	Solid	Sublimation	Melting	Transformation[b]

[a]*Also, boiling.*
[b]*For example, change in crystal structure.*

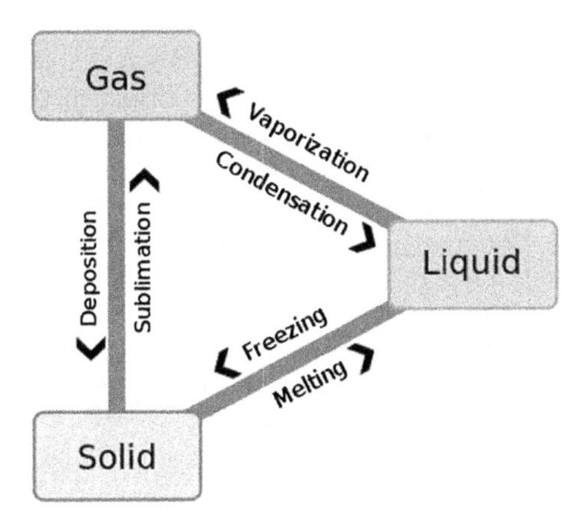

FIGURE 4.2 Representation of phase changes.

ammonia (NH_3) and hydrogen chloride (HCl) is *not* sublimation but a chemical reaction:

$$NH_4Cl \rightarrow NH_3 + HCl$$

Sublimation requires additional energy and is an endothermic change and the enthalpy of sublimation (also referred to as the *heat of sublimation*) can be calculated by adding the enthalpy of fusion and the enthalpy of vaporization.

5. DISTRIBUTION IN THE ENVIRONMENT

Pollution by chemicals and the distribution of these contaminants in the environment is a major concern. However, the movement of chemicals through the environment is complex and is difficult to predict. Different types of chemicals react differently with soils, sediments, and other geologic materials and commonly travel along different flow paths and at different velocities. For example, using soil as the example and which can vary considerably in properties (Table 4.9), contaminants that are highly soluble, such as salts (e.g., sodium chloride, NaCl) move readily from surface soils to saturated materials below the water table. This often occurs during and after rainfall events. Other contaminants that are not highly soluble may have considerably longer residence times in the soil zone. Some contaminants adsorb readily onto soil particles and slowly dissolve during precipitation events, resulting in dissolve fraction concentrations of contaminants migrating to groundwater. Once below the water table, chemicals are also subject to dispersion (mechanical mixing with uncontaminated water) and diffusion (dilution by concentration gradients).

Typically, chemicals released from various sources are ultimately dispersed among, and can at times accumulate in, various environmental compartments (such as soil as well as various floral and faunal species). Some contaminants

TABLE 4.9 General Characteristics of Soil

Property	Characteristic	Comments
Composition	Porous body	Surface horizons have approximately equal volumes of solids and pores containing water or air
	Weathering	Weatherable minerals decompose and more resistant minerals form
	Transported chemicals	Small particles may be transported by water, wind, or gravity
Physical properties	Texture	Surface area per unit volume and particle size; influence on properties and reactivity
	Structure	Influence on porosity and large and small pore ratio; affects adsorption-absorption
	Consistency	Physical behavior of soil changes as moisture content changes
Chemical properties	Cation exchange	Ability of soil clay minerals to reversibly adsorb and exchange cations
	pH	A measure of the acidity or alkalinity; can change soil chemistry
Environment	Water	Can change soil chemistry; hydrolysis
	Temperature	Can change soil chemistry; decomposition
	Aeration	Can change soil chemistry; redox reactions

may contribute primarily to environmental compartments on a local scale but other contaminants that are more persistent in the environment can be distributed over much greater distances—even up to a regional scale, a national scale, or an international scale.

Thus, understanding the potential environmental impact of dispersed chemicals requires an understanding of the relative contribution of the various sources of the pollutants as well as the types of environments and the potential for a pollutant to undergo chemical (or physical) transformation in the environment (Chapters 5–8). Therefore, an investigation of a potential contaminant must account for transport of the contaminant through ecosystems. The required characterization of concentrations of contaminants in an environmental medium, such as air, involves accounting for the gains (or inputs to) and losses from that medium, and transport through it.

Some examples of atmospheric pollutants include nitrogen dioxide (NO_2), sulfur dioxide (SO_2), and carbon monoxide (CO). The first two pollutants combine with water to form acids, which not only irritate the lungs but also contribute to the long-term destruction of the environment due to the generation of acid rain (Chapter 1).

$$2[C]_{\text{fossil fuel}} + O_2 \rightarrow 2CO$$

$$[C]_{\text{fossil fuel}} + O_2 \rightarrow CO_2$$

$$2[N]_{\text{fossil fuel}} + O_2 \rightarrow 2NO$$

$$[N]_{\text{fossil fuel}} + O_2 \rightarrow NO_2$$

$$[S]_{\text{fossil fuel}} + O_2 \rightarrow SO_2$$

$$2SO_2 + O_2 \rightarrow 2SO_3$$

Carbon monoxide, generated by the incomplete combustion of hydrocarbons, displaces and prevents oxygen from binding to hemoglobin and causes asphyxiation. Also, it binds with metallic pollutants and causes them to be more mobile in air and water.

Fresh, clean, and drinkable water is a necessary but limited resource on the planet. Industrial, agricultural, and domestic wastes can contribute to the pollution of this valuable resource, and water pollutants can damage human and animal health. Three important classes of water pollutants are heavy metals, inorganic pollutants, and organic pollutants. Heavy metals include transition metals such as cadmium, mercury, and lead, all of which can contribute to brain damage. Pollutants like hydrochloric acid, sodium chloride, and sodium carbonate change the acidity, salinity, or alkalinity of the water, making it undrinkable or unsuitable for the support of floral and faunal species.

The use of pesticides (such as phosphates) in agriculture contributes to environmental pollution. Pesticides are used to control the growth of insects, weeds, and fungi, which compete with humans in the consumption of crops.

This use not only increases crop yields and decreases grocery prices, but also controls diseases such as malaria and encephalitis. However, the spraying of crops and the water runoff from irrigation transports these harmful chemicals to the habitats of nontarget animals. Chemicals can build up in the tissues of these animals, and when humans consume the animals the increased potency of the pesticides manifests as health problems and in some cases death. Chemists have recently developed naturally occurring pesticides that are toxic only to their particular targets and are benign to birds and mammals. The most significant pesticide of the 20th century was DDT (Chapter 1), which was highly effective as an insecticide but did not break down in the environment and led to the death of birds, fish, and some humans.

To effectively monitor changes in the environmental behavior of chemicals that are of most concern, it is extremely important to understand how these chemicals typically behave in natural systems and in specific ecosystems. Equally important is an understanding of how these chemicals might respond to specific best management practices. Some of the discharged chemicals only become problems at high concentrations that impair the beneficial uses of these ecosystems. An effective monitoring program explicitly considers how these chemicals may change as they move from a source into the groundwater, surface water, or into the soil. This includes an understanding of how a specific chemical may be introduced or mobilized within an ecosystem, how the chemical moves through an ecosystem, and the transformations that may occur during this process.

As an example, groundwater will normally look clear and clean because the ground naturally filters out particulate matter. But, natural and human-induced chemicals can be found in groundwater (Tables 4.10–4.12). As groundwater flows through the ground, metals such as iron and manganese are dissolved and may later be found in high concentrations in the water. Industrial discharges, urban activities, agriculture, groundwater pumpage, and disposal of waste all can affect groundwater quality. Contaminants can be human-induced, as from leaking fuel tanks or toxic chemical spills. Pesticides and fertilizers applied to lawns and crops can accumulate and migrate to the water table. Leakage from septic tanks and/or waste-disposal sites also can introduce bacteria to the water, and pesticides and fertilizers that seep into farmed soil can eventually end up in water drawn from a well. Or, a well might have been placed in land that was once used for something like a garbage or chemical dump site.

However, the entry of chemicals into the environment and the distribution of these chemicals within the environment is often complex and there have been many occasions when a significant amount of a chemical (or a mixture of chemicals) has entered an ecosystem and the effects of contamination are well-defined. Generally, the assumption is that the chemical (or a mixture thereof) does not rapidly diffuse away but remains in the immediate vicinity at a noticeably high concentration or perhaps moves, but in such a way that

TABLE 4.10 Physical Properties of Groundwater

Contaminant	Sources to Groundwater	Potential Health and Other Effects
Turbidity	Caused by the presence of suspended matter such as clay, silt, and fine particles of inorganic and organic matter, plankton, and other microscopic organisms. A measure how much light can filter through the water sample.	Objectionable for aesthetic reasons. Indicative of clay or other inert suspended particles in drinking water. May not adversely affect health but may cause need for additional treatment. Following rainfall, variations in groundwater turbidity may be an indicator of surface contamination.
Color	Can be caused by decaying leaves, plants, organic matter, copper, iron, and manganese, which may be objectionable. Indicative of large amounts of chemicals, inadequate treatment, and high disinfection demand. Potential for production of excess amounts of disinfection byproducts.	Suggests that treatment is needed. No health concerns. Aesthetically unpleasing.
pH	Indicates, by numerical expression, the degree to which water is alkaline or acidic. Represented on a scale of 0 −14 where 0 is the most acidic, 14 is the most alkaline, and 7 is neutral.	High pH causes a bitter taste; water pipes and water-using appliances become encrusted; depresses the effectiveness of the disinfection of chlorine, thereby causing the need for additional chlorine when pH is high. Low-pH water will corrode or dissolve metals and other substances.
Odor	Certain odors may be indicative of contaminants that originate from municipal or industrial waste discharges or from natural sources.	
Taste	Some substances such as certain salts produce a taste without an odor and can be evaluated by a taste test. Many other sensations ascribed to the sense of taste actually are odors, even though the sensation is not noticed until the material is taken into the mouth.	

TABLE 4.11 Changes in Physicochemical Properties Related to a Reduction in Aquatic Toxicity

Physicochemical Property	Changes
Molecular size and weight	Generally, as molecular weight increases, aquatic bioavailability and toxicity decrease. At MW > 1000, bioavailability is negligible. Caution must be taken, however, to consider possible breakdown products that may have MW < 1000 and exert toxicity.
Octanol–water partition Coefficient (log P) and octanol–water distribution coefficient at biological pH ($logD_{7.4}$)	log *P* usually correlates exponentially with acute aquatic toxicity by narcosis for nonionic organic chemicals up to a value of about 5–7. Chemicals with log *P* < 2 have a higher probability of having low acute and chronic aquatic toxicity. For ionizable organic chemicals, log $D_{7.4}$ is a more appropriate measure: ionizable compounds with log $D_{7.4} < 1.7$ have been shown to have increased probability of being safe to freshwater fish than those with log $D_{7.4} > 1.7$.
Water solubility	Generally, compounds with higher log *P* have lower water solubility. Very poorly water-soluble chemicals (<1 ppb) generally have low bioavailability and are less toxic.

TABLE 4.12 Contaminants Found in Groundwater

Contaminant	Sources	Effects
Aluminum	Occurs naturally in some rocks and drainage from mines.	Can precipitate out of water after treatment, causing increased turbidity or discolored water.
Antimony	Enters environment from natural weathering, industrial production, municipal waste disposal, and manufacturing of flame retardants, ceramics, glass, batteries, fireworks, and explosives.	Decreases longevity, alters blood levels of glucose and cholesterol in laboratory animals exposed to high levels over their lifetime.
Arsenic	Enters environment from natural processes, industrial activities, pesticides, and industrial waste, smelting of copper, lead, and zinc ore.	Causes acute and chronic toxicity, liver and kidney damage; decreases blood hemoglobin. A carcinogen.
Barium	Occurs naturally in some limestones, sandstones, and	Can cause a variety of cardiac, gastrointestinal, and neuromuscular effects. Associated

TABLE 4.12 Contaminants Found in Groundwater—cont'd

Contaminant	Sources	Effects
	soils in the eastern United States.	with hypertension and cardiotoxicity in animals.
Beryllium	Occurs naturally in soils, groundwater, and surface water. Often used in electrical industry equipment and components, nuclear power, and space industry. Enters the environment from mining operations, processing plants, and improper waste disposal. Found in low concentrations in rocks, coal, and petroleum and enters the ground.	Causes acute and chronic toxicity; can cause damage to lungs and bones. Possible carcinogen.
Cadmium	Found in low concentrations in rocks, coal, and petroleum and enters the groundwater and surface water when dissolved by acidic waters. May enter the environment from industrial discharge, mining waste, metal plating, water pipes, batteries, paints and pigments, plastic stabilizers, and landfill leachate.	Replaces zinc biochemically in the body and causes high blood pressure, liver and kidney damage, and anemia. Destroys testicular tissue and red blood cells. Toxic to aquatic biota.
Chloride	May be associated with the presence of sodium in drinking water when present in high concentrations. Often from saltwater intrusion, mineral dissolution, industrial and domestic waste.	Deteriorates plumbing, water heaters, and municipal water-works equipment at high levels. Above secondary maximum contaminant level, taste becomes noticeable.
Chromium	Enters environment from old mining operations runoff and leaching into groundwater, fossil-fuel combustion, cement-plant emissions, mineral leaching, and waste incineration. Used in metal plating and as a cooling-tower water additive.	Chromium III is a nutritionally essential element. Chromium VI is much more toxic than Chromium III and causes liver and kidney damage, internal hemorrhaging, respiratory damage, dermatitis, and ulcers on the skin at high concentrations.
Copper	Enters environment from metal plating, industrial and domestic waste, mining, and mineral leaching.	Can cause stomach and intestinal distress, liver and kidney damage, anemia in high doses. Imparts an adverse taste and significant staining to clothes and fixtures.

Continued

TABLE 4.12 Contaminants Found in Groundwater—cont'd

Contaminant	Sources	Effects
		Essential trace element but toxic to plants and algae at moderate levels.
Cyanide	Often used in electroplating, steel processing, plastics, synthetic fabrics, and fertilizer production; also, from improper waste disposal.	Poisoning is the result of damage to spleen, brain, and liver.
Dissolved solids	Occur naturally but also enter environment from manmade sources such as landfill leachate, feedlots, or sewage. A measure of the dissolved "salts" or minerals in the water. May also include some dissolved compounds.	May have an influence on the acceptability of water in general. May be indicative of the presence of excess concentrations of specific substances not included in the Safe Water Drinking Act, which would make water objectionable. High concentrations of dissolved solids shorten the life of hot water heaters.
Fluoride	Occurs naturally or as an additive to municipal water supplies; widely used in industry.	Decreases incidence of tooth decay but high levels can stain or mottle teeth. Causes crippling bone disorder (calcification of the bones and joints) at very high levels.
Hardness	Result of metallic ions dissolved in the water; reported as concentration of calcium carbonate. Calcium carbonate is derived from dissolved limestone or discharges from operating or abandoned mines.	Decreases the lather formation of soap and increases scale formation in hot-water heaters and low-pressure boilers at high levels.
Iron	Occurs naturally as a mineral from sediment and rocks or from mining, industrial waste, and corroding metal.	Imparts a bitter astringent taste to water and a brownish color to laundered clothing and plumbing fixtures.
Lead	Enters environment from industry, mining, plumbing, gasoline, coal, and as a water additive.	Affects red blood cell chemistry; delays normal physical and mental development in babies and young children. Causes slight deficits in attention span, hearing, and learning in children. Can cause slight increase in blood pressure in some adults. Probable carcinogen.

TABLE 4.12 Contaminants Found in Groundwater—cont'd

Contaminant	Sources	Effects
Manganese	Occurs naturally as a mineral from sediment and rocks or from mining and industrial waste.	Causes aesthetic and economic damage and imparts brownish stains to laundry. Affects taste of water and causes dark brown or black stains on plumbing fixtures. Relatively nontoxic to animals but toxic to plants at high levels.
Mercury	Occurs as a salt and as mercury compounds. Enters the environment from industrial waste, mining, pesticides, coal, electrical equipment (batteries, lamps, switches), smelting, and fossil-fuel combustion.	Causes acute and chronic toxicity. Targets the kidneys and can cause nervous system disorders.
Nickel	Occurs naturally in soils, groundwater, and surface water. Often used in electroplating, stainless steel and alloy products, mining, and refining.	Damages the heart and liver of laboratory animals exposed to large amounts over their lifetime.
Nitrate (as nitrogen)	Occurs naturally in mineral deposits, soils, seawater, freshwater systems, the atmosphere, and biota. More stable form of combined nitrogen in oxygenated water. Found in the highest levels in groundwater under extensively developed areas. Enters the environment from fertilizer, feedlots, and sewage.	Toxicity results from the body's natural breakdown of nitrate to nitrite. Causes "bluebaby disease," or methemoglobinemia, which threatens oxygen-carrying capacity of the blood.
Nitrite (combined nitrate/nitrite)	Enters environment from fertilizer, sewage, and human or farm-animal waste.	Toxicity results from the body's natural breakdown of nitrate to nitrite. Causes "bluebaby disease," or methemoglobinemia, which threatens oxygen-carrying capacity of the blood.
Selenium	Enters environment from naturally occurring geologic sources, sulfur, and coal.	Causes acute and chronic toxic effects in animals—"blind staggers" in cattle. Nutritionally essential element at low doses but toxic at high doses.
Silver	Enters environment from ore mining and processing, product fabrication, and	Can cause argyria, a blue–gray coloration of the skin, mucus membranes, eyes, and organs in

Continued

TABLE 4.12 Contaminants Found in Groundwater—cont'd

Contaminant	Sources	Effects
	disposal. Often used in photography, electric and electronic equipment, sterling and electroplating, alloy, and solder. Because of great economic value of silver, recovery practices are typically used to minimize loss.	humans and animals with chronic exposure.
Sodium	Derived geologically from leaching of surface and underground deposits of salt and decomposition of various minerals. Human activities contribute through deicing and washing products.	Can be a health risk factor for those individuals on a low-sodium diet.
Sulfate	Elevated concentrations may result from saltwater intrusion, mineral dissolution, and domestic or industrial waste.	Forms hard scales on boilers and heat exchangers; can change the taste of water and has a laxative effect in high doses.
Thallium	Enters environment from soils; used in electronics, pharmaceuticals manufacturing, glass, and alloys.	Damages kidneys, liver, brain, and intestines in laboratory animals when given in high doses over their lifetime.
Zinc	Found naturally in water, most frequently in areas where it is mined. Enters environment from industrial waste, metal plating, and plumbing, and is a major component of sludge.	Aids in the healing of wounds. Causes no ill-health effects except in very high doses. Imparts an undesirable taste to water. Toxic to plants at high levels.

concentration levels of the chemical remain high as it moves. Such cases would normally occur when large quantities of a substance were being stored, transported, or otherwise handled in concentrated form. Thus, due to leakages, spills, improper disposal and accidents during transport, compounds have become subsurface contaminants that threaten various.

Background knowledge of the sources, chemistry, and potential risks of toxic heavy metals in contaminated soils is necessary for the selection of appropriate remedial options. Remediation of soil contaminated by heavy metals is necessary to reduce the associated risks, make the land resource available for agricultural production, enhance food security, and scale down land tenure problems. Immobilization, soil washing, and phytoremediation are frequently listed among the best available technologies for cleaning up heavy metal contaminated soils. These technologies are recommended for field applicability and commercialization in the developing countries also where agriculture, urbanization, and industrialization are leaving a legacy of environmental degradation (Wuana and Okieimen, 2011).

REFERENCES

Bahadori, A., 2014. Natural Gas Processing: Technology and Engineering Design. Gulf Professional Publishing, Elsevier, Amsterdam, Netherlands.

Kidnay, A.J., Parrish, W.R., McCartney, D.G., 2011. Fundamentals of Natural Gas Processing, second ed. CRC Press, Taylor & Francis Group, Boca Raton, Florida.

Levy, D.B., Barbarick, K.A., Siemer, E.G., Sommers, L.E., 1992. Distribution and partitioning of trace metals in contaminated soils near Leadville, Colorado. Journal of Environmental Quality 21 (2), 185–195.

Miller, D.R., 1984. Effects of Pollutants at the Ecosystem Level. SCOPE 22. In: Sheehan, P.J., Miller, D.R., Butler, G.C., Bourdeau, P. (Eds.), Scientific Committee on Problems of the Environment (SCOPE) of the International Council of Scientific Unions (ICSU). John Wiley & Sons Inc., Hoboken, New Jersey. http://dge.stanford.edu/SCOPE/SCOPE_22/SCOPE_22_front%20material.pdf.

Mokhatab, S., Poe, W.A., Speight, J.G., 2006. Handbook of Natural Gas Transmission and Processing. Elsevier, Amsterdam, Netherlands.

Shiowatana, J., McLaren, R.G., Chanmekha, N., Samphao, A., 2001. Fractionation of arsenic in soil by a continuous flow sequential extraction method. Journal of Environmental Quality 30 (6), 1940–1949.

Speight, J.G., 2007. Natural Gas: A Basic Handbook. GPC Books. Gulf Publishing Company, Houston, Texas.

Speight, J.G., 2014. The Chemistry and Technology of Petroleum, fifth ed. CRC Press, Taylor & Francis Group, Boca Raton, Florida.

Speight, J.G., 2017. Handbook of Petroleum Refining. CRC Press, Taylor & Francis Group, Boca Raton, Florida.

Wuana, R., Okieimen, F.E., 2011. Heavy metals in contaminated soils: a review of sources, chemistry, risks and best available strategies for remediation. ISRN Ecology 2011, 402647. https://doi.org/10.5402/2011/402647. https://www.hindawi.com/journals/isrn/2011/402647/.

Part II

Transformation Processes

Chapter 5

Sorption, Dilution, and Dissolution

1. INTRODUCTION

Chemicals interact with the environment in different ways—two such ways are adsorption and absorption—both are important phenomena but with significant differences in the outcome (Table 5.1). *Adsorption* is a word that describes removal of a solute from solution to a contiguous solid phase and also refers to the two-dimensional accumulation of an adsorbate at a solid surface. The word *adsorbate* refers to the solute that adsorbs on to the solid phase (the *adsorbent*). The word *absorption* is used to describe the relationship between a solute and solid when there is diffusion of the sorbate *into* the solid phase and, in addition, absorption processes usually show a significant time dependency. The word *sorption* and the related words *sorbate and sorbent* are more general terms that are often applied to the processes of adsorption and absorption (Chapter 10) (Jenne, 1986).

When a discharged chemical is associated with the solid phase, it is not known if the chemical is adsorbed on to the surface of a solid, absorbed into the structure of a solid, precipitated as a 3-dimensional molecular structure on the surface of the solid, or partitioned into the organic matter (Sposito, 1984, 1989, 1994). Dissolution/precipitation and adsorption/desorption are considered the most important processes affecting metal and radionuclide interactions with soils and are also important chemical processes. Moreover, dissolution/precipitation is more likely to be the key process when no chemical equilibrium exists, such as in an area where high contaminant concentrations exist, or where steep pH or redox gradients exist. On the other hand, adsorption/desorption will likely be the key process controlling inorganic contaminant migration in areas where the naturally present constituents are already in equilibrium and only the anthropogenic constituents (contaminants) are out of equilibrium, such as in areas far from the point source.

The concentration of chemical contaminants in an ecosystem is controlled primarily by (1) the amount of contaminant present at the source, (2) rate of release from the source, (3) hydrologic factors such as dispersion and dilution in the case of water systems; and (4) several geochemical processes including

Reaction Mechanisms in Environmental Engineering. https://doi.org/10.1016/B978-0-12-804422-3.00005-5

TABLE 5.1 Comparison of Adsorption and Absorption

	Adsorption	Absorption
Definition	Accumulation of the molecular species at the surface rather than in the bulk of the solid or liquid	Assimilation of molecular species throughout the bulk of the solid or liquid
Characteristic	A surface phenomenon	A bulk phenomenon
Reaction type	Exothermic process	Endothermic process
Temperature	Unaffected by temperature	Not affected by temperature
Reaction rate	Increases to equilibrium	Occurs at a uniform rate
Concentration	Different at surface and bulk	Same throughout

aqueous geochemical processes, adsorption/desorption, precipitation, and diffusion. The precipitation reaction of dissolved species is a special case of complexation reaction in which the complex formed by two or more aqueous species is a solid. Precipitation is particularly important to the behavior of heavy metals (e.g., nickel and lead) in soil/groundwater systems.

To accurately predict contaminant transport through the subsurface, it is essential that the important geochemical processes affecting contaminant transport be identified and, perhaps more importantly, accurately described in a mathematically defensible manner. Dissolution/precipitation and adsorption/desorption are usually the most important processes affecting contaminant interaction with soils. In addition, the structure (physical and electronic) of the chemical contaminants plays a role in both phenomena as well as such properties as water solubility and (in the case of mixtures) the composition, the latter of which is particularly important. Sorption (adsorption and desorption) reactions, involving minerals and chemicals are an important control on the transport and fate of many chemicals in the environment and understanding the controls is essential for a number of applications such as the transport of contaminants and bioavailability. Evaluation of adsorption or absorption can be obtained through either laboratory measurements or by use of several property correlations. However, any deductions from laboratory measurement must also consider the potential for transformation of the chemicals (or chemical mixtures) in the environment, as well as degradation of these chemicals.

Adsorption (Chapter 10) occurs when a chemical attaches to particles in the soil or in a geologic formation, thus immobilizing the chemical and limiting the availability for further transport. The most common host materials for adsorption are silicate minerals, aluminosilicate minerals, carbonate minerals, and especially clay minerals. In fact, due to the unique polarity, pore-size distribution, and high surface areas, pillared and delaminated clays are potentially useful materials for the adsorption and removal of discharged chemicals from the environment (Zielke and Pinnavaia, 1988).

Adsorption is typically considered to be a physical phenomenon rather than a chemical phenomenon, but as chemical contaminants migrate through the subsurface they encounter a range of geologic surfaces that may retard the mobility of the chemicals through a variety of chemical reactions. Adsorption reactions are one type of interaction to consider when examining the migration of contaminants through the subsurface. The mobility of chemical contaminants in the subsurface can be highly influenced by adsorption onto geological surfaces (Stipp et al., 2002; Stewart et al., 2003; Garelick et al., 2005).

The adsorption properties of silica minerals result from their significant porosity and the presence of surface hydroxyl groups. These groups are created during the natural formation of opal, and as a result of chemical reactions between the mineral surface and substances present in the environment. Diatomaceous earth and diatomite minerals are formed from the exoskeletons of unicellular algae known as diatoms, which collected at the bottom of water bodies over many millions of years. The exoskeletons of these microorganisms form a unique structure which provides the mineral with a significant contribution of free spaces (between the exoskeletons), as well as mesoporosity and macroporosity. Among the silica adsorbents, diatomite minerals exhibit a great diversity in terms of sorption capacities which may arise from their different particle sizes or from any thermal treatment. In addition, diatomite minerals may have a different mineral composition depending on the geological occurrence, which directly influences their sorption properties. Diatomaceous minerals are widely used as sorbents of petroleum chemicals that are in the form of liquids and vapors including benzene, toluene, ethylbenzene, and xylene (BTEX).

Calcite, $CaCO_3$, is one of the most common minerals on the surface of the earth, comprising about 4% by weight of the crust of the earth crust and is formed in many different geological environments. Limestone, a sedimentary rock, becomes marble from the heat and pressure of metamorphic events. Calcite is even a major component in the igneous rock called carbonatite. Dolomite, $CaCO_3 \cdot MgCO_3$, is a common sedimentary rock—forming mineral that can be found in massive beds several hundred feet thick. They are found all over the world and are quite common in sedimentary rock sequences. All dolomite rock was initially deposited as calcite/aragonite rich limestone, but during a process called diagenesis the calcite and/or aragonite is altered to dolomite.

Of the aluminosilicate mineral adsorbents, other than the complex aluminosilicate silica, bauxite is a naturally occurring, heterogeneous material composed primarily of one or more aluminum hydroxide minerals, plus various mixtures of silica, iron oxide, titania, aluminosilicate, and other impurities in minor or trace amounts. The principal aluminum hydroxide minerals found in

TABLE 5.2 The Various Clay Mineral Groups

Group	Layer Type	Layer Charge (x)	Types of Chemical Formula
Kaolinite	1:1	<0.01	$[Si_4]Al_4O_{10}(OH)_8.nH_2O$ (n = 0 or 4)
Illite	2:1	1.4–2.0	$M_x[Si_{6.8}Al_{1.2}]Al_3Fe.025Mg_{0.7}5O20(OH)_4$
Vermiculite	2:1	1.2–1.8	$M_x[Si_7Al]AlFe.05Mg0.5O_20(OH)_4$
Smectite	2:1	0.5–1.2	$M_x[Si_8]Al_{3.2}Fe_{0.2}Mg_{0.6}O_20(OH)_4$
Chlorite	2:1:1	Variable	$(Al(OH)_{2.55})4[Si_{6.8}Al0_{1.2}]Al_{3.4}Mg_{0.6})20(OH)_4$

varying proportions with bauxite are gibbsite and the polymorphs boehmite and diaspore. The bauxite mineral group is typically classified according to the intended commercial application: abrasive, cement, chemical, metallurgical, and refractory material. On the other hand, the aluminosilicate minerals are composed of aluminum, silicon, and oxygen, plus countercations.

Clay minerals (Table 5.2) are common weathering products and products of low-temperature hydrothermal alteration processes. Aluminosilicates are major components of clay minerals. Clay minerals are very common in soils, in fine-grained sedimentary rocks such as shale and mudstone and siltstone. Various types of clay minerals are abundant throughout the world and, in the current context, in the southern half of the United States. The color, texture, density, opacity, refractive index, and the slaking and swelling properties of adsorbent clays all have a bearing on their utility. Refractive indices vary with water content and with the removal of bases by acid and interplanar distances vary stepwise with moisture variation. Generally, clay minerals are used in the laboratory for the chromatographic separation of crude oil and crude oil products by the selective adsorption of the darker higher molecular weight more basic and less saturated constituents (Speight, 2014). Water is associated with adsorbent clays in at least four different ways, which are evaluated by vapor pressure and thermal tests.

In composition the adsorbent clays have a silica-alumina ratio ranging from 1:2 to 1:8, but the chemical composition bears no apparent relation to activity. Base-exchange properties are pronounced. All bases are soluble in strong acids, leaving isotropic silica. Bentonite clay minerals attain maximum sorption efficiency on removal of about half their bases by leaching in strong acid. In dilute acid the various types of adsorbent clays lose bases in ways characteristic of the type. It is worthy of note for laboratory use that the time of storage can affect the sorption capacity of some clay minerals. The principal use of adsorbent clays is in decolorizing oils. Special uses are in insecticides and fungicides, fertilizers, water softeners, adsorbent carbon, ceramic materials, drilling muds, molding sands, cements, and as catalysts in cracking and reforming oil products.

Sorption is most effective for a chemical that is present in low concentration and which do not overwhelm the potential sites for sorption. With sorption, the contaminants are mobilized, but not destroyed and, while sorption is always reversible, it is conventionally termed irreversible if the sorbent (such as soil) encloses (traps) the chemical. When this happens, the sorbed chemical (the sorbate) is less likely to be affected by changes in the environment that might desorb or dissolve the chemical, such as changes in acidity (or alkalinity) or exposure of the chemical to an additional mutually reactive chemical.

Briefly, an irreversible reaction is one in which the reactant(s) proceed to product(s), but there is no significant backward reaction. In a generalized form, irreversible reactions can be represented as:

$$nA + mB \rightarrow \text{Products}$$

In this reaction, the products do not recombine or change to form reactants in any appreciable amount. An example of an irreversible reaction is hydrogen and oxygen combining to form water in a combustion reaction. A reversible reaction is a reaction in which the products can revert to the starting materials (A and B). Thus:

$$nA + mB \leftrightarrow \text{Products}$$

As part of the sorption process, adhesion must also be recognized and considered. Adhesion is the tendency of similar and/or dissimilar particles and/or surfaces to cling to one another—on the other hand, cohesion is the tendency of similar or identical particles/surfaces to cling to one another. The forces that cause adhesion and cohesion can be divided into several types: (1) chemical adhesion, (2) dispersive adhesion, and (3) diffusive adhesion.

There are times when the sorption (or adhesion) of a chemical can be enhanced through a chemical transformation reaction or chemical stabilization. That is, before the metals attach to soil particles, they form solid compounds with chemicals, such as oxygen and sulfur that are found in the soil. This has two advantages: (1) the chemical will not travel as easily with the flow of groundwater, and (2) the new chemical formed by reaction may be more likely to adhere to the soil, especially the clay minerals in the soil. However, over a period of time that is dependent upon the properties of the chemical and the environment into which it is released, changes in acidity, or the concentration of charged particles, and reactive chemicals in the soil can destabilize the sorptive capacity of the soil and the ability of the chemical to be sorbed.

Furthermore, clay minerals are very important industrial minerals; they have been in use as raw materials for hundreds of industrial applications due to their abundant availability and inexpensive nature (Nutting, 1943; Srinivasan, 2011).

Clay minerals are usually classified according to their structure and layer type. The classification of Grim becomes the basis for outlining the nomenclature and the differences between the various clay minerals. A simple classification of clay minerals is also available in the literature; in this classification, clay minerals are divided into four main groups: (1) kaolinite group, (2) illite group, (3) smectite group, and (4) vermiculite. Because of the sorbent properties, clay minerals play an important role in the environment protection and have been used in the disposal and storage of hazardous chemicals, as well as for remediation of polluted water. The use of clay minerals as for the adsorption of various hazardous substances (heavy metals, dyes, antibiotics, biocide chemicals, and other chemicals) has been widely acknowledged.

Clay minerals belong to the family of phyllosilicate or sheet silicate family of minerals, which are distinguished by layered structures composed of polymeric sheets of SiO_4 tetrahedra linked into sheets of $(Al, Mg, Fe)(O,OH)_6$ octahedra. Clay minerals are layer-type aluminosilicates that are formed as products of chemical weathering of other silicate minerals at the earth's surface (Sposito et al., 1999). These minerals have a platy morphology because of the arrangement of atoms in the structure. Clay minerals are utilized in agricultural applications, in engineering and construction applications, in environmental remediation, in geology, pharmaceuticals, food processing, and many other industrial applications (Murray, 2007).

Unfortunately, sorption is more difficult to measure than the degradation of chemicals. With degradation, the byproducts of chemical reactions and of biological reactions can be measured in order to estimate the extent of any degradation. In contrast, chemicals are sorbed by displacing ions that are abundant in the soil and measuring relatively small variations in the concentration of those ions is very difficult, and the results are often less reliable.

In evaluating sorption as a natural attenuation technique, the following three factors must be considered: (1) based upon available soil data, such as the sorptive capacity of the soil for the specific contaminant and whether all of the chemical be sorbed by soil particles, (2) if sorption of the chemical can reduce the contamination to acceptable levels, it is necessary to determine the length of time it will take and whether or not any contamination reach receptors before sorption is complete, and (3) the stability of the sorbed contaminants and whether or not any changes in the soil chemistry will cause the contaminant to desorb.

Sorption, like other natural attenuation processes, occurs whether or not it has been approved as a remedy. In fact, sometimes *irreversible sorption* (*fixed sorption*) makes active removal of the contamination-now-bound-to-the-soil much more difficult. Furthermore, sorption can be subdivided into two categories: (1) adsorption and (2) absorption (Table 5.1). Chemicals interact with the environment in different ways and once a chemical is released into the environment, there are two physical effects that can influence the distribution of the chemicals: (1) sorption and (2) dilution.

2. SORPTION

Sorption occurs when a chemical is able to attach to underground particles, immobilizing and thus limiting the availability and movement of the chemical. The most common host materials for sorption are iron hydroxides, clay minerals, and carbonate minerals. Initially chromatographic techniques were used to separate chemicals based on their color as was the case with herbal pigments. With time the application of the technique has been extended considerably. It is now accepted as an extremely sensitive and effective separation method for many uses and is, in fact, one of the most useful separation and determination methods. As part of the sorption process, adhesion must also be recognized and considered. Unfortunately, sorption is more difficult to measure than the degradation of chemicals. With degradation, the byproducts of chemical reactions and biological reactions can be measured in order to estimate the extent of any degradation. In contrast, chemicals are sorbed by displacing ions that are abundant in the soil and measuring relatively small variations in the concentration of those ions is very difficult, and the results are often less reliable.

In evaluating sorption as a natural attenuation technique, the following three factors must be considered: (1) based upon available soil data, such as the sorptive capacity of the soil for the specific contaminant and whether all of the chemical can be sorbed by soil particles, (2) if sorption of the chemical can reduce the contamination to acceptable levels, it is necessary to determine the length of time it will take and whether or not any contamination will reach receptors before sorption is complete, and (3) the stability of the sorbed contaminants and whether or not any changes in the soil chemistry will cause the contaminant to desorb. Sorption, like other natural attenuation processes, occurs whether or not it has been approved as a remedy. In fact, sometimes *irreversible sorption* (*fixed sorption*) makes active removal of the contamination-now-bound-to-the-soil much more difficult.

Sorption is a mass-transfer process that results in the movement of a chemical species from one phase to another, that is, from an aqueous phase to a solid phase. At times this process also may not involve addition of a different chemical species, but instead may be aided by mixing to enhance mass transfer rates. However, it should be noted that different forms of a chemical differ in their susceptibility to volatilization or sorption.

For example, carbon dioxide (CO_2) is a volatile gas, while bicarbonate $\left(HCO_3^-\right)$ is not, just as hydrogen sulfide (H_2S) is a volatile gas, while bisulfide (HS^-) is not. The sorption characteristics of the zinc cation (Zn^{2+}) are different from the cation $ZnOH^+$. The pH affects the relative proportions of these different species, and thus by implementing pH control, the potential for volatilization or sorption can be made to vary considerably. This again illustrates the importance that pH control can have on the movement and fate of chemicals in groundwater.

2.1 Adsorption

As it pertains to a soil, a sediment, and their constituents in nature, adsorption is the passage of a chemical (typically from an aqueous phase or liquid phase) to the surface of a solid adsorbent—desorption is the reverse process. The solute may be a neutral molecule, or an ionic species, and the process can take place either in the macropores or in the micropores of the medium. The role of the solute structure and of molecular diffusion must also be considered. In fact, microbial adsorption reactions have the potential to control the mobility of contaminants in the subsurface. Understanding the reactivity of microbe surfaces is a key factor in determining the potential impact bacterial adsorption reactions may have on a system of interest.

In many ecosystems, the extent of the adsorption phenomenon is controlled by the electrostatic surface charge of the mineral phase. Most soils have net negative charges which originate from permanent and variable charges—the permanent charge results from the substitution of a lower valence cation for a higher valence cation in the mineral structure, whereas the variable charge results from the presence of surface functional groups, each of which can influence the interaction with constituents of an ecosystem (Table 5.3). The permanent charge is the dominant charge of clay minerals, such as biotite and montmorillonite. On the other hand, the permanent positive charge is essentially nonexistent in natural rock and soil systems and the variable charge is the dominant charge of aluminum, iron, and manganese oxide solids and organic matter. The magnitude and polarity of the net surface charge changes with a number of factors, including pH—as the pH increases, the surface becomes increasingly more negatively charged. The pH where the surface has a net charge of zero is referred to as the pH of zero-point-of-charge. At the pH of the majority of natural soils (pH = 5.5 to 8.3), calcite, gibbsite, and goethite, if present, would be expected to have some, albeit little, positive charge and therefore some anion sorption capacity.

In order to assess adsorption, it is necessary to obtain information about (1) the relationships at equilibrium between the amount of the chemical adsorbed

TABLE 5.3 Potential Reactions of Organic Functional Groups on the Environment

Functional Group	Interaction
Carboxylic acid, $-COOH$	Ion exchange, complexation
Alcohol, phenol, $-OH$	Hydrogen bonding, complexation
Carbonyl, $>C{=}O$	Reduction-oxidation
Hydrocarbon, $[-CH_2-]_n$	Hydrophobic

and the concentration of the bulk solution in contact with the adsorbent; this is given by isotherm curves of adsorption and desorption, (2) the energy that characterizes the equilibrium between the solid surface and the liquid phase, and (3) thermodynamic treatments of adsorption data allow their values to be obtained; and (4) the speed at which equilibrium is attained and the magnitude of energy involved.

Adsorption depends on surface energy—the surface atoms of the adsorbent are partially exposed, so they can chemically attract the adsorbate molecules. Adsorption may result from electrostatic attraction, chemisorption, or physisorption—in the process, molecules (the adsorbate) bind to the top layer of a material by any one or more several chemical forces which, typically, are weak intermolecular bonds that are responsible for molecular ordering of the adsorbate on the surface of the adsorbent and which include, for example, ionic interactions, dipolar interactions, van der Waals forces, as well as hydrogen bonds.

Overall, adsorption is the physical accumulation of material (usually a gas or liquid) on the surface of a solid adsorbent and is a *surface phenomenon* (Calvet, 1989). The process of adsorption arises due to presence of unbalanced or residual chemical forces at the surface of liquid or solid phase. These unbalanced residual forces have tendency to attract and retain the molecular species with which it comes in contact with at the surface. Typically, adsorption processes remove solutes from liquids based on their mass transfer from liquids to porous solids.

Typically, adsorption processes remove solutes from liquids based on their mass transfer from liquids to porous solids. Functional groups on mineral surfaces (a chemically reactive molecular unit bound into the structure of a solid) are the binding sites for adsorption reactions. When a discharged chemical comes into contact with the mineral, various bonding options come into play by which the chemical adheres to the mineral (Chapter 10).

Ion exchange is the exchange of dissolved ions for ions on solid media and can take place either between (1) cations and negatively charged surfaces or (2) between anions and positively charged surfaces. Ion exchange involves the movement of cations (positively charged elements such as calcium, magnesium, and sodium) and anions (negatively charged elements such as chloride, and chemicals like nitrate) through the soils. More specifically, *cation exchange* is the interchanging between a cation in the solution of water around the soil particle and another cation that is on the surface of the clay mineral.

A cation is a positively charged ion and most contaminants are cations, such as: ammonium $\left(NH_4^+\right)$ calcium (Ca^{2+}), copper (Cu^{2+}), magnesium (Mg^{2+}), manganese (Mn^{2+}), potassium (K^+), and zinc (Zn^{2+}). These cations are in the soil solution and are in dynamic equilibrium with the cations adsorbed on the surface of clay and matter. It is the cation exchange interaction of ions with clay minerals in the soil that gives soil the ability to adsorb and exchange cations with those in soil solution (water in soil pore space).

The adsorption capacity of a clay mineral can cause a mixture of pollutants to physically transform when one or more constituents of the mixture adsorb on to the clay mineral. In addition, the total amount of positive charges that the clay can absorb (the cation exchange capacity, CEC) impacts the rate of movement of the pollutants through the soil and the physical and chemical changes that can occur to the pollutants. For example, soil (i.e., clay) with a low CEC is much less likely to retain pollutants and, furthermore, the soil (i.e., the clay) is less able to retain chemicals with the potential that the chemicals are released into the groundwater.

Thus, an ion-exchange reaction is a reaction by which ions of one substance are replaced by similarly charged ions of another substance. As such, a previously sorbed ion of weaker affinity is exchanged by the soil for an ion in aqueous solution. Most metals in aqueous solution occur as charged ions and thus metal species adsorb primarily in response to electrostatic attraction. The term *ion-exchange* denotes purification, separation, and decontamination of aqueous and other ion-containing solutions with solid polymeric or mineral ion exchangers. Many organic chemical contaminants are ionic and can be removed through specialized ion exchange media. The media are often derived from organic polymers or plastic (often referred to as *resins*) and regeneration of ion exchange resins is accomplished with a chemical flush, typically a highly acidic or basic solution, brine solution, or solvent-brine solution, rendering the resin reusable.

Ion exchange for groundwater remediation is virtually always carried out by passing the water downward under pressure through a fixed bed of granular medium (either cation exchange media and anion exchange media) or spherical beads. Cations are displaced by specific cations from the solutions and ions are displaced by certain anions from the solution. Ion exchange media most often used for remediation are zeolite (both natural and synthetic) and synthetic resins (Hayman and Dupont, 2001).

The adsorption process creates a film of the *adsorbate* on the surface of the *adsorbent* and the process differs from the absorption process in which a fluid (the *absorbate*) is dissolved by a liquid or permeates into a solid (the *absorbent*), respectively. Thus, adsorption is a surface-based process while absorption involves the whole volume of the material. The term *sorption* encompasses both processes, while *desorption* is the reverse of sorption. In the environment, chemicals will collect on the surfaces of particles, such as soil or suspended sediment. Most of these particles are covered with a layer of material; thus, the adsorption results from the attraction of two materials to one another.

Generally, adsorption is not recognized as having any chemical interactions but that is not true. Adsorption can be understood by considering a simple example. In case of liquid state, a water molecule present on the surface is attracted inwards by chemical forces from the molecules of water present in the bulk liquid phase, which gives rise to surface tension. While the molecule of water present within the bulk is equally attracted from all the sides and the net force experienced by the water molecule in bulk is zero and this is due to particles at surface and particles at the bulk being in different chemical (or physical) surroundings. In case of the solid state, the residual forces arise because of unbalanced valence forces of atoms at the surface. Due to cleavage of a big crystal into smaller units, residual forces or vacancies gets generated on the surface of the solid. Occupancy of these vacancies by some other molecular species results in adsorption.

In nature, a variety of potential natural adsorbents exists in the soil—adsorption occurs in many natural, physical, biological, and chemical systems (especially in the environment) where molecules can adsorb on to minerals (such as clay) or on to charred wood that remains after a forest fire. In fact, clay minerals are particularly good adsorbents and have a high adsorption capacity for chemicals that have been released into the environment (Ismadji et al., 2015).

A natural clay mineral is not composed of one clay mineral only. Impurities such as calcite ($CaCO_3$), quartz (SiO_2), feldspar ($KAlSi_3O_8$/$NaAlSi_3O_8$/$CaAl_2Si_2O_8$), iron oxides (FeO and Fe_2O_3), and humic acids (degradation products of materials) are the most common components in addition to the pure clay mineral. Calcite, iron oxides, and humic acids can be removed by chemical treatments. Quartz and feldspar can be removed by sedimentation if the particle size is bigger than that of clay minerals but traces of quartz are often found in the purified samples. Physically, clay minerals are typically ultrafine-grained (normally considered to be less than 2 μm (<2 microns, $<2 \times 10^{-6}$ m) in size on standard particle size classifications). In the present context, clay minerals, which can be classified into various chemical groups, such as the silicate clay mineral groups (Table 5.2) are an important part of many soils thus rendering the soil capable of having a high adsorption capacity for chemicals. Generally, no two clay minerals are the same and the adsorption capacity will vary accordingly.

In terms of adsorption, the most active constituent of soils is the clay fraction. Clays not only have large surface areas, but also have mineral structures that bear negative charges that are balanced by exchangeable or charge-balancing cations. Among environmental chemical contaminants, cations of heavy elements such as cadmium (Cd^{2+}), mercury (Hg^{2+}), or lead (Pb^{2+}) can easily exchange innocuous elements such as sodium (Na^+) or calcium (Ca^{2+}). In addition, clay minerals can bind a large number of chemicals because of their large specific surface areas and van der Waals, ion–dipole, and/or dipole–dipole interactions. These interactions can result in a synergistic effect whereby a pollutant is "clathrated" by the chemical pollutant. The clays in soils are never saturated by heavy elements, but they are always hydrated. The intermolecular interactions of pollutants with clay mineral surfaces can be expected to play a crucial role in the subsequent chemical/biological transformations, transport, and retention of these contaminants. This is especially true when considering volatile chemicals for which diffusion in the vapor phase may be significant (Siantar et al., 1994).

Thus, adsorption of a chemical on to a solid adsorbent, such as a clay mineral or any other mineral, is measured by a partition coefficient, which is the ratio of the concentration of the chemical on the solid to the concentration of the chemical in the fluid (usually water) surrounding the solid:

$$K_d = C_{solid}/C_{water}$$

The concentration on the solid has units of mol/kg, and the concentration in the water is mol/L and, thus, the adsorption coefficient (K_d) has units of L/kg. Assuming a solid density of 1 kg/L, these units are often ignored. K_d will often depend on how much of the total mass of the particle is organic material. Thus, K_d can be corrected by the fraction of organic material (f_{om}) in the particles:

$$K_{om} = K_d/f_{om}$$

Adsorbed molecules are those that are resistant to washing with the same solvent medium in the case of adsorption from solutions. The washing conditions can thus modify the measurement results, particularly when the interaction energy is low. The exact nature of the bonding depends on the details of the chemical species involved, but the adsorption process is generally classified as physisorption (which is characteristic of weak van der Waals forces) or chemisorption (which is characteristic of covalent bonding). It may also occur due to electrostatic attraction (Table 5.4).

The interactions involved in the sorption of heavy metal cations and anions to the surfaces of materials such as soil are complex. Specific adsorption/surface precipitation onto various mineral phases present on composite materials explains the capacity of such materials to adsorb metals. However, differences in operational parameters make a comparison of the adsorption capacity between materials difficult to occur and are not fully defined or explained. The ease of desorption of the chemical is also an important consideration, because in the treatment of waste waters, materials are used primarily as ion exchangers, while for in situ immobilization, the metals need to be irreversibly bound to the added adsorbent.

In an industrial setting, adsorption processes are used to remove certain components from a mobile phase (i.e., a gas phase or a liquid phase) or to separate mixtures. The applications of adsorption can be production-related or abatement-related and may include the removal of water from natural gas or the removal of constituents from flue gas, such as is often witnessed in refinery processes and/or in natural gas processing operations and/or coal gas processing operations (Mokhatab et al., 2006; Speight, 2007, 2013, 2014, 2016). The most preferential adsorbents are characterized by varying size pores and, accordingly, activated carbon, zeolites, silica gel, and aluminum oxide are the most commercially important adsorbents. This enables the adsorbent to accommodate types and sizes of the various molecular species that occur in gas or liquid streams. Zeolites (molecular sieves) have a very narrow distribution of micropores and preferentially adsorb polar or polarizable materials

TABLE 5.4 Characteristics of Chemical Adsorption and Physical Adsorption

Chemical Adsorption (Chemisorption)
When the gas molecules or atoms are held to the solid surface via chemical bonds, this type of adsorption is chemical adsorption or chemisorption
Characteristics
Occurs only if there is formation of chemical bonds between the adsorbate and adsorbent Irreversible An exothermic process but the process occurs slowly at low temperature Accompanied by increase in temperature Promoted by high pressure Increases with increase in surface area Enthalpy is high because the chemical bond formation of chemisorption is high Activation energy is needed Results in an unimolecular layer
Physical Adsorption
Involves adsorption of gases on solid surface via weak van der Waals forces
Characteristics
No specificity in case of physical adsorption Every gas is adsorbed on the surface of the solid Easily liquefiable gases are strongly adsorbed physically Reversible If pressure is increased volume of gas decreases; as a result more gas is adsorbed By decreasing the pressure, gas can be removed from the solid surface Low temperature promotes physical adsorption High temperature decreases the rate of adsorption More the surface area, more is the rate of adsorption Porous substances and finely divided metals are good adsorbents Exothermic process No activation energy is needed

(e.g., water or carbon dioxide). By contrast, activated carbon has a hydrophobic character and is especially suitable for the removal of substances.

In nature, it is different—a variety of potential natural adsorbents exists in the soil—adsorption occurs in many natural, physical, biological, and chemical systems (especially in the environment) where molecules can adsorb on to minerals (such as clay) or on to charred wood that remains after a forest fire. In fact, clay minerals are particularly good adsorbents and have a high adsorption capacity for chemicals that have been released into the environment.

Ion exchange can be used to remove water hardness and toxic metals during wastewater treatment. Disinfection is the removal or inactivation of pathogenic organisms in wastewater prior to be discharged to the receiving body of water.

The treatment of an ecosystem where the discharged chemical is adsorbed on to soil or on to a sediment can be achieved by either physical separation or destruction of the chemical(s). Physical separation or treatment technologies transfer the chemical contaminants from the soil/sediment matrix to another medium (such as from soil to water or from soil to air). The physical separation process requires a capture step for the chemical contaminant, which can result in concentrating the contaminant for recovery and reuse or for more cost-effective ultimate disposal.

In summary, for a given system, the description of adsorption mechanisms is a difficult task because of the wide range of solute chemical structures and of adsorbent properties of the soil and sediment constituents. In fact, the adsorption of discharged chemicals on soils, sediments, and their constituents reveals a great variety of systems and system behavior. It is difficult to derive a general rule to describe adsorption that is applicable to the various chemical–adsorbent systems that relate to the wide range of variation of molecular properties of the discharged chemical and of adsorbent substrates. For any given system or incident, there needs to be knowledge of the structure and surface properties of amorphous mineral and organic adsorbents and of their association with clay minerals (Calvet, 1989).

2.2 Absorption

Absorption is another phenomenon that can be a beneficial or adverse influence of the environment and involves the uptake of one substance into the inner structure of another; most typically a gas into a liquid solvent. Furthermore, absorption is a physical or chemical phenomenon or a process in which atoms, molecules or ions enter some bulk phase—gas, liquid, or solid material. This is a different process from *adsorption*, since molecules undergoing absorption are taken up by the volume, not by the surface (as in the case for adsorption). A more general term is *sorption*, which covers absorption and adsorption—the former (absorption) is a condition in which something takes in another substance. In many processes important in technology, the chemical absorption is used in place of the physical process. It is possible to extract from one liquid phase to another a solute without a chemical reaction. The process of absorption means that a substance captures and transforms energy and the absorbent distributes the material it captures throughout whole and adsorbent only distributes it through the surface.

In chemical absorption (sometimes referred to in the shortened word form as *chemisorption*), the absorbed material is generally converted to a product different to the starting material. Thus, chemical absorption or *reactive absorption* involves a chemical reaction between the absorbent (the absorbing substance) and the absorbate (the absorbed substance) and may be combined with the physical absorption phenomenon. This type of absorption depends upon the stoichiometry of the reaction and the concentration of the potential reactants.

Physical absorption or nonreactive absorption occurs between two phases of matter: a liquid absorbs a gas, or a solid absorbs a liquid. When a liquid solvent absorbs a gas mixture or part of it, a mass of gas moves into the liquid. For example, water may absorb oxygen from the air. This mass transfer takes place at the interface between the liquid and the gas, at a rate depending on both the gas and the liquid. This type of absorption depends on the solubility of gases, the pressure, and the temperature. The rate and amount of absorption also depend on the surface area of the interface and its duration in time. For example, when the water is finely divided and mixed with air, as may happen in a waterfall or a strong ocean surf, the water absorbs more oxygen. When a solid absorbs a liquid mixture or part of it, a mass of liquid moves into the solid. This mass transfer takes place at the interface between the solid and the liquid, at a rate depending on both the solid and the liquid. Absorption is essentially a molecule attaching itself to a substance and will not be attracted by other molecules.

2.3 Effects of the Various Parameters on Adsorption–Adsorption

2.3.1 Sorbent

Knowledge of sorbent properties such as particle size, particle shape, particle-size distribution, crystallinity, chemical composition, surface area, porosity, and numbers and types of surface functional groups is necessary for the interpretation of the mechanistic aspects of sorption data. Furthermore, it must be recognized that the chemical and physical properties of a bulk material may be different from the chemical and physical properties of its surface of the sorbent. Also, the general rule is that the more crystalline a sorbent material, the less sorption capacity it has for chemicals. Furthermore, the adsorption of chemicals by soils and sediments generally increases with decreasing grain size because of the increasing surface area of the sorbent surface with decreasing grain size.

By way of definition, the specific surface area is the amount of reactive surface area available for adsorbing a chemical on a unit weight basis of the sorbent. Knowledge of the specific surface area is necessary for calculation of layer charges and potentials in many surface-complexation models. The specific surface area can be determined by a number of techniques including (1) gas adsorption and (2) adsorption from solution. Surface area measurements frequently are used to estimate adsorption binding site densities of minerals (White and Peterson, 1990; Chiou and Rutherford, 1993).

2.3.2 Chemical Composition

The type of the discharged chemical or the composition of the chemical mixture can influence adsorption reactions on adjacent solid surfaces. For example, the various functional groups (Chapter 3) in dissolved constituents can compete with the solid surface for interaction and adsorption. In such cases, sorbate ions compete with other sorbate ions for surface-binding sites and bind to different surface sites and alter the surface properties. In addition, the interaction of functional groups with the mineral surface can seriously hinder the potential for absorption of a chemical within the pore system of the mineral. Consequently, it is essential to perform thorough chemical analyses to characterize the system and determine the presence and concentration of sorbate solutes and potential complexing species.

2.3.3 Acidity and Alkalinity

In the case of surface-binding reactions of chemicals, metals on oxide minerals, the acidity or alkalinity of the system (measured as the pH of the system) can be a controlling variable. Typically, cation (a positively changed ion) adsorption increases with increasing pH from near zero to nearly 100% over a pH range of 1–2 units (Davis and Hayes, 1986; Smith, 1999). This pH region (adsorption edge) is related to and its placement seems to be characteristics of the chemical (the adsorbate) and, to a lesser extent, to the adsorbent, as well as to the concentrations of surface binding sites and adsorbate (Smith, 1999).

As expected, the adsorption of an anion (a negatively changed ion) is the mirror opposite of cation adsorption in that anion adsorption tends to decrease with increasing pH. Thus, for a given chemical concentration, increasing the amount of the adsorbent material will shift down the pH of the adsorption edge for cations and shift up the pH of the adsorption edge for anions. Conversely, increasing the concentration of a cationic adsorbate for a given amount of adsorbent will shift the adsorption edge to higher pH values.

However, because sorption of monovalent ions (not including H^+) is usually nonspecific, selectivity differences between monovalent ions is usually small, and monovalent-ion selectivity is generally much lower than that for multivalent ions on metal oxide minerals. Divalent transition–metal cations typically have much higher selectivity than do alkaline earth metal cations. There is also an approximate relationship between the selectivity of metal cations and the onset of hydrolysis. For alkali and alkaline earth cations, the sorption selectivity increases with the ionic radius of the ion. Thus:

Alkali metal cations:

Cesium (Cs^+) > Rubidium (Rb^+) > Potassium (K^+) > Sodium (Na^+) > Lithium (Li^+)

Alkaline earth metal cations:

Barium (Ba^{2+}) > Strontium (Sr^{2+}) > Calcium (Ca^{2+}) > Magnesium (Mg^{2+})

Also, the pH at which metal sorption becomes significant varies with the particular metal cation, the particular sorbent, the solid–solution ratio, the specific surface area of the sorbent, the total metal cation concentration, and the concentration of other competing or interacting species. Consequently, it is very difficult to generalize about metal sorption on different oxide minerals.

2.3.4 Temperature

Several factors can affect the temperature dependence of sorption reactions, such as: (1) the aqueous speciation of the sorbate solutes is temperature dependent, which will in turn affect sorption reactions, (2) the surface charge of the sorbent material is temperature dependent, which can produce the effect of allowing more positively charged surface binding sites to be present at a given pH. Generally, the specific adsorption of anions tends to increase with temperature while the adsorption of cations tends to decrease with decreasing temperature (Smith, 1999). Thus, seasonal variations in temperature have the potential to impact the adsorption reactions in ecosystems.

2.3.5 Sorption Rate

The importance of sorption reactions in controlling the concentrations of chemicals in ecosystems is partly due to the initial rapid equilibration of most sorption reactions in contrast to the typically longer periods of time required for precipitation reactions. The sorption may be a two-step process (Smith, 1999). In the first step, a chemical may sorb on to an external site and then surface sites and rapidly equilibrate with the surrounding solution. In the second step, the metal slowly diffuses to interior sites allowing this fraction to become isolated from the bulk solution. Also, it is important to recognize that sorption processes may be limited by mass transfer in natural systems.

2.3.6 Partition Coefficient

A partition coefficient is typically derived for the partition of a chemical between nonmiscible liquids, such as a solvent and water. Similarly, a partition coefficient can be derived for the partition of a chemical between a liquid and a solid adsorbent.

In the case of adsorption, the partition coefficient, K_d, is the ratio of the amount of adsorbate adsorbed per mass of adsorbent solid to the amount of the adsorbate remaining in solution per solution volume. Thus:

$$K_d = q/C$$

K_d is the partition coefficient (distribution coefficient, C is the adsorbate concentration remaining in solution and q is the amount of adsorbate adsorbed on the adsorbent solid). For derivation of the partition coefficient, it is generally assumed that available adsorption sites are in ample excess compared with C.

The K_d derived in this manner is valid only under the conditions it is measured and cannot be extrapolated to other adsorbents, adsorbates, or aqueous conditions (such as pH and electrolyte concentrations) (Reardon, 1981; Balistrieri and Murray, 1984; Tessier et al., 1989).

It is worthy of note, that K_d decreases with an increase in the amount of total suspended solids. This effect, frequently referred to as the *particle concentration effect* (PCE) or the *solids concentration effect* can have a noticeable influence on solid–solution partitioning (Honeyman and Santschi, 1988).

The PCE is an unexpected decline in K_d as suspended particulate matter increases. This anomaly has been attributed to a variety of causes, such as reaction kinetics, irreversible adsorption, incomplete desorption, variations in surface chemistry, filtration artifacts, and particle–particle interactions. However, a prominent cause of the PCE is believed to be the existence of colloidal forms of the adsorbate, which are included, in error, in the dissolved fraction when calculating K_d (Benoit and Rozan, 1999).

2.4 Ion Exchange

Ion exchange occurs in an ecosystem when an ion in a solid phase in contact with a liquid is exchanged (replaced) by another ion (SenGupta, 2017). More specifically, ion exchange is an exchange of ions between two electrolytes or between an electrolyte solution and a complex. In most cases when considering ecosystems, typical ion exchangers are clay minerals. Ion exchange media are either cation exchangers, which exchange positively charged ions (cations), or anion exchangers, which exchange negatively charged ions (anions). There are also amphoteric exchangers that are able to exchange both cations and anions simultaneously. However, the simultaneous exchange of cations and anions can be more efficiently performed in *mixed beds* (or *mixed mineral strata*) which contain a mixture of anion- and cation-exchange clay minerals or if the discharge chemical or discharged solution has to pass through several different ion exchange materials.

Competition for ion exchange sites on a clay mineral resin can greatly impact the effect of the clay minerals' efficiency in interacting with discharged chemicals. Generally, ions with higher valence, greater atomic weights, and smaller radii are said to have a greater affinity for certain types of clay minerals. Relative affinities of common ions are:

Monovalent cations:

$$\text{Silver } (Ag^{+}) > \text{Cesium } (Cs^{+}) > \text{Potassium } (K^{+}) > \text{Sodium } (Na^{+}) > \text{Lithium } (Li^{+})$$

Divalent cations:

$$\text{Barium } (Ba^{+2}) > \text{Strontium } (Sr^{+2}) > \text{Calcium } (Ca^{+2}) > \text{Magnesium } (Mg^{+2})$$

Monovalent anions:

$$\text{Iodide}(I^{-}) > \text{Nitrate}\left(NO_3^{-}\right)\text{Cyanide}(> CN)\text{Bisulfate}\left(HSO_4^{-}\right) > \text{Nitrite}\left(NO_2^{-}\right) > \text{Chloride}(Cl^{-}) > \text{Bicarbonate}\left(HCO_3^{-}\right)$$

Most clay minerals have a permanently negative charge which is compensated for by interlayer cations or accumulation of counterions. Clay minerals generally favor polyvalent cations over monovalent cations. Among the preferred monovalent cations are the larger, less hydrated cations such as cesium (Cs^+). Typically, clay minerals do not show much selectivity among multivalent cations of the same charge. Furthermore, understanding of exchange reactions is a starting point for a general understanding of soil sorption phenomena.

2.5 Desorption

The desorption process is an extremely important aspect of the adsorption–absorption process, for understanding the mobility, bioavailability, and fate of chemicals in the environment. When the chemicals desorb from the mineral, the mineral can act as a constant source of chemicals to the surrounding solution. In fact, the sorbed chemicals are not always readily or completely desorbed since some sorption reactions are partially irreversible as is often the case when nitrogen-containing (or basis chemicals) chemicals are adsorbed on to clay minerals. In such cases, the desorption process may require drastic (1) destruction of the adsorbate, (2) destruction of the adsorbent, or (3) both.

The reversibility of chemical adsorption can be influenced by the transformation of the adsorbent phases to a thermodynamically more stable phase. During this process, some of the adsorbed chemical may be incorporated into the structure and other chemicals may be excluded from the structure (Smith, 1999).

2.6 Mechanism

2.6.1 Adsorption

During the migration of chemical contaminants through the subsurface, they encounter a range of geologic surfaces that may retard the mobility of the contaminants through a variety of chemical reactions. The mobility of chemical contaminants in the subsurface can be highly influenced by adsorption on to geological surfaces. In order to monitor such migration, the approach of using the K_d (distribution coefficient) for describing adsorption of chemical contaminants introduced into the environment is appropriate.

This technique is most useful when adsorption behavior is relatively simple, that is, there is no pH dependence, or when the surface properties of the sorbent are too heterogeneous to be described with balanced chemical equations. The distribution coefficient approach can typically describe adsorption accurately, but the predictive properties provided by the method for adsorption in systems not directly studied in the laboratory could be weakened due to its lack of grounding in thermodynamics.

More generally, there are two types of adsorption mechanism: (1) physical adsorption, also known as physisorption and (2) chemical adsorption, also

TABLE 5.5 Factors Influencing the Mechanism of Adsorption

	Chemisorption	Physisorption
Material specificity	Substantial variation between materials	Slight dependence upon substrate composition
Crystallographic specificity	Marked variation between crystal planes	Virtually independent of surface atomic geometry
Temperature range	Virtually unlimited	Near or below the condensation point of the gas
Adsorption enthalpy	Related to the chemical bond strength; in the order of 40–800 kJ/mol	Related to factors such as molecular mass and polarity; in the order of 5–40 kJ/mol
Nature of adsorption	Often dissociative; may be irreversible	Nondissociative; reversible
Saturation uptake	Limited to a monolayer	Multilayer uptake possible
Kinetics of adsorption	Variable; an activated process	Rapid; a nonactivated process

known as chemisorption and there are several factors that influence the mechanism of adsorption (Table 5.5). The former (physical adsorption) involves the participation of weak Van der Waals forces and there is no significant redistribution of electron density in either the molecule or at the substrate surface. The latter (chemical adsorption) involves the formation of a chemical bond between the adsorbate and substrate in which there is a substantial rearrangement of the electron density. The nature of this bond may lie anywhere between the extremes of being completely ionic in character or completely covalent in character.

The chemical structure as well as the polarity of the adsorbed chemical and the nature of adsorbent surface allow hydrogen bonds to be formed. Hydrogen bonds have been assumed to be responsible for adsorption in various systems, either directly through associations with functional surface groups or indirectly through associations with hydration water molecules of exchangeable metallic cations. There are also interactions with metallic cations.

Adsorbents in nature (soils, sediment, clay minerals) contain various cations as exchangeable ions or as constitutive units of crystalline and amorphous minerals; so there are many opportunities for adsorbate–adsorbent interactions. Two types of interactions are possible: (1) dipole interactions, likely to occur with sodium (Na^{+}), potassium (K^{+}), calcium (Ca^{2+}), or magnesium (Mg^{2+}); and (2) coordination bonds with transition metallic cations.

Charge transfer is also a possible adsorption mechanism. Formation of a donor—acceptor complex between an electron donor molecule and an electron acceptor involves partial overlap of the respective molecular orbitals that allow electron exchanges. Chemical characteristics of both organic solutes and soil constituents may explain such interactions in the adsorption process.

London—van der Waals Dispersion Forces, particularly for hydrophobic compounds, also play a role in the adsorption of discharged chemicals on to natural adsorbents.

Several factors affect chemical adsorption on soils and sediments: (1) molecular properties of adsorbed compounds, (2) properties of adsorbents, and (3) liquid-phase characteristics (composition and water content of adsorbing medium). Adsorption also depends on temperature, as shown by thermodynamic and kinetic descriptions. Increases in temperature generally cause adsorbed amounts to decrease and adsorption and desorption rates to increase. Furthermore, since the mobility of adsorbed molecules is very low, the transport of chemicals in soils and sediments only occurs in the fluid phase. Thus, the mass transport of chemicals decreases as adsorption increases. In water-containing soils, the relative contribution of the transport in the gas phase and the transport in the liquid phase depends on the partitioning of the compounds between the two phases.

Thus, chemical adsorption is specific as compared to physical adsorption and occurs only if there is formation of chemical bonds between the adsorbate and adsorbent. Chemical adsorption is generally irreversible, and it is an exothermic process that occurs slowly at low temperature. Due to the formation of a chemical bond, the enthalpy of chemisorption is high and activation energy is needed. The result is an unimolecular layer on the surface of the adsorbent.

The rate of adsorption, R_{ads}, of a molecule onto a surface can be expressed in the same manner as any kinetic process. If the rate constant is expressed in an Arrhenius form, the kinetic equation is:

$$R_{ads} = A\ \exp\ (-E_a/RT) \cdot P^x$$

E_a is the *activation energy for adsorption*, and A the *preexponential (frequency) factor*, x is the kinetic order, k' is the rate constant, and P is the partial pressure.

It may be much more informative, however, to consider the factors controlling this process at the molecular level and the rate of adsorption is governed by (1) the rate of arrival of molecules at the surface and (2) the proportion of incident molecules which undergo adsorption.

2.6.2 Absorption

For absorption, the mechanism is somewhat different to the mechanism of adsorption. Assuming that the absorption process is a physical process and is not accompanied by any other physical or chemical process, it usually follows the Nernst distribution law in which the ratio of concentrations of some solute species in two bulk phases, when it is in equilibrium and in contact, is constant for a given solute and bulk phases. Thus:

$$X_1/X_2 = \text{constant} = K_{N(x,\ 12)}$$

The value of constant K_N (the partition coefficient) depends on the temperature and the equation is valid if concentrations are not too large and if the species X does not change its form in any of the two phases, 1 or 2.

If such a molecule undergoes association or dissociation, this equation still describes the equilibrium between X in both phases, but only for the same form and the concentrations of all remaining forms must be calculated by taking into account all the other equilibria. In the case of gas absorption, the concentration can be calculated by using the Ideal gas law ($c = p/RT$), or partial pressures can be used instead of concentrations.

In summary, mineral-based and carbon-based sorption and stabilization techniques vary in their effectiveness according to site conditions and the type of the target molecule(s). An example of a site condition that can affect sorption is high organic matter in soil, which can foul carbon sorbents with competing compounds. The type of target molecule affects sorption in that many discharged chemicals often occur as mixtures and include constituents with varying sorption characteristics. Clay minerals are often used because they are (1) environmentally benign, (2) have a high sorption capacity, and (3) can be easily modified to enhance their sorption capacity with mesopores.

3. DILUTION

Dilution is the process in which a chemical in an ecosystem becomes less concentrated and there is a decrease in the concentration of a solute in solution, usually simply by mixing with more solvent (such as water). Dilution is also a reduction in the acidity or alkalinity of a chemical (gas, vapor, solution). To dilute a solution means to add more solvent without the addition of more solute. The resulting solution is thoroughly mixed to ensure that all parts of the solution are identical. Mathematically dilution can be represented by a simple equation:

$$C_1 \times V_1 = C_2 \times V_2$$

In this equation, C_1 is the initial concentration of the solute, V_1 is the initial volume of the solution, C_2 is the final concentration of the solute, and V_1 is the final volume of the solution. However, this type of simple equation does not consider any solute–solvent interactions, solute–solute interactions, and solvent–solvent interactions.

Dilution of the chemical may reduce the risk to the floral and faunal species because the potential individual receptors are likely to be exposed to lower, less toxic concentrations of the hazard. However, the chronic effects associated with the dilution of a toxic chemical (or any chemical for that matter) to a less concentrated form are extremely difficult to measure. By itself, however, dilution does not reduce the mass of the chemical but, rather, spreads the area of potential exposure to the chemical. In addition, some contaminants are believed to be hazardous (such as the always lethal potassium cyanide, KCN) even at levels that may be too dilute to be detected with standard field characterization equipment and techniques.

Dilution is one of the main processes for reducing the concentration of substances away from the discharge point. Thus, dilution is more important for reducing the concentration of conservative substances (those that do not undergo rapid degradation, e.g., metals) than for nonconservative substances (those that do undergo rapid degradation). For example, when a chemical (or a mixture of chemicals, such as crude oil) is introduced into the aqueous environment, the chemical(s) is subject to a number of physical processes which result in the dilution in the receiving water. The process of dilution can be separated into (1) initial dilution and (2) secondary mixing.

3.1 Dilution Capacity

The dilution capacity of the receiving water can be defined as the effective volume of receiving water available for the dilution of the effluent. The effective volume can vary according to tidal cycles and transient physical phenomena such as stratification. In estuaries, in particular, the effective volume is much greater at high spring tides than at low neap tides. It is important to consider concentrations of substances in worst-case scenarios (usually low neap tides except, for example, when pollutants might be carried further into a sensitive location by spring tides) when calculating appropriate discharge consent conditions. Stratification can reduce the effective volume of the receiving water by reducing vertical mixing and constraining the effluent to either the upper or lower layers of the water column.

For the discharge of a chemical from a pipe, the effluent is principally fresh water, containing a mixture of pollutants and the discharge point is generally located below mean low water springs (MLWS) such that the effluent is released under seawater. Initial dilution occurs as the buoyant discharge rises to the surface because of the density differential between the saline receiving water and the fresh water effluent. Under certain circumstances of stratification or where the effluent comprises seawater (in a cooling water discharge, for instance), the effluent may not rise to the surface but may be trapped in the lower layers of the water column. For example, the design of sewage outfalls including the use of diffusers can maximize the initial dilution by entraining as much receiving water into the effluent as possible.

For the many buoyant discharges, the effluent rises to the surface where it can form a "boil." The plume then forms and spreads and secondary mixing takes place. Eventually, the plume disperses both vertically and horizontally in the water column as the density differential becomes inconsequential and the concentration of pollutants in the water column approaches uniformity. Further dilution occurs as a result of the action of tide, wind, and wave driven currents.

3.2 Stratification

Stratification occurs as a result of a density differential between two water layers and can arise as a result of the differences in salinity, temperature, or a combination of both. Stratification is more likely when the mixing forces of wind and wave action are minimal and this occurs more often in the summer months.

In estuaries, the prime reason for stratification is restricted mixing of freshwater river flows and saltwater tidal incursions. In summer, the effects of temperature can reinforce the salinity stratification. In some estuaries, this can divide the water into two distinct layers which do not mix and are kept separate by a sharp change in density. In stratified estuaries, the lower, saline water is more rapidly replaced, since less dense freshwater inputs (at lower volumes during summer) do not mix but float above the pycnocline. The saline water is more strongly mixed with seawater by tidal flow and so most exchange with the open coast occurs through the bottom layer. In such estuaries, discharge plumes usually rise to the surface and dilution/dispersion occurs primarily in the surface waters which, during stratification, have an increased retention time. This provides an argument for using summer "upper layer" retention times to assess environmental impact, rather than whole estuary or annual mean retention times.

3.3 Mixing Zone

The concept of the mixing zone was developed to allow a sound basis for the derivation of discharge consents which can be readily related to enforceable end of pipe effluent concentrations and outfall design criteria. A mixing zone is an area of sea surface surrounding a surface boil. It comprises an early part of the secondary mixing process and is prescribed to ensure that no significant environmental damage occurs outside its boundaries. The relation of the mixing zone to the location of dilution European marine site features will be a key consideration for determining the acceptability of dilution criteria.

Dilution within the mixing zone consists of initial dilution (the dilution received as a plume rises from the discharge point to the water surface) and secondary dilution (a slower rate of dilution, occurring between the surface "boil" and the edge of the mixing zone). To ensure that the integrity of a European marine site is not affected, the minimum size of a discharge to be consented (in terms of flow or load) should be assessed on a site-specific basis. This will depend on the substances and/or physicochemical parameters associated with the discharge, together with the positioning of the discharge in relation to the biotopes(s) or species for which it was designated. The initial dilution of discharges also needs to be considered. For example, an initial dilution of 50 times may be considered appropriate for secondary treated sewage effluent and low toxicity industrial effluents, but highly toxic industrial chemical effluents may require a minimum initial dilution of 100 times.

3.4 Flushing Time

The concept of flushing time (retention or residence time) is important in the consideration of the impact of polluting substances because it is one of the factors determining the exposure period to marine organisms. In relation to estuaries, flushing time is defined as the time required to replace the existing freshwater accumulated in the estuary by the river discharge. Flushing time is influenced primarily by fresh water flow from the river. Estuaries with shorter flushing times are better able to accept effluents because mixing of fresh and saltwater is greater and dilution of polluting substances is better. Also, the flushing times can be influenced by stratification.

The importance of flushing time for nontoxic substances is that shorter flushing times reduce the risk of primary or secondary effects occurring. For example, the risk of deoxygenation of the water column is reduced if the oxygen depleting substances are better mixed in the water column and are washed out of an estuary quickly. Similarly, the risk of the secondary effects of eutrophication are reduced if the phytoplankton are rapidly removed from an estuary and do not die off causing oxygen depletion. For toxic substances, shorter flushing times result in reduced contamination of the water column and sediments and reduced exposure periods for organisms in the waterbody.

When setting consent conditions, it is important that the longest estimated flushing time is used (i.e., in an estuary, during minimum river flow and neap tides) in calculations to maximize the potential for protecting organisms in a European marine site. In considering the effect of flushing time within a site, it is important that the effect is considered over the entire site, for example, including its effect on residence time within offshore/coastal parts of the site (where relevant) and not considering flushing time of the estuary alone.

3.5 Sediment Type

Sediment is present in the marine environment as both suspended and deposited particles and comprises both inorganic and organic components. The extent to which particles are suspended or deposited is a function of their density and the hydrodynamics of the water column.

Suspended sediment influences the effects of some nontoxic substances such as nutrients by reducing light penetration and preventing uptake by primary producers. This allows nutrient concentrations to increase while eutrophication effects are not apparent. Some toxic substances have an affinity for sediments, in particular, organic particles, and the presence of suspended sediments can reduce water column concentrations here, but the toxic substance may still be biologically available, for example, to sediment feeders/dwellers.

Sediment type is critical, not only because of its role in determining the faunal community, but also because of its role in adsorbing pollutants. In general terms, fine mud/silt/clay sediments with high chemical content retain more contaminants than relatively coarse sandy sediments. The smaller the particle size, the more likely sediments are to retain contaminants which may subsequently be released back into the water column.

3.6 Biodilution

Biodilution, sometimes referred to as bloom dilution, is the decrease in concentration of a chemical with an increase in the trophic level (Pickhardt et al., 2002). This effect is primarily observed during algal blooms whereby an increase in algal biomass reduces the concentration of pollutants in organisms higher up in the food chain, like zooplankton or daphnia (Pickhardt et al., 2002).

The primary elements and pollutants of concern are heavy metals such as mercury, cadmium, and lead. These toxins have been shown to bioaccumulate in a food chain. In some cases, metals, such as mercury, can biomagnify (Campbell et al., 2005). This is a major concern since methylmercury, the most toxic mercury species, can be found in high concentrations in human-consumed fish and other aquatic organisms. Persistent pollutants, such as carcinogenic polycyclic aromatic hydrocarbons and alkylphenols, have also been shown to biodilute in the marine environment.

While many chemicals (especially the heavy metals) bioaccumulate, under certain conditions, heavy metals and other chemicals have the potential to biodilute, making a higher organism less exposed to the chemical. By way of explanation, a heavy metal may be defined on the basis of density, atomic number, or behavior. Density criteria range from above 3.5 g/cm^3 to above 7 g/cm^3. Using atomic number as the basis for the definition can range from greater than sodium (atomic weight 22.98) to an atomic number higher than 40 or even higher than 200. Criteria based on chemical behavior or periodic table position have been used or suggested.

4. SOLUBILITY

The solubility of a chemical is the property of a solid, liquid, or gaseous chemical (the solute) to dissolve in a solid, liquid, or gaseous solvent. The phenomenon is also called dissolution, which affects the transportation or mobility of a chemical in the environment, especially in the aquasphere. The solubility of a chemical depends on the chemical and physical properties of the solute and the solvent, as well as on temperature, pressure, and the pH of the solution. The extent of the solubility of a substance in a specific solvent is measured as the saturation concentration, where adding more solute does not increase the concentration of the solution and begins to precipitate the excess amount of solute.

At the molecular level, solubility is controlled by the energy balance of the intermolecular forces between solute–solute, solvent–solvent, and solute–solvent molecules. But without getting too far into a study of such forces, the simple, very useful and practical empirical rule that is quite reliable is *like dissolves like* which is based on the polarity of the chemical systems insofar as polar chemicals typically dissolve in polar solvents (such as water, alcohols) and nonpolar chemicals typically dissolve in nonpolar solvents (such as nonpolar hydrocarbon solvents). The polarity of molecules is determined by the presence of polar bonds due to the presence of electronegative atoms (such as nitrogen and oxygen) in polar functional groups such as amine derivatives (RNH_2) and alcohol derivatives (ROH) (Chapter 3). Furthermore, since the polarity of a chemical is related to the presence of a functional group, the solubility characteristics of an unknown contaminant can provide evidence (as well as evidence from other analytical techniques) for the presence (or absence) of a functional group.

In the case of organic chemicals, relative to inorganic chemicals (Speight, 2017a,b), most molecules are relatively nonpolar and are usually soluble in solvents (such as diethyl ether, dichloromethane, chloroform, and hexane) but not in polar solvents (such as water). If the chemical is soluble in water, this denotes that the chemical is polar and therefore soluble in water which denotes a high ratio of polar group(s) to the nonpolar hydrocarbon chain, such as a low molecular–weight chemical that contains a hydroxy function (–OH) or an amino function ($-NH_2$) or a carboxylic acid function ($-CO_2H$) group, or a higher molecular–weight chemical that contains two or more functional (polar) groups. The presence of an acidic carboxylic acid or basic amino group in a water-soluble compound can be detected by measurement of the pH of the solution—a low pH ($pH = <7$) indicates the present of an acidic function while a high pH ($pH = >7$) indicates the presence of a basic function. Thus, chemicals that are insoluble in water can become soluble in an aqueous environment if they form an ionic species when treated with an acid or a base—the ionic form (for example, the sedum salt of a carbocyclic acid, $RCO_2^-\ Na^+$) is much more polar than the carboxylic acid.

However, *solubility* is not to be confused with the ability to dissolve a chemical because the solution might also occur because of a chemical reaction. For example, zinc dissolves in hydrochloric acid as a result of a chemical reaction releasing hydrogen gas in a displacement reaction:

$$Zn + 2HCl \rightarrow ZnCl_2 + H_2$$

The zinc ions (Zn^{2+}) are soluble in the acid.

The solubility of a substance is an entirely different property from the rate of solution, which is how fast it dissolves. The smaller a particle is, the faster it dissolves although there are many factors to add to this generalization. In fact, solubility applies to all areas of chemistry, geochemistry, inorganic chemistry, physical chemistry, organic chemistry, biochemistry, and environmental chemistry. In all cases, the solubility of the chemical will depend on the physical conditions (temperature, pressure, and concentration) and the enthalpy and entropy directly relating to the solvents and solutes concerned. By far the most common solvent in chemistry is water (especially in the current context) which is a suitable solvent for most ionic compounds, as well as a wide range of organic substances. This is a crucial factor in acidity/alkalinity as it relates to the environment.

Dissolution kinetics (rate of solubility) and equilibrium solubility (amount of dissolved material) of a chemical will influence its environmental fate and toxicity. Although solubility is material-dependent, the solubility of a solid material is not an inherent property as such but also depends on the media composition (e.g., ionic strength, ligands, pH, and temperature). For many chemicals additional parameters play a role in the dissolution process, including particle size, state of aggregation, particle coating, water chemistry, and presence of natural organic matter. Most of the currently available models to predict dissolution agree that the dissolution rate increases with decreasing particle diameter. Upon dissolution, the dissolved ions or molecules may form dissolved complexes with, for example, anions or organic matter (complexation) in the media or the ions may form a solid phase and sediment out (precipitation). For example, precipitation is commonly used for waste water treatment to remove heavy metals and reduce hardness. In general, the process involves addition of agent to an aqueous waste stream in a stirred reaction vessel, either batchwise or with continuous flow. Most metals can be converted to insoluble compounds by chemical reactions between the agent and the dissolved metal ions.

The opposite effect of dissolution is agglomeration and aggregation. Briefly, agglomerated particles are held together by weaker forces that do not exclude reversible processes, while aggregates are defined as clusters of particles held together by strong chemical bonds or electrostatic interactions. Either of these effects may occur as a result of attractive forces between particles, causing them to cluster together. This can happen after discharge to the environment and does not rely upon whether the chemical is in solution or in the gas phase at the time of discharge.

The direction of such processes will depend on the conditions of the surrounding media. The processes of aggregation and agglomeration significantly influence the fate and behavior of chemicals in the environment, with a dependency on particle properties (e.g., size, chemical composition, surface charge) as well as environmental conditions (e.g., mixing rates, pH, and natural organic matter). In some cases, aggregation and agglomeration may lead to changes in bioavailability of discharged chemicals and serve as a starting point for sedimentation of the chemicals in the environment.

A common example that is often treated as a physical transformation is the settling of suspended sediment particles (Chapter 10). Although settling of the particles does not actually transform the sediment into another chemical, it does remove sediment from a water course (such as a river) and depositing the particle on, say, the river bed. This process can be expressed mathematically by heterogeneous transformation equations at the river bed and, hence, is typically considered to be a transformation.

Sedimentation is linked to aggregation/agglomeration as the velocity of the settling of particles in water depends on both the viscosity and density of water, as well as particle radius and density. This implies that larger agglomerates and aggregates will settle more quickly compared to more disperse particles. This also means that agglomeration is the rate limiting factor for sedimentation of discharged chemicals in the aquatic environment. As for agglomeration and aggregation, sedimentation is considered relevant for many chemicals and as a potential key removal mechanism for chemicals from the water phase in the aquatic environment.

Sedimentation is linked to aggregation as larger particles tend to settle more rapidly, and the gravitational settling implies that larger agglomerates and aggregates will settle more quickly compared to more disperse particles. This also means that agglomeration is the rate limiting factor for sedimentation of discharged chemicals, in the aquatic environment.

Interactions with other substances (such as macromolecules, surfactants, humic acids) will take place after chemicals are released to the environment. These interactions can be described as an adsorption process on the minerals or the binding of the chemical with natural organic matter such as humic acid. The association of the discharged chemical with natural organic matter will alter the surface properties and behavior of the chemical and influence its interactions with other particles and surfaces (e.g., agglomeration) and the surrounding media (e.g., dissolution), and in turn be a determining factor for the environmental transport and fate (e.g., sedimentation) of the chemical.

The interaction with a discharged chemical has been dealt with under Section 2 Sorption which is a factor of great importance for the transport and fate of chemicals in the environment. In this chapter the deposition mechanism by which a chemical can attach to other materials (sometimes referred to as heteroaggregation) is described with a focus on soil and mineral particles. However, it can be difficult to distinguish between sorption and other retention mechanisms for some chemicals. Retention can be strongly linked to aggregation and agglomeration as larger particle sizes are more likely to become immobilized in the micropores of the soil. Under the prevailing conditions in an ecosystem, particle-to-particle agglomeration (homoagglomeration) may play a minor role and that the interaction with soil colloids (hetero-agglomeration) will have the largest impact on the overall sorption behavior.

Thus, soils with a higher fraction of clay minerals will generally show the highest retention of discharged chemicals. Also changes in pH affect the sorption process.

Particle dissolution kinetics (rate of solubility) and equilibrium solubility (amount of dissolved material) of a chemical are both influenced by media constituents and the properties of the chemical. Upon dissolution into the aqueous media, dissolved ions or molecules will interact with other media components, for example, anions such as bicarbonate (HCO_3^-), sulfate (SO_4^{2-}), or chloride (Cl^-) and organic matter, to form dissolved metal complexes or precipitates.

Precipitation, the process by which dissolved chemicals form a solid phase, is the case when metal ions, released from a discharged chemical, reassemble into a solid (insoluble) phase. The processes of dissolution and precipitation are governed by the solubility product (K_{sp}), which expresses the equilibrium between a solid material and the ion concentration in solution. However, this is not an inherent property of the material but depends also on media composition (ionic strength, ligands, pH, and temperature) and different chemicals will behave differently in terms of these processes and the solubility product for a chemical may be different from that of the bulk materials.

The rate by which dissolution takes place governs to which material form different environmental compartments will be exposed. For example, it is possible that for metal-containing discharged chemicals with a slow dissolution rate, benthic (sediment) organisms will have a higher exposure to metal ions than pelagic (water column) organisms because (some of) the chemicals will distribute preferentially into the sediment in the longer term. Also, in a toxicity test system it is important to determine the dissolution kinetics and degree of solubility in order to correctly interpret the test results.

Thus, dissolution kinetics (rate of solubility) and equilibrium solubility (amount of dissolved material) of a chemical will influence the environmental fate and toxicity of the chemical. Although solubility is material-dependent, the solubility of a solid material is not only an inherent property as such but depends also on the media composition (e.g., ionic strength, ligands, pH, and temperature). Additional parameters that can also play a role in the dissolution process, include (1) particle size, (2) state of aggregation, (3) water chemistry, and (4) the presence of natural organic matter. Although no single kinetic model exists that takes consideration of all the parameters that are expected to play a role in the dissolution of chemicals, most models agree that the dissolution rate increases with decreasing particle diameter. Upon dissolution the dissolved ions or molecules may form dissolved complexes with, for example, anions or organic matter (complexation) in the solution or the ions may form a solid phase and sediment out (precipitation) of the solution.

In order for a chemical (the solute) to be soluble in a particular solvent, three things need to be considered: (1) the intermolecular forces holding the solvent molecules together must be broken to make room for the solute, which requires energy and (2) the intermolecular forces holding the solute together must be broken, and (3) the solute and solvent must interact through whatever forces are available. The general adage that *like dissolves like* refers to the fact that if the intermolecular forces in the solvent and the solute are similar (H-bonding, dipole–dipole, or London forces), solute/solvent interactions are probably similar as well.

Generally speaking, the water solubility of a liquid or solid will increase with increasing temperature. However, there are some exceptions to this. Some chemicals such as cerium sulfate [$Ce_2(SO_4)_3$] are less soluble with increasing temperature because these salts exhibit an exothermic heat of solution—the thermodynamics of the solution is responsible (Chapter 9). Since a solution that has reached the solubility limit is in equilibrium, the various laws of equilibrium apply to that system. Since increasing the temperature will always favor the endothermic process, the dissolution (solution breakdown) process will be favored and the solubility will decrease. Gaseous solutes always have an exothermic heat of solution. Consequently, the solubility of all gases in water decrease with increasing temperature. In any remediation process that relies upon the water solubility of the target chemical, this is a particular property that must be considered carefully before application of the process.

The effect of pressure is another parameter that can or cannot affect the solubility of a chemical in water. Pressure changes above the solution do not affect the solubility limits of solids or liquids dissolved in water. However, gaseous solutes are affected and if the pressure of the gas is increased above the gaseous solution, then the solubility will be increased in a linear fashion.

On the other hand, colligative properties must also be given consideration—these are the properties of a liquid that may be altered by the presence of a solute because the magnitude of the change is due to the number of particles in the solution and not their chemical identity. Examples of properties that fall under this category are (amongst others) the vapor pressure, melting point, and boiling point. In a solution, a fraction of the molecules with energy in excess of the intermolecular force are nonvolatile solute molecules, or another way of saying this is that the number of solvent molecules with energies above the intermolecular force has been reduced. Because of this the solution has a lower vapor pressure than the pure solvent at all temperatures, and, therefore, a higher normal boiling point. The presence of solute particles interferes with the crystallization process and thus the normal melting point is lower (more energy needs to be removed to favor crystallization).

In conclusion, a chemical that is very soluble in water will tend to not accumulate in soil or biota because of its strong polar nature. This suggests that the chemical will degrade via hydrolysis which is the reaction that is favored in water. In addition, a chemical with a high solubility and a high vapor pressure generally will vaporize and be transported by air.

5. VAPOR PRESSURE

Vapor pressure (Chapter 3) controls the volatility of a chemical from soil, and along with its water solubility, determines evaporation from water. Predicting the volatility of a chemical in soil systems is important for estimating chemical partitioning in the subsurface between sorbed phase, dissolved phase, and gas phase. Vapor pressure is an important parameter in estimating vapor transport of an organic chemical in air. It is often useful to determine the vapor pressure and other properties of a chemical by assuming that it is a liquid or supercooled liquid at a temperature less than the melting point. At very low environmental concentrations such as in liquid solutions or in aerosol particles, pure chemical behavior relates to the liquid rather than the solid state.

Because changes of state depend on vapor pressure, the presence of a solute affects the freezing point and boiling point of a solvent. The normal boiling point of a liquid occurs at the temperature where the vapor pressure is equal to 1 atm. A nonvolatile solute elevates the boiling point of the solvent. The magnitude of the boiling point elevation depends on the concentration of the particles of solute. Also, when a solute is dissolved in a solvent, the freezing point of the solution is lower than that of the pure solvent.

In a solution, the fraction of the molecules with energy in excess of the intermolecular force is nonvolatile solute molecules—the number of solvent molecules with energies above the intermolecular force has been reduced. Because of this the solution has a lower vapor pressure than the pure solvent. Yet, some strong solute–solvent interactions do not behave ideally and gives a vapor pressure lower than that predicted by thermodynamics.

Soil vapor extraction removes harmful chemicals, in the form of vapors, from the soil above the water table. The vapors are extracted from the ground by applying a vacuum to pull it out.

Like soil vapor extraction, *air sparging* uses a vacuum to extract the vapors. Air sparging uses air to help remove harmful vapors like the lighter gasoline constituents (i.e., benzene, ethylbenzene, toluene, and xylene [BTEX]), because they readily transfer from the dissolved to the gaseous phase but is less applicable to diesel fuel and kerosene. When air is pumped underground, the chemicals evaporate faster, which makes them easier to remove. Methane can be used as an amendment to the sparged air to enhance cometabolism of chlorinated organics. Soil vapor extraction and air sparging are often used at the same time to clean up both soil and groundwater.

Biosparging is used to increase the biological activity in soil by increasing the oxygen supply via sparging air or oxygen into the soil. In some instances, air injections are replaced by pure oxygen to increase the degradation rates. However, in view of the high costs of this treatment in addition to the limitations in the amount of dissolved oxygen available for microorganisms, hydrogen peroxide (H_2O_2) was introduced as an alternative, and it was used on a number of sites to supply more oxygen (Schlegel, 1977) and is more efficient in enhancing microbial activity during the biodegradation of contaminated soil and ground water (Brown and Norris, 1994; Flathman et al., 1991; Lee et al., 1988; Lu, 1994; Lu and Hwang, 1992; Pardieck et al., 1992),but it can be a disadvantage if the toxicity is sufficiently high to microorganisms even at low concentrations (Brown and Norris, 1994; Scragg, 1999).

Soil vapor extraction requires drilling extraction wells within the polluted area. The necessary equipment to create a vacuum is attached to the well, which pulls air and vapors through the soil and up to the surface. Once the extraction wells pull the air and vapors out of the ground, special air pollution control equipment collects them. The equipment separates the harmful vapors from the clean air.

Air sparging works very much like soil vapor extraction. However, the wells that pump air into the ground are drilled into water-soaked soil below the water table. Air pumped into the wells disturbs the groundwater. This helps the pollutant change into vapors. The vapors rise into the drier soil above the groundwater and are pulled out of the ground by extraction wells. The harmful vapors are removed in the same way as soil vapor extraction. The air used in soil vapor extraction and air sparging also helps clean up pollution by encouraging the growth of microorganisms. In general, the wells and equipment are simple to install and maintain and can reach greater depths than other methods that involve digging up soil. Soil vapor extraction and air sparging are effective at removing many types of pollutants that can evaporate.

Air sparging should not be used if free products are present. Air sparging can create groundwater mounding which could potentially cause free product to migrate and contamination to spread. Also, it is not suitable around basements, sewers, or if other subsurface confined spaces are present at the site. Potentially dangerous constituent concentrations could accumulate in basements unless a vapor extraction system is used to control vapor migration. If the contaminated groundwater is located in a confined aquifer system, air sparging is not advisable because the injected air would be trapped by the saturated confining layer and could not escape to the unsaturated zone. Anaerobic sparging, an innovative technique in biodegradation, depends on the delivery of an inert gas (nitrogen or argon) with low ($<2\%$) levels of hydrogen. Cometabolic air sparging is the delivery of oxygen containing gas with enzyme-inducing growth substrate (such as methane or propane).

Bioventing is a technology that stimulates the natural in situ biodegradation of any aerobically degradable compounds in soil by providing oxygen to existing soil microorganisms. In contrast to soil vapor vacuum extraction, bioventing uses low airflow rates to provide only enough oxygen to sustain microbial activity. Two basic criteria have to be satisfied for successful bioventing (1) the air must be able to pass through the soil in sufficient quantities to maintain aerobic conditions, and (2) natural hydrocarbon–degrading microorganisms must be present in concentrations large enough to obtain reasonable biodegradation rates.

Bioventing is a medium to long-term technology—cleanup ranges from a few months to several years—and it is applicable to any chemical that can be aerobically biodegraded. The technique has been successfully used to remediate soils contaminated by hydrocarbons, nonchlorinated solvents, some pesticides, wood preservatives, and other organic chemicals. Though there are limitations, this technology does not require expensive equipment and relatively few personnel are involved in operation and maintenance; therefore, bioventing is receiving increased exposure in the remediation consulting community. Potential improvements on the current bioventing methods that have been taking place are the use of electrochemical oxygen gas sensors, detailed characterization of *nonaqueous phase liquid* (NAPL) distribution, and neutron probe logging. Another bioventing enhancement is the use of this technique with bioslurping and soil vapor extraction (Baker, 1999).

Bioslurping is an in situ remediation technology that combines the two remedial approaches of bioventing and vacuum-enhanced free-product recovery. It is faster than the conventional remedy of product recovery followed by bioventing. The system is made to minimize groundwater recovery and drawn down in the aquifer. Bioslurping was designed and is being tested to address contamination by hydrocarbons with a floating lighter nonaqueous phase liquids (LNAPL) layer.

Bioslurping efficiently recovers free product and extracts less groundwater for treatment, which speeds up remediation and reduces water handling and treatment costs. It enhances natural in situ biodegradation of vadose zone (which extends from the top of the ground surface to the water table) soils and may be the only feasible remediation technology at low permeability sites. But for bioslurping to even be considered at a contamination site, free product must be present, and the product must be biodegradable; also the soil must respond to bioventing.

REFERENCES

Baker, R.S., 1999. Bioventing systems: a critical review. In: Adriana, D.C., Bollag, J.M., Frankenberger, W.T., Sims, R.C. (Eds.), Bioremediation of Contaminated Soils, Agronomy Monograph No. 37. Amer. Soc. Agron., Crop Sci. Soc. Amer., Soil Sci. Soc. Amer. Madison, Wisconsin, pp. 595–630.

Balistrieri, L.S., Murray, J.W., 1984. Marine scavenging: trace metal adsorption by interfacial sediment from MANOP site H. Geochimica et Cosmochimica Acta 48, 921–929.

Benoit, G., Rozan, T.F., 1999. The influence of size distribution on the particle concentration effect and trace metal partitioning in rivers. Geochimica et Cosmochimica Acta 63 (1), 113–127.

Brown, R.A., Norris, R.D., 1994. The Evolution of a Technology: Hydrogen Peroxide in In Situ Bioremediation. In: Hinchee, R.E., Alleman, B.C., Hoeppel, R.E., Miller, R.N. (Eds.), Hydrocarbon Bioremediation. CRC Press, Boca Raton, Florida, pp. 148–162.

Calvet, R., 1989. Adsorption of organic chemicals in soils. Environmental Health Perspectives 83, 145–177.

Campbell, L.M., Norstrom, R.J., Hobson, K.A., Muir, D.C.G., Backus, S., Fisk, A.T., 2005. Mercury and other trace elements in a pelagic Arctic marine food web (Northwater Polynya, Baffin Bay). The Science of the Total Environment 351–352, 248–263.

Chiou, C.T., Rutherford, D.W., 1993. Sorption of N_2 and EGME vapors on some soils, clays, and mineral oxides and determination of sample surface areas by use of sorption data. Environmental Science and Technology 27, 1587–1594.

Davis, J.A., Hayes, K.F., 1986. Geochemical processes at mineral surfaces: an overview. In: Davis, J.A., Hayes, K.F. (Eds.), Geochemical Processes at Mineral Surfaces, ACS Symposium Series No. 323. American Chemical Society, Washington, D.C., pp. 2–18

Flathman, P.E., Carson Jr., J.H., Whitenhead, S.J., Khan, K.A., Barnes, D.M., Evans, J.S., 1991. Laboratory Evaluation of the Utilization of Hydrogen Peroxide for Enhanced Biological Treatment of Crude Oil Hydrocarbon Contaminants in Soil. In: Hinchee, R.E., Olfenbuttel, R.F. (Eds.), In Situ Bioreclamation: Applications and Investigations for Hydrocarbon and Contaminated Site Remediation. Butterworth-Heinemann, Stoneham, Massachusetts, pp. 125–142.

Garelick, H., Dybowska, A., Valsami-Jones, E., Priest, N.D., 2005. Remediation technologies for arsenic contaminated drinking waters. Journal of Soils and Sediments 5 (3), 182–190.

Hayman, M., Dupont, R.R., 2001. Groundwater and Soil Remediation: Process Design and Cost Estimating of Proven Technologies. ASCE Press, Reston, Virginia.

Honeyman, B.D., Santschi, P.H., 1988. Metals in aquatic systems. Environmental Science and Technology 22, 862–871.

Ismadji, S., Soetaredjo, F.E., Ayucitra, R.A., 2015. Clay minerals for environmental remediation. In: Green Chemistry for Sustainability. Springer Briefs, Springer, Heidelberg, Germany.

Jenne, E.A., 1968. Controls on Mn, Fe, Co, Ni, Cu, and Zn concentrations in soils and water – the significant role of hydrous Mn and Fe oxides. In: Gould, R.F. (Ed.), Traces in Water, Advances in Chemistry Series No. 73. American Chemical Society, Washington, D.C., pp. 337–387

Lee, M.D., Thomas, J.M., Borden, R.C., Bedient, P.B., Ward, C.H., 1988. Biorestoration of aquifers contaminated with organic compounds. CRC Critical Reviews in Environmental Control 18, 29–89.

Lu, C.J., Hwang, M.C., 1992. Effects of hydrogen peroxide on the in situ biodegradation of chlorinated phenols in groundwater. In: Proceedings. Water Environ. Federation 65th Annual Conference, New Orleans, Louisiana. September 20–24.

Lu, C.J., 1994. Effects of hydrogen peroxide on the in situ biodegradation of organic chemicals in a simulated groundwater system. In: Hinchee, R.E., Alleman, B.C., Hoeppel, R.E., Miller, R.N. (Eds.), Hydrocarbon Bioremediation. CRC Press, Inc., Boca Raton, Florida, pp. 140–147.

Mokhatab, S., Poe, W.A., Speight, J.G., 2006. Handbook of Natural Gas Transmission and Processing. Elsevier, Amsterdam, Netherlands.

Murray, H.H., 2007. Applied Clay Mineralogy: Occurrences, Processing and Application of Kaolins, Bentonites, Palygorskite-Sepiolite, and Common Clays. Elsevier, Amsterdam, Netherlands.

Nutting, P.G., 1943. Adsorbent Clays: Their Distribution, Properties Production, and Uses. Bulletin 928-C. United States Geological Survey, Reston, Virginia.

Pardieck, D.L., Bouwer, E.J., Stone, A.T., 1992. Hydrogen peroxide use to increase oxidant capacity for in situ bioremediation of contaminated soils and aquifers: a review. Journal of Contaminant Hydrology 9, 221–242.

Pickhardt, P.C., Folt, C.L., Chen, C.Y., Klaue, B., Blum, J.D., 2002. Algal blooms reduce the uptake of toxic methylmercury in freshwater food webs. Proceedings of the National Academy of Sciences 99, 4419–4424.

Reardon, E.J., 1981. K_d's – can they Be used to describe reversible ion sorption reactions in contaminant transport? Ground Water 19, 279–286.

Schlegel, H.G., 1977. Aeration without air: oxygen supply by hydrogen peroxide. Biotechnology and Bioengineering 19, 413.

Scragg, A., 1999. Environmental Biotechnology. Pearson Education Limited, Harlow, Essex, England.

SenGupta, A.K., 2017. Ion Exchange in Environmental Processes: Fundamentals, Applications and Sustainable Technology. John Wiley & Sons Inc., Hoboken, New Jersey.

Siantar, D.P., Feinberg, B.A., Fripiat, J.J., 1994. Interaction between clays and pollutants in the clay interlayer. Clays and Clay Minerals 42 (2), 187–196.

Smith, K.S., 1999. Metal sorption on mineral surfaces: an overview with examples relating to mineral deposits. In: The Environmental Geochemistry of Mineral Deposits – Part A: Processes, Techniques, and Health Issues and Part B: Case Studies and Research Topics. Reviews in Economic Geology, Volumes 6A and 6B. Society of Economic Geologists, Littleton, Colorado.

Stewart, M.A., Jardine, P.M., Brandt, C.C., Barnett, M.O., Fendorf, S.E., McKay, L.D., Mehlhorn, T.L., Paul, K., 2003. Effects of contaminant concentration, aging, and soil properties on the bioaccessibility of Cr(III) and Cr(VI) in soil. Soil and Sediment Contamination 12 (1), 1–21.

Stipp, S.L.S., Hansen, M., Kristensen, R., Hochella, M.F., Bennedsen, L., Dideriksen, K., Balic-Zunic, T., Leonard, D., Mathieu, H.J., 2002. Behavior of Fe-Oxides relevant to contaminant uptake in the environment. Chemical Geology 190 (1–4), 321–337.

Speight, J.G., 2007. Natural Gas: A Basic Handbook. GPC Books. Gulf Publishing Company, Houston, Texas.

Speight, J.G., 2013. The Chemistry and Technology of Coal, third ed. CRC Press, Taylor & Francis Group, Boca Raton, Florida.

Speight, J.G., 2014. The Chemistry and Technology of Petroleum, fifth ed. CRC Press, Taylor & Francis Group, Boca Raton, Florida.

Speight, J.G., 2016. Deep Shale Oil and Gas. Gulf Professional Publishing, Elsevier, Oxford, United Kingdom.

Speight, J.G., 2017a. Environmental Organic Chemistry for Engineers. Butterworth-Heinemann, Elsevier, Oxford, United Kingdom.

Speight, J.G., 2017b. Environmental Inorganic Chemistry for Engineers. Butterworth-Heinemann, Elsevier, Oxford, United Kingdom.

Sposito, G., 1984. The Surface Chemistry of Soils. Oxford University Press, Oxford, United Kingdom.

Sposito, G., 1989. The Chemistry of Soils. Oxford University Press, Oxford, United Kingdom.

Sposito, G., 1994. Chemical Equilibria and Kinetics in Soils. Oxford University Press, Oxford, United Kingdom.

Sposito, G., Skipper, N.T., Sutton, R., Park, S.H., Soper, A.K., Greathouse, J.A., 1999. Surface geochemistry of the clay minerals. Proceedings of the National Academy of Sciences 96, 3358–3364.

Srinivasan, R., 2011. Advances in application of natural clay and its composites in removal of biological chemicals and contaminants from drinking water. Advances in Materials Science and Engineering, 872531. https://doi.org/10.1155/2011/872531. https://www.hindawi.com/journals/amse/2011/872531/.

Tessier, A., Carignan, R., Dubreuil, B., Rapin, F., 1989. Partitioning of zinc between the water column and the oxic sediments in lakes. Geochimica et Cosmochimica Acta 53, 1511–1522.

White, A.F., Peterson, M.L., 1990. Role of reactive-surface-area characterization in geochemical kinetic models. In: Melchior, D.C., Bassett, R.L. (Eds.), Chemical Modeling of Aqueous Systems II, ACS Symposium Series No. 416. American Chemical Society, Washington, D.C., pp. 461–475

Zielke and Pinnavaia, 1988. Modified clays for the adsorption of environmental toxicants: binding of chlorophenols to pillared, delaminated, and hydroxy-interlayered smectites. Clays and Clay Minerals 36 (5), 403–408.

Chapter 6

Hydrolysis

1. INTRODUCTION

By way of definition, a chemical transformation is the conversion of a substrate (or reactant) to a product. In more general terms, a chemical transformation involves a (or is) chemical reaction which is characterized by a chemical change, and yields one or more products, which usually have properties substantially different from the properties of the individual reactants. Reactions often consist of a sequence of individual substeps that can be described by means of chemical equations, which symbolically present the starting materials, end products, and sometimes intermediate products and reaction conditions.

Thus, a chemical transformation involves changing a chemical (or chemicals) from a beginning chemical (the reactants or reactants) mass to a different resulting (the product or products) chemical substance. The transformation produces new chemicals which have different physical and chemical properties when compared to the physical and chemical properties of the starting material(s). Such changes are usually irreversible in environment because the newly formed chemical(s) cannot easily change back into the original chemical(s). In the laboratory, chemical changes can easily be identified with the help of change in color, odor, energy level, and physical state. Any chemical change can easily be represented with the help of chemical equations which is a simple representation of a chemical reaction and which involves the molecular or atomic formulas of reactants and products.

Except for spontaneous reactions such as radioactive decay, chemical transformations often require that two or more substances be brought together for the transformation to occur. Examples of interest in groundwater is the oxidation of both inorganic and organic chemicals, which require the presence of some oxidant, such as diatomic oxygen (O_2), nitrate, sulfate, or ferric iron (Fe^{3+}).

In biological reactions (Chapter 8), three entities generally are required, the chemical being oxidized (electron donor), the oxidant (electron acceptor), and the microorganism carrying out the transformation. At times, the required entities are already present together, and then transformation occurs, based simply on normal reaction kinetics. However, this is often not the case in groundwater remediation, and then the missing reactants must be supplied

Reaction Mechanisms in Environmental Engineering. https://doi.org/10.1016/B978-0-12-804422-3.00006-7

through some means and mixed with the substance or substances targeted for removal. The speed of the reaction is then likely to be governed primarily by the rate at which the required substances can be brought together. Natural attenuation for transformation of materials may require mixing brought about by the diffusion of oxygen into an aquifer from the vadose zone above, or from an adjacent groundwater flow stream. The process of adding and mixing needed substances for desired transformation is one of the most challenging and costly aspects of in situ remediation of contaminated groundwater and soil. This is a much more difficult process than with an aboveground reactor because of complex and often undefined hydrogeology and the general uncertainty of the exact location of the contaminants.

Some form of mixing may also be required for processes other than chemical oxidations. Included are the addition of reducing chemicals for chemical reductions; acids or bases for pH control; chemicals that promote precipitation for in-place stabilization; detergents, solvents, or other chemicals that promote solubilization of the chemical of interest for easier removal; addition of a separate phase such as air; use of thermal treatment to enhance vaporization; as well as chemical changes resulting from groundwater–surface water interactions that are driven by variability in rates of precipitation, extraction, and aquifer recharge. All such processes involve mixing in one form or another. The emphasis in this chapter is not on the mixing processes themselves, but on the chemical and biological requirements for contaminant transformation, destruction, or removal (McCarty and Criddle, 2012).

Chemical reactions occur at a characteristic rate (the reaction rate) at a given temperature and chemical concentration. Typically, reaction rates increase with increasing temperature because there is more thermal energy available to reach the activation energy necessary for breaking bonds between atoms. The general rule of thumb (see above) is that for every 10°C (18°F) increase in temperature the rate of a chemical reaction is doubled and there is no reason to doubt that this would not be the case for chemicals discharged into the environment.

Chemicals can enter the environment (air, water, and soil) when they are produced, used, or disposed and the impact on the environment is determined by the amount of the chemical that is released, the type and concentration of the chemical, and where it is found, as well as through any chemical transformation that occurs after the chemical has entered the environment, whether it is in the atmosphere, the aquasphere, or the terrestrial biosphere. Some chemicals can be harmful if released to the environment even when there is not an immediate, visible impact. Some chemicals are of concern as they can work their way into the food chain and accumulate and/or persist in the environment for prolonged periods, including years which is in direct contradiction of the earlier *conventional wisdom* (or unbridled optimism) that assumed that chemicals would either (1) degrade into harmless byproducts as a result of microbial or chemical reactions, (2) immobilize completely by binding to soil solids, or (3) volatilize to the atmosphere where dilution to harmless levels was assured. This false assurance led to years of agricultural chemical use and chemical waste disposal with no monitoring of atmosphere, or groundwater (the aquasphere), or soil (the terrestrial biosphere) near to the discharge. Thus, the volatility of a chemical is of concern predominantly for surface-located chemicals and is affected by (1) temperature of the soil, (2) the water content of the soil, (3) the adsorptive interaction of the chemical and the soil, (4) the concentration of the chemical in the soil, (5) the vapor pressure of the chemical, and (6) the solubility of the chemical in water, which is the predominant liquid in the soil.

To predict the persistency of a chemical in the environment, the physical-chemical properties of the chemical need to be known or at least estimated which lead to the reactivity of the chemical and, in the current context, the reactivity of the chemical in the environment. In fact, knowledge of the chemical properties and the physical properties is also an advantage (Rahm et al., 2005). The partitioning of a chemical (Chapter 9) can be described based on (1) the water solubility of the chemical, (2) the octanol–water partitioning coefficient of the chemical, and (3) the vapor pressure of the chemical. Furthermore, chemicals that can undergo elimination reactions are rapidly transformed, as are perhalogenated chemicals that can undergo substitution reactions. These chemicals are not likely to persist in the environment, while those that did not show any observable reactivity under similar hydrolytic conditions may persist for prolonged periods (Rahm et al., 2005).

In terms of chemical transformation, water occurs on all parts of the earth—a substantial proportion of the earth's surface is covered by the oceans, and in the terrestrial environment there are large fresh water lakes, rivers, and streams. The soil retains water because of the hydrophilic nature of clay surfaces (Chapter 1). Water is a major constituent of all cells. Any chemical discharged into the environment will, therefore, encounter water and it is necessary to consider the extent to which and the conditions under which the chemical and water will react. Such a process is usually classified as a hydrolysis reaction and the net effect is the exchange of functional groups in the chemical with the hydroxyl groups of water.

This chapter covers the possible hydrolysis reactions and the related nucleophilic substitution reactions of discharged chemicals.

2. NUCLEOPHILIC SUBSTITUTION REACTIONS

By way of definition, a *nucleophile* is an atom, ion, or molecule that has an electron pair (giving the atom, ion, or molecule an overall negative charge). The electron pair may be donated in forming a covalent bond to an electrophile (or Lewis acid). Conversely, an *electrophile* is an electron deficient atom, ion, or molecule that has an affinity for an electron pair and will bond to a base or nucleophile. In general terms, the nucleophilic character of a chemical depends on the availability of the electrons in the nucleophile. The more available the electrons, the more readily the electrons can be donated and the more nucleophilic is the system.

To study the effect and behavior of a chemical in the environment without any foreknowledge of the chemistry that is involved is unwise and can lead to serious environmental consequences. In addition, the partitioning of a chemical can be described based on water solubility (especially the octanol–water partitioning coefficient) or by its adherence to minerals, chemical or physical–chemical interactions with minerals (water–rock partitioning coefficient) (Chapters 5 and 9), as well as the vapor pressure of the chemical (which indicates the degree of volatilization into the atmosphere) (Chapter 5) (McCall et al., 1983; Cole and Mackay, 2000; Delle Site, 2001; Green and Bergman, 2005). As part of this preknowledge, the mechanisms by which a chemical can be transformed may be categorized as being hydrolysis reactions, oxidation–reduction (redox) reactions, photolysis reactions, as well as any potential biological transformations (Rahm et al., 2005).

Water occurs throughout the earth and is a necessary part of life as evidenced by the oceans, and in the terrestrial environment there are large fresh water lakes, rivers, and streams. The soil retains water because of the hydrophilic nature of clay surfaces. Water is a major constituent of all cells. Even though a small molecule like water has an unusually high boiling point, it does have appreciable vapor pressure at ambient conditions (Chapter 5) and thus the atmosphere also contains substantial amounts of water. Any compound introduced into the environment, therefore, will encounter water and the potential chemistry involves the conditions under which they might react. Such a process is usually classified as a hydrolysis reaction and the net effect is the exchange of a functional group (X) (to differentiate the group for a hydrogen atom) with the hydroxyl group (OH). Thus:

$$RX + H_2O \rightarrow ROH + HX$$

This type of reaction can involve a variety of intermediates depending on the compound and conditions but the overall reaction is generally classified as a nucleophilic reaction.

In inorganic chemistry and organic chemistry, nucleophilic substitution is a fundamental class of reactions in which an electron rich nucleophile selectively bonds with or attacks the positive or partially positive charge of an atom or a group of atoms to replace a leaving group; the positive or partially positive atom is referred to as an electrophile. The whole molecular entity of which the electrophile and the leaving group are part of is usually called the substrate (Rossi and De Rossi, 1983; March, 1992; Imyanitov, 1990; Wade, 2003; House, 2008; Strohfeldt, 2016). Thus:

$$\text{Nuc} + \text{R(LG)} \rightarrow \text{R(Nuc)} + \text{LG}$$

The electron pair (:) from the nucleophile (Nuc) attacks the substrate (R-LG) forming a new bond, while the leaving group (LG) departs with an electron pair. The principal product in this case is R-Nuc. The nucleophile may be electrically neutral or negatively charged, whereas the substrate is typically neutral or positively charged.

There are two fundamental events in a nucleophilic substitution reaction: (1) the formation of the new σ bond to the nucleophile and (2) breaking of the σ bond to the LG. Depending on the relative timing of these events, two different mechanisms are possible: (1) bond breaking to form a carbocation precedes the formation of the new bond, S_N1 reaction, and (2) simultaneous bond formation and bond breaking, S_N2 reaction. A third possibilityis where the nucleophile adds, then the LG departure cannot occur because it would require that the electrophilic carbon atoms become pentavalent. However, it is also useful to appreciate that the overall outcome of the transformation (that is, the substitution) is often the same regardless of whether it is S_N1 or S_N2, though they may be differences in stereochemistry, which can provide some evidence as to which mechanistic path is occurring (Table 6.1).

2.1 S_N1 and S_N2 Reactions

The S_N1 reaction, although common in organic chemistry, is more likely to occur in inorganic chemistry in which ionization of the starting chemical is the rate-determining step. In the second step this cation is attacked by the

TABLE 6.1 Some Parameters Relating to Nucleophilic Substitution Reaction of RX (Alkyl Halide or Equivalent Target Molecule)

Reaction type: $R{-}X + OH^- \rightarrow R{-}OH + X^-$

Factor	*S_N1*	*S_N2*
Kinetics	Rate = k[RX]	Rate = k[RX][Nucleophile]
Leaving group	Important	Important
Nucleophilicity	Unimportant	Important
Preferred solvent	Polar, protic[a]	Polar, aprotic[a]
Rearrangements	Common	Rare
Eliminations	Common with basic nucleophiles	Only with heat and basic nucleophiles

[a]A protic solvent is a solvent that has a hydrogen atom bound to an oxygen (as in a hydroxyl group) or a nitrogen (as in an amine group)—such solvents readily donate protons (H^+) to reagents. In general terms, any solvent that contains a labile H^+ is a protic solvent. Conversely, an aprotic solvent cannot donate hydrogen.

nucleophile to form the product. In the S_N1 reaction, the barrier to ready reaction is the formation of the cation or the stability of the carbocation and, thus, the rate of the reaction is proportional to the stability of the cation or the carbocation.

On the other hand, the S_N2 reaction is a type of reaction mechanism that is common in organic chemistry. The reaction type is so common that it has other names, such as *bimolecular nucleophilic substitution*, or, among chemists, *associative substitution* or *interchange mechanism*. In the S_N2 reaction, being predominantly a reaction of organic chemicals, a barrier to the reaction is steric hindrance. The reaction can only proceed if there is ready access to the carbocation. The more groups that are present around the vicinity of the LG, the slower the reaction will be.

In organic chemistry, nucleophilic substitution reactions are common occurrences and are an important class of reactions that allow the interconversion of functional groups. Of importance, are the reactions of alcohols (R−OH) and alkyl halide derivatives (R−X). In the case of alcohols, the range of substitution reactions possible can be increased by utilizing the derivative such as the tosyl derivative (p-toluene sulfonic acid derivative, R-OTs), an alternative method of converting the −OH to a group that can be more readily displaced for the starting chemical.

In the case of the alkyl halide derivatives, the functional group is a carbon−halogen bond, involving fluorine (C−F), chlorine (C−Cl), bromine (C−Br), and iodine (C−I). Except for iodine; the halogen atoms each have an electronegativity significantly greater than carbon. Consequently, this functional group is polarized so that the carbon is electrophilic, and the halogen is nucleophilic. Two characteristics other than electronegativity also have an important influence on the chemical behavior of these compounds. The first of these is the covalent bond strength. The strongest of the carbon−halogen covalent bonds is that to fluorine, which is the strongest common single bond to carbon, being roughly 30 kcal/mole stronger than a carbon−carbon bond and about 15 kcal/mole stronger than a carbon−hydrogen bond.

Thus, alkyl fluorides and fluorocarbons in general are chemically and thermodynamically quite stable, and do not share any of the reactivity patterns shown by the other alkyl halides. The carbon−chlorine covalent bond is slightly weaker than a carbon−carbon bond, and the bonds to the other halogens are weaker still, the bond to iodine being about 33% weaker. The second factor to be considered is the relative stability of the corresponding halide anions, which is likely the form in which these electronegative atoms will be replaced. This stability may be estimated from the relative acidities of the H−X acids, if the strongest acid releases the most stable conjugate base (halide anion). Except for hydrogen fluoride (HF, $pK_a = 3.2$), all the hydrohalogen acids (also called hydrohalic acids) are very strong, small differences being in the direction $HCl < HBr < HI$.

2.2 Reactive Nucleophiles

The most reactive nucleophiles are said to be more nucleophilic than less reactive members of the group. The nucleophilicity of some common nucleophilic anion reactants (Nu^-) vary as shown in the following:

$$\text{Nucleophilicity: } CH_3CO_2^- < Cl^- < Br^- < N_3^{(-)} < CH_3O^- < CN^- \approx SCN^- < I^- < CH_3S^-$$

The cumulative results of studies of this kind has led to useful empirical rules pertaining to nucleophilicity: (1) for a given element, negatively charged species are more nucleophilic (and basic) than are equivalent neutral species, (2) for a given period of the periodic table (Fig. 6.1), nucleophilicity (and basicity) decreases on moving from left to right, and (3) for a given group of the periodic table, nucleophilicity increases with increasing size, although there is a solvent dependence due to hydrogen bonding. Conversely, basicity varies in the opposite manner.

Nucleophilic substitution reactions are commonplace in chemistry and they can be broadly categorized as taking place at a saturated aliphatic carbon or at (less often) an aromatic carbon center or other unsaturated carbon center (Wade, 2003). An example of nucleophilic substitution is the hydrolysis of an alkyl bromide, RBr, under basic conditions, where the *attacking* nucleophile is the hydroxyl ion (OH^-) and the LG is bromide ion (Br^-):

$$RBr + OH^- \rightarrow ROH + Br^-$$

The functional group of alkyl halides is a carbon–halogen bond, the common halogens being fluorine, chlorine, bromine, and iodine. Except for iodine, a halogen atom has an electronegativity significantly greater than

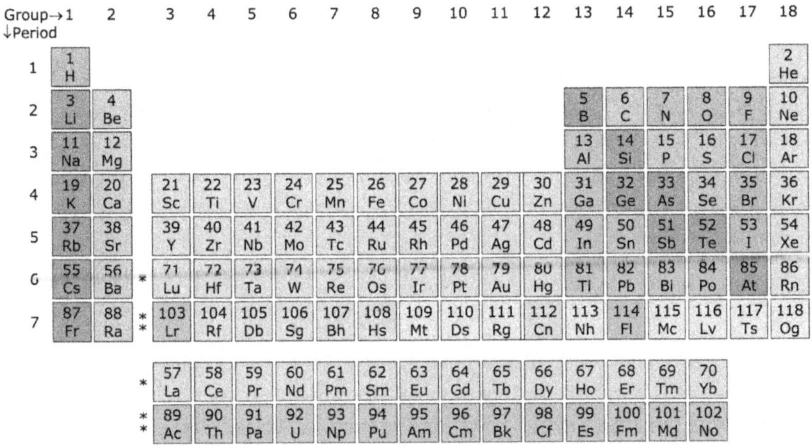

Group→ ↓Period	1	2		3	4	5	6	7	8	9	10	11	12	13	14	15	16	17	18
1	1 H																		2 He
2	3 Li	4 Be												5 B	6 C	7 N	8 O	9 F	10 Ne
3	11 Na	12 Mg												13 Al	14 Si	15 P	16 S	17 Cl	18 Ar
4	19 K	20 Ca		21 Sc	22 Ti	23 V	24 Cr	25 Mn	26 Fe	27 Co	28 Ni	29 Cu	30 Zn	31 Ga	32 Ge	33 As	34 Se	35 Br	36 Kr
5	37 Rb	38 Sr		39 Y	40 Zr	41 Nb	42 Mo	43 Tc	44 Ru	45 Rh	46 Pd	47 Ag	48 Cd	49 In	50 Sn	51 Sb	52 Te	53 I	54 Xe
6	55 Cs	56 Ba	*	71 Lu	72 Hf	73 Ta	74 W	75 Re	76 Os	77 Ir	78 Pt	79 Au	80 Hg	81 Tl	82 Pb	83 Bi	84 Po	85 At	86 Rn
7	87 Fr	88 Ra	**	103 Lr	104 Rf	105 Db	106 Sg	107 Bh	108 Hs	109 Mt	110 Ds	111 Rg	112 Cn	113 Nh	114 Fl	115 Mc	116 Lv	117 Ts	118 Og
			*	57 La	58 Ce	59 Pr	60 Nd	61 Pm	62 Sm	63 Eu	64 Gd	65 Tb	66 Dy	67 Ho	68 Er	69 Tm	70 Yb		
			**	89 Ac	90 Th	91 Pa	92 U	93 Np	94 Pu	95 Am	96 Cm	97 Bk	98 Cf	99 Es	100 Fm	101 Md	102 No		

FIGURE 6.1 The periodic table of the elements showing the groups and periods including the Lanthanide elements and the Actinide elements.

carbon. Consequently, this functional group is polarized so that the carbon is electrophilic, and the halogen is nucleophilic.

2.3 Bond Strength and Anion Stability

Two characteristics other than electronegativity also have an important influence on the chemical behavior of these chemicals. The first of these is the covalent bond strength. The strongest of the carbon–halogen covalent bonds is that to fluorine. Remarkably, this is the strongest common single bond to carbon, being roughly 30 kcal/mole stronger than a carbon–carbon bond and about 15 kcal/mole stronger than a carbon–hydrogen bond. Because of this, alkyl fluorides and fluorocarbons in general are chemically and thermodynamically quite stable, and do not share any of the reactivity patterns shown by the other alkyl halides. The carbon–chlorine covalent bond is slightly weaker than a carbon–carbon bond, and the bonds to the other halogens are weaker still, the bond to iodine being about 33% weaker.

The second factor to be considered is the relative stability of the corresponding halide anions, which is likely the form in which these electronegative atoms will be replaced. This stability may be estimated from the relative acidities of the H-X acids, assuming that the strongest acid releases the most stable conjugate base (halide anion). With the exception of HF ($pK_a = 3.2$), all the hydrohalic acids are very strong, small differences being in the direction $HCl < HBr < HI$.

Solvation of nucleophilic anions markedly influences their reactivity. The nucleophilicities cited above were obtained from reactions in methanol solution. Polar, protic solvents such as water and alcohols solvate anions by hydrogen bonding interactions. These solvated species are more stable and less reactive than the nonsolvated anions. Polar, aprotic solvents such as DMSO (dimethyl sulfoxide), DMF (dimethylformamide), and acetonitrile do not solvate anions nearly as well as methanol but provide good solvation of the accompanying cations. Consequently, most of the nucleophiles discussed here react more rapidly in solutions prepared from these solvents. These solvent effects are more pronounced for small basic anions than for large weakly basic anions. Thus, for reaction in dimethyl sulfoxide solution, the following order of reactivity is observed, which is approximately equivalent to the order of increasing basicity:

$$\text{Nucleophilicity: } I^- < SCN^- < Br^- < Cl^- \approx N_3^- < CH_3CO_2^- < CN^- \approx CH_3S^- < CH_3O^-$$

Some of the most important information concerning nucleophilic substitution and elimination reactions of alkyl halides has come from studies in which the structure of the alkyl group has been varied. For example, using the strong nucleophile thiocyanide (SCN) in ethanol solvent, there are large decreases in the rates of reaction as alkyl substitution of the alpha-carbon increases. Methyl bromide reacts 20–30 times faster than simple primary alkyl bromides, which in turn react about 20 times faster than simple secondary alkyl bromides, and tertiary alkyl bromides are essentially unreactive or undergo elimination reactions. Furthermore, β-alkyl substitution also decreases the rate of substitution, as witnessed by the failure of neopentyl bromide, $(CH_3)_3CCH_2{-}Br$ (a primary alkyl bromide), to react.

It is necessary to point out that substitution and elimination reactions often compete with each other because it is a question of nucleophilic or basic properties. Thus:

```
                     Nu⁻        |   |
                   ------>   H-C-C-Nu        LG⁻
      |   |                     |   |
   H-C-C-LG      or
      |   |          B⁻         \     /
                   ------>        C=C          H-LG
                                 /     \
```

Both substitution and elimination reactions are strongly influenced by many experimental factors (Table 6.2).

At any pH the observed rate constant for hydrolysis of a chemical structure in water is given by the equation (Mill, 1982):

$$\text{Rate} = k_h[C] = (k_A[H_3O^+] + k_B[OH^-] + k_N)\cdot[C]$$

In this equation [C] is the concentration of chemical; k_h is the observed rate constant; k_A, k_B and k_N are the specific rate constants for the acid-catalyzed processes, the base-promoted process and the neutral process, respectively.

TABLE 6.2 Comparison of Substitution Reactions (S_N1, S_N2) and Elimination Reactions (E1, E2)

Reaction	Solvent	Nu or Base	Leaving Group	Substrate	Example Conditions
S_N1	Polar solvent	Good Nu and weak base	Good LG	Strong	Alkyl halide/ $AgNO_3$/aq. EtOH
S_N2	Polar aprotic solvent	Good Nu and weak base	Good LG	Strong	Alcohol/$SOCl_2$ or PX_3
E1	Polar solvents	Weak base	Good LG	Strong	Alcohol/ H_2SO_4/heat
E2	Polar aprotic solvent	Poor Nu and strong base	Good LG	Strong	Alkyl halide/ KOH/heat

The rate constants k_A, k_B, and k_N, for any specific chemical exhibit different dependence on the temperature which leads to factors in the order of 2–4 in the value of k_h, with a 10°C (18°F) increase in the temperature of the reaction. In addition to temperature, pH, ionic strength, and the interaction with humic substances in the natural aquatic bodies, all have strong influences on the hydrolysis of organic pollutants (Jiang et al., 1994).

2.4 Kinetic Order and Steric Effects

The kinetic order of a reaction is a description of the rate of the reaction as it depends upon the concentrations of various species. It includes a rate constant, k (small k, not an equilibrium constant, k), which is equal to the rate of the reaction at unit concentrations of all reagents. Thus, it is a measure of the inherent rate of the reaction, independent of concentrations.

A *first-order reaction* is a reaction that depends upon the concentration of a single chemical and the rate of the reaction is linearly dependent upon the concentration of that chemical (the rate-determining chemical) (S_N1). In contrast, a *second-order reaction* is dependent upon the concentration of both chemicals (in a two-chemical system) and the rate determining step is bimolecular (S_N2). In this type of reaction, both reactants must collide in order to react and the number of collisions between the two reagents is directly proportional to the concentrations of both chemicals.

Like S_N2 reactions, there are quite a few factors that affect the reaction rate of S_N1 reactions. Instead of having two concentrations that affect the reaction rate, there is only one substrate. Since the rate of a reaction is only determined by its slowest step, the rate at which the LG leaves the target chemical determines the speed (rate) of the reaction. Thus, the better the LG, the faster the reaction rate. A general rule for the type of group that constitutes a good LG is the weaker the conjugate base, the better the leaving group. In this case, for example, halogen atoms are the best LGs, while compounds such as amines (RNH_2) and alkane derivatives are poor LGs. An important aspect of the S_N2 reaction is that the new bond does not directly replace the old one in space. Instead, the nucleophile approaches from the opposite side (or face) of the molecule from which the LG departs, resulting in inversion of configuration.

As S_N2 reactions were affected by steric effects, S_N1 reactions are determined by bulky groups attached to the carbocation [such as $(CH_3)_3C^+$, which is typically unstable]. The carbocation is a molecule having a carbon atom bearing three bonds and a positive formal charge. Since there is an intermediate that contains a positive charge, bulky groups attached are going to help stabilize the charge on the carbocation through resonance and distribution of charge. In this case, tertiary carbocation will react faster than a secondary one which will react much faster than a primary. It is also due to this carbocation intermediate that the product does not have to have inversion. The nucleophile can attack from the top or the bottom and therefore create a racemic product.

It is important to use a protic solvent, water, and alcohols, since an aprotic solvent could attack the intermediate and cause unwanted product. It does not matter if the hydrogens from the protic solvent react with the nucleophile since the nucleophile is not involved in the rate determining step.

The chemistry of nucleophilic reactions represents an innovative approach to the problem of introducing nucleophiles onto aromatic ring systems in a way that greatly reduces the negative environment. The most common method of producing substituted benzene rings involves the well-known reaction of electrophilic aromatic substitution. This allows facile substitution for hydrogen of nitro groups, halogens, sulfonic acid, alkyl and acyl groups. While the applications of this reaction for multiple substitutions are limited by the ortho/para or meta directing properties of groups on the ring as well as their activating/deactivating properties, it is usually possible to prepare the desired compound if the group to be introduced to the ring is an *electrophile*. Thus, unless the reaction parameters are changed, a whole family of substituents, which are nucleophilic in nature, cannot be introduced to the ring by electrophilic aromatic substitution.

In general, reactions involving the substitution of nucleophiles are difficult. Under the usual conditions for nucleophilic substitution reactions they produce vanishingly small amounts of product. Recalling the two mechanisms for nucleophilic substitution, S_N1 and S_N2 demonstrates the reason: with an aromatic ring there is no readily available approach to the carbon. In addition, hydrogen is a very poor LG and the aromatic system will not readily ionize by loss of hydrogen. Therefore, to introduce important nucleophiles onto the aromatic ring, novel approaches (not readily available) are necessary so that under the reaction conditions, it becomes possible to incorporate nucleophiles onto the aromatic ring (Haag and Mill, 1988a,b).

Thus, in addition, to hydrolytic reactions by water (Mabey and Mill, 1978), other naturally occurring nucleophiles and redox agents are known from the older literature to contribute to the abiotic transformations of compounds in ground water environments. For example, under aerobic conditions, sediment and clay surfaces can, in certain circumstances, alter hydrolysis kinetics and cause oxidation (Saltzman et al., 1976; El-Amamy and Mill, 1984; Stone and Morgan, 1984; Haag and Mill, 1988a,b). In anaerobic sediments, abiotic reduction reactions have been reported for a variety of compound classes, including haloalkanes, nitroaromatics, and azo compounds (Wolfe et al., 1986; Jafvert and Wolfe, 1987; Macalady et al., 1986).

Generally, the catalytic or redox agents in natural systems have not been identified. Reduction in pond sediments are slowed by sterilization but can still occur, suggesting that some of the reductants are microbial exudates or low-valency metals that react abiotically (Macalady et al., 1986; Jafvert and Wolfe, 1987). Extracellular iron porphyrins are plausible reductants that have been shown to catalyze the transformation of chlorinated compounds (Klecka and Gonsior, 1984; Macalady et al., 1986). Also, natural metal sulfides are reductants for alkyl iodides, releasing metal ions, mercaptans, and sulfides (Thayer et al., 1984). Reactions on sediments and metal sulfides are heterogeneous; however, homogeneous reactions also can occur as, for instance, the reaction of bromoalkene derivatives with bisulfide to form mercaptans and alkyl sulfides in an anaerobic aquifer and in model studies as well as the reactions of 1,2-dibromoethane (DBE) and 1,2-dichloroethane with HS^- (Schwarzenbach et al., 1985; Barbash and Reinhard, 1987; Haag and Mill, 1987; Weintraub and Moye, 1987). Other homogeneous nucleophiles exist in natural waters that potentially contribute to abiotic transformation. Carbonates are present in high concentrations in some waters (up to 10 mM in ground waters and 400 mM in soda lakes) and may enhance hydrolysis rates, as HCO is reported to be 5000 times more nucleophilic than water toward methyl bromide (Swain and Scott, 1953). Particularly powerful nucleophiles are the sulfite and thiosulfate dianions, which are more nucleophilic than water toward methyl bromide (Swain and Scott, 1953). Sulfite and thiosulfate occur, in salt marsh pore waters, along with other potential reactants such as sulfides, polysulfides, and thiols (Boulegue et al., 1982; Luther et al., 1986). Partially oxidized forms of sulfur can occur in sediment systems through microbial activity or by contact with aerated water (Chen and Morris, 1972).

Thus, to summarize, in nucleophilic processes (including hydrolysis reactions), a reactant (the nucleophile) bonds with the target chemical and displaces an atom or group of atoms from the target molecule. The efficiency of the process depends upon the injection of the nucleophilic agent in a manner to attack the target molecule. On the other hand, a combined redox (Chapter 7) and nucleophilic attack almost completely mineralizes certain target molecules.

2.5 S_E1 and S_E2 Reactions

Electrophilic substitution reactions (S_E1 and S_E2 reactions) are chemical reactions in which an electrophilic chemical displaces a functional group in another chemical, which is typically, but not always, a hydrogen atom. This reaction is included here because of the similarity to nucleophile substitution, where the reactant is a nucleophile rather than an electrophile. The possible electrophilic substitution reaction mechanisms of interest here are the S_E1 and S_E2 reactions that are similar to the nucleophile counterparts S_N1 and S_N2. In the S_E1 course of action the substrate first ionizes into a carbanion (C^-) and a positively charged remainder after which the carbanion then quickly recombines with the electrophile. On the other hand, the S_E2 reaction mechanism has involved the formation of a single transition state in which the old bond and the newly formed bond are both present. In addition, if substitution occurs (i.e., there is a similar polarized covalent bond on the electronegative carbon, which breaks up during the reaction, so that the reagent substitutes the original group or the LG); this specific reaction is electrophilic substitution.

2.6 Effect of Substrate

The nature of the substrate (in particular the degree of substitution of the α-carbon) is of principal importance in determining which mechanism predominates. Generally, S_N1, S_N1, and the S_E1 are less common than the S_N2 and S_E2 reactions because of the first-order kinetics and the stability constraints imposed by the charged intermediate species the mechanisms generate.

In general, elimination reactions require higher temperatures than substitution reactions. Elimination reactions, at the most simplified level, involve two reactant molecules (that is, a base and the target chemical) and yield three products: (1) the protonated base, (2) the target chemical that has undergone elimination, and (3) the LG of the target chemical. Consequently, elimination reactions are entropically favorable but are also endothermic because heat energy is required to break the two σ-bonds of the abstracted proton and the LG, while the resulting products form only one σ- and one π-bond, the latter of which is less stable. The implication of this is that, in order for the overall change in free energy to be negative—as a condition required for the spontaneity of the reaction—the temperature needs to be sufficiently high if the reaction is to take place.

The solvent is of only slightly less importance. Polar protic solvents (water, alcohols, weak carboxylic acids, etc.) have partially positive hydrogens, which serve to stabilize negatively charged species. Consequently, the mechanisms of the S_N2 and S_E2 reactions, which are aided by having negatively charged bases or nucleophiles in high concentration, are somewhat retarded by protic solvents which allow competing reactions (S_N1, S_E1) to occur.

Thus, because elimination reactions require heat, substitution reactions will be heavily favored and immediately take place unless the reaction vessel is heated prior to mixing of the reagents. The problem is compounded by the fact that, with hydroxide as the nucleophile, an alcohol would form, which would be stabilized by hydrogen bonding with water molecules. Further, because many organic target chemicals (such as the alkyl halides) have a poor solubility in water and immiscible liquids in the same container will boil at lower temperatures than they would unmixed, it could become very difficult in practice to achieve the temperatures necessary to favor elimination reactions while also maintaining the agitation necessary to keep the target chemical and the aqueous phase from completely separating. In an in situ remediation process, this latter effect could be the proverbial *game-changer.*

3. HYDROLYSIS REACTIONS

Many kinds of chemical structures can undergo hydrolysis under conditions typically found in freshwater and marine systems with rates that are fast enough to make these reactions the dominant loss processes. In the most general sense, hydrolysis refers to a reaction with water in which old bonds are broken and new bonds formed. In fact, hydrolysis processes provide the baseline loss rate for any chemical in an aqueous environment. Both inorganic and organic chemicals react with water, although the reactions of inorganic chemicals are the lesser focused group in the environment (Table 6.3).

The common types of hydrolysis are: (1) salts, which occur when salt from a weak base or acid dissolves in liquid and, in this reaction, water spontaneously ionizes into hydroxide anions and hydronium cations, (2) acid hydrolysis in which water can act as an acid or a base and, in this case, the water molecule would give away a proton, (3) base hydrolysis in which the reaction is very similar to the hydrolysis for base dissociation.

Although the various hydrolytic pathways account for significant degradation of certain classes of chemicals, other chemical structures are completely inert. Moreover, hydrolysis is an important abiotic transformation process for many types of discharged synthetic chemicals in aquatic freshwater or marine environments (Mill, 1982). More specifically, hydrolysis should involve only the reactant species water provides—the proton (H^+), the hydroxyl ion (OH^-), and water (H_2O)—but the complete picture includes analogous reactions and thus the equivalent effects of other chemical species present in the local environment, such as the hydrosulfide anion (SH^-) in anaerobic bogs and the chloride ion (Cl^-) in sea water.

Hydrolysis is a double decomposition nucleophilic reaction insofar as the operative entity, the hydroxyl anion (HO^-) as does, in many cases, use water as one of the reactants. Hydrolysis reactions occur when (or) chemicals react with water and the reactions are characterized by the splitting of a water molecule into a hydrogen and a hydroxide group with one or both of these becoming attached to the starting chemical. Hydrolysis usually requires the use of an acid or base catalyst and is used in the synthesis of many useful compounds. Temperature, pH, ionic strength, and humic substances in the natural aquatic bodies all have strong influences on the hydrolysis of pollutants (Mabey and Mill, 1978; Jiang et al., 1994; Katagi, 2002; Tolosa et al., 2013; Tebes-Stevens et al., 2017). The essential parameters of the hydrolysis reaction are (1) pH and (2) temperature.

In a simple representation of the hydrolysis reaction, if a chemical is represented by the formula *AB* in which *A* and *B* are atoms or groups and water is represented by the formula HOH, the hydrolysis reaction may be represented by the reversible chemical equation:

$$AB + \mathrm{HOH} \rightleftharpoons A\mathrm{H} + B\mathrm{OH}$$

The reactants other than water, and the products of hydrolysis, may be neutral molecules (as in most hydrolysis reactions involving chemicals) or ionic molecules (charged species), as in hydrolysis reactions of salts, acids, and bases.

TABLE 6.3 Examples of Water-Reactive Inorganic Chemicals (Listed Alphabetically)

Chemical	Reaction With Water
Aluminum bromide	Violent hydrolysis
Aluminum chloride	Violent decomposition forming hydrogen chloride gas
Boron tribromide	Violent or explosive reaction when water added
Butyl lithium	Ignites on contact with water
Calcium carbide	Gives off explosive acetylene gas
Calcium hydride	Hydrogen gas liberated
Chlorosulfonic acid	Highly exothermic violent reaction
Chlorotrimethylsilane	Violent reaction
Dichlorodimethylsilane	Violent reaction
Lithium aluminum hydride	Releases and ignites hydrogen gas
Lithium hydride	Violent decomposition
Lithium metal	Powder reacts explosively with water
Phosphorus pentachloride	Violent reaction with water
Phosphorus pentachloride	Violent reaction
Phosphorus pentoxide	Violent exothermic reaction
Phosphorus tribromide	Reacts violently with limited amounts of warm water
Phosphorus trichloride	Violent reaction releasing flammable diphosphane
Phosphoryl chloride	Slow reaction which may become violent
Potassium amide	Violent reaction which may cause ignition
Potassium hydride	Releases hydrogen gas
Potassium metal	Forms potassium hydroxide and hydrogen gas
Potassium hydroxide	Highly exothermic reaction
Silicon tetrachloride	Violent reaction producing silicic acid
Sodium amide	Generates sodium hydroxide and ammonia
Sodium azide	Violent reaction with strongly heated azide
Sodium hydride	Reacts explosively with water
Sodium hydrosulfite	Heating and spontaneous ignition

Continued

TABLE 6.3 Examples of Water-Reactive Inorganic Chemicals (Listed Alphabetically)—cont'd

Chemical	Reaction With Water
Sodium hydroxide	Highly exothermic reaction
Sodium metal	Generates flammable hydrogen gas
Sodium peroxide	Reacts violently or explosively
Strontium metal	Violent reaction
Sulfuric acid	May boil and spatter
Tetrachlorosilane	Violent reaction
Thionyl chloride	Violent reaction which forms hydrochloric acid and sulfur dioxide
Titanium tetrachloride	Violent reaction that produces hydrogen chloride gas
Trichlorosilane	Releases toxic and corrosive fumes
Triethyl aluminum	Explodes violently in water
Tri-isobutyl aluminum	Violent reaction with water
Zirconium tetrachloride	Violent reaction with water

3.1 Acid Hydrolysis

In chemistry, acid hydrolysis is a process in which a protic acid is used to catalyze the cleavage of a chemical bond via a nucleophile substitution reaction, with the addition of the elements of water (H_2O). A common type of hydrolysis occurs when a salt of a weak acid or weak base (or both) is dissolved in water. The water spontaneously ionizes into hydronium cations (H_3O^+, for simplicity usually represented as H^+) and hydroxide anions ($-OH^-$). The salt (for example), using sodium acetate (CH_3COONa) as the example, dissociates into the constituent cations (Na^+) and anions ($CH3COO^-$). The sodium ions tend to remain in the ionic form (Na^+) and react very little with the hydroxide ions (OH^-) whereas the acetate ions combine with hydronium ions to produce acetic acid (CH^3COOH). In this case the net result is a relative excess of hydroxide ions, and the solution has basic properties. On the other hand, strong acids also undergo hydrolysis and when sulfuric acid (H_2SO_4) is dissolved in (mixed with) water the dissolution is accompanied by hydrolysis to produce hydronium ion (H_3O^+) ion and a bisulfate ion (HSO_4^-), which is the conjugate base of sulfuric acid.

An example of hydrolysis from chemistry is the hydrolysis of iron chloride ($FeCl_3$) with water ions which occur because the salt (iron chloride) is formed from a weak base [iron hydroxide, ferric hydroxide, $Fe(OH)_3$] and a strong acid (hydrochloric acid, HCl):

$$FeCl_3 + 3H_2O \rightarrow Fe(OH)_3 + 3HCl$$

The reaction is difficult to reverse because the ferric hydroxide separates from the aqueous solution as a precipitate.

Thus, a hydrolysis reaction is the cleavage of chemical bonds by the addition of water or a base that supplies the hydroxyl ion ($-OH^-$). A chemical bond is cleaved, and two new bonds are formed, each one having either the hydrogen component (H) or the hydroxyl component (OH) of the water molecule. This produces the strong dependence on the acidity or alkalinity (pH) of the solution often observed but, in some cases, hydrolysis can occur in a neutral ($pH = 7$) environment. Adsorption on to a mineral sediment (such as a clay sediment that has strong adsorptive powers) generally reduces the rates of hydrolysis for acid-catalyzed or base-catalyzed reactions.

For example, silicone polymers which are often difficult to remediate (Rücker and Kümmerer, 2015) undergo clay-catalyzed hydrolytic degradation in soil (Xu, 1998). In this example, the effects of moisture levels and exchangeable cations on degradation of a poly(dimethylsiloxane) fluid on clay minerals was investigated. Kaolinite, talc, and Arizona montmorillonite saturated with sodium ions (Na^+), calcium ions (Ca^{2+}), or aluminum ions (Al^{3+}) were used to scission of its silicon-oxygen-silicon (Si—O—Si) backbone. The hydrolytic degradation has two stages—both are zero-order reactions. Although high humidity may result in the formation of some volatile cyclic methyl siloxane derivatives on an artificial catalyst (the aluminum-saturated montmorillonite), the ultimate degradation product was otherwise water-soluble dimethylsilanediol. Thus, the degradation rates and products of silicone polymers were influenced by (1) the exchangeable cation type, (2) the moisture level, and (3) the type of clay.

There is also evidence that the rate of enzymatic hydrolysis of glucose-1-phosphate (G1P) adsorbed on goethite by acid phosphatase (AcPase) can be of the same order of magnitude as in aqueous solution. The surface process releases carbon to the solution whereas orthophosphate remains adsorbed on goethite. This hydrolysis reaction is strictly an interfacial process governed by the properties of the interface. A high surface concentration of substrate mediates the formation of a catalytically active layer and although adsorption likely reduces the catalytic efficiency of the enzyme, this reduction is almost balanced by the fact that enzyme and substrate are concentrated at the mineral surfaces. Thus, mineral surfaces with appropriate surface properties can be very effective in concentrating substrates and enzymes thereby creating microchemical environments of high enzymatic activity. Hence, strongly adsorbed molecules in soils and aquatic environments also may be subjected to biodegradation by extracellular enzymes (Olsson et al., 2012).

In addition, neutral reactions appear to be unaffected by adsorption (Chapter 5) although there is always the possibility that the mineral sediment can cause catalyzed chemical transformation reactions. Many chemicals can be altered by a direct reaction with water. The rate of a hydrolysis reaction is typically expressed in terms of the acid-catalyzed, neutral-catalyzed, and base-catalyzed hydrolysis rate constants.

Many types of chemicals undergo hydrolysis under conditions typically found in the various ecosystems. In fact, there are four possible general rules involving the reactions of salts in the aqueous environment: (1) if the salt is formed from a *strong* base and *strong* acid, then the salt solution is neutral, indicating that the bonds in the salt solution will not break apart (indicating no hydrolysis occurred) and the solution is basic, (2) if the salt is formed from a *strong* acid and *weak* base, the bonds in the salt solution break and the solution becomes acidic, (3) if the salt is formed from a *strong* base and *weak* acid, the salt solution is basic and hydrolyzes, and (4) if the salt is formed from a weak base and weak acid, it will hydrolyze but the acidity or basicity depends on the equilibrium constants of K_a and K_b. If the K_a value is greater than the K_b value, the resulting solution will be acidic and if the K_b value is greater than the K_a value, the resulting solution will be basic. Moreover, there is the need to estimate the relative susceptibility of some potential environmental pollutants to undergo hydrolysis reactions in order to determine the potential beneficial effects or the potential adverse effects that the product can have on the environment (Rahm et al., 2005).

Typically, in a hydrolysis reaction the hydroxyl group (OH^-) replaces another chemical group in the molecule and hydrolysis reactions are usually catalyzed by hydrogen ions or hydroxyl ions. This produces the strong dependence on the acidity or alkalinity (pH) of the solution often observed but, in some cases, hydrolysis can occur in a neutral ($pH = 7$) environment. Adsorption on to a mineral sediment (such as a clay sediment that has strong adsorptive powers) generally reduces the rates of hydrolysis for acid- or base-catalyzed reactions. Neutral reactions appear to be unaffected by adsorption although there is always the possibility that the mineral sediment can cause catalyzed chemical transformation reactions.

Hydrolytic processes provide the baseline loss rate for any chemical in an aqueous environment. Although various hydrolytic pathways account for significant degradation of certain classes of chemicals, other structures are completely inert. Strictly speaking, hydrolysis should involve only the reactant species water provides, that is, H+, OH^-, and H_2O, but the complete picture includes analogous reactions and thus the equivalent effects of other chemical species present in the local environment, such as SH^- in anaerobic bogs, the chloride ion (Cl^-) in sea water, and various ions in laboratory buffer solutions (Jiang et al., 1994; Wolfe and Jeffers, 2000).

Many chemicals can be altered by a direct reaction of the chemical with water (hydrolysis) in which a chemical bond is cleaved and two new bonds are formed, each one having either the hydrogen component (H^+) or the hydroxyl component (OH^-) of the water molecule. Typically, the hydroxyl replaces another chemical group on the molecule and hydrolysis reactions are usually catalyzed by hydrogen ions or hydroxyl ions. This produces the strong dependence on the acidity or alkalinity (pH) of the solution often observed but, in some cases, hydrolysis can occur in a neutral (pH = 7) environment. Adsorption on to a mineral sediment (such as a clay sediment that has strong adsorptive powers) generally reduces the rates of hydrolysis for acid- or base-catalyzed reactions. Neutral reactions appear to be unaffected by adsorption although there is always the possibility that the mineral sediment can cause catalyzed chemical transformation reactions.

Furthermore, the hydrolysis of chemicals is influenced by the composition of the solvent and the rate constants may be much higher in water than in solvents. In fact, the introduction of a complex mixture of chemicals into a water body can be expected to produce a significant shift in acidity or alkalinity of the medium and, therefore, it would not be surprising to anticipate that hydrolysis would be affected in complex mixtures.

3.2 Alkaline Hydrolysis

As an example of alkaline hydrolysis (base-catalyzed hydrolysis), the method has been used to dispose of animal waste (Krička et al., 2014). Alkaline hydrolysis gives positive results in harmless destruction of prions and as such it proved very efficient in the disposal of high-risk animal waste. In the process, hydroxide is added in the proportional ratio with the quantity of waste and the entire mixture is continuously heated and mixed from three to 6 h at a pressure of 4 bars in adequate stainless-steel reactors. The process can be catalyzed by enzymes, metal salts, acids and alkalis—potassium and sodium hydroxide water solutions. The hydrolysis process is accelerated when the system is heated, and the optimal results are achieved at a temperature of 150°C (302°F). Amino acids, as the basic building units of all proteins, are interconnected by peptide links creating the polypeptide chains. During the alkaline hydrolysis process, the peptide links are broken down leaving small peptide chains and amino acids in the form of potassium and sodium salts.

Predictive methods can be applied for assessing hydrolysis for simple one-step reactions where the product distribution is known. Generally, however, pathways are known only for simple molecules. Often, for environmental studies, the investigator is interested in not only the parent chemical but also the intermediates and products. Therefore, estimation methods may be required for several reaction pathways. Two approaches have been used extensively to obtain estimates of hydrolytic rate constants for use in environmental systems (Wolfe and Jeffers, 2000). The first and potentially more precise method is to apply quantitative structure/activity relationships. To develop such predictive methods, one needs a set of rate constants for a series of chemicals that have systematic variations in structure and a database of molecular descriptors related to the substituents on the reactant molecule. The second and more widely used method is to compare the target chemical with an analogous chemical or chemicals containing similar functional groups and structure, to obtain an estimate of the rate constant.

As an example of the need for hydrolysis, and of importance, are the pesticides which are fairly resilient in the environment (Crosby, 1973; Katagi, 2002) and the potential for pesticides to be permanent pollutants . For example, the reactions of five organophosphorus insecticides (chlorpyrifos-methyl, parathion-methyl, fenchlorphos, chlorpyrifos, and parathion) with hydrogen sulfide/bisulfide (H_2S/HS^-) and polysulfides (Sn_{2-}) were examined in well-defined aqueous solutions over a pH range from 5 to 9 (Wu et al., 2006). The rates are first-order in the concentration of the different reduced sulfur species. Experiments at 25°C (77°F) demonstrated that the reactions of the five organophosphorus insecticides with the reduced sulfur species follow a S_N2 mechanism. The activation parameters of the reaction of the organophosphorus insecticides with bisulfide have been reported from the measured second-order rate constants over a temperature range of 5−60°C (41−140°F). The determined second-order rate constants show that the reaction of organophosphorus insecticides with polysulfides is 15 to 50 times faster than the reaction of the same organophosphorus insecticides with bisulfide. The dominant transformation products are dealkylated organic pollutants, which indicate that the nucleophilic substitution of reduced sulfur species occurs at the carbon atom of the alkoxy groups. And also, the results show that these reduced sulfur species are much better nucleophiles, and thus degrade these pesticides faster than the well-studied base hydrolysis by the hydroxyl ion (OH^-). When the determined second-order rate constants are multiplied with the concentration of the hydrosulfide anion (HS^-) and the polysulfide anion (Sn_2^-) reported in salt marshes and porewater of sediments, predicted half-lives show that abiotic degradation by sulfide species may be of comparable importance to microbially mediated degradation in anoxic environments.

3.3 Reaction Profiles

The hydrolytic profiles of the reaction depend on the chemical structure and functional group(s) of a pesticide molecule, but they are not always consistent within a chemical class of pesticides. For example, organophosphorus pesticides are primarily susceptible to alkaline hydrolysis with less acidic catalysis, but some of phosphorodithioates are found to be acid labile. In the case of carbamates, the pK_a value (a measure of the acid strength) of a LG control the hydrolysis mechanism. In addition, it is advantageous to estimate the hydrolysis profiles of pesticides because of pH and temperature dependency of the hydrolysis reaction. Therefore, it would still be practical when investigating abiotic hydrolysis of a new pesticide that a laboratory study be effectively designed on the basis of accumulated knowledge of the profiles of the hydrolysis reactions for the essential chemical structure and functional groups and conducted to obtain pH-rate and temperature-rate profiles.

The reaction rate for hydrolysis reaction is variable. For simple reactions, which consist of a single step, the order equals the molecularity as predicted by collision theory. For example, a bimolecular elementary reaction (such as $A + B \rightarrow$ products) will be second order overall and first order in each reactant. For multistep reactions, the order of each step equals the molecularity, but this is *not* generally true for the overall rate. The rate equation of a reaction with an assumed multistep mechanism can often be derived theoretically using quasi steadystate assumptions from the underlying elementary reactions and be compared with the experimental rate equation as a test of the assumed mechanism. In addition, the kinetic equation may depend on the concentration of an intermediate ion or species.

Two approaches have been used extensively to obtain estimates of hydrolytic rate constants for use in environmental systems: (1) application of quantitative structure/activity relationships which requires a set of rate constants for a series of compounds that have systematic variations in structure and a database of molecular descriptors related to the substituents on the reactant molecule, and (2) comparison of the target chemical with an analogous compound or compounds containing similar functional groups and structure, to obtain a less quantitative estimate of the rate constant.

Predictive methods can be applied for assessing hydrolysis for simple one-step reactions where the product distribution is known but, typically, reaction pathways are known only for simple molecules. For environmental studies, there is interest not only in reactivity of the parent chemical but also any intermediate product(s) (that might be preserved by the environmental conditions) and the final product(s). Therefore, estimation methods may be required for several reaction pathways. Some preliminary examples of hydrolysis reactions illustrate the very wide range of reactivity of discharge chemicals. Example are the triesters of phosphoric acid which hydrolyze in near-neutral solution at ambient temperatures with half-lives ranging from several days to several years, whereas the halogenated alkane derivatives (pentachloroethane, carbon tetrachloride, and hexachloroethane) have "environmental effects" (pH = 7, 25°C, 77°F) with half-lives in the order of 2 h, 50 years, and 1000 millennia, respectively (Wolfe and Jeffers, 2000). On the other hand, pure hydrocarbons from methane through the polynuclear aromatic hydrocarbon derivatives are not hydrolyzed under any circumstances that are environmentally relevant.

Hydrolysis can explain the attenuation of contaminant plumes in aquifers where the ratio of rate constant to flow rate is sufficiently high. For example, 1,1,1-trichloroethane (TCA) has been observed to disappear from a mixed halocarbon plume over time, while trichloroethylene and 1,2-dichloroethylene (the biodegradation product of trichloroethylene) persist. Also, trichloropropane (1,2,3-trichloropropane) is an emerging contaminant because of increased recognition of its occurrence in groundwater, potential carcinogenicity, and resistance to natural attenuation. The physical and chemical properties of trichloropropane make it difficult to remediate, with all conventional options being relatively slow or inefficient. Treatment that results in alkaline conditions (e.g., permeable reactive barriers containing zerovalent iron) favor base-catalyzed hydrolysis of trichloropropane, but high temperature (e.g., conditions of in situ thermal remediation) is necessary for this reaction to be significant. Common reductants (sulfide, ferrous iron adsorbed to iron oxides, and most forms of construction-grade or nano-Fe^0) produce insignificant rates of the reductive dechlorination of trichloropropane. However, quantifiable rates of trichloropropane reduction can be obtained with several types of activated nano-Fe^0, but the surface area normalized rate constants (kSA) for these reactions were lower than is generally considered useful for in situ remediation applications. Much faster rates of the degradation of trichloropropane can be obtained with granular zero-valent zinc (Zn^0) and potentially problematic dechlorination intermediates (1,2- or 1,3-dichloropropane, 3-chloro-1-propene) are not detected. The advantages of Zn^0 over Fe^0 are somewhat peculiar to trichloropropane and suggest a practical application for Zn^0 even though it has not found favor for remediation of contamination with other chlorinated solvents (Sarathy et al., 2010).

The hydrolytic loss of organophosphate pesticides in sea water, as determined from both laboratory and field studies, suggests that these compounds will not be long-term contaminants despite runoff into streams and, eventually, the sea (Wolfe and Jeffers, 2000). The oceans also can provide a major sink for atmospheric species ranging from carbon tetrachloride to methyl bromide. Loss of methyl bromide in the oceans by a combination of hydrolysis and exchange of chloride (Cl^-) for bromide (Br^-) constitutes a significant contribution to the total degradation and must be given consideration in any predictive and/or modeling efforts. In addition, sulfur chemistry is an important aspect of the transformation of chemical species in the atmosphere.

Sulfur compounds have been observed in the atmosphere and are known to be important to the evolution of the atmospheric chemistry and the climate of the earth. In an oxidizing atmosphere, sulfur dioxide (SO_2) is transformed (typically by of the transient sulfur trioxide, SO_3) to highly hygroscopic sulfuric acid (H_2SO_4) (Chapter 1). Because of the hygroscopic nature and ubiquitous nature in atmosphere of the earth, gas phase sulfuric acid is considered to be a critical agent in much of the chemistry leading to article formation in the atmosphere. The gas-phase formation is via the reaction of hydroxyl radicals with sulfur dioxide, with the resulting complex hydroxyl sulfur dioxide species ($HOSO_2$) rapidly reacting with the atmospherically abundant oxygen to form sulfur trioxide, thereby leading (through hydrolysis) to sulfuric acid.

Two water molecules, perhaps as a water dimer, are required for this last process, giving rise to a strong temperature-dependence and relative humidity-dependent high hydrolysis rate (Lovejoy et al., 1996; Morokuma and Muguruma, 1994; Larson et al., 2000; Donaldson et al., 2016). Once formed, the very hygroscopic sulfuric acid molecule provides a good nucleus for the formation of small water clusters, which then may grow by rapid condensation of further water molecules onto this nucleus. The presence of ammonia or small amines in the atmosphere of the earth is believed to further aid the water condensation process by reducing the water activity in the growing condensation nucleus.

As the temperature increases the molecules in solution have more energy, causing them to move and react faster. This causes hydrolysis reactions to occur at a faster rate. In addition, some hydrolysis reactions work better in slightly basic or acidic environments. The previously shown mechanism needs a slightly basic solution to start, because it is the hydroxyl ion (HO^-) that is attracted to the carboxyl center (carbon double bonded to the oxygen). The next reaction needs a slightly acidic solution, so the nitrogen can be doubly protonated (add two hydrogens to nitrogen).

The hydrolysis half-life ($t_{1/2}$) is determined by adding a known amount of chemical into a solution and then measuring the amount of the original chemical present at various time intervals. Then a graph of the natural log of chemical present versus time is plotted and the slope of the line will equal the value of K_T (hydrolysis rate constant) from which the half-life ($t_{1/2}$) can be then determined:

$$t_{1/2} = 0.693/_T$$

The hydrolysis half-life is an aid for the estimation of the time that a chemical will persist in an aqueous environment. If the chemical resists hydrolysis, then it may degrade via some other pathway such as microbial metabolism (Chapter 8).

A chemical with a high vapor pressure may cause an environmental problem because the chemical can volatilize and disperse over a large area and such chemicals with a high vapor pressure need to be handled in a way so that the vapors do not escape into the atmosphere (Chapter 5). A chemical with a low vapor pressure does not move into air so there is a potential for accumulation in water if it is water-soluble. If it is not water-soluble, the chemical may accumulate in soil or biota.

The Henry's law constant (HLC) is a measure of the concentration of a chemical in air over its concentration in water. Thus, the constant can be conveniently represented by:

$$\text{HLC} = (\text{Liquid vapor pressure})/(\text{Solubility of the chemical})$$

A chemical with a high Henry's law constant will volatilize from water into air and be distributed over a large area. A chemical with a low Henry's law constant will tend to persist in water and may be adsorbed on to soil. The value of the Henry's law constant value is an integral part in calculating the volatility of a chemical.

Chemicals in the air can partition (move) into water droplets in clouds and fog. If the Henry's law constant is low, substantial amounts of the volatilized chemical will dissolve in the water droplets and be transported back to the surface of the earth by rain. This process of a chemical moving from the gas phase into water droplets and being deposited onto the earth's surface is called wet deposition. Dry deposition is another process that occurs when the chemical is adsorbed onto soil particles in air which is deposited on the surface of the earth.

In summary, the reactivity of a chemical is the most important factor determining whether hydrolysis will occur. This reactivity is largely determined by the substituents that are bound to the chemical. Substituents are atoms or groups of atoms bonded to the substrate (main body of chemical). Some substituents are readily displaced from the substrate by hydrolysis reactions since the products formed are very stable in water. There are other substituents that pull electron density away from the substrate, making a partially positive center that readily accepts hydrolysis.

REFERENCES

Barbash, J.E., Reinhard, M., 1987. Nucleophilic substitution reactions between HS^- and halogenated aliphatic compounds in homogeneous aqueous solution. In: Proceedings. 194th National Meeting, American Chemical Society, New Orleans, Louisiana, August 30–September 4. Division of Environmental Chemistry, vol. 27, pp. 334–337.

Boulegue, J., Lord, J.C., Church, T.M., 1982. Sulfur speciation and associated trace metals (Fe, Cu) in the pore waters of Great Marsh, Delaware. Geochimica et Cosmochimica Acta 461, 453–464.

Chen, K.Y., Morris, J.C., 1972. Kinetics of oxidation of aqueous sulfide by O_2. Environmental Science & Technology 6, 529–537.

Cole, J.G., Mackay, D., 2000. Correlating environmental partitioning properties of compounds: the three solubility approach. Environmental Toxicology & Chemistry 19 (2), 265–270.

Crosby, D.G., 1973. The fate of pesticides in the environment. Annual Review of Plant Physiology 24 (1), 467–492.

Delle Site, A., 2001. Factors affecting sorption of compounds in natural sorbent/water systems and sorption coefficients for selected pollutants. A review. Journal of Physical and Chemical Reference Data 30 (1), 187–439.

Donaldson, D.J., Kroll, J.A., Vaida, V., 2016. Gas-phase hydrolysis of triplet SO_2: a possible direct route to atmospheric acid formation. Scientific Reports 6 (30000). https://www.nature.com/articles/srep30000.pdf?origin=ppub.

El-Amamy, M.M., Mill, T., 1984. Hydrolysis kinetics of chemicals on montmorillonite and kaolinite surfaces as related to moisture content. Clays and Clay Minerals 32, 67–73.

Green, N., Bergman, A., 2005. Chemical reactivity as a tool for estimating persistence. Environmental Science & Technology 39 (23), 480A–486A.

Haag, W.R., Mill, T., 1987. Reactions of sulfur nucleophiles with haloalkanes. In: Proceedings. 194th National Meeting, American Chemical Society, New Orleans, LA, August 30–September 4. Division of Environmental Chemistry, vol. 27, pp. 332–333.

Haag, W.R., Mill, T., 1988a. Effect of a subsurface sediment on hydrolysis of haloalkanes and epoxides. Environmental Science & Technology 22, 658–663.

Haag, W.R., Mill, T., 1988b. Some reactions of naturally occurring nucleophiles with haloalkanes in water. Environmental Toxicology & Chemistry 7 (11), 917–924.

House, J.E., 2008. Chemistry. Academic Press, Elsevier, Burlington, Massachusetts.

Imyanitov, N.S., 1990. Electrophilic bimolecular substitution as an alternative to nucleophilic monomolecular substitution in and chemistry. Journal of General Chemistry of the USSR 60 (3), 417–419 (English translation).

Jafvert, C.T., Wolfe, N.L., 1987. Degradation of selected halogenated ethanes in anoxic sediment-water systems. Environmental Chemistry 6, 827–837.

Jiang, L., Han, S., Wang, L., Li, C., Du, D., 1994. The influential factors of the hydrolysis of pollutants in the environment. Chemosphere 28 (10), 1749–1756.

Katagi, T., 2002. Abiotic hydrolysis of pesticides in the aquatic environment. Reviews of Environmental Contamination and Toxicology 175, 79–261.

Klecka, G.M., Gonsior, S.J., 1984. Reductive dechlorination of chlorinated methanes and ethanes by reduced iron (II) porphyrins. Chemosphere 13, 391–402.

Krička, T., Toth, I., Kalambura, S., Jovičić, N., 2014. Efficiency of alkaline hydrolysis method in environment protection. Collegium Antropologicum 38 (2), 487–492.

Larson, L.J., Kuno, M., Tao, F.M., 2000. Hydrolysis of sulfur trioxide to form sulfuric acid in small water clusters. The Journal of Chemical Physics 112, 8830–8838.

Lovejoy, E.R., Hanson, D.R., Huey, L.G., 1996. Kinetics and products of the gas-phase reaction of SO_3 with water. Journal of Physical Chemistry 100, 19911–19916.

Luther, G.W., Church, T.M., Scudlark, J.R., Cosman, M., 1986. Inorganic and organic sulfur cycling in salt-marsh pore waters. Science 232, 746–749.

March, J., 1992. Advanced Chemistry, fourth ed. John Wiley & Sons, Inc., Hoboken, New Jersey.

Mabey, W.R., Mill, T., 1978. Critical review of hydrolysis of compounds in water under environmental conditions. Journal of Physical and Chemical Reference Data 7 (2), 383–415.

Macalady, D.L., Tratnyek, P.G., Grundl, T.J., 1986. Abiotic reduction of anthropogenic systems: a critical review. Journal of Contaminant Hydrology 1, 1–28.

McCall, P.J., Laskowski, D.A., Swann, R.L., Dishburger, H.J., 1983. Estimation of environmental partitioning of chemicals in model ecosystems. In: Gunther, F.A., Gunther, J.D. (Eds.), Residue Reviews, vol. 85. Springer, New York, pp. 231–244.

McCarty, P.L., Criddle, C.S., 2012. In: Kitanidis, P.K., McCarty, P.L. (Eds.), Delivery and Mixing in the Subsurface: Processes and Design Principles for In Situ Remediation. Springer Science, New York, pp. 7–52 (Chapter 2).

Mill, T., 1982. Hydrolysis and oxidation processes in the environment. Environmental Toxicology & Chemistry 1, 135–141.

Morokuma, K., Muguruma, C., 1994. Ab-initio molecular-orbital study of the mechanism of the gas-phase reaction $SO_3 + H_2O$ – importance of the 2nd water molecule. Journal of the American Chemical Society 116, 10316–10317.

Olsson, R., Giesler, R., Loring, J.S., Persson, P., 2012. Enzymatic hydrolysis of organic phosphates adsorbed on mineral surfaces. Environmental Science & Technology 46 (1), 285–291.

Rahm, S., Green, N., Norrgran, J., Bergman, A., 2005. Hydrolysis of environmental contaminants as an experimental tool for indication of their persistency. Environmental Science and Technology 39 (9), 3128–3133.

Rossi, R.A., De Rossi, R.H., 1983. Aromatic Substitution by the $S_{RN}1$ Mechanism. ACS Monograph Series No. 178. American Chemical Society, Washington, DC.

Rücker, C., Kümmerer, K., 2015. Environmental chemistry of organosiloxanes. Chemical Reviews 115 (1), 466–524.

Saltzman, S., Mingelgrin, U., Yaron, B., 1976. Role of water in the hydrolysis of parathion and methylparathion on kaolinite. Journal of Agricultural and Food Chemistry 24, 739–743.

Sarathy, V., Salter, A.J., Nurmi, J.T., Johnson, G.O., Johnson, R.L., Tratnyek, P.G., 2010. Degradation of 1,2,3-trichloropropane (TCP): hydrolysis, elimination, and reduction by iron and zinc. Environmental Science & Technology 44 (2), 787–793.

Schwarzenbach, R.P., Giger, W., Schaffner, C., Wanner, O., 1985. Groundwater contamination by volatile halogenated alkanes: abiotic formation of volatile sulfur compounds under anaerobic conditions. Environmental Science & Technology 19, 322–327.

Stone, A., Morgan, J.J., 1984. Reduction and dissolution of manganese (III) and manganese (IV) oxides by organics: 2. Survey of the reactivity of organics. Environmental Science & Technology 18, 617–624.

Strohfeldt, K.A., 2016. Essentials of Chemistry. John Wiley & Sons, Inc., Chichester, United Kingdom.

Swain, C.G., Scott, C.B., 1953. Quantitative correlation of relative rates. Comparison of hydroxide ion with other nucleophilic reagents toward alkyl halides, esters, epoxides and acyl halides. Journal of the American Chemical Society 75, 141–147.

Tebes-Stevens, C., Patel, J.M., Jones, W.J., Weber, E.J., 2017. Prediction of hydrolysis products of chemicals under environmental pH conditions. Environmental Science & Technology 51 (9), 5008–5016.

Thayer, J.S., Olson, G.J., Brickman, F.E., 1984. Iodomethane as a potential metal mobilizing agent in nature. Environmental Science & Technology 18, 726–729.

Tolosa, S., Hidalgo, A., Sansón, J.A., 2013. Thermodynamic study of hydrolysis reactions in aqueous solution from *ab initio* potential and molecular dynamics simulations. Hindawi Journal of Chemistry, 265958.

Wade, L.G., 2003. Chemistry, fifth ed. Prentice Hall, Upper Saddle River, New Jersey.

Weintraub, R.A., Moye, H.A., 1987. Ethylene dibromide (EDB) transformations in abiotic-reducing aqueous solutions in the presence of hydrogen sulfide. In: Proceedings. 194th National Meeting, American Chemical Society, New Orleans, Louisiana, August 30–September 4. Division of Environmental Chemistry, vol. 27, pp. 236–240.

Wolfe, N.L., Kitchens, B.E., Macalady, D.L., Grundl, T.J., 1986. Physical and chemical factors that influence the anaerobic degradation of methyl parathion in sediment systems. Environmental Toxicology & Chemistry 5, 1019–1026.

Wolfe, N.L., Jeffers, P.M., 2000. Hydrolysis. In: Boethling, R.S., Mackay, D. (Eds.), Handbook of Property Estimation Methods for Chemicals: Environmental and Health Sciences. Lewis Publishers, Boca Raton, Florida, pp. 311–333 (Chapter 13).

Wu, T., Gan, Q., Jans, U., 2006. Nucleophilic substitution of phosphorothionate ester pesticides with bisulfide (HS^-) and polysulfides (S_n^{2-}). Environmental Science & Technology 40 (17), 5428–5434.

Xu, S., 1998. Hydrolysis of poly(dimethylsiloxanes) on clay minerals as influenced by exchangeable cations and moisture. Environmental Science & Technology 32 (20), 3162–3168.

FURTHER READING

Baulch, D.L., Cox, R.A., Hampson Jr., R.F., Kerr, J.A., Roe, J., Watson, R.T., 1980. Evaluation of kinetic and photochemical data for atmospheric chemistry. Journal of Physical and Chemical Reference Data 9 (2), 295–471.

Gaffney, J.S., Marley, N.A., 2003. Atmospheric chemistry and air pollution. The Scientific World Journal 3, 199–234.

McNeill, V.F., 2015. Aqueous chemistry in the atmosphere: sources and chemical processing of aerosols. Environmental Science & Technology 49 (3), 1237–1244.

NRC, 2014. Physicochemical Properties and Environmental Fate: A Framework to Guide Selection of Chemical Alternatives. National Research Council, Washington, DC.

Prather, M.J., 2007. Lifetimes and time-scales in atmospheric chemistry. Philosophical Transactions of the Royal Society A 365, 1705–1726.

Seinfeld, J.H., Pandis, S.N., 2006. Atmospheric chemistry and physics: from air pollution to climate change, second ed. John Wiley & Sons, Hoboken, New Jersey.

Speight, J.G., Singh, K., 2014. Environmental Management of Energy from Biofuels and Biofeedstocks. Scrivener Publishing, Beverly, Massachusetts.

Speight, J.G., 2017. Environmental Chemistry for Engineers. Butterworth-Heinemann, Elsevier, Oxford, United Kingdom.

Wayne, R.P., 2000. Chemistry of Atmospheres, third ed. Oxford University Press, Oxford, United Kingdom.

Chapter 7

Redox Transformations

1. INTRODUCTION

Oxidation–reduction (redox) reactions are among the most important and interesting chemical reactions that occur in the various environmental systems (the atmosphere, the aquasphere, and the terrestrial biosphere). Redox reactions are central to major element cycling, many sorption processes, trace element mobility and toxicity, most remediation schemes, and life itself. In fact, one of the important questions in the chemistry of pollutant degradation is the identity and distribution of chemical agents that are responsible for reduction reactions in the environment.

Historically, the terms oxidation and reduction arose from experimental observations: oxidation reactions consumed oxygen (O_2) by incorporating oxygen (as oxygen functions) into products and reduction reactions reduced the mass or volume of products by expelling oxygen. A similarly empirical definition of reduction is that it usually involves incorporation of hydrogen, and, therefore, oxidation can be regarded as dehydrogenation (e.g., dehydrogenase enzymes catalyze oxidation). More specifically, oxidation–reduction (redox) reactions are commonly understood to occur by the exchange of electrons between reacting chemical species. Electrons are lost (or donated) in oxidation and gained (or accepted) in reduction. The reaction processes of oxidation and reduction involve a loss or gain of electrons, respectively. Redox reactions are the basis for chemical transformation processes for inorganic substances, including dissolution, and are relevant for discharged chemicals that participate in electron transfer or uptake.

All life on earth derives energy from redox processes. Biomass production further requires the transfer of electrons to bring carbon, macronutrients (N and S), and micronutrients (e.g., Fe, Mn) into the proper oxidation states for incorporation in biomolecules. During the history of the earth, biological redox activity has caused the appearance of a highly oxidizing surface environment overlying a sedimentary cover rich in highly reduced biogenic products, such as organic matter, sulfide minerals, and methane. Along this global redox gradient, the numerous potential combinations of electron donors, electron acceptors, and carbon sources have given rise to a tremendous ecological and metabolic diversity of microorganisms (DeLong and Pace, 2001). Redox gradients in time

Reaction Mechanisms in Environmental Engineering. https://doi.org/10.1016/B978-0-12-804422-3.00007-9

and space can be found, for example, in poorly drained or groundwater-fed soils, riparian zones, coastal marshes, shallow aquifers, contaminant plumes, lakes, and estuaries. The sequence of redox reactions that occurs along redox gradients in time or space depends on chemical composition, microbial activity, and the reduction potentials of relevant redox couples.

In the context of this book, oxidation–reduction (redox) reactions are commonly understood to occur by the exchange of electrons between reacting chemical species. Electrons are lost (or donated) in oxidation and gained (or accepted) in reduction. Oxidation of a species is caused by an oxidizing agent (or oxidant), which accepts electrons (and is thereby reduced). Similarly, reduction results from reaction with a reducing agent (or reductant), which donates electrons (and is oxidized). Reduction occurs primarily in water-saturated environments, such as sediments, soils, and sludge. Redox indicators can be used (1) as chemical probes to obtain fundamental insights into biogeochemical processes and (2) as the basis for demonstrations suitable for teaching aspects of environmental chemistry.

A redox reaction (reduction–oxidation reaction) is a chemical reaction in which the oxidation states of the atoms are changed. Any such reaction involves both a reduction process and a complementary oxidation process, both of which involve electron transfer processes. The chemical species from which the electron is stripped is said to have been oxidized, while the chemical species to which the electron is added is said to have been reduced. Thus: (1) oxidation is the loss of electrons or an *increase* in oxidation state by a molecule, atom, or ion, while (2) reduction is the *gain* of electrons or a *decrease* in oxidation state by a molecule, atom, or ion. The reaction can occur relatively slowly, as in the case of rust, or more quickly, as in the case of combustion. Thus, oxidation refers to the loss of electrons while reduction refers to the gain of electrons. There are simple redox processes, such as the oxidation of carbon to carbon dioxide (CO_2) or the reduction of carbon by hydrogen to yield methane (CH_4).

$$C + O_2 \rightarrow CO_2$$

$$C + 2H_2 \rightarrow CH_4$$

In conjunction with redox reactions, it is necessary to know the oxidation number of the compound. The oxidation number is the effective charge on an atom in a compound, calculated according to a prescribed set of rules. An increase in oxidation number corresponds to oxidation, and a decrease to reduction. Also, the oxidation number suggests the strength or tendency of the compound to be oxidized or reduced, to serve as an oxidizing agent or reducing agent.

These above given definitions of oxidation and reduction are adequate for most purposes, but not all. Just as acid–base concepts have proton-specific definitions (the Brönsted model) and more general definitions (e.g., the

Lewis model), redox concepts can be extended from electron transfer specific definitions to more general definitions that are based on electron density of chemical species. The latter allows for redox reactions that occur by atom transfer, as well as electron transfer mechanisms. While often ignored, the role of atom transfer mechanisms can be important, particularly in redox reactions involving organic compounds. Redox reactions, defined inclusively, are central to many priority and emerging areas of research in the aquatic sciences. This scope includes all aspects of the aquatic sciences: not just those involving the hydrosphere, but also aquatic (i.e., aqueous) aspects of environmental processes in the atmosphere, lithosphere, biosphere, etc. As a field of study, aquatic redox chemistry has multidisciplinary roots (spanning mineralogy to microbiology) and interdisciplinary applications (e.g., in removal of contaminants from water, sediment, or soil).

The general rules relating to oxidation number are (1) for atoms in their elemental form, the oxidation number is 0, (2) for ions, the oxidation number is equal to their charge, (3) for single hydrogen, the number is usually +1 but in some cases it is −1, (4) for oxygen, the number is usually −2, and (5) the sum of the oxidation number (ONs) of all the atoms in the molecule or ion is equal to its total charge.

An example of a redox reaction that occurs in the environment is the oxidation of sulfur dioxide as it pertains to the formation of acid rain and its effects on the environment (Wondyfraw, 2014). Thus, sulfur dioxide (SO_2) is released into the atmosphere naturally through volcanic activity as well as from the combustion of sulfur-containing fuels. The sulfur dioxide is carried in the atmosphere by the prevailing winds during which time the sulfur dioxide dissolves in water, ultimately forming sulfuric acid which represents a change in the oxidation state of sulfur from +4 to +6:

$$[S]_{\text{fossil fuel}} + O_2 \rightarrow SO_2$$

$$2SO_2 + O_2 \rightarrow 2SO_3$$

$$SO_2 + H_2O \rightarrow H_2SO_3 \text{ (sulfurous acid)}$$

$$SO_3 + H_2O \rightarrow H_2SO_4 \text{ (sulfuric acid)}$$

In situ chemical oxidation involves the introduction of a chemical oxidant into the subsurface for the purpose of transforming ground water or soil contaminants into less harmful chemical species.

The four most commonly used oxidants: (1) permanganate (MnO_4^-), (2) hydrogen peroxide (H_2O_2) and iron (Fe) (Fenton-driven, or hydrogen peroxide–derived oxidation), (3) persulfate ($S_2O_8^{2-}$), and (4) ozone (O_3). The type and physical form of the oxidant indicates the general material handling and injection requirements. The persistence of the oxidant in the subsurface is important since this affects the contact time for advective (the transport of a chemical by bulk motion) and diffusive transport and ultimately the delivery of oxidant to targeted zones in the subsurface. For example, permanganate persists for long periods of time, and diffusion into low-permeability materials and greater transport distances through porous media are possible. Hydrogen peroxide has been reported to persist in soil and aquifer material for minutes to hours, and the diffusive and advective transport distances will be relatively limited. Radical intermediates formed using some oxidants (hydrogen peroxide, H_2O_2, persulfate, $S_2O_8^{2-}$, and ozone, O_3) that are largely responsible for various contaminant transformations react very quickly and persist for very short periods of time (typically, less than 1 s).

Anaerobic sediments and permeable reactive barriers of granular iron metal are two examples of environmental systems where the characteristic pathway for contaminant transformation is reduction. Redox indicator dyes can be used to probe the physicochemical properties of these reducing systems in research on the environmental fate of contaminants, but also to form educational exercises that illustrate some of the many facets of environmental chemistry. Key concepts that can be highlighted with the methods described in this manuscript include (1) thermodynamic versus kinetic control on reactivity, (2) mass transport versus chemical control on reaction kinetics, (3) initial rates and pseudo first-order kinetics, (4) the relationship between chemical structure and reactivity, (5) reversibility, (6) stability analysis with E and pH as master variables, (7) adsorption versus reaction at surfaces, (8) chemical versus biologically mediated reactions in natural systems, (9) spatial and temporal variability in environmental samples, and (10) the chemical nature of both pristine and contaminated environments.

Once environmental conditions are established, however, many important redox reactions proceed without further mediation by organisms. These reactions are considered to be abiotic when it is no longer practical (or possible) to link them to any particular biological activity. Thus, the overall favorability of these processes is not necessarily affected by microbiological mediation (i.e., the redox ladder applies either way). However, currently systems where biotic and abiotic controls on contaminant fate are closely coupled are frontier areas of research.

Also, microbial respiration indirectly influences the speciation of trace metals via changes in sorption and precipitation equilibria, as well as solid–solution partitioning of metals. For example, the reductive dissolution of metal oxides may result in the loss of sorption capacity. At the same time, the released Fe^{2+} may compete with trace metal cations for sorption to mineral and organic sorbents. Such mobilizing effects may be counteracted by trace metal adsorption or coprecipitation with newly formed Fe(II)-bearing precipitates, whose formation is promoted by increasing Fe^{2+}, carbonate, or phosphate concentrations resulting from microbial respiration.

A fundamental reason for complexity in assessing the redox conditions of many environmental systems is that most aqueous redox reactions in such systems are kinetically limited and therefore not in equilibrium with each other

(Borch et al., 2010; Grundl et al., 2011). Under these circumstances, potential measurements made with an inert electrode (e.g., Pt) are mixed potentials in which each redox couple in contact with the electrode exchanges electrons independently and the electrode response is the sum of anodic (reductive) and cathodic (oxidative) currents, each weighted by the corresponding exchange current density (a measure of the sensitivity of the electrode to particular species). This mixed potential does not necessarily reflect equilibrium among the couples or between the electrode and any particular couple; so the relationship between mixed potentials measured on complex environmental sample and specific redox active species in the sample is not well defined. The theoretical and practical difficulties with direct potentiometry as a means to define the redox intensity of aquatic systems were a major issue in the early literature on aquatic redox chemistry.

It is important to recognize that the need for a process-level understanding is not limited to the specifics of a given electron transfer reaction but also extends to external constraints affecting the progress of the reaction. External constraints include field-scale effects such as mass transfer limitations and heterogeneity in both the flow regime and the reactivity that cause competition between the residence time and the characteristic reaction time of solutes within a moving parcel of water or between aquifer and aquitard. Whether the reaction is driven biotically or abiotically will affect the rates of reaction directly or in the case of smectite reduction, completely change the mechanism of electron transfer. Obviously, the microbial consortia that drive biotic processes are subject to electron donor/acceptor and nutrient availability, as well as to the buildup of metabolic waste products.

In terms of the chemical cycles of the earth, the biogeochemical cycles of many major and trace elements are driven by redox processes. Examples include the cycles of carbon (C), nitrogen (N), sulfur (S), iron (Fe), and manganese (Mn), as well as those of redox-sensitive trace elements such as arsenic (As), chromium (Cr), copper (Cu), and uranium (U). The chemical speciation, bioavailability, toxicity, and mobility of these elements in the environment are directly affected by reduction and oxidation. In addition, the biogeochemical behavior of other not redox-active elements and compounds may be indirectly coupled to redox transformations of natural organic matter (NOM) and mineral phases, in particular (hydr)oxides of iron (Fe) and manganese (Mn), Fe-bearing clay minerals, and Fe sulfides. Redox-active functional groups associated with humic substances and mineral surfaces can further catalyze the oxidation or reduction of ions and molecules, including many organic chemical contaminants (Borch et al., 2010). Redox processes may also provide new opportunities for engineered remediation strategies, such as the in situ microbial degradation of organic solvents and reductive sequestration of uranium from groundwater. Thus, understanding the biogeochemical processes at environmental redox interfaces is crucial for predicting and protecting the various ecosystems.

For example, the carbon cycle is driven by oxygenic photosynthesis, which fixes carbon dioxide by removing electrons from water, thereby producing oxygen. Anoxygenic photosynthesis and chemolithoautotrophic carbon fixation may be locally important sources of biomass, especially in more extreme environments (D'Hondt et al., 2002; Overmann and Manske, 2006). Sequestration of organic matter and other biologically derived reduced materials, for example, the mineral pyrite (FeS_2), in sedimentary deposits and their subsequent uplift and oxidative weathering at the surface of the earth are key processes controlling the composition of the atmosphere, and hence the climate of the earth, on a geological time scale. The behavior of nitrogen and phosphorus in the environment is closely linked to redox processes of carbon. Nitrogen exhibits a variety of oxidation states; many of its redox transformations, for example, nitrogen fixation, nitrification, denitrification, dissimilatory nitrate reduction to ammonium and anammox, are carried out by microorganisms. These microbial processes affect nitrogen availability, and therefore organic matter production and recycling, from local to global scales.

Links also exist between the redox cycle of nitrogen and those of iron, sulfur, phosphorus, and trace elements. For example, denitrification coupled to pyrite oxidation can be an important nitrate removal pathway in fertilizer-impacted aquifers (Postma et al., 1991). However, the sulfate produced by this process may fuel microbial sulfate reduction, which, in turn, may lead to the reductive dissolution of ferric (hydr)oxides by sulfide, thereby releasing sorbed phosphate. When the phosphate-rich groundwater subsequently discharges into a surface water body, it may cause eutrophication (Lucassen et al., 2004). This example illustrates the highly coupled nature of biogeochemical redox cycles.

As the most abundant transition metal on the earth's surface, iron plays a particularly important role in environmental biogeochemistry. The oxidized form of iron (Fe^{3+}) is soluble under extremely acidic conditions but ferric iron precipitates as ferric (hydr)oxides in near-neutral pH environments. Many nutrients, trace elements, and contaminants strongly sorb to these ferric minerals. The surfaces of the ferric (hydr)oxides also catalyze many environmentally important redox transformations (Table 7.1).

Under reducing conditions, ferric (hydr)oxides can be reduced abiotically, for example, by sulfide (Afonso and Stumm, 1992), thereby releasing potentially harmful sorbates. In addition, ferric iron minerals may serve as terminal electron acceptors to dissimilatory iron-reducing bacteria. These organisms, which are ubiquitous in freshwater and marine environments, are able to couple the cytoplasmic oxidation of organic compounds or hydrogen to the extracellular reduction of poorly soluble ferric (Fe^{3+}) minerals, thus gaining energy for growth via electron transport phosphorylation. Reduction of ferric minerals produces soluble ferrous (Fe^{2+}) products and a wide range of secondary minerals, including ferrous minerals (e.g., vivianite [$Fe(PO_4)_2$] and siderite [$FeCO_3$]), ferric minerals (e.g., goethite), and mixed ferrous–ferric

TABLE 7.1 Examples of the Advantages and Disadvantages of In Situ Oxidation Processes

Advantages
Contaminant mass can be destroyed in situ.
Rapid destruction/degradation of contaminants (measurable reduction in weeks or months).
Produces no significant wastes (volatile organic chemical off-gas is minimal), except Fenton reagent method.
Compatible with posttreatment monitored natural attenuation.
Can enhance aerobic and anaerobic biodegradation of residual organic contaminants.
May cause only minimal disturbance to site operations.
Disadvantages
Contamination in low permeability soils may not be readily contacted by oxidant.
Fenton reagent can produce significant quantity of explosive off-gas.
Dissolved contaminant concentrations may rebound weeks or months after treatment.
Contaminant plume configuration may be altered by chemical oxidation application.
May not be able to reduce contaminants to background or very low concentrations.
Significant losses of chemical oxidants may occur as they react with soil/bedrock material.
May alter aquifer flow patterns by precipitation of minerals in pore spaces.

minerals (such as magnetite [Fe_3O_4] and green rusts [ferrous–ferric layered double hydroxides]). Dissolved, adsorbed, and solid state ferrous iron can act as powerful reductants in a variety of abiotic redox processes.

Oxidation of ferrous iron may be mediated by aerobic and anaerobic bacteria. Microbial oxidation of ferrous iron is common in low pH environments where chemical oxidation is not favorable. Under near-neutral pH conditions, aerobic ferrous iron-oxidizing bacteria have to compete with the rapid chemical oxygenation of ferrous iron. Therefore, they thrive mainly at the low levels of oxygen typically encountered at oxic–anoxic interfaces, such as around plant roots in waterlogged soils. Phototrophic and nitrate-dependent ferrous iron-oxidizing bacteria mediate ferrous iron oxidation in neutrophilic, anoxic environments, where nitrite and manganese (Mn^{4+}) oxides can serve as chemical oxidants for ferrous iron. Manganese oxide minerals commonly occur in the environment as coatings and fine-grained aggregates.

Because of the large surface area, many metal oxides exert chemical influences far out of proportion to their abundance. Manganese oxides are potent sorbents of heavy metals and nutrients, serving as natural sinks for contaminants. Additionally, they participate in a wide variety of environmental redox reactions; for example, birnessite (δ-MnO_2) directly oxidizes quadrivalent selenium (Se^{4+}) to hexavalent selenium (Se^{6+}), trivalent chromium (Cr^{3+}) to quadrivalent chromium (Cr^{4+}), and trivalent arsenic (As^{3+}) to pentavalent arsenic (As^{5+}). The oxidation of manganese (Mn^{2+}) occurs in a wide array of environments and may be catalyzed by a variety of bacteria and fungi. The initial products of biological Mn(II) oxidation are generally poorly crystalline, layered manganese (Mn^{4+}) oxide minerals, although the final mineral form is often determined by the geochemical conditions during and after manganese (Mn^{2+}) oxidation. Biologically mediated manganese (Mn^{2+}) oxidation is a possible major source of environmental manganese oxides. The structure and reactivity of ferric mineral surface and manganese (Mn^{3+}, Mn^{4+}) mineral surfaces in natural waters are influenced by inorganic sorbates and natural organic matter (NOM).

In fact, NOM is an inherently complex and inseparable group of molecules primarily resulting from the partial decay of senescent plant materials and microorganisms. NOM is ubiquitous in natural waters and in soils and, in addition, to redox chemistry of NOM, it plays significant roles in metal transport; microbial, photochemical, and water treatment processes; and in soil fertility. Because of its complexity and wide range of molecular sizes and physical properties, NOM has often been conceptually divided into separate classes of molecules. This classification has generally been based on solubility and pH properties of the NOM. NOM is conveniently subdivided into three types: (1) humin, which is water insoluble at all pH values, (2) humic acid, which is base-soluble but reprecipitates at slightly acid pH, and (3) fulvic acids are solubilized by base and remain in solution under acid conditions (Eaton et al., 1985). All three types can influence the adsorption and mobility of chemicals in soils.

Humic substances are redox-active and can be reduced by microorganisms. They can stimulate microbial reduction of poorly soluble ferric minerals by acting as electron shuttles between cell and mineral. Strong adsorption of NOM phosphate, and bicarbonate to ferric and manganese (Mn^{4+}) mineral surfaces may cause a release of sorbed pollutants but may also restrict access of the minerals to microorganisms, thereby protecting the solid phases against enzymatic reduction. Humic substances further influence the formation (biomineralization) of iron and manganese minerals, through complexation and solubilization of the metal ions and by sorption to the mineral surface, generally causing less crystalline minerals to form.

Biogeochemical redox processes strongly influence the environmental fate of metalloids such as arsenic (As) and antimony (Sb). Whereas As has by far received the most attention during the last two decades due to the worldwide health impacts of arsenic-contaminated drinking water and soils. Antimony may represent an important contaminant, for instance, in the vicinity of copper and lead ore smelters or at shooting range sites due to the weathering of bullets. Oxidation of zero valent antimony (Sb^0) to multivalent antimony (Sb^{3+} or Sb^{5+}), as well as sorption to ferric (hydr)oxides, controls the toxicity and mobility of antimony at these sites.

The mobility and toxicity of arsenic is not only influenced by the presence of suitable sorbents, but also by its redox speciation, with arsenite generally considered to be more mobile and toxic than its oxidized counterpart arsenate. Microbial reduction of pentavalent arsenic (As^{5+}) to lower-valency arsenic (As^{3+}) has been proposed to contribute to arsenic mobilization although this may not be universally true. Several biogeochemical processes can directly or indirectly lead to redox transformations of arsenic. Abiotic oxidation of trivalent arsenic (As^{3+}) has, for example, been observed at reactive iron barriers, in ferrous–goethite systems, or in the presence of higher valent iron formed by hydrogen peroxide–dependent Fenton reactions.

Bacteria can also control arsenic mobility and toxicity directly by changing the redox state of arsenic via pentavalent (As^{5+}) reduction and oxidation of trivalent arsenic (As^{3+}). These microbially catalyzed arsenic redox transformations can be part of dissimilatory processes or reflect microbial detoxification mechanisms. Microorganisms may further indirectly induce trivalent arsenic (As^{3+}) oxidation or reduction of pentavalent arsenic (As^{5+}). In particular, the microbial species can produce reactive organic or inorganic compounds that subsequently undergo redox reactions with pentavalent arsenic (As^{5+}) or trivalent arsenic (As^{3+}). Also, the semiquinone radicals and hydroquinone derivatives in humic substances and humic-quinone model compounds can oxidize trivalent arsenic (As^{3+}) to pentavalent arsenic (As^{5+}) and reduce pentavalent arsenic (As^{5+}) to trivalent arsenic (As^{3+}), respectively. Such semiquinone and hydroquinone species are produced during microbial reduction of humic substances, a process that plays an important role in electron shuttling during microbial reduction of ferric minerals. Additionally, it is conceivable that the arsenic redox state can be changed by reactive ferrous–ferric mineral systems as observed, for example, in the oxidation of trivalent arsenic by ferrous iron–goethite systems. These reactive iron phases can form when ferrous iron produced during microbial reduction of ferric-iron minerals sorbs on to remaining ferric minerals. Such reactive systems have also been shown to be able to reductively transform metal ions such as hexavalent uranium (U^{6+}).

2. OXIDATION REACTIONS

The effective removal of chemical contaminants from contaminated media, such as soil, is an essential problem because of environmental concerns. One of the major groups of organic chemical contaminants, which causes harmful consequences, endangers natural environment and human health, is crude oil and oil products. It is explained by their ability to form toxic compounds in soils and superficial and ground water. The remediation of the environment from crude oil contaminants is especially difficult and most essential for enterprises and regions involved in crude oil production.

An effective solution of this problem is in situ chemical oxidation (ISCO) (Karpenko et al., 2009), which is a technique used to remediate contaminated soil and groundwater systems and utilizes chemical agents capable of oxidizing organic chemical contaminants (i.e., active chemical oxidants). Physicochemical processes, which are able to destroy persistent contaminants, are affected less by environmental and contaminant factors. The oxidants used are readily available, and treatment time is usually measured in months (or even days) rather than years, making the process economically feasible. In situ chemical oxidation (ISCO) is accomplished by introducing the agents into the soil or aquifer at a contaminated site using injection and mixing apparatuses.

Although chemical oxidation processes have been studied and publicized for decades, the use of chemical oxidation for remediation applications is limited to the past decade. The effectiveness of oxidation depends upon the geological conditions, the residence time of the oxidant, the amount of oxidant used and the effective contact between oxidant and the contaminant(s). Chemical oxidation offers several advantages (as well as disadvantages) over other in situ or ex situ remediation technologies: (1) rapid treatment time and treatment of concentrated contaminants, and (2) effectiveness for a diverse group of contaminants. The disadvantages are: (1) nonselective oxidation, that is, the oxidant may react not only with the definite contaminants but with substances in the soil which can be readily oxidized, as well, (2) control of pH, (3) control of the temperature, and (4) control of the contact time is important to ensure the desired extent of oxidation; (3) the costs necessary for the implementation of this process are rather high (Table 7.1).

Chemical oxidation is one half of a redox reaction, which results in the loss of electrons. One of the reactants in the reaction becomes oxidized, or loses electrons, while the other reactant becomes reduced, or gains electrons. In an ISCO process, oxidizing compounds, compounds that give electrons away to other compounds in a reaction, are used to change the contaminants into harmless compounds (Siegrist et al., 2001).

ISCO processes are chemical oxidation processes that occur at the site of the contamination and are a form of controlled oxidation coupled with an advanced oxidation technology, which can be employed as remediation techniques for application to polluted soil and/or groundwater, to reduce the concentrations of targeted environmental contaminants to acceptable levels. ISCO is accomplished by injecting or otherwise introducing strong chemical oxidizing agents directly into the contaminated medium (soil or groundwater) to destroy chemical contaminants in place. It can be used to remediate a variety of organic compounds, including some that are resistant to natural degradation.

A wide range of soil contaminants and groundwater contaminants react either moderately or highly with the ISCO method and, furthermore, the ISCO can be applied in a variety of different situations (e.g., unsaturated vs. saturated ground, above ground or underground, etc.); so it is a popular method to use. For example, the remediation of sites contaminated with chlorinated solvents, such as trichloroethylene and tetrachloroethylene, as well as low-boiling gasoline constituents (such as benzene, toluene, ethylbenzene, and the xylene isomers [BTEX]) is also possible using the ISCO method (Stroo et al., 2010). In addition, other chemicals contaminants that are not completely removable can be made less toxic through chemical oxidation.

Fenton's reagent (hydrogen peroxide catalyzed with iron) and potassium permanganate are the oxidants that have been used the most frequently over the past decades and are remain the most widely using oxidants, now used the most widely. For example, hydrogen peroxide (H_2O_2) was first used in 1985 to treat a formaldehyde spill. In this process 10% v/v aqueous solution of hydrogen peroxide was injected into a formaldehyde (HCHO) plume. Fenton reagent was initially used to treat hydrocarbon sites where benzene, toluene, and ethylbenzene were present (Pignatello et al., 2006).

As the industry shifted its focus to the remediation of chlorinated solvents, hydrogen peroxide was found to be effective in both the hydrocarbon industry and the chlorinated solvent industry. Scientists also found that permanganate could be used on chlorinated solvents.

Hydrogen peroxide has such a short life that it cannot be transported correctly or safely. Sodium (or potassium) permanganate only treats chlorinated solvents with double bonds and is easily used up by organic material in soil. Sodium persulfate is more stable, treats a wider range of contaminants, and is not used up by soil organics as easily.

Common oxidants used in the in situ oxidation process are (1) permanganate, both sodium permanganate ($NaMnO_4$) and potassium permanganate ($KMnO_4$), Fenton reagent (a solution of hydrogen peroxide with ferrous iron, Fe^{2+}, as that is used to oxidize contaminant chemicals), persulfate, and ozone. Other oxidants can be used, but these four are the most commonly used.

2.1 Permanganate

As an oxidizer, permanganate has a unique affinity for destroying organic compounds containing carbon–carbon double bonds. The permanganate (MnO^{4-}) ion is strongly attracted to the negative charge associated with electrons in the *pi*-cloud of carbon–carbon double bonds (>C=C<) of chlorinated alkenes and borrows electron density from the *pi*-bond (π-bond), which disturbs the carbon–carbon double bond, thus forming hypomanganate diester. This intermediate product is unstable and further reacts by a number of mechanisms including hydroxylation, hydrolysis, or cleavage. The final oxidation product is carbon dioxide, chloride salt, and manganese dioxide. Other contaminants may be oxidized by free radical oxidation.

Although the reaction of the permanganate ion with contaminants is well understood, these reactions have not been commonly used for in situ remediation due to complications resulting from precipitate of manganese dioxide. Plugging of the soil matrix with MnO_2 has been observed when permanganate salts are injected into the soils through injection wells or points, which results in the poor contact of the oxidant with the contaminants. However, the manganese dioxide precipitates are unstable as they are readily reduced by chemical or biological in situ processes to form soluble divalent manganese ions (Karpenko et al., 2009).

Sodium permanganate ($NaMnO_4$) or potassium permanganate ($KMnO_4$) is used in groundwater remediation. Both compounds have the same oxidizing capabilities and limitations and react similar to contaminants. The biggest difference between the two chemicals is that potassium permanganate is less soluble than sodium permanganate. Because the temperature in the aquifer is usually less than the temperature in the area that the solution is mixed, the potassium permanganate may revert to the solid form which does not react readily with the contaminants but over time, depending on the temperature, the potassium permanganate will become soluble again. This compound (potassium permanganate) has the ability to oxidize many different contaminants but is notable for oxidizing chlorinated solvents such as perchloroethylene (tetrachloroethylene, CCl_2CCl_2, PCE), trichloroethylene ($CHClCCl_2$, TCE), and vinyl chloride (CH_2CHCl, VC). However, potassium permanganate is unable to efficiently oxidize diesel, gasoline, or the BTEX mixtures.

Sodium permanganate is more expensive than potassium permanganate, but because sodium permanganate is more soluble than potassium permanganate, it can be applied to the site of contamination at a much higher concentration. This shortens the time required for the contaminant to be oxidized. Sodium permanganate is also useful in that it can be used in places where the potassium ion cannot be used. Another advantage that sodium permanganate has over potassium permanganate is that sodium permanganate, due to its high solubility, can be transported above ground as a liquid, decreasing the risk of exposure to granules or skin contact with the substance.

The reaction stoichiometry of permanganate salts in natural systems is rather complex. Due to its multiple valence states and mineral forms, manganese can participate in numerous reactions. The reactions proceed according to the second-order kinetics and at a much slower rate compared with the reactions of hydrogen peroxide or ozone. Depending on pH, the reaction can include destruction by direct electron transfer or free radical advanced oxidation. Permanganate reactions are effective over a pH range in the order of 3.5–12. The stoichiometric reaction for the complete destruction of trichloroethylene by potassium permanganate ($KMnO_4$) is:

$$2KMnO_4 + C_2HCl_3 \rightarrow 2CO_2 + 2MnO_2 + 2KCl + HCl$$

The primary redox reactions for permanganate are given by the equations:

$$MnO_4^- + 8H^+ + 5e^- \rightarrow Mn^{2+} + 4H_2O \qquad (\text{at pH} < 3.5)$$

$$MnO_4^- + 2H_2O + 3e^- \rightarrow MnO_2(s) + 4OH^- \qquad (\text{at pH} < 3.5\text{–}12)$$

$$MnO_4^- + e^- \rightarrow MnO_4^{2-} \qquad (\text{at pH} > 12)$$

The typical reactions that occur under common environmental conditions, the reactions at pH 3.5−12, when the product is solid manganese dioxide (MnO_2).

The advantage of using sodium or potassium permanganate in an ISCO process is that it reacts comparatively slowly in the subsurface, which allows the compound to move further into the contaminated space and oxidize more contaminants. The permanganate-based process can also help with the cleanup of materials that have a low permeability. In addition, because both sodium permanganate and potassium permanganate solutions have a density greater than density of water, permanganate can travel through the contaminated area through density-driven diffusion.

As noted above, the use of permanganate creates the byproduct manganese dioxide (MnO_2) which is naturally present in the soil and is, therefore, often classified as a safe byproduct and is relatively benign. However, the relative toxicity and safety of chemical is dependent upon the amount present compared to the indigenous amount of the chemical in an ecosystem. Unfortunately, manganese dioxide can act as a mastic in and cementation of sand particles to form a low-permeability rock-like material. As the rock-like materials build up, the permanganate is prevented (blocked) from contacting the remainder of the chemical contaminant thereby lowering the efficiency of the permanganate-based process. This effect can be mitigated by extracting the manganese dioxide from the contaminated area.

Permanganate as an oxidant is weaker than hydrogen peroxide. Its inability to oxidize benzene can lead to the early elimination of permanganate as a candidate for oxidation technology at petroleum cleanup sites. However, permanganate has several advantages over other oxidants: (1) oxidation of organic chemicals over a wider pH range, (2) reactivity over a prolonged period in the subsurface allowing the oxidant to permeate soil more effectively and contact adsorbed chemical contaminants, and (3) absence of heat, steam, and vapors or associated health and safety concerns.

As with other chemical oxidation technologies, the success of the permanganate use relies heavily on its ability to come into contact with the site contaminants. The delivery mechanism must be capable of dispersing the oxidant throughout the treatment zone. To accomplish this, permanganate may be delivered in a solid or liquid form in a continuous or cyclic application schedule using injection probes, soil fracturing, soil mixing, or groundwater recirculation.

In situ permanganate reactions can yield low pH (e.g., pH = 3) which can temporarily mobilize naturally occurring metals and metal contaminants, which may also be present in the treatment area. The release of these metals from the aquifer formation, however, may be offset by sorption of the metals onto strongly sorbent manganese dioxide (MnO_2) solids that are precipitated as a byproduct of permanganate oxidation.

Trivalent chromium (Cr^{3+}) in the form of chromium trioxide [$Cr(OH)_3$] in soil may be oxidized to hexavalent chromium (Cr^{6+}), which may persist for some time. This may be dangerous if the aquifer is used for drinking water. Thus, regardless all advantages of the application of permanganate salts the questions remain about their efficiency as well as about MnO2 that is generated, and the effect, which it may have on subsurface permeability and remediation performance.

2.2 Fenton Reagent

Fenton reagent is a mixture of hydrogen peroxide (H_2O_2) and ferrous (Fe^{2+}) iron salts which act as a catalyst for the oxidation process. Hydrogen peroxide is a strong oxidant that can be injected into a contaminated zone to destroy organic chemical contaminants. When injected to groundwater, hydrogen peroxide is unstable, and reacts with organic chemical contaminants and subsurface materials. It has long been known that hydrogen peroxide decomposes to oxygen and water within hours of its introduction into groundwater generating heat in the process (Stewart, 1984).

When the peroxide is catalyzed by soluble iron, hydroxyl radicals (OH•) are formed that serve to oxidize contaminants such as chlorinated solvents, fuel oil, and BTEX mixtures. When the Fenton reaction is applied to ISCO, the collective reaction results in the degradation of contaminants in the presence of ferrous iron (Fe^{2+}) as a catalyst and the overall simplified reaction for the process:

$$H_2O_2 + \text{carbonaceous contaminant} \rightarrow H_2O + CO_2 + O_2$$

Oxidation using liquid hydrogen peroxide in the presence of native or supplemental ferrous iron (Fe^{+2}) produces Fenton reagent:

$$H_2O_2 + Fe^{2+} \rightarrow HO\bullet + OH^{+} + Fe^{3+}$$

This optimal reaction occurs under relatively low pH (e.g., pH = 2–4; high acidity). Using a ferrous sulfate solution ($FeSO_4$ aq) simultaneously adjusts aquifer pH and adds the iron catalyst needed for Fenton reagent. Because of the low pH requirement, Fenton reagent treatment may not be effective in limestone geology or sediments with elevated pH levels. In addition, the reaction between hydrogen peroxide and ferric iron can consume hydrogen peroxide, reducing the effectiveness of the oxidant dose. The same effect may also occur in soils with high ferric iron (Fe^{3+}) content.

However, the actual chemistry of the Fenton reaction is complex and has many steps and is typically represented by the equations:

$$Fe^{2+} + H_2O_2 \rightarrow Fe^{3+} + \bullet OH + OH^{-}$$

$$Fe^{3+} + H_2O_2 \rightarrow Fe^{2+} + \bullet OOH + H^{+}$$

The typical process requires (1) adjusting the waste water to pH 3–5, (2) adding the iron catalyst as a solution of ferrous sulfate, $FeSO_4$, and (3) adding the hydrogen peroxide slowly. If the pH is too high, the iron precipitates as ferric hydroxide, $Fe(OH)_3$, and catalytically decomposes the hydrogen peroxide to oxygen thereby creating a potentially hazardous situation.

The reaction rates of the process with Fenton reagent are generally limited by the rate of generation of the hydroxyl radical (HO•) which depends upon the concentration of iron catalyst rather than by the specific waste water being treated. Typical iron peroxide ratios are in the order of 1:5 to 10 w/w, though iron levels that are less, 25–50 mg/L, can require excessive reaction times (10–24 h). This is particularly true where the oxidation products (organic acids) sequester the iron and remove it from the catalytic cycle. Fenton reagent is most effective as a pretreatment method, where the chemical oxygen demand is >500 mg/L. This is due to the loss in selectivity as pollutant levels decrease:

$$HO\bullet + H_2O_2 \rightarrow HOO\bullet + H_2O$$

$$HO\bullet + Fe^{3+} \rightarrow Fe^{2+} + OH^-$$

There are a number of subsequent reactions that can lead to the production of the perhydroxyl radical ($HO_2\bullet$), as well as the scavenging of the hydroxyl radical (•OH). Both hydrogen peroxide and the free radicals will oxidize contaminants of concern, though their reaction rates vary. Hydrogen peroxide alone and Fenton reagent react with the long-chain carbon sources prior to oxidizing the lower-boiling hydrocarbon derivatives but Fenton reagent reacts faster and is much more efficient due to the genesis of free hydroxyl radicals. The hydroxyl radical rapidly reacts nonselectively by attacking and cleaving the carbon–hydrogen bonds of contaminant organic compounds and are used to treat many organics such as chlorinated and nonchlorinated solvents, and polynuclear aromatic hydrocarbon derivatives. Intermediate compounds formed because of the reaction with the hydroxyl radical can then be subsequently oxidized to carbon dioxide and water in the presence of excess oxidant (Kluck and Achari, 2015).

In addition to free radical scavengers, the process is inhibited by (iron) chelants such as phosphates, ethylenediaminetetraacetic acid, formaldehyde, citric acid, and oxalic acid. Because of the sensitivity of Fenton reagent to different waste waters, it is recommended that the reaction always be characterized through laboratory treatability tests before proceeding to plant scale.

As noted above, the Fenton reagent requires a significant pH reduction of the soil or the groundwater in the treatment zone for the process to be successful and to allow for the introduction and distribution of aqueous iron because the iron will oxidize and precipitate at a pH greater than 3.5. Unfortunately, the contaminated groundwater that needs to be treated has a pH level that is at or near neutral (pH = 7). Due to this, there have been questions relating to whether or not the use of Fenton reagent for ISCO processes is a manifestation of the Fenton reaction. Instead, scientists call these reactions Fenton-like. However, some ISCO processes successfully apply pH neutral

Fenton reagent by chelating the iron which keeps the iron in solution and mitigates the need for acidifying the treatment zone (Karpenko et al., 2009).

The advantages of using the Fenton reagent process method include that the hydroxyl radicals are very strong oxidants and react very rapidly with contaminants and impurities in the ground water. Moreover, the chemicals needed for this process are inexpensive and abundant.

Traditional Fenton reagent applications can be highly exothermic when significant iron, manganese, or contaminant (that is, nonaqueous phase liquids or nonaqueous polar liquids often represented by the acronym NAPL) are present in an injection zone. Over the course of the reaction, the groundwater heats up and, in some cases, reagent and vapors can surface out of the soil. Stabilizing the peroxide can significantly increase the residence time and distribution of the reagent while reducing the potential for excessive temperatures by effectively isolating the peroxide from naturally occurring divalent transition metals in the treatment zone. However, the NAPL contaminant concentrations can still result in rapid oxidation reactions with an associated temperature increase and more potential for surfacing even with reagent stabilization. The hydroxyl radicals can be scavenged by carbonate, bicarbonate, and naturally occurring organic matter, in addition to the targeted contaminant; so it important to evaluate a site's soil matrix and apply additional reagent when these soil components are present in significant abundance.

The hydroxyl radical, a major participant in the Fenton reaction, is one of the most reactive chemical species known, second only to elemental fluorine. The chemical reactions of the hydroxyl radical in water are of four types:

Addition reaction, in which the hydroxyl radical adds to an unsaturated compound, aliphatic or aromatic, to form a free radical product:

$$HO\bullet + C_6H_6 \rightarrow (OH)C_6H_6$$

Hydrogen abstraction, in which organic free radical and water are formed:

$$HO\bullet + CH_3OH \rightarrow \bullet CH_2OH + H_2O$$

Electron transfer, in which ions of a higher valence state are formed, or an atom or free radical is formed if a mono-negative ion is oxidized:

$$HO\bullet + [Fe(CN)_6]^{4-} \rightarrow [Fe(CN)_6]^{3-} + OH^-$$

Radical interaction, in which the hydroxyl radical reacts with another hydroxyl radical, or with an unlike radical, to form a stable product.

$$HO\bullet + \bullet OH \rightarrow H_2O_2$$

When the Fenton reaction is applied to the treatment of industrial waste treatment, the conditions of the reaction are adjusted so that first two mechanisms (hydrogen abstraction and oxygen addition) predominate.

In the absence of iron, there is no evidence of hydroxyl radical formation when, for example, hydrogen peroxide is added to a phenol-containing waste water (that is, no reduction in the level of phenol occurs). As the concentration of iron is increased, phenol removal accelerates until a point is reached where further addition of iron becomes inefficient. This feature (an optimal dose range for iron catalyst) is characteristic of the Fenton reaction, although the definition of the range varies between waste waters. Three factors typically influence its definition: (1) a minimal threshold concentration of 3–15 mg/L iron which allows the reaction to proceed within a reasonable period of time regardless of the concentration of organic material, (2) a constant ratio of iron to the substrate above the minimal threshold, typically 1 part iron per 10 to 50 parts substrate, which produces the desired end products—the ratio of iron to the substrate may affect the distribution of reaction products, and (3) a supplemental aliquot of iron which saturates the chelating properties in the waste water, thereby availing the nonsequestered iron to catalyze the formation of hydroxyl radicals. The iron dose may also be expressed as a ratio to hydrogen peroxide dose and typical ranges are 1 part iron per 5–25 parts of hydrogen peroxide on a weight-weight basis.

For most applications, there does not seem to be a preference whether ferrous (Fe^{2+}) or ferric (Fe^{3+}) salts are used to catalyze the reaction since the catalytic cycle initiates rapidly if the hydrogen peroxide and the organic substances are present in plentiful quantities. However, if a low dose of the Fenton reagent is being used (e.g., <10–25 mg/L hydrogen peroxide), there are some indications that ferrous iron is the preferred form of iron to be present. In terms of the anions, there is little difference whether a chloride salt or a sulfate salt of the iron is used, although with the former, chlorine may be generated at high rates of application.

It is also possible to recycle the iron following the reaction by (1) increasing the pH, (2) separating the iron-containing precipitate, often referred to as the iron floc, and (3) reacidifying the iron sludge. Furthermore, because of the indiscriminate nature by which hydroxyl radicals oxidize organic materials, it is important to profile the reaction in the laboratory for each waste to be treated.

The rate of reaction with Fenton reagent increases with increasing temperature, with the effect more pronounced at temperatures <20°C (<68°F) and, in practice, most commercial applications of Fenton chemistry occur at temperatures between 20 and 40°C (69 and 104°F) but as the temperature increases above 40–50°C (104–122°F), the efficiency of hydrogen peroxide utilization declines because of the accelerated decomposition of hydrogen peroxide into oxygen and water.

For Fenton chemistry, the optimal pH is between 3 and 6. The decrease in efficiency as the pH approaches the basic value (pH = 7–14) is attributed to the transition of iron from a hydrated ferrous ion [represented simply as ($[Fe^{2+} \cdot H_2O]_n$) to a colloidal ferric species in which the hydrated iron catalytically decomposes the hydrogen peroxide into oxygen and water without the formation of hydroxyl radicals (•OH) (Karpenko et al., 2009).

The time needed to complete the Fenton reaction will depend on the many variables discussed above, most notably, catalyst dose and waste water strength. For simple phenol oxidation (less than ca. 250 mg/L), typical reaction times are 30–60 min. For more complex or more concentrated wastes, the reaction may take several hours. In such cases, performing the reaction in steps by adding both iron and hydrogen peroxide may be more effective (and safer) than increasing the initial charges. In addition, determining the completion of the reaction may not be straight forward. The presence of residual hydrogen peroxide will interfere with many waste water analyses. Any residual hydrogen peroxide may be removed by raising the pH to a basic regime (e.g., pH 7–10) or by neutralizing with bisulfite solution.

The difficulty in addressing contamination in low permeability soils may be alleviated to some degree by controlled pneumatic or hydraulic fracturing of the soil. In any case, long-term postinjection monitoring of contaminant levels in groundwater is critical in evaluating the success of putting Fenton reagent into contact with adsorbed contaminants. Controlled oxidation is increasingly being practiced using solid peroxides, pH modifiers, and catalysts which promote the generation of free radicals (Karpenko et al., 2009).

The main advantage of hydrogen peroxide treatment is the speed at which cleanup can be achieved. Due to oxidation kinetics small sites can be possibly remediated in days or weeks. The main disadvantage of hydrogen peroxide treatment is the cost of the hydrogen peroxide. If the aim may be achieved in a much shorter time period compared to bioremediation, the cost of the peroxide may be offset by reduced labor and monitoring costs. Being highly reactive, hydrogen peroxide is classified as a strong oxidant. The widespread availability of hydrogen peroxide is another advantage. To weaken its potential for reactivity and decomposition, commercially available hydrogen peroxide is stabilized through the addition of stannite [$Sn(OH)^{3-}$] derivatives.

2.3 Persulfate

Activated persulfate ($Na_2S_2O_8$) is an oxidation agent which is used for in situ and ex situ destruction of a wide range of chemical contaminants. Field pilot or full-scale application of activated persulfate now has been performed at over 60 sites in more than 20 states, treating effectively chlorinated ethylene derivatives, chlorinated ethane derivatives, chlorinated methane derivatives, polyaromatic hydrocarbon derivatives, petroleum hydrocarbons, and benzene, toluene, ethylbenzene, and xylene singly or in mixtures.

Oxidants such as sodium persulfate can be thermally activated to promote the formation of sulfate-free radicals. A potential alternative to thermal activation is chemical activation. Transition metal activators such as ferrous ion (Fe^{2+}) can activate persulfate decomposition at ambient temperature (approximately 20°C, 68°F). The ferrous ion activated persulfate reaction requires activation energy of 12 kcal/mol, which is lower than the value of 33.5 kcal/mol required for thermal rupture of the oxygen–oxygen bond.

Citric acid is a natural multidentate organic complexing agent that is environmentally friendly, and readily biodegradable, and has the ability to extract toxic metals from contaminated soils and sediments. Therefore, citric acid is often used in environmental applications and is recommended for application in combination with the source of ferrous ions for sodium persulfate activation.

Activated persulfate forms sulfate radicals, which are primary oxidizing agents. Sulfate radicals are one of the strongest oxidants available, with an oxidation potential of 2.6 V, as compared to the hydroxyl radical (2.7 V), the permanganate ion (1.4 V), and ozone (2.2 V) and are effective in oxidizing a broad range of chemical substances. In addition, the persulfate anion is also formed during oxidation, which itself is a strong oxidant (2.1 V), relatively stable and can persist in the subsurface for up to several months before decomposing. Proper choice of activation chemistry can tailor the kinetics of radical formation, thus allowing the applicator to be flexible in balancing between persulfate distribution and speed of contaminant destruction. The soil oxidant demand for persulfate is reported to be less than that for hydrogen peroxide and permanganate, which makes this method more effective economically.

The persulfate compound that is used in groundwater remediation is in the form of the peroxidisulfate ($S_2O_2^{8-}$) but is generally called a persulfate ion in the field of environmental engineering (Tsitonaki et al., 2010). More specifically, sodium persulfate is used because it has the highest water solubility and its reaction with contaminants leaves least harmful byproducts. Although sodium persulfate by itself can degrade many environmental contaminants, the sulfate radical is usually derived from the persulfate because sulfate radicals can degrade a wider range of contaminants at a faster pace (several orders of magnitude depending upon the site conditions and the type of chemical contaminant) than the persulfate ion. Various agents, such as heat, ultraviolet light, high pH, hydrogen peroxide, and transition metals, are used to activate persulfate ions and generate sulfate radicals.

The sulfate radical is an electrophile insofar as it can accept electrons and reacts by accepting an electron pair in order to bond to a nucleophile. Therefore, the performance of sulfate radicals is enhanced in an area where there are many electron-donating chemicals. The sulfate radical reacts with the organic compounds to form a radical cation, particularly when reacting with chemicals such as amino ($-NH_2$) derivatives, hydroxyl ($-OH$) derivatives, and alkoxy ($-OR$) derivatives. On the other hand, the sulfate radical does not react readily with chemicals that contain electron attracting groups such as nitro ($-NO_2$) derivatives and carbonyl ($>CO$) derivatives, as well as chemicals containing chlorine atoms. Also, as the number of ether bonds (R^1-O-R^2, R^1-O-R^2 can be the same or different groups) increases, the reaction rate shows a marked decrease.

In practice, the persulfate must first be activated (by conversion to the sulfate radical) to be effective as a remediating agent. This is commonly achieved using a catalyst containing ferrous iron (Fe^{2+}). The combination of the persulfate and the ferrous iron results in the production of ferric iron (Fe^{3+}) and two types of sulfate radicals, one with a charge of -1 and the other with a charge of -2. However, the persulfate and the iron should not be mixed beforehand but are injected into the area of contamination together. The persulfate and iron react underground to produce the sulfate radicals.

The advantage of using persulfate is that it (the persulfate) is much more stable than either hydrogen peroxide or ozone above the surface. Thus, there are fewer transportation limitations and the persulfate can be injected into the site of contamination at high concentrations after which it is transported through porous media by density-driven diffusion.

2.4 Ozone

Ozone (O_3) is a strong oxidant and can oxidize contaminants directly or through the formation of hydroxyl radicals. Like peroxide, ozone reactions are most effective in systems with acidic pH. The oxidation reactions proceed with extremely fast, pseudo first-order kinetics. For decades it has been used in the waste water industry as a disinfectant as well as a chemical oxidant, to oxidize or destroy organic chemicals in the waste stream.

Due to the high reactivity and instability, ozone should be produced onsite, and it requires closely spaced delivery points (e.g., air sparging wells). The gas may be transferred to the dissolved phase onsite by sparging upgradient water with ozone. Groundwater which is extracted upgradient from the area to be treated may be amended with ozone, then reinjected or reinfiltrated into the subsurface. More commonly, gaseous ozone is injected or sparged directly into contaminated groundwater. In situ decomposition of the ozone can lead to beneficial oxygenation and biostimulation. The simplified stoichiometric reaction of ozone with trichloroethylene in water is:

$$O_3 + H_2O + C_2HCl_3 \rightarrow 2CO_2 + 3HCl$$

The innovative in situ remedial technology using ozone is a form of chemical oxidation, which destroys organic chemical contaminants leaving behind harmless end products. ISCO becomes a desirable remedial alternative due to the potential for large cost savings, as well as limited site disturbance. However, ozone is a strong oxidant with an oxidation potential about 1.2 times greater than hydrogen peroxide. It can be used to destroy petroleum contamination in situ. Delivery concentrations and rates vary depending on the object and the type of contaminations and because of the high reactivity of ozone and associated free radicals.

The unique aspect of the use of ozone (O_3) as an oxidant for ISCO remediation project is that it is typically injected into the contamination site from the bottom rather than the top of the contamination area. Thus, well-bores are drilled into the ground to transport the ozone to lower regions of the contamination.

The major advantage in using ozone in ISCO remediation projects is that ozone does not leave any residual chemical as happens with persulfate and permanganate projects. The processes involved with ozonization (treating water with ozone) only leave behind oxygen. In addition, because ozone is a gas, adding ozone to the bottom of the contaminant pool allows the ozone to rise up through the contaminants and react.

The advantages of ozone application for ISCO are the reactivity and the wide range of chemicals which can be treated. The disadvantages are the same high reactivity, which results in complications during its application and its distribution in soil after injection. For example: (1) ozone reacts with a variety of contaminants, (2) ozone reacts quickly with many other substances such as minerals and in-place organic matter, such as humic materials, that are not the targeted substances, (3) ozone is not very soluble and stays in gas form in the water, which makes ozone prone to nonuniform distribution and rising up to the top of contamination site by the shortest routes rather than traveling through the entire material.

3. REDUCTION REACTIONS

In situ chemical reduction (ISCR) is a type of environmental remediation technique used for soil and/or ground water remediation to reduce the concentrations of chemical contaminants to acceptable levels (Brown et al., 2006). The process is usually applied in the environment by injecting chemically reductive additives in liquid form into the contaminated area or placing a solid medium of chemical reductants in the path of a contaminant plume. It can be used to remediate a variety of chemical contaminants, especially organic chemical contaminants that are resistant to natural degradation.

By way of definition and explanation, the chemical reduction reaction is one-half of a redox reaction, which results in the gain of electrons. One of the reactants in the reduction reaction is oxidized (loses electrons) while the other reactant is reduced (gains electrons). In the ISCR process, reducing chemicals (i.e., chemicals that accept electrons donated given by other chemicals in a reaction) are used to change the contaminants into harmless compounds.

In the ISCR process, there are many reductive reaction options processes that can take place, such as (1) hydrogenolysis, (2) β-elimination, (3) hydrogenation, (4) α-elimination, and (5) electron transfer. The specific combination of reductive processes that actually take place in the subsurface depends on the species of contaminant that is present and also the type of reduction being used. The natural and biological processes that take place in the substratum also affect the kind of reductive processes that are found.

The reactions that occur with permeable reactive barriers and ferrous iron are surface based. The surface reactions take three different forms: (1) direct reduction, (2) electron shunting through ferrous iron, and (3) reduction by production and reaction of hydrogen. Also, the reductive process can be enhanced in two ways. One method is to increase the amount of usable iron in the subsurface to increase the rate of the reduction by chemical or biological means. The second method is to enhance the reducing ability of the iron by a coupling reaction with other chemical reductants or by using biological reduction. Combining bacterial action and biological processes with iron is also known to be effective. The biological processes (Chapter 8) have iron within some biological matrix (such as iron suspended in vegetable oil) and use microbial organisms to enhance the reduction zone and to create a more anaerobic environment for the reactions to take place in.

The chemical agents used for the ISCR remediation process are: (1) zero valent metals, (2) iron minerals, (3) polysulfides, (4) dithionite, and (5) bimetallic materials.

3.1 Zero Valent Metals

Zero valent metals are the main reductants used in ISCR processes. The most common metal used is iron, in the form of zero valent iron, zero valent zinc, which can be more effective at eradicating the contaminants than zero valent iron. The zero valent metal technology is usually implemented by a permeable reactive barrier such as iron that has been embedded in a swellable, organically modified silica (SiO_2) that creates a permanent soft barrier underground to capture and reduce small, organic compounds as groundwater passes through it.

The reactivity of particulate zero valent metals in solution is affected by the metal type (such as iron or zinc), particle size (nanoparticles or microparticles), surface conditions (passivation by coatings of oxides), and solution conditions (including the type and concentration of oxidants). Comparing the reactivity of various types of zero valent iron (Fe^0) and zero valent zinc (Zn^0) with carbon tetrachloride (CCl_4) shows that the intended effect of properties engineered to give enhanced reactivity can be obscured by effects of environmental factors. In this case, the rate of the reduction of carbon tetrachloride (by Zn^0) is more strongly affected by solution chemistry than particle size or surface morphology. Under favorable conditions, however, Zn^0 reduces carbon tetrachloride more rapidly and more completely than Fe^0, regardless of particle size (Tratnyek et al., 2010).

Hydrolysis can explain the attenuation of contaminant plumes in aquifers where the ratio of rate constant to flow rate is sufficiently high. For example, 1,1,1-trichloroethane (TCE) has been observed to disappear from a mixed halocarbon plume over time, while trichloroethylene and 1,2-dichloroethylene (the biodegradation product of trichloroethylene) persist. Also, trichloropropane (1,2,3-trichloropropane) is an emerging contaminant because of increased recognition of its occurrence in groundwater, potential carcinogenicity, and resistance to natural attenuation. The physical and chemical properties of trichloropropane make it difficult to remediate, with all conventional options being relatively slow or inefficient. Treatment that results in alkaline conditions (e.g., permeable reactive barriers containing zero valent iron) favor base-catalyzed hydrolysis of trichloropropane, but high temperature (e.g., conditions of in situ thermal remediation) is necessary for this reaction to be significant. Common reductants (sulfide, ferrous iron adsorbed to iron oxides, and most forms of construction-grade or nano-Fe^0) produce insignificant rates of the reductive dechlorination of trichloropropane. However, quantifiable rates of trichloropropane reduction can be obtained with several types of activated nano-Fe^0, but the surface area normalized rate constants (k_{SA}) for these reactions were lower than is generally considered useful for in situ remediation applications. Much faster rates of the degradation of trichloropropane can be obtained with granular Zn^0 and potentially problematic dechlorination intermediates (1,2- or 1,3-dichloropropane, 3-chloro-1-propene) are not detected. The advantages of Zn^0 over Fe^0 are somewhat peculiar to trichloropropane and suggest a practical application for Zn^0 even though it has not found favor for remediation of contamination with other chlorinated solvents (Sarathy et al., 2010).

3.2 Iron Minerals

There are also many iron minerals that can actively be used in dechlorination. These minerals use ferrous iron (Fe^{2+}). Minerals that can be used include green rust, magnetite, pyrite, and glauconite. The most reactive of the iron minerals are the iron sulfide minerals and the iron oxide minerals. Pyrite, an iron sulfide (FeS_2), can dechlorinate carbon tetrachloride in suspension. These substances are very interesting because they are naturally present and learning about how they produce reductive zones could lead to the development of better reductants for ISCR.

3.3 Polysulfides

Polysulfide derivatives are chemicals that have chains of sulfur atoms. The use of polysulfide derivatives is a type of abiotic reduction and works best in anaerobic conditions where ferric iron (Fe^{3+}) is available. The benefit of using polysulfide derivatives is that they do not produce any biological waste products; however, the reaction rates are slow, and they require more time to create the dual valent iron minerals that are needed for the reduction to occur.

3.4 Dithionite

The dithionite anion $(S_2O_4)^{2-}$ (a derivative of dithionous acid, $H_2S_2O_4$) is an oxoanion of sulfur that acts as a reducing agent. At neutral pH (pH = 7) the redox reaction occurs with formation of sulfite:

$$S_2O_4^{2-} + 2H_2O \rightarrow 2HSO_3^{-} + 2e^{-} + 2H^{+}$$

Dithionite undergoes acid hydrolysis to thiosulfate and bisulfite and undergoes alkaline hydrolysis at high pH (pH > 7) to sulfite and sulfide:

$$2S_2O_4^{2-} + H_2O \rightarrow \underset{\text{thiosulfate}}{S_2O_3^{2-}} + \underset{\text{bisulfite}}{2HSO_3^-}$$

$$3Na_2S_2O_4 + 6NaOH \rightarrow \underset{\text{sulfite}}{5Na_2SO_3} + \underset{\text{sulfide}}{Na_2S} + 3H_2O$$

Dithionite derivatives can also be used as a reductant and several reactions take place and eventually the contaminant is removed. For example, dithionite is used in conjunction with a complexing agent (such as citric acid) to reduce ferric oxyhydroxide into soluble iron (Fe^{2+}) compounds and to remove amorphous iron-bearing (Fe^{3+}) mineral phases in soil analyses (selective extraction). In the process, the dithionite is consumed and the final product of all the reactions is two sulfur dioxide anions. The dithionite is not stable for a prolonged period.

3.5 Bimetallic Chemicals

Bimetallic catalysis has been known for decades in petroleum refining, especially in hydrotreating processes and hydrocracking processes (Speight, 2014, 2017c). The move to these types of catalyst complements the more traditional approach that uses single-metal (single-site) catalysis. Bimetallic catalysts do not only reveal the combination of the properties related to the presence of two individual metals, but also generate new and distinctive properties due to synergetic effects between the two metals present. The structure of the bimetallic particles can be oriented in random, for instance, alloy or intermetallic compound, cluster-in-cluster or core-shell structures etc. However, their final structure strongly depends upon the composition, the synthesis method and conditions, relative strengths of metal–metal bond, surface energies of bulk elements.

In the ISCR process, bimetallic materials are small pieces of metals that are coated lightly with a catalyst such as palladium, silver, or platinum. The catalyst increases the rate of the reaction and the small size of the particles allows them to effectively move into and remain in the target zone.

The in situ chemical reduction process can involve any one or more of several possible reactions, such as (1) hydrogenolysis, (2) hydrogenation, (3) α-elimination, (4) β-elimination, and (5) electron transfer. The specific combination of reductive processes that actually take place in the subsurface depends on the species of contaminant that is present and also the type of reduction being used. The natural and biological processes that take place in the substratum also affect the kind of reductive processes that are found (Brown et al., 2006).

The surface catalyzed reactions occur with permeable reactive barriers and ferrous iron and these reactions take three different forms: (1) direct reduction,

Reaction 1, (2) reduction by production and reaction of hydrogen, Reaction 2, and (3) electron shunting through ferrous iron, Reaction 3. In terms of the mechanisms of these different reactions, Reaction 1 involves electron transfer for Fe^0 to the adsorbed chemical at the metal/water point of contact, resulting in reduction of the substrate chemical and production of ferrous iron (Fe^{2+}). Reaction 2 involves the production of hydrogen from the anaerobic corrosion of ferrous iron (Fe^{2+}) that can then react with the discharged chemical if a catalyst is present. Reaction 3 involves the production of ferrous iron (Fe^{2+}) resulting from corrosion of Fe^0 that can react with the discharged chemical and produce ferric iron (Fe^{3+}).

In addition, these reductive processes can be enhanced by increasing the amount of usable iron in the subsurface to increase the rate of the reduction by chemical or by biological means. The second method is to enhance the reducing ability of the iron by coupling it with other chemical reductants or using a bioreduction process (Brown et al., 2006). Combining bacterial action and biological processes (Chapter 8) along with iron to enhance the reduction zone creates a more anaerobic environment for the reactions to take place.

4. PHOTOCHEMICAL AND PHOTOCATALYTIC TRANSFORMATIONS

Photochemistry is the branch of chemistry concerned with the chemical effects of light and, in a general sense, is used to describe a chemical reaction caused by absorption of ultraviolet light (wavelength: 100–400 nm), visible light (wavelength: 400–750 nm), or infrared radiation (wavelength: 750–2500 nm). Thus, photochemistry is the branch of chemistry that deals with the chemical processes that are caused by the absorption of light energy (Klán and Wirz, 2009). Environmentally relevant (photo)chemical transformations can cause photo-induced changes of surface properties, which can influence aggregation–agglomeration and adsorption to/of other surfaces/contaminants.

In the current context, an example, of a photochemical reaction in the environment is the formation of ozone in the upper atmosphere resulting from the action of light on oxygen molecules, represented simply (but not fully) as:

$$O_2 \rightarrow O_2{}^* \text{ (absorption of energy)}$$

$$O_2 \rightarrow 2O\bullet \text{ (dissociation)}$$

$$O_2 + + O\bullet \rightarrow 2O_3 \text{ (reaction)}$$

The photochemical process (photolysis) is usually initiated by infrared, visible, or ultraviolet light. A primary photochemical reaction is the immediate consequence of the absorption of light but there are subsequent chemical changes due to the participation of secondary processes. It is often the secondary processes that produce an array of intermediates that can interfere with the production of the desired product(s).

In the case of photochemical reactions, light provides the activation energy required for many reactions. If laser light is employed, it is possible to selectively excite a molecule by using light of a specific wavelength to produce a desired electronic and vibrational state. Equally important, the energy emission (typically in the form of light) from a particular state may be selectively monitored, thereby providing a measure of the character of that intermediate reaction state. If the chemical system is at low pressure, the energy distribution of the products of a chemical reaction can be observed before the differences in energy have been masked by repeated collisions between the reactive intermediates.

Photocatalysis is the acceleration of a photoreaction in the presence of a catalyst. In catalyzed photolysis, light is absorbed by an adsorbed substrate. In photogenerated catalysis, the photocatalytic activity depends on the ability of the catalyst to create electron–hole pairs, which generate free radicals (such as hydroxyl radicals: •OH) able to undergo secondary reactions. Metal oxides are of great technological importance in environmental remediation and electronics because of their capability to generate charge carriers, when stimulated with required amount of energy. Also, photochemical reactions proceed differently than temperature-driven reactions because photochemical reaction pathways may include high-energy intermediate species that cannot be generated thermally. This overcomes the potentially large activation barriers in a short period of time and allows the reactions otherwise inaccessible by thermal processes.

Many photochemical reactions are several orders of magnitude faster than thermal reactions. For example, reactions as fast as 10^{-9} s and associated processes as fast as 10^{-15} s are often observed. Most photochemical transformations occur through a series of simple steps known as primary photochemical processes.

4.1 Photochemistry

Photolysis (also called photodissociation and photodecomposition) is a chemical reaction in which an inorganic chemical (or an organic chemical) is broken down by photons and is the interaction of one or more photons with one target molecule. The photolysis reaction is not limited to the effects of visible light but any photon with sufficient energy can cause the chemical transformation of the inorganic bonds of a chemical. Since the energy of a photon is inversely proportional to the wavelength, electromagnetic waves with the energy of visible light or higher, such as ultraviolet light, X-rays, and gamma rays can also initiate photolysis reactions. In the current context, photolysis should not be confused with photosynthesis which is a two-part process in which natural chemicals (typically organic chemicals) are synthesized by a living organism as part of the life cycle (life chemistry) of the organism.

To begin a photochemical process, an atom or molecule must absorb a quantum of light energy from a photon. This reaction causes an atom or molecule to undergo a transient excited state, consequently changing the physical and chemical properties from those of the original atom or molecule of the substance. When this occurs, the recipient molecule tends to form a new structure, or combines with other molecules, and transfers electrons, atoms, protons, or excitation energy to other molecules, thus causing a prolonged chemical chain reaction.

During the reaction, the energy of the atom or molecule increases above its normal level—at this stage, the atom or molecule is now in an excited (or activated) state. If a quantum of visible or ultraviolet light is absorbed, then an electron in a relatively low energy state of the atom or molecule is excited into a higher energy state. If infrared radiation is absorbed by a molecule, then the excitation energy affects the motions of the nuclei in the molecule. After the initial absorption of a quantum of energy, the excited molecule can undergo a number of primary photochemical processes. A secondary process may occur after the primary step.

The absorption step can be represented by where the molecule M absorbs a quantum of light of appropriate energy to yield the excited molecular, M*, which can then react further to produce a range of products:

$$M + \text{light energy}, h\nu \rightarrow M^*$$

$$M^* \rightarrow M + \text{light (luminescence)}$$

$$M^* + C \rightarrow M + C^* \text{ (energy transfer)}$$

$$M^* \rightarrow M^+ + e^- \text{ (photoionization)}$$

$$M^* \rightarrow A + B \text{ (photodissociation)}$$

$$M^* \rightarrow N \text{ (rearrangement)}$$

$$M^* + C \rightarrow \text{products (reaction)}$$

Thus, the general form of photolysis reaction is:

$$X + h\nu \rightarrow Y + Z$$

The rate of reaction is:

$$-d/dt\,[X] = d/dt[Y] = d/dt[Z] = k[X]$$

K is the photolysis rate constant for this reaction in units of s^{-1}.

Thus, photolysis is a chemical process by which chemical bonds are broken as the result of transfer of light energy (direct photolysis) or radiant energy (indirect photolysis) to these bonds. The rate of photolysis depends upon numerous chemical and environmental factors including the light adsorption properties and reactivity of the chemical, and the intensity of solar radiation. In the process, the photochemical mechanism of photolysis is divided into three stages: (1) the adsorption of light which excites electrons in the molecule, (2) the primary photochemical processes which transform or deexcite the excited molecule, and (3) the secondary ("dark") thermal reactions which transform the intermediates produced in the previous step (step 2).

An example of a secondary photochemical reaction in the atmosphere is the dissociation of a molecule into radicals (unstable fragments of molecules). The secondary process may involve a chain reaction, which is a process in which a reactive radical attacks a molecule to produce another unstable radical. This new radical can now attack another molecule, thereby reforming the original radical, which can now begin a new cycle of events.

A well-known example of a chain reaction is the hydrogen–chlorine reaction in which hydrogen and chlorine gases (in the presence of ultraviolet light) form hydrogen chloride; it is given by

$$Cl_2 + h\nu \rightarrow 2Cl\bullet$$

$$Cl\bullet + H_2 \rightarrow HCl + H\bullet$$

$$H\bullet + Cl_2 \rightarrow HCl + Cl\bullet$$

Indirect photolysis or sensitized photolysis occurs when the light energy captured (absorbed) by one molecule is transferred to the inorganic molecule of concern. The donor species (the sensitizer) undergoes no net reaction in the process but has an essentially catalytic effect. Moreover, the probability of a sensitized molecule donating its energy to an acceptor molecule is proportional to the concentration of both chemical species. Thus, complex mixtures may, in some cases, produce enhancement of photolysis rates of individual constituents through sensitized reactions.

There are two fundamental principles (sometimes referred to as the *first law of photochemistry* and the *second law of photochemistry*) are the foundation for understanding photochemical transformations: (1) light must be absorbed by a compound in order for a photochemical reaction to take place, and (2) for each photon of light absorbed by a chemical system, only one molecule is activated for subsequent reaction, sometime referred to as the *photo-equivalence law*.

The efficiency with which a photochemical process occurs is given by the *quantum yield* (Φ) which is the number of moles of a stated reactant disappearing, or the number of moles of a stated product produced, per mole of photons of monochromatic light absorbed:

$$\text{Quantum yield, } \Phi = \text{(product molecules)/(photons absorbed)}$$

Since many photochemical reactions are complex, and may compete with unproductive energy loss, the quantum yield is usually specified for a particular reaction. As an example, the irradiation of acetone with 313 nm light (3130 Å) gives a complex mixture of products, as shown in the following diagram:

$$H_3C{-}\overset{O}{\overset{\|}{C}}{-}CH_3 \xrightarrow[\text{vapor phase 313 nm light}]{h\nu} \left[H_3C{-}\overset{O}{\overset{\|}{C}}{-}CH_3\right]^* \longrightarrow H_3C{-}\overset{O}{\overset{\|}{C}}\bullet + \bullet CH_3$$

$$H_3C{-}\overset{O}{\overset{\|}{C}}\bullet \longrightarrow H_3C\bullet + CO$$

$$2\ H_3C\bullet \longrightarrow H_3C{-}CH_3$$

$$2\ H_3C{-}\overset{O}{\overset{\|}{C}}\bullet \longrightarrow H_3C{-}\overset{O}{\overset{\|}{C}}{-}\overset{O}{\overset{\|}{C}}{-}CH_3$$

$$H_3C{-}\overset{O}{\overset{\|}{C}}\bullet + \bullet CH_3 \longrightarrow H_2C{=}C{=}O + CH_4$$

The quantum yield of these products is less than 0.2, indicating there are radiative (fluorescence and phosphorescence) and nonradiative return pathways. The primary photochemical reaction is the homolytic cleavage of a carbon–carbon bond shown in the top equation in which the asterisk represents an electronic excited state.

The molecule in the excited state (that is, $CH_3COCH_3{}^*$) may return to its initial state according to any of three physical processes: (1) the molecule can release its excitation energy by emitting luminescent radiation through fluorescence or phosphorescence, (2) the molecule may transfer its energy to some other molecule with which it collides, without emitting light, in which the energy transfer process results in a normal molecule and an excited molecule, and (3) within the molecule, as a result of the initial light absorption step, an electron (e^-) may absorb so much energy that it may escape from the molecule, leaving behind a positive molecular ion as a result of a process known as photoionization. If the excited species M* (molecule or atom) does react, then it may undergo any of the following chemical processes: photodissociation, intramolecular (or internal) rearrangement, and reaction with another molecule. Photodissociation may result when the excited molecule breaks apart into atomic and/or molecular fragments. A rearrangement reaction (or photoisomerization reaction) involves the conversion of molecule the into an isomer, which has same numbers and types of atoms but with a different structural arrangement of the atoms.

If the primary photochemical process involves the dissociation of a molecule into radicals (unstable fragments of molecules), then the secondary process may involve a chain reaction. A chain reaction is a cyclic process whereby a reactive radical attacks a molecule to produce another unstable radical. This new radical can now attack another molecule, thereby reforming the original radical, which can now begin a new cycle of events. The hydrogen–chlorine reaction is an example of a chain reaction.

The rate-determining step in a photochemical reaction is determined by the number of photons present that can react with the chemical(s). At midday, the sun is at its apex and light has to travel through the least amount of atmosphere to reach the ground. Thus, a lower number of photons will be filtered out by ozone causing more photochemical reactions to occur. In early morning and evening the sun is at an angle, so light must travel through more of the atmosphere causing more photons to be filtered out. Photochemical degradation processes increase with temperature, so the maximum degradation rates will occur at midday.

Some chemicals are not prone to photochemical reactions and, thus, are not degraded by light because the necessary wavelength needed for the reaction is not present. When a molecule is absorbed onto a solid particle in air, the binding of the molecule onto the surface can change the bond strengths within the molecule. This can affect the absorbance wavelength needed for degradation. Particulate matter can have a negative effect upon photolysis. The particles in air can scatter light, preventing photons from reaching molecules which serves to decrease the number of intermediate frees radicals formed and the amount of pesticide that is degraded by light directly.

Photochemical and free radical reactions are major degradation pathways in the atmosphere, so an understanding of the products that are formed is important. The products formed by photochemical reactions may or may not be more toxic than the parent compound. Once a pesticide, for example, has been degraded, a major removal process for chemicals is to precipitate out of air and return to the earth's surface. Another removal process is for the products to be dissolved in rain and fall back to the surface of the earth.

4.2 Photochemistry in the Environment

Photochemical reactions play a major role in the environment including a wide range of reactions in the atmosphere, natural waters, soil and living organisms. For example, *photochemical smog* is a typical form of pollution of all the main urban and industrial areas of the world. It occurs in or near areas with a high traffic density and in the presence of specific climatic conditions (no wind or weak winds, high temperatures, etc.) that cause the concentration of polluting gases to increase and prevent them from dispersing. In these areas, the concentrations of some gases (tropospheric ozone, carbon monoxide, particulate, volatile chemicals (typically volatile organic chemicals [VOC], nitrogen oxides, etc.) very often exceed the threshold values, above which they are risks to human health, farming, and natural vegetation.

Indirect photolysis or sensitized photolysis occurs when the light energy captured (absorbed) by one molecule is transferred to the inorganic molecule of concern. The donor species (the sensitizer) undergoes no net reaction in the process but has an essentially catalytic effect. Moreover, the probability of a sensitized molecule donating its energy to an acceptor molecule is proportional to the concentration of both chemical species. Thus, complex mixtures may, in some cases, produce enhancement of photolysis rates of individual constituents through sensitized reactions.

Photolysis occurs in the atmosphere as part of a series of reactions by which primary pollutant nitrogen oxides react to form secondary pollutants such as peroxyacyl nitrate derivatives. The two most important photolysis reactions in the troposphere are:

$$O_3 + h\nu \rightarrow O_2 + O^*$$

The excited oxygen atom (O^*) can react with water to give the hydroxyl radical:

$$O^* + H_2O \rightarrow 2OH\cdot$$

The hydroxyl radical is central to atmospheric chemistry because it can initiate the oxidation of hydrocarbons in the atmosphere.

Another reaction involves the photolysis of nitrogen dioxide to produce nitric oxide and an oxygen radical that is a key reaction in the formation of tropospheric ozone:

$$NO_2 + h\nu \rightarrow NO + O$$

The formation of the ozone layer is also caused by photolysis. Ozone in the stratosphere is created by ultraviolet light striking oxygen molecules (O_2) and splitting the oxygen molecules into individual oxygen atoms (atomic oxygen). The atomic oxygen then combines with an unbroken oxygen molecule to create ozone (O_3):

$$O_2 \rightarrow 2O$$

$$O_2 + O \rightarrow O_3$$

In addition, the absorption of radiation in the atmosphere can cause photodissociation of nitrogen (as one of several possible reactions) that can lead to the formation of nitric oxide (NO) and nitrogen dioxide (NO_2) that can act as a catalyst to destroy ozone:

$$N_2 \rightarrow 2N$$

$$O_2 \rightarrow 2O$$

$$CO_2 \rightarrow C + 2O$$

$$H_2O \rightarrow 2H + O$$

$$2NH_3 \rightarrow 3H_2 + N_2$$

$$N + 2O \rightarrow NO_2$$

As shown in the above equations, most photochemical transformations occur through a series of simple steps (Wayne and Wayne, 2005).

Photochemical (or photooxidant) pollution involves a series of complex phenomena leading to the formation of ozone (O_3) and other oxidizing compounds (such as hydrogen peroxide, aldehyde derivatives, peroxyacetyl nitrate, PAN) from primary pollutants: nitrogen oxides (NO_x), non-methane volatile organic compounds (MVOC), carbon monoxide (CO), and methane (CH_4) and the energy from solar ultraviolet radiation. This type of ozone-laden atmospheric pollution (smog) occurs in the lower layer of the atmosphere (the troposphere, which is 0 to 8–10 km above the surface of the earth) (Speight, 2017a,b). The lifetime of ozone has been estimated to be on the order of 22 days, but it is much shorter—only 1 or 2 days because of rapid reactions—in the layer 0–2 km above the surface of the earth. The chemical reactions involved are complex but can be represented fairly simply. In a polluted atmosphere, they involve nitrogen dioxide non-MVOCs as primary pollutants or precursors. Thus:

$$NO_2 + h\nu \rightarrow NO + O\bullet$$

$$O\bullet + O_2 \rightarrow O_3$$

$$O_3 + NO \rightarrow NO_2 + O_2$$

Since the nitric oxide (NO) reduces ozone concentrations by consuming ozone, it (the nitric oxide) is described as an ozone sink. Thus, the concentration of ozone in the atmosphere depends on the ratio between the concentrations of nitrogen dioxide (NO_2) and nitric oxide (NO). If no volatile organic chemicals are present, the concentration of ozone in the troposphere is low.

When volatile organic chemicals are present, a complex series of reactions occurs that causes ozone to accumulate. These reactions increase the nitrogen dioxide sink in the atmosphere by consuming nitric oxide, which then ceases to function as an ozone sink and nitrogen dioxide forms with no ozone destruction. The decomposition of any volatile organic chemicals is triggered by the presence of the hydroxyl radical (•OH), which is highly reactive and naturally present in the atmosphere. Thus:

$$RH + \bullet OH \rightarrow R\bullet + H_2O$$

$$R\bullet + O_2 \rightarrow RO_2$$

$$RO_2 + NO \rightarrow NO_2 + RO\bullet$$

These reactions generate a great many organic gaseous species, especially organic nitrogen compounds such as PAN.

An important characteristic of atmospheric chemistry is its nonlinear nature. This means that ozone formation is not proportional to concentrations of its precursors. The reactions involving the different compounds will either form or destroy ozone, depending on their relative abundance. This accounts for the fact that ozone concentrations measured at a distance from the source of ozone precursors (an urban area, for example) are higher than those measured near the actual sources of emission. Thus, peak ozone concentrations affect suburban and rural areas more than cities. Over an area of high population, for example, emissions of nitric oxide (especially from traffic) are high. Any ozone likely to form is quickly destroyed by the high concentrations of nitric oxide. On the other hand, if the pollutant cloud that is over a city moves away to a low-population area (the countryside), where emissions of nitric oxide are lower, the concentration of ozone will increase since it is no longer consumed by nitric oxide.

The environmental fate and behavior of discharged chemicals may change due to a photochemical transformation of the properties of the chemical itself. Such a transformation may occur if the photochemical energy changes and/or degrades the chemical or, in the context of a mixture, changes component of the mixture. Also, the photoactivation process may alter the binding of the chemicals to any natural adsorbent which in turn may influence the behavior of the chemical in the ecosystem. At the same time, a photocatalytic reaction could also cause the degradation (by photochemical surface modification) of the surface of the chemical thereby affecting the stability, agglomeration, and dissolution behavior of the chemical.

4.3 Photocatalysis in the Environment

Photocatalysis is the acceleration of a photoreaction in the presence of a catalyst in which light is absorbed by a substrate that is typically adsorbed on a (solid) catalyst (Shen et al., 2017). In photogenerated catalysis, the photocatalytic activity depends on the ability of the catalyst to generate free radicals that are able to undergo secondary reactions. Thus, a photocatalyst is defined as a material that can absorb light, producing electron–hole pairs that enable chemical transformations of the reaction participants and regenerate its chemical composition after each cycle of such interactions.

Photocatalysis is an emerging technology that enables a wide variety of applications, including degradation of organics and dyes, antibacterial action, and fuel generation through water splitting and carbon dioxide reduction. Numerous inorganic semiconducting materials have been explored as photocatalysts, and the versatility of these materials and reactions has been expanded (Shen et al., 2017). Understanding the relationship between the physicochemical properties of photocatalytic materials and their performances, as well as the fundamentals in catalytic processes, is important to design and synthesis photocatalytic materials (Shen et al., 2017).

There are two types of photocatalytic reactions: (1) homogeneous photocatalysis and (2) heterogeneous photocatalysis.

4.3.1 Homogeneous Photocatalysis

In homogeneous photocatalysis, the reactants and the photocatalysts exist in the same phase. The most commonly used homogeneous photocatalysts include ozone and photo-*Fenton* systems (Fe^+ and Fe^+/H_2O_2). The reactive species is the •OH which is used for different purposes. The mechanism of hydroxyl radical production by ozone can follow two paths:

$$O_3 + h\nu \rightarrow O_2 + O\bullet$$

$$O\bullet + H_2O \rightarrow \bullet OH + \bullet OH$$

$$O\bullet + H_2O \rightarrow H_2O_2$$

$$H_2O_2 + h\nu \rightarrow \bullet OH + \bullet OH$$

Similarly, the Fenton system produces hydroxyl radicals by the following mechanism:

$$Fe^{2+} + H_2O_2 \rightarrow HO\bullet + Fe^{3+} + OH^-$$

$$Fe^{3+} + H_2O_2 \rightarrow Fe^{2+} + HO\bullet 2 + H^+$$

$$Fe^{2+} + HO\bullet \rightarrow Fe^{3+} + OH^-$$

In a photo Fenton–type process, additional sources of hydroxyl radicals (•OH) should be considered: (1) through photolysis of hydrogen peroxide and (2) through the reduction of ferric (Fe^{3+}) ions under ultraviolet light:

$$H_2O_2 + h\nu \rightarrow HO\bullet + HO\bullet$$

$$Fe^{3+} + H_2O + h\nu \rightarrow Fe^{2+} + HO\bullet + H^+$$

The efficiency of Fenton-type processes is influenced by several operating parameters like concentration of hydrogen peroxide, pH, and intensity of UV. The main advantage of this process is the ability of using sunlight with light sensitivity up to 450 nm, thus avoiding the high costs of UV lamps and electrical energy. These reactions have been proven more efficient than the other photocatalysis, but the disadvantages of the process are the low pH values which are required, since iron precipitates at higher pH values and the fact that iron has to be removed after treatment.

4.3.2 Heterogeneous Photocatalysis

Heterogeneous catalysis has the catalyst in a different phase from the reactants. Heterogeneous photocatalysis is a discipline which includes a large variety of reactions: mild or total oxidations, dehydrogenation, hydrogen transfer, $^{18}O_2$– $^{16}O_2$ and deuterium–alkane isotopic exchange, metal deposition, water detoxification, gaseous pollutant removal, etc.

In this process, a metal oxide is activated with either ultraviolet light, visible light, or a combination of both, and photoexcited electrons are promoted from the valence band to the conduction band, forming an electron/hole pair (e^-/h^+). The photogenerated pair (e^-/h^+) can reduce and/or oxidize a compound adsorbed on the photocatalyst surface. The photocatalytic activity of metal oxide comes from two sources: (1) generation of hydroxyl radicals by oxidation of hydroxyl anions, OH^-, (2) generation of oxygen radicals by reduction of molecular oxygen. Both the radicals and anions can react with pollutants to degrade or otherwise transform them to lesser harmful byproducts (Hoffmann et al., 1995; Hernández-Ramírez and Medina-Ramírez, 2015).

Most common heterogeneous photocatalysts are transition metal oxides and semiconductors, which have unique characteristics. Unlike the metals which have a continuum of electronic states, semiconductors possess a void energy region where no energy levels are available to promote recombination of an electron and hole produced by photoactivation in the solid. Thus, metal oxides are of technological importance in environmental remediation because of the arrangement of electronic structure, light absorption properties, and charge transport characteristics.

However, the physicochemical properties of the metal oxides are crucial for the virtuous photocatalytic performance, which are typically size, shape, morphology, and composition dependent. The synthetic procedure employed can control the size, shape and morphology of the materials prepared, which can contribute towards the development of certain properties of the photoactive materials. This can facilitate the formation of powders or thin films with the required characteristics that improve the performance of the catalyst. Furthermore, the source and type of the light used can affect the performance of the material as a photocatalyst (Khan et al., 2015).

Heterogeneous photocatalysis employing metal oxides such as titanium dioxide (TiO_2), zinc oxide (ZnO), stannic oxide (SnO_2), and cerium oxide (CeO_2) has proved its efficiency in degrading a wide range of distinct pollutants into biodegradable compounds and eventually mineralizing them to harmless carbon dioxide and water (Khan et al., 2015). Among oxidation reactions, most of the photocatalytic processes, are focused on the conversion of highly toxic chemicals to either less toxic chemicals or carbon dioxide and water. These properties make metal oxides an effective photocatalyst for degradation of environment pollutants. When the photocatalytic oxidation reaction is carried out in the presence of oxygen, the catalyst not only plays a role in scavenging the photogenerated electrons but also produces active oxygen species. Apart from the UV-light activated photocatalyst, it was recently reported that the defected metal oxides could also respond to visible light, a process which is termed as visible light–induced photocatalysis, widely employed for environmental remediation (Khan et al., 2015).

REFERENCES

Afonso, M.D., Stumm, W., 1992. Reductive dissolution of iron(III) (hydr)oxides by hydrogen sulfide. Langmuir 8, 1671–1675.

Borch, T., Kretschmar, R., Kappler, A., Van Cappellen, P., Ginder-Vogel, A., Voegelin, A., Campbell, K., 2010. Biogeochemical redox processes and their impact on contaminant dynamics. Environmental Science and Technology 44, 15–23.

Brown, R.A., Lewis, R.L., Fiacco, J., Leahy, M.C., 2006. The technical basis for in situ chemical reduction. In: Remediation of Chlorinated and Recalcitrant Compounds. Battelle Press, Columbus, Ohio.

DeLong, E.F., Pace, N.R., 2001. Environmental diversity of bacteria and archaea. Systematic Biology 2001 (50), 470–478.

D'Hondt, S., Rutherford, S., Spivack, A.J., 2002. Metabolic activity of subsurface life in deep-sea sediments. Science 295, 2067–2070.

Eaton, A.D., Clesceri, L.S., Greenberg, A.E., 1985. Humic Substances in Soil, Sediment, and Water: Geochemistry, Isolation and Characterization. John Wiley & Sons Inc., Hoboken, New Jersey.

Grundl, T.J., Haderlein, S.B., Nurmi, J.T., Tratnyek, P.G., 2011. Introduction to aquatic redox chemistry. ACS Symposium Series No. 1071. In: Tratnyek, P.G., Grundl, T.J., Haderlein, S.B. (Eds.), Aquatic Redox Chemistry. American Chemical Society, Washington, DC.

Hernández-Ramírez, A., Medina-Ramírez, I., 2015. Photocatalytic Semiconductors: Synthesis, Characterization, and Environmental Applications. Springer International Publishing AG, Heidelberg, Germany.

Hoffmann, M.R., Martin, S.T., Choi, W., Bahnemann, D.W., 1995. Environmental applications of semiconductor photocatalysis. Chemical Reviews 95, 69–96.

Karpenko, O., Lubenets, V., Karpenko, E., Novikov, V., 2009. Chemical oxidants for remediation of contaminated soil and water. A review. Chemistry and Chemical Technology 3 (1), 41–49.

Khan, M.M., Adil, S.F., Al-Mayouf, A., 2015. Metal oxides as photocatalysts. Journal of Saudi Chemical Society 19, 462–464.

Klán, P., Wirz, J., 2009. Photochemistry of Organic Compounds: From Concepts to Practice. John Wiley & Sons, Hoboken, New Jersey.

Kluck, C., Achari, G., 2015. Chemical Oxidation Techniques for In Situ Remediation of Hydrocarbon Impacted Soils. Center for Environmental Engineering Research and Education (CEERE) and Department of Civil Engineering, University of Calgary, Calgary, Alberta, Canada. http://www.esaa.org/wp-content/uploads/2015/06/04-29Chu.pdf.

Lucassen, E., Smolders, A.J.P., Van der Salm, A.L., Roelofs, J.G.M., 2004. High groundwater nitrate concentrations inhibit eutrophication of sulphate-rich freshwater wetlands. Biogeochemistry 2004 (67), 249–267.

Overmann, J., Manske, A.K., 2006. Anoxygenic phorotrophic bacteria in the black sea chemocline. In: Neretin, L.N. (Ed.), Past and Present Water Column Anoxia. Springer, Dordrecht, Netherlands.

Pignatello, J.J., Oliveros, E., MacKay, A., 2006. Advanced oxidation processes for organic contaminant destruction based on the Fenton reaction and related chemistry. Critical Reviews in Environmental Science and Technology 36 (1), 1–84.

Postma, D., Boesen, C., Kristiansen, H., Larsen, F., 1991. Nitrate reduction in an unconfined sandy aquifer – water chemistry, reduction processes, and geochemical modeling. Water Resources Research 27, 2027–2045.

Sarathy, V., Salter, A.J., Nurmi, J.T., Johnson, G.O., Johnson, R.L., Tratnyek, P.G., 2010. Degradation of 1,2,3-trichloropropane (TCP): hydrolysis, elimination, and reduction by iron and zinc. Environmental Science and Technology 44 (2), 787–793.

Shen, S., Kronawitter, C., Kiriakidis, G., 2017. An overview of photocatalytic materials. Journal of Materiomics 3, 1–2.

Siegrist, R.L., Urynowicz, M.A., West, O.R., Crimi, M.L., Lowe, K.S., 2001. Principles and Practices of In Situ Chemical Oxidation Using Permanganate. Battelle Press, Columbus, Ohio.

Speight, J.G., 2014. The Chemistry and Technology of Petroleum, fifth ed. CRC Press, Taylor & Francis Group, Boca Raton, Florida.

Speight, J.G., 2017a. Environmental Organic Chemistry for Engineers. Butterworth-Heinemann, Elsevier, Oxford, United Kingdom.

Speight, J.G., 2017b. Environmental Inorganic Chemistry for Engineers. Butterworth-Heinemann, Elsevier, Oxford, United Kingdom.

Speight, J.G., 2017c. Handbook of Petroleum Refining. CRC Press, Taylor & Francis Group, Boca Raton, Florida.

Stewart, R., 1984. Oxidation Mechanisms. W.A. Benjamin Inc., New York.

Stroo, H., Ward, C.H., Alleman, B.C., 2010. In Situ Remediation of Chlorinated Solvent Plumes. Springer Verlag, Berlin, Germany, pp. 481–535.

Tratnyek, P.G., Salter, A.J., Nurmi, J.T., Sarathy, V., 2010. Environmental applications of zerovalent metals: iron vs. zinc. ACS Symposium Series No. 1045. In: Nanoscale Materials in Chemistry: Environmental Applications. American Chemical Society, Washington, DC, pp. 165–178 (Chapter 9).

Tsitonaki, A., Petri, B., Crimi, M., Mosbaek, H., Siegrist, R.L., Bjerg, P.L., 2010. In situ chemical oxidation of contaminated soil and groundwater using persulfate: a review. Critical Reviews in Environmental Science and Technology 40 (1), 55–91.

Wayne, C.E., Wayne, R.P., 2005. Photochemistry. Oxford University Press, Oxford, United Kingdom.

Wondyfraw, M., 2014. Mechanisms and effects of acid rain on environment. Journal of Earth Science & Climatic Change 5 (6), 204–206.

FURTHER READING

Kornblum, Z.C., 2010. Photochemistry. Encyclopedia Americana. Grolier Inc., Danbury, Connecticut.

Tratnyek, P.G., 2010. Chemical reductants for ISCR: the potential for improvement. In: Proceedings of the 7th International Conference on Remediation of Chlorinated and Recalcitrant Compounds. Monterey, California.

Chapter 8

Biological Transformations

1. INTRODUCTION

The terms biodegradation, mineralization, or transformation are used interchangeably, which is partially unjustified and at the same time may cause confusion since there are obvious differences in the meaning of the terms (Speight and Lee, 2000; Speight, 2017a,b). The term mineralization is usually understood as the complete decomposition of a the discharged (inorganic or organic) chemical into inorganic products (such as carbon dioxide and water) while biodegradation is a process taking place with the participation of living organisms that also involves the decomposition of organic compounds into inorganic products, but with the simultaneous buildup and accumulation of biomass (or biomass byproducts). The term biotransformation, on the other hand, is understood as the process leading to the change of the structure of the original chemical compound to such a degree that its original characteristic properties change as well. Also, the biotransformation process modifies not only the chemical and physical properties of chemicals, such as solubility or (bio) availability, but also the toxicity level of the chemical.

In the context of this book, a transformation is the conversion of a chemical (or reactant chemical) to a chemical product. In more general terms, a transformation is a chemical reaction which is characterized by a chemical change and yields one or more products. These products usually have properties substantially different from the properties of the individual reactants. Reactions often consist of a sequence of individual substeps that can be described by means of chemical equations, which are symbolic representations of the starting materials, intermediate products, and the end products, as well as the reaction parameters (such as temperature, pressure, and time).

Following on from the general definition of transformation, biotransformation is a process by which a discharged environmental pollutant chemical is transformed from one form to another to reduce its persistence and toxicity, aided by major range of microorganisms, such as bacteria and fungi, and their products, enzymes. Biotransformation can also be used to synthesize compounds or materials, if synthetic approaches are challenging. Natural transformation process is slow, nonspecific and less productive. Microbial biotransformation or microbial biotechnology is gaining importance and

Reaction Mechanisms in Environmental Engineering. https://doi.org/10.1016/B978-0-12-804422-3.00008-0

extensively utilized to generate metabolites in bulk amounts with more specificity.

In this context, biotransformation techniques have the potential to respond to the needs and offers environmental protection, sustainability, and management (Koenigsberg et al., 2005; Hatti-Kaul et al., 2007; Dowling, 2009; Azadi, 2010). While some applications such as bioremediation are direct applications of biotechnology, there are many which are indirectly beneficial for environmental remediation, pollution prevention, and waste treatment. Large-scale pollution due to anthropogenic chemicals is of global concern now. Seepage and run-offs due to the mobile nature, and continuous cycling of volatilization and condensation of many organic chemicals such as pesticides, have even led to their presence in rain, fog, and snow.

Biological transformations are unique in that the reactions are catalyzed by enzymes and the cell can generate the energy to drive the reaction(s) to the product(s). The catalytic action of an enzyme reduces the activation energy to give significant reaction rates at ambient temperatures. Thus, the process of biotransformation (often called bioremediation or biodegradation) uses microorganisms to degrade contaminants in soil, groundwater, sludge, and solids. The microorganisms break down contaminants by using them as an energy source or cometabolizing them with an energy source. More specifically, the process involves the production of energy in a redox reaction (Chapter 7) within microbial cells. These reactions include respiration and other biological functions needed for cell maintenance and reproduction. A delivery system that provides one or more of the following is generally required: an energy source (electron donor), an electron acceptor, and nutrients. Different types of microbial electron acceptor classes can be involved in bioremediation, such as oxygen-, nitrate-, manganese-, iron (III)-, sulfate-, or carbon dioxide-reducing, and their corresponding redox potentials. Redox potentials provide an indication of the relative dominance of the electron acceptor classes. Generally, electron acceptors and nutrients are the two most critical components of any delivery system.

Finally, evidence for the effectiveness of biodegradation should include: (1) faster disappearance of chemicals in treated areas than in untreated areas, and (2) a demonstration that biodegradation was the main reason for the increased rate of disappearance of the chemical(s). To obtain such evidence, the analytical procedures must be chosen carefully, and careful data interpretation is essential, but there are disadvantages and errors when the method is not applied correctly (Speight, 2005; Speight and Arjoon, 2012).

The concentration of discharged chemicals changes in response to a variety of processes which are (1) physical, (2) chemical, or (3) biological in nature (Senesi et al., 2009). The previous chapters have dealt with the first two categories of chemical removal and it is the third category—biological removal—that is the subject of this chapter.

2. BIOTRANSFORMATION

Biotransformation (sometimes called biomodification) is a biologically mediated transformation processes. Such processes include intraorganism processes that occur after uptake or are "indirectly" mediated by an organism, for example, by release of exudates that bind to the discharged chemical and changes the properties of the chemical. In such case, the organism can also be biomodified.

Microbial biotransformation is widely used in the transformation of various pollutants or a large variety of compounds including hydrocarbons, pharmaceutical substances, and metals. These transformations can be congregated under the categories: oxidation, reduction, hydrolysis, isomerization, condensation, formation of new carbon bonds, and introduction of functional groups. In fact, the biogeochemical cycles of many major and trace elements are driven by redox processes. Examples include the geochemical cycles of carbon (C), nitrogen (N), sulfur (S), iron (Fe), and manganese (Mn), as well as those of redox-sensitive trace elements such as arsenic (As), chromium (Cr), copper (Cu), and uranium (U). The chemical speciation, bioavailability, toxicity, and mobility of these elements in the environment are directly affected by reduction and oxidation.

Some of the metals that anaerobic organisms use as electron acceptors are considered contaminants and are not always organic chemicals. For example, some microorganisms can use soluble uranium (U^{6+}) as an electron acceptor, reducing it to insoluble uranium (U^{4+}). Under this circumstance the organisms cause the uranium to precipitate, decreasing its concentration and mobility in the ground water. In addition to organisms that use inorganic chemicals as electron acceptors for anaerobic respiration, other organisms can use inorganic molecules as electron donors.

Examples of inorganic electron donors are ammonium (NH^{4+}), nitrite (NO^{2-}), reduced iron (Fe^{2+}), reduced manganese (Mn^{2+}), and hydrogen sulfide (H_2S). When these inorganic molecules are oxidized (for example, to NO^{2-}, NO^{3-}, Fe^{3+}, Mn^{4+}, and $SO_4{}^{2-}$, respectively), the electrons are transferred to an electron acceptor (usually oxygen, O_2) to generate energy for cell synthesis. In most cases, microorganisms whose primary electron donor is an inorganic molecule must obtain their carbon from atmospheric carbon.

In some cases, microorganisms can transform these types of contaminants, even though the transformation reaction yields little or no benefit to the cell. The general term for such nonbeneficial biotransformations is *secondary utilization*, and an important special case is called cometabolism. In cometabolism, the transformation of the contaminant is an incidental reaction catalyzed by enzymes involved in normal cell metabolism or special detoxification reactions.

Microbes are also capable of reducing the chemical contaminant. Reductive dehalogenation is potentially important in the detoxification of halogenated organic chemical contaminants, such as chlorinated solvents. In reductive dehalogenation, microbes catalyze a reaction in which a halogen atom (such as chlorine) on the contaminant molecule gets replaced with a hydrogen atom. The reaction adds two electrons to the contaminant molecule, thus reducing the contaminant. For reductive dehalogenation to proceed, a substance other than the halogenated contaminant must be present to serve as the electron donor. Possible electron donors are hydrogen and low molecular–weight organic compounds (lactate, acetate, methanol, or glucose). In most cases, reductive dehalogenation generates no energy but is an incidental reaction that may benefit the cell by eliminating a toxic material. However, researchers are beginning to find examples in which cells can obtain energy from this metabolic process.

Also, the biogeochemical behavior of other nonredox active elements and compounds may be indirectly coupled to redox transformations of natural organic matter and mineral phases, oxides and hydroxides of iron (Fe) and manganese (Mn), iron-bearing clay minerals, and iron sulfides. Redox-active functional groups associated with humic substances and mineral surfaces can further catalyze the oxidation or reduction of ions and molecules, including many organic chemical contaminants. Redox processes (Chapter 7) may also provide new opportunities for engineered remediation strategies, such as the in situ microbial degradation of organic solvents and reductive sequestration of uranium from groundwater. Thus, understanding the biogeochemical processes at environmental redox interfaces is crucial for predicting and protecting an ecosystem, especially an aquatic ecosystem (Kappler and Haderlein, 2003; Ginder-Vogel et al., 2005; Borch and Fendorf, 2008; Polizzotto et al., 2008; Moberly et al., 2009; Borch et al., 2010).

Biotransformation is looked upon as an environmentally friendly technique used to restore soil and water to its original state by using indigenous microbes to break down and eliminate contaminants. The microorganisms used for biodegradation may be indigenous to a contaminated area or they may be isolated from elsewhere and brought to the contaminated site. Chemical contaminants are transformed by living organisms through reactions that take place as a part of their metabolic processes. The mechanisms of biodegradation of a compound are often a result of the actions of multiple organisms and microorganisms, which can be imported to a contaminated site to enhance degradation through biostimulation and bioaugmentation. It is also the result of the prevalent conditions (aerobic or anaerobic) at the site (Tables 8.1 and 8.2).

However, the distribution of microbe populations is not homogeneous in soil. There are different populations of microbes which are close to the surface because these microbes naturally live off material excreted from the roots of a plant. Pesticides are generally degraded by microbes along with material excreted by the roots of plants. Bacteria account for 65% of the total biomass in soil, so they generally account for most degradation processes. Bacterial

TABLE 8.1 Contaminant Type and the Preferred Conditions for Biodegradation

Contaminant	Preferred Conditions
Mineral Oil Hydrocarbons	
Short-chain hydrocarbons derivatives	Aerobic
Long-chain/branched hydrocarbons derivatives	Aerobic
Cycloalkane derivatives	Aerobic
Monoaromatic Hydrocarbons	
(Mono)aromatic hydrocarbon derivatives	Aerobic
Phenol derivatives	Aerobic
Cresol derivatives	Aerobic
Catechol derivatives	Aerobic
Polycyclic Aromatic Hydrocarbons	
2- to 3-ring-polyaromatic hydrocarbons (PAHs) (e.g., naphthalene)	Aerobic
4- to 6-membered ring PAHs (e.g., benzo(a)pyrene)	Aerobic
Chlorinated Aliphatic Hydrocarbons	
Tetrachloroethene, trichloroethane	Anaerobic
Trichloroethylene	Anaerobic
Dichloroethane, dichloroethylene, vinyl chloride	Anaerobic/aerobic
Chlorinated Aromatic Hydrocarbons	
Chlorophenol derivatives (highly chlorinated)	Anaerobic
Chlorophenol derivatives (low chlorinated)	Anaerobic/aerobic
Chlorobenzene derivatives (highly chlorinated)	Anaerobic
Chlorobenzene derivatives (low chlorinated)	Anaerobic/aerobic
Chloronaphthalene	Anaerobic/aerobic
Polychlorinated biphenyls (PCBs) (highly chlorinated)	Anaerobic
Polychlorinated biphenyls (low chlorinated)	Anaerobic/aerobic
Nitroaromatic Compounds	
Mono- and dinitro-aromatic derivatives	Anaerobic/aerobic
Trinitrotoluene (TNT)	Anaerobic/aerobic

Continued

TABLE 8.1 Contaminant Type and the Preferred Conditions for Biodegradation—cont'd

Contaminant	Preferred Conditions
Trinitrophenol (picric acid)	Anaerobic/aerobic
Nitro-aliphatic Compounds	
Glycerol trinitrate	Aerobic
Pesticides	
g-hexachlorocyclohexane (lindane)	Anaerobic/aerobic
b-hexachlorocyclohexane (lindane)	Anaerobic/aerobic
Atrazine derivatives	Aerobic
Dioxins	
PCDD/F (several)	Anaerobic
Inorganic Compounds	
Cyanide derivatives	Aerobic
Ammonium	Anaerobic/aerobic
Nitrate	Anaerobic
Sulfate	Anaerobic

populations dominate the degradation process in alkaline soils and water ($pH > 5.5$) while fungal populations dominate the degradation process in acidic soils.

In addition, the surrounding conditions will determine whether an aerobic or an anaerobic metabolism will occur in the degradation of a chemical contaminant. Aerobic metabolism is a process where oxygen is utilized to oxidize a chemical. On the other hand, anaerobic metabolism is a process where oxygen is not present, so degradation occurs via some other pathway. Aerobic metabolism generally occurs in soil but not as much in water because less oxygen is available for use in water. In soils, the first layer degrades chemicals via aerobic metabolism because air (oxygen) flows over the surface, while below this layer anaerobic metabolism occurs because of the lack of oxygen.

To stimulate and enhance microbial activity, microorganisms (bio-augmentation) or amendments (biostimulation), such as air, organic substrates, or other electron donors/acceptors, nutrients, and other compounds that affect and can limit treatment in their absence, can be added. Biostimulation can be

TABLE 8.2 Factors That Affect the Mechanism of Biodegradation

Contaminant Concentrations
Directly influence microbial activity At high concentration, the contaminants may have toxic effects on the present bacteria Low concentration may prevent induction of bacterial degradation enzymes
Contaminant Bioavailability
Depends on the degree to which they sorb to solids or are sequestered by molecules in contaminated media, are diffused in macropores of soil or sediment, and other factors such as whether contaminants are present in nonaqueous phase liquid form. Bioavailability for microbial reactions is lower for contaminants that are more strongly sorbed to solids, enclosed in matrices of molecules in contaminated media, more widely diffused in macropores of soil and sediments, or are present in nonaqueous phase liquid form.
Site Characteristics
Have a significant impact on the effectiveness of any bioremediation strategy Site environmental conditions important to consider for bioremediation applications include pH, temperature, water content, nutrient availability, and redox potential
pH
Affects the solubility and biological availability of nutrients, metals, and other constituents; for optimal bacterial growth Should remain within the tolerance range for the target microorganisms Bioremediation processes preferentially proceed at a pH of 6–8
Redox Potential
With oxygen content typify oxidizing or reducing conditions\ Influenced by the presence of electron acceptors such as nitrate, manganese oxides, iron oxides, and sulfate
Nutrients
Necessary for microbial cell growth and division Suitable amounts of trace nutrients for microbial growth are usually present Can be added in a useable form or via an organic substrate amendment, which also serves as an electron donor, to stimulate bioremediation
Temperature
A direct effect on the rate of microbial metabolism and microbial activity The biodegradation rate, to an extent rises with increasing temperature and slows with decreasing temperature

used where the bacteria necessary to degrade the contaminants are present but conditions do not favor their growth (e.g., anaerobic bacteria in an aerobic aquifer, aerobic bacteria in an anaerobic aquifer, lack of appropriate nutrients or electron donors/acceptors). Bioaugmentation can be used when the bacteria necessary to degrade the contaminants do not occur naturally at a site or occur at a too low population to be effective. Biostimulation and bioaugmentation can be used to treat soil and other solids, groundwater, or surface water.

For example, if a treatability study shows no degradation (or there is an extended period before significant degradation is achieved) in contamination contained in an ecosystem (such as groundwater), inoculation with strains known to be capable of degrading the contaminants may be helpful. This process increases the reactive enzyme concentration within the bioremediation system and contaminant degradation rates over the nonaugmented rates (Hayman and Dupont, 2001; Karigar and Rao, 2011).

For biodegradation to be effective, microorganisms must convert the pollutants to harmless products. As biodegradation can be effective only where environmental conditions permit microbial growth and activity, its application often involves the manipulation of environmental parameters to allow microbial growth and degradation to proceed at a faster rate. However, as is the case with other technologies, biodegradation has its limitations and there are several disadvantages that must be recognized (Table 8.3).

The control and optimization of biotransformation processes is a complex system of many factors, which include: (1) the existence of a microbial

TABLE 8.3 Advantages and Disadvantages of Bioremediation

Advantages
1. A natural process. 2. Toxic chemicals are destroyed or removed from environment. 3. Can affect complete destruction of a wide variety of contaminants. 4. Can transform contaminants to harmless products. 5. Can be carried out onsite without causing a major disruption of normal activities. 6. if performed in situ, no excavation or transport costs. 7. Less energy is required as compared to other technologies. 8. Microbes do not need manual supervision.
Disadvantages
1. Not all contaminants are biodegradable. 2. The microbial action may be highly specific to one site only. 3. Heavy metals remain on the site. 4. The products may be more toxic than the original contaminant(s). 5. For in situ bioremediation, the site must have highly permeable soil. 6. Process may not remove all quantities of contaminants. 7. Extrapolation from laboratory scale to field scale may be subject to inaccuracies. 8. The contaminant may be a complex mixture that is not evenly dispersed. 9. The contaminant may be present as solids, liquids, and gases.

population capable of degrading the pollutants, (2) the availability of contaminants to the microbial population, (3) the environment factors—the type of soil, the temperature, the pH, and (4) the presence of oxygen or other electron acceptors, and nutrients. Moreover, one of the important factors in biological removal of hydrocarbons from a contaminated environment is their bioavailability to an active microbial population, which is *the degree of interaction of chemicals with living organisms* or the degree to which a contaminant can be readily taken up and metabolized by a bacterium (Harms et al., 2010). It is also important to note that the bioavailability of a contaminant is controlled by factors such as the physical state of the hydrocarbon in situ, its hydrophobicity, water solubility, sorption to environmental matrices such as soil, and diffusion out of the soil matrix. When contaminants have a low solubility in water, as in the case of *n*-alkane derivatives and polynuclear aromatic hydrocarbon derivatives, the organic phase components will not partition efficiently into the aqueous phase supporting the microbes.

In the case of soil, a chemical contaminant will also partition or adsorb on the natural organic matter in the soil (for example, humic acid) and become even less bioavailable. Two-phase bioreactors containing an aqueous phase and a nonaqueous phase liquid (NAPL) have been developed and used for biodegradation of hydrocarbon-contaminated soil to address this very problem, but the adherence of microbes to the NAPL–water interface can still be an important factor in reaction kinetics. Similarly, two-phase bioreactors, sometimes with silicone oil as the nonaqueous phase, have been proposed for biocatalytic conversion of hydrocarbons like styrene (Osswald et al., 1996) to make the substrate more bioavailable to microbes in the aqueous phase. When the carbon source is in limited supply, its availability will control the rate of metabolism and hence biodegradation, rather than catabolic capacity of the cells or availability of oxygen or other nutrients.

In the case of the bioremediation of the aquasphere (waterways), similar principles apply (Vidali, 2001). Under enhanced conditions (1) certain fuel hydrocarbons can be removed preferentially over others, but the order of preference is dependent upon the geochemical conditions, and (2) augmentation and enhancement via electron acceptors to accelerate the biodegradation process. For example, with regard to the aromatic benzene-toluene-ethyl benzene-xylenes mixture (BTEX) or individual component of this mixture: (1) toluene can be preferentially removed under intrinsic biodegradation conditions, (2) biodegradation of benzene is relatively slow, (3) augmentation with sulfate can preferentially stimulate biodegradation of *o*-xylene, and (4) ethylbenzene may be recalcitrant under sulfate-reducing conditions but readily degradable under denitrifying conditions (Cunningham et al., 2000).

In the current context, biodegradation is a collection of chemical reactions for dealing with chemical contaminants and the process typically occurs through the degradation of the chemical through the action of microorganisms (biodegradation). The method utilizes indigenous bacteria (microbes) compared to the customary (physical and chemical) remediation methods. Also, the microorganisms engaged can perform almost any detoxification reaction. Biodegradation studies provide information on the fate of a chemical or mixture of chemicals (such as crude oil spills or process wastes) in the environment thereby opening the scientific doorway to develop further methods of cleanup by (1) analyzing the contaminated sites, (2) determining the best method suited for the environment, (3) optimizing the cleanup techniques which lead to the emergence of new processes.

2.1 Natural Methods

Natural biodegradation typically involves the use of molecular oxygen (O_2), where oxygen (the *terminal electron acceptor*) receives electrons transferred from an organic contaminant:

$$\text{Organic substrate} + O_2 \rightarrow \text{biomass} + CO_2 + H_2O + \text{other products}$$

In the absence of oxygen, some microorganisms obtain energy from fermentation and anaerobic oxidation of organic carbon. Many anaerobic organisms (*anaerobes*) use nitrate, sulfate, and ferric (Fe^{3+}) salts as practical alternates to an oxygen acceptor as, for example, in the anaerobic reduction process of nitrates, sulfates, and ferric salts:

$$2NO_3^- + 10e^- + 12H^+ \rightarrow N_2 + 6H_2O$$

$$SO_4^{2-} + 8e^- + 10H^+ \rightarrow H_2S + 4H_2O$$

$$Fe(OH)_3 + e^- + 3H^+ \rightarrow Fe^{2+} + 3H_2O$$

2.2 Traditional Methods

Methods for the cleanup of chemical pollutants have usually involved removal of the polluted materials, and their subsequent disposal by land filling or incineration (so called *dig*, *haul*, *bury*, or *burn* methods) (Speight, 1996, 2005; Speight and Lee, 2000). Furthermore, available space for landfills and incinerators is declining. However, one of the greatest limitations to traditional cleanup methods is the fact that these methods do not always ensure that contaminants are destroyed.

Conventional biodegradation methods that have been, and are still, used are (1) composting, (2) land farming, (3) biopiling and (4) use of a bioslurry reactor (Speight, 1996; Speight and Lee, 2000; Semple et al., 2001).

Composting is a technique that involves combining contaminated soil with nonhazardous organic materials such as manure or agricultural wastes. The presence of the organic material promotes the development of a rich microbial population and the elevated temperature characteristic of composting. *Land farming* is a simple technique in which contaminated soil is excavated and spread over a prepared bed and periodically tilled until pollutants are degraded. *Biopiling* is a hybrid of land farming and composting, it is essentially engineered cells that are constructed as aerated composted piles. A *bioslurry reactor* can provide rapid biodegradation of contaminants due to enhanced mass transfer rates and increased contaminant-to-microorganism contact. These units are capable of aerobically biodegrading aqueous slurries created through the mixing of soils or sludge with water. The most common state of bioslurry treatment is in batches; however, continuous-flow operation is also possible.

The technology selected for a site will depend on the limiting factors present at the location. For example, where there is insufficient dissolved oxygen, bioventing or sparging is applied; biostimulation or bioaugmentation is suitable for instances where the biological count is low. On the other hand, application of the composting technique, if the operation is unsuccessful, will result in a greater quantity of contaminated materials. Land farming is only effective if the contamination is near the soil surface or else bed preparation is required. The main drawback with slurry bioreactors is that high-energy input is required to maintain suspension and the potential needed for volatilization.

Other techniques are also being developed to improve the microbe–contaminant interactions at treatment sites to use biodegradation technologies at their fullest potential. These biodegradation technologies consist of monitored natural attenuation, bioaugmentation, biostimulation, surfactant addition, anaerobic bioventing, sequential anaerobic/aerobic treatment, soil vapor extraction, air sparging, enhanced anaerobic dechlorination, and bioengineering (Speight, 1996; Speight and Lee, 2000).

2.3 Enhanced Methods

An *enhanced biodegradation* process is a process in which indigenous or inoculated microorganisms (e.g., fungi, bacteria, and other microbes) degrade (metabolize) chemical contaminants (especially chemical contaminants that are organic in character) that have been discharged into the soil and/or into groundwater and convert the contaminants to innocuous end products. The process involves the addition of microorganisms (e.g., fungi, bacteria, and other microbes) or nutrients (e.g., oxygen, nitrates) to the subsurface environment to accelerate the natural biodegradation process.

The method relies on the general availability of naturally occurring microbes to consume contaminants as a food source or as an electron acceptor (chlorinated solvents, which may be waste materials from chemical processes). In addition to microbes being present, to be successful, these processes require nutrients such as carbon, nitrogen, and phosphorus.

2.4 Biostimulation and Bioaugmentation

Biostimulation is the method of adding nutrients such as phosphorus and nitrogen to a contaminated environment to stimulate the growth of the microorganisms that break down chemicals. Additives are usually added to the subsurface through injection wells, although injection well technology for biostimulation purposes is still emerging. Limited supplies of these necessary nutrients usually control the growth of native microorganism populations. Thus, addition of nutrients causes rapid growth of the indigenous microorganism population thereby increasing the rate of biodegradation.

The primary advantage of biostimulation is that biodegradation will be undertaken by already present native microorganisms that are well suited to the subsurface environment and are well distributed spatially within the subsurface, but the main disadvantage is that the delivery of additives in a manner that allows the additives to be readily available to subsurface microorganisms is based on the local geology of the subsurface.

It is to be anticipated that the success of biostimulation is case-specific and site-specific, depending on the properties of the chemicals, the nature of the nutrient products, and the characteristics of the contaminated environments. When oxygen is not a limiting factor, one of the keys for the success of biostimulation is to maintain an optimal nutrient level in the interstitial pore water. Several types of commercial biostimulation agents are available for use in biodegradation (Zhu et al., 2004).

Bioaugmentation is the addition of pregrown microbial cultures to enhance microbial populations at a site to improve contaminant cleanup and reduce cleanup time and cost. Indigenous or native microbes are usually present in very small quantities and may not be able to prevent the spread of the contaminant. In some cases, native microbes do not have the ability to degrade a contaminant. Therefore, bioaugmentation offers a way to provide specific microbes in sufficient numbers to complete the biodegradation (Atlas, 1991). Bioaugmentation adds highly concentrated and specialized populations of specific microbes to the contaminated area, while biostimulation is dependent on appropriate indigenous microbial population and organic material being present at the site. Therefore, it might be that bioaugmentation is more effective than biostimulation, but most cleanup programs have a site specificity that cannot be matched from one site to another.

Mixed cultures have been most commonly used as inocula for seeding because of the relative ease with which microorganisms with different and complementary biodegradative capabilities can be isolated (Atlas, 1977). Different commercial cultures were reported to degrade various hydrocarbons

(Compeau et al., 1991; Leavitt and Brown, 1994; Chhatre et al., 1996; Mishra et al., 2001a,b; Vasudevan and Rajaram, 2001).

Microbial inocula (the microbial materials used in an inoculation) are prepared in the laboratory from soil or groundwater either from the site where they are to be used or from another site where the biodegradation of the chemicals of interest is known to be occurring. Microbes from the soil or groundwater are isolated and are added to media containing the chemicals to be degraded. Only microbes capable of metabolizing the chemicals will grow on the media. This process isolates the microbial population of interest. One of the main environmental applications for bioaugmentation is at sites with chlorinated solvents. Microbes (such as *Dehalococcoides ethenogenes*) usually perform reductive dechlorination of solvents such as perchloroethylene and trichloroethylene (TCE).

Thus, bioaugmentation adds highly concentrated and specialized populations of specific microbes to the contaminated area, while biostimulation is dependent on appropriate indigenous microbial population and organic material that is present at the site.

2.5 In Situ and Ex Situ Methods

Biotransformation processes can be used as a cleanup method for both contaminated soil and water and the applications of this technology fall into two broad categories: in situ or ex situ. In situ biodegradation treats the contaminated soil or groundwater in the location in which it was found while ex situ biodegradation processes require excavation of contaminated soil or pumping of groundwater before they can be treated. In situ biotransformation is used when physical and chemical methods of remediation may not completely remove the contaminants, leaving residual concentrations that are above regulatory guidelines.

Biotransformation applications fall into two broad categories: (1) in situ or (2) ex situ. In situ *biotransformation processes* are used to treat the contaminated soil or groundwater in the location in which it is found. Ex situ *biotransformation processes* require excavation of contaminated soil or pumping of groundwater before they can be treated. In situ techniques do not require excavation of the contaminated soils, so may be less expensive, create less dust, and cause less release of contaminants than ex situ techniques. Also, it is possible to treat a large volume of soil at once. In situ techniques, however, may be slower than ex situ techniques, difficult to manage, and are most effective at sites with permeable soil.

In situ biotransformation process applied to groundwater speeds the natural biodegradation processes that take place in the water-soaked underground region that lies below the water table. One limitation of this technology is that differences in underground soil layering and density may cause reinjected conditioned groundwater to follow certain preferred flow paths. On the other hand, ex situ techniques can be faster, easier to control, and used to treat a wider range of contaminants and soil types than in situ techniques. However, they require excavation and treatment of the contaminated soil before and, sometimes, after the actual biodegradation step.

In situ technologies do not require excavation of the contaminated soils so may be less expensive, create less dust, and cause less release of contaminants than ex situ techniques. Also, it is possible to treat a large volume of soil at once. In situ techniques, however, may be slower than ex situ techniques, may be difficult to manage, and are only most effective at sites with permeable soil.

3. MECHANISMS AND METHODS

Biodegradation is not the only degradation process in soil, but it is the main one. Nature provides enzymes to degrade natural products, so that pollutant chemicals that are similar to natural products are more likely to be degraded since enzymes will have the correct active sites. A chemical pollutant that cannot be degraded by microbes is likely to accumulate in soils and contaminate ground water.

Thus, interest in the microbial biodegradation of pollutants has intensified in recent years as mankind strives to find sustainable ways to clean up contaminated environments. These bioremediation and biotransformation methods endeavor to harness the astonishing, naturally occurring, microbial catabolic diversity to degrade, transform, or accumulate a huge range of compounds including hydrocarbons (e.g., oil), polychlorinated biphenyls (PCBs), polyaromatic hydrocarbons (PAHs), pharmaceutical substances, radionuclides, and metals. Major methodological breakthroughs in recent years have enabled detailed genomic, metagenomic, proteomic, bioinformatics, and other high-throughput analyses of environmentally relevant microorganisms providing unprecedented insights into key biodegradative pathways and the ability of organisms to adapt to changing environmental conditions.

In fact, the study of the fate of persistent chemicals in the environment has revealed a large reservoir of enzymatic reactions with a large potential in preparative organic synthesis that has already been exploited for several oxygenase derivatives on pilot scale and even on an industrial scale. Novel catalysts can be obtained from metagenomic libraries and DNA-sequence based approaches. Our increasing capabilities in adapting the catalysts to specific reactions and process requirements by rational and random mutagenesis broaden the scope for application in the fine chemical industry, but also in the field of biodegradation. In many cases, these catalysts need to be exploited in whole cell bioconversions or in fermentations, calling for system-wide approaches to understanding strain physiology and metabolism and rational approaches to the engineering of whole cells as they are increasingly put forward in systems biotechnology and synthetic biology.

The elimination of a wide range of pollutants and wastes from the environment is an absolute requirement to promote a sustainable development of our society with low environmental impact. Biological processes play a major role in the removal of contaminants and they take advantage of the astonishing catabolic versatility of microorganisms to degrade/convert such compounds. New methodological breakthroughs in sequencing, genomics, proteomics, bioinformatics, and imaging are producing vast amounts of information. In the field of environmental microbiology, genome-based global studies open a new era providing unprecedented in silico views of metabolic and regulatory networks, as well as clues to the evolution of degradation pathways and to the molecular adaptation strategies to changing environmental conditions. Functional genomic and metagenomic approaches are increasing our understanding of the relative importance of different pathways and regulatory networks to carbon flux environments and for inorganic or organic chemicals which can promote the acceleration of the development of bioremediation technologies and biotransformation processes.

The most effective means of implementing an in situ biotransformation process depends on the structure of the subsurface area, the extent of the contaminated area, and the nature (type) of the contamination, as well as the hydrology of the subsurface. In general, this method is effective only when the subsurface soils are highly permeable, the soil horizon to be treated falls within a relatively shallow range (in the order of 25–35 ft). The depth of contamination plays an important role in determining whether an in situ biotransformation project should be employed. If the contamination is near the groundwater but the groundwater is not yet contaminated, then it would be unwise to set up a hydrostatic system. It would be safer to excavate the contaminated soil and apply an onsite method of treatment away from the groundwater. However, due to the inherently poor mixing at some sites, it may be necessary to treat for long periods of time to ensure that all the pockets of contamination have been treated.

3.1 Aerobic Biodegradation

The burgeoning amount of bacterial genomic data provides unparalleled opportunities for understanding the genetic and molecular bases of the degradation of organic pollutants—inorganic chemicals are somewhat more difficult to degrade by this method. Aromatic compounds are among the most recalcitrant of these pollutants and lessons can be learned from the recent genomic studies of *Burkholderia xenovorans LB400* and *Rhodococcus* sp. strain *RHA1*, two of the largest bacterial genomes completely sequenced to date. These studies have helped expand our understanding of bacterial catabolism, noncatabolic physiological adaptation to organic compounds, and the evolution of large bacterial genomes.

First, the metabolic pathways from phylogenetically diverse isolates are very similar with respect to overall organization. Thus, as originally noted in pseudomonads, a large number of *peripheral aromatic pathways* funnel a range of natural and xenobiotic compounds into a restricted number of *central aromatic pathways*.

3.2 Anaerobic Biodegradation

Anaerobic microbial mineralization of recalcitrant organic chemical pollutants (which, by definition, resist degradation) is of great environmental significance and involves intriguing novel biochemical reactions. In particular, hydrocarbon derivatives and halogenated compounds have long been doubted to be degradable in the absence of oxygen, but the isolation of hitherto unknown anaerobic, hydrocarbon-degrading bacteria capable of dehalogenation of the chemical substrate during the last decades has provided ultimate proof for these processes in nature. Many novel biochemical reactions were discovered enabling the respective metabolic pathways, but progress in the molecular understanding of these bacteria was rather slow, since genetic systems are not readily applicable for most of them. However, with the increasing application of genomics in the field of environmental microbiology, a new and promising perspective is now at hand to obtain molecular insights into these new metabolic properties.

Several complete genome sequences were determined during the last few years from bacteria capable of anaerobic organic pollutant degradation. The ~4.7 Mb genome of the facultative denitrifying *Aromatoleum aromaticum* strain EbN1 was the first to be determined for an anaerobic hydrocarbon degrader (using toluene or ethylbenzene as substrates). The genome sequence revealed about two dozen gene clusters (including several paralogs) coding for a complex catabolic network for anaerobic and aerobic degradation of aromatic compounds. The genome sequence forms the basis for current detailed studies on regulation of pathways and enzyme structures.

3.3 Microbe Selection

Microbes are widely diverse organisms and, thus, can make excellent agents in biodegradation and bioremediation. There are few universal toxins to bacteria, so there is likely an organism able to break down any given substrate, when provided with the right conditions (such as: anaerobic vs. aerobic environment and sufficient electron donors or acceptors). Microorganisms use a wide range of processes to transform chemicals in their environment. In some cases, pollutants serve as the carbon and energy source for microbial growth, while in other cases pollutants serve as the terminal electron acceptor. This becomes evident when consideration is given to the diverse ability of microbes to transform and degrade toxic molecules.

In the process, it is essential that the microbe chosen for the remediation is able to convert the chemical pollutant into a less harmful (more benign) one. Furthermore, it is also essential to ensure that a relatively benign chemical contaminant is not converted into a more toxic product. Thus, it is important that the microbe has been previously tested (in the laboratory) and shown to be adequate to the task. It is also important to know if the product toxins are more or less mobile than its predecessor and to estimate the potential for the toxic product(s) to remain at the site for an extended period of time before being transported out of the site.

Traditional molecular analyses have led to great understanding of the microbial diversity in natural systems. These approaches can tell the presence of a particular group of bacteria but does not address the activity. Molecular methods, including microautoradiography, mRNA analysis, growth assays, and incorporation of stable isotopes, can be used to determine which bacteria are involved in biodegradation of chemical pollutants. This information leads to a greater understanding of the role of microbial community structure and function with respect to bioremediation.

Although fungi demonstrate significant biochemical and ecological useful qualities, they are hardly utilized for biotechnological purposes. Instead, bacteria are most commonly used because they usually produce superior results in their numerous advantages ranging from their highly specific biochemical reactions to their capabilities of breaking down pollutants efficiently. Fungi are underused primarily because of the costs that come from providing oxygen to fungi in polluted environments. However, filamentous fungi could be highly valuable in situations where bacteria cannot perform. For example, fungi are useful in situations where contaminants are physically blockaded and bacteria cannot reach or in circumstances of environmental extremes such as high acidity or dryness that prevent bacteria from functioning (Harms et al., 2011).

3.4 Extracellular Electron Transfer

Geobacter species are often the predominant organisms when extracellular electron transfer is an important bioremediation process in subsurface environments. Therefore, a systems biology approach to understanding and optimizing bioremediation with *Geobacter* species has been initiated with the ultimate goal of developing in silico models that can predict the growth and metabolism of *Geobacter* species under a diversity of subsurface conditions. It is expected that this system's approach to bioremediation with *Geobacter* will provide the opportunity to evaluate multiple *Geobacter*-catalyzed bioremediation strategies in silico prior to field implementation, thus providing substantial savings when initiating large-scale in situ bioremediation projects for groundwater polluted with inorganic chemical contaminants (such as uranium) and/or by organic chemical contaminants.

3.5 Bioavailability and Transport of Pollutants

Bacterial pathways for the degradation of organic pollutants have been the subject of intense study for decades. However, important physiological events that precede the catabolism of these compounds have recently been receiving significant scientific attention. Bioavailability, or the amount of a substance that is physicochemically accessible to microorganisms, is a key factor in the efficient biodegradation of pollutants. Chemotaxis, the directed movement of motile organisms toward or away from chemicals in the environment, is an important physiological response that may contribute to effective catabolism of molecules in the environment. In addition, mechanisms for the intracellular accumulation of aromatic molecules via various transport mechanisms are also important.

3.6 Crude Oil Biodegradation

Petroleum oil is toxic for most life forms and episodic and chronic pollution of the environment by oil causes major ecological perturbations. Marine environments are especially vulnerable since oil spills of coastal regions and the open sea are poorly containable and mitigation is difficult. In addition to pollution through human activities, millions of tons of petroleum enter the marine environment every year from natural seepages. Despite its toxicity, a considerable fraction of petroleum oil entering marine systems is eliminated by the hydrocarbon-degrading activities of microbial communities, in particular, by a remarkable recently discovered group of specialists, the so-called hydrocarbonoclastic bacteria (HCB).

Alcanivorax borkumensis, a paradigm of HCB and, probably the most important global oil degrader, was the first to be subjected to a functional genomic analysis. This analysis has yielded important new insights into its capacity for (1) *n*-alkane degradation including metabolism, biosurfactant production, and biofilm formation, (2) scavenging of nutrients and cofactors in the oligotrophic marine environment, as well as (3) coping with various habitat-specific stresses. The understanding thereby gained constitutes a significant advance in efforts toward the design of new knowledge-based strategies for the mitigation of ecological damage caused by oil pollution of marine habitats. HCB also have potential biotechnological applications in the areas of bioplastics and biocatalysis.

3.7 Emerging Technologies

Natural attenuation is one of several cost-saving options for the treatment of polluted environment, in which microorganisms contribute to pollutant degradation. For risk assessments and endpoint forecasting, natural attenuation sites should be carefully monitored (monitored natural attenuation). When site assessments require rapid removal of pollutants, bioremediation, categorized into biostimulation (introduction of nutrients and chemicals to stimulate indigenous microorganisms) and bioaugmentation (inoculation with exogenous microorganisms) can be applied. In such a case, special attention should be paid to its influences on indigenous biota and the dispersal and outbreak of inoculated organisms. Recent advances in microbial ecology have provided molecular technologies, for example, detection of degradative genes, community fingerprinting, and metagenomics, which are applicable to the analysis and monitoring of indigenous and inoculated microorganisms in polluted sites. Scientists have started to use some of these technologies for the assessment of natural attenuation and bioremediation in order to increase their effectiveness and reliability.

4. ADVANTAGES AND DISADVANTAGES

Bioremediation has advantages over traditional cleanup methods of chemical spills. One of the major advantages of bioremediation is the savings in cost and also the savings in the time put forth by workers to clean a contaminated site. The financial savings of bioremediation, when used properly, have tremendous benefits compared to traditional cleanup processes. After the Exxon Valdez spill, the cost to clean 75 miles of shoreline by bioremediation was less than the cost to provide physical washing of the shore for 1 day (Atlas, 1991; Zhu et al., 2004). In addition to traditional methods of site cleanup, bioremediation continues to clean the contaminated site without the constant need of workers.

Bioremediation is also advantageous due to its environmentally friendly approach since no foreign or toxic chemicals are added to the site. It is also environmentally friendly because it does not require any disruption to the natural habitat which often occurs from physical and chemical methods of cleanup. Bioremediation allows for natural organisms to degrade the toxic hydrocarbons into simple compounds which pose no threat to the environment, and this also eliminates the need to remove and transport the toxic compounds to another site. This loss of a need to transport the oil and contaminated soils lowers further risk of additional oil spills, and also saves energy and money which would be put forth in the transportation process. These environmental benefits also make bioremediation a positively viewed method by the general public. With the limited resources in today's world, this is a very much supported technology, which pleases the public and hence is given political support and funding for further research.

One of the greater downsides of bioremediation for petroleum-related spills is that it is a slow process. Such spills can pose a great threat to many different habitats, environments, and industries, and depending upon the urgency of cleanup, bioremediation may not always be the best available option. Also, there are many variables that affect whether bioremediation is capable and practical for the cleanup of different spills. Depending on where the spill takes place and the conditions of the soil or water, it may be very difficult to provide proper nutrient concentrations to the oil degrading microorganisms (Delille et al., 2009). If an oil spill occurs offshore, there is typically much more energy and waves, and this can cause the quick loss and dilution of nutrients provided by biostimulation. In the case of bioaugmentation, there are other issues—particularly the competition that will develop between the native and foreign microbes, which has the potential to render the bioremediation method unsuccessful.

However, it must be recognized that all of the spilled petroleum or petroleum product is removed by microbes in a linear timeframe, since the residual fractions (especially those that contain high molecular−weight polar constituents) that are not consumed initially will be more refractory to microbial attack (Speight, 2005, 2007). Conditions in any ecosystem are rarely favorable for maximum biodegradation.

Furthermore, a major issue relates to whether or not degradation that has been demonstrated to occur in the laboratory will occur in the soil or water ecosystems. Biodegradation phenomena reported in laboratory studies of pure culture of microorganisms reflect only the *potential degradation* that *may* occur in any natural ecosystem. Physical−chemical properties of petroleum, petroleum products, and any of the related constituents, the concentration of the petroleum related contaminant(s), as well as the concentration and diversity of the microbial flora of a specific ecosystem are variable factors in the biodegradation process.

Furthermore, there have been various suggestions that when the concentration of the petroleum-related contaminants passes below a threshold level (which is typically not defined), biodegradation ceases because bacterial growth is limited. Surely this relates to the amount of recalcitrant species (such as resin constituents and asphaltene constituents or related polynuclear aromatic systems) remaining in the nonbiodegraded material.

The qualitative and quantitative differences in the hydrocarbon and the nonhydrocarbon content of petroleum (Speight, 2005, 2007) influence the susceptibility of petroleum and certain petroleum products to biodegradation. This must be acknowledged as a major consideration in determining ecotoxicological effects of petroleum constituents.

Frequently, contaminated sites contain a combination of several manmade organic chemical contaminants and naturally occurring organic chemicals from decayed plant and animal matter. When such mixtures of organics are present, microbes may selectively degrade the compound that is easiest to digest or that provides the most energy. Moreover, the complex mechanisms regulating microbial metabolism may cause some carbon compounds to be ignored while others are selectively used. This phenomenon could have serious implications for bioremediation efforts if the targeted contaminant is accompanied by substantial quantities of preferred growth substrates. However, mixtures do not always cause problems and sometimes can promote bioremediation. For example, biomass that grows primarily to degrade one type of organic compound may also degrade a second compound present at a concentration too low to support bacterial growth by itself.

In some cases, contaminants may not be fully degraded by the organisms. Partial degradation may diminish the concentration of the original pollutant but create metabolic intermediates that in some cases are more toxic than the parent compound. There are two main reasons why intermediates build up. In one case a so-called deadend product is produced. Deadend products may form during cometabolism, because the incidental metabolism of the contaminant may create a product that the bacterial enzymes cannot transform any further. For example, in the cometabolism of chlorinated phenol derivatives, deadend products such as chlorocatechol derivatives, which are toxic, sometimes build up. In the second case the intermediate builds up even though the compound can be fully degraded, because some of the key bacterially mediated reactions are slow. For example, vinyl chloride, a cancer-causing agent, may build up during TCE biodegradation. The bacteria can convert TCE to vinyl chloride relatively quickly, but the subsequent degradation of vinyl chloride usually occurs slowly.

Microorganisms may sometimes be physiologically incapable, even when environmental conditions are optimal, of reducing pollutant concentrations to very low, health-based levels, because the uptake and metabolism of organic compounds sometimes stops at low concentrations. This may be caused by internal mechanisms for regulating what reactions they perform or by an inability of the capable microbial populations to survive given inadequate sustenance. Regardless of the mechanism, if the final contaminant concentration fails to meet the cleanup goal, other cleanup strategies (microbiological or other) may have to be implemented to effectively reduce the concentration to acceptable levels.

Stimulating the growth of enough microorganisms to ensure contaminant degradation is essential to in situ bioremediation. However, if all the organisms accumulate in one place, such as near the wells that supply growth-stimulating nutrients and electron acceptors, microbial growth can clog the aquifer. Clogging can interfere with effective circulation of the nutrient solution, limiting bioremediation in places that the solution does not reach. Protozoan predators may help mitigate clogging. In addition, two engineering strategies can help prevent clogging: (1) feeding nutrients and substrates in alternating pulses and (2) adding hydrogen peroxide as the oxygen source. Pulse feeding prevents excessive biomass growth by ensuring that high concentrations of all the growth-stimulating materials do not accumulate near the injection point. Hydrogen peroxide prevents excessive growth because it is a strong disinfectant, until it decomposes to oxygen and water.

Finally, it is important to realize that no single set of site characteristics will favor bioremediation of all chemical contaminants. For example, certain compounds can only be metabolized under anaerobic conditions, while metabolism of others requires oxygen. When the degradation mechanisms for two cooccurring contaminants are mutually exclusive, difficult choices need to be made or sequential treatment schemes need to be devised.

5. THE FUTURE

The increasingly long list of chemical contaminants released into the environment on a large scale includes numerous aliphatic and aromatic compounds, such as petroleum hydrocarbons. The local concentration of such contaminants depends on the amount present and the rate at which the compound is released, its stability in the environment under both aerobic and anaerobic conditions, the extent of its dilution in the environment, the mobility of the compound in a particular environment, and its rate of biological or nonbiological degradation (Ellis, 2000; Janssen et al., 2001; Dua et al., 2002).

Remediation technologies that can reduce or remove a contaminant can be classified into four categories based on the process acting on the contaminant. These categories are removal, separation, destruction, and containment, which can be either a physical, chemical, or biological process. Physical removal (Speight, 1996, 2005; Speight and Lee, 2000), isolation, microbial remediation, which includes phytoremediation for the purposes of this text, are the most commonly used remediation techniques.

The physical removal of contaminated soil and groundwater is the most common form of remediation, but the process does not eliminate the contamination, but rather transfers it to another location. In ideal cases, the other location will be a facility that is specially designed to contain the contamination for a sufficient period of time or treat it as necessary. In this way, proper removal reduces the risk of exposure to the environmental contaminants.

Isolation technology is typically carried out using clay, concrete, manmade liners, or a combination of these and is often used as a contaminant is difficult or extremely expensive to remove or destroy. The process essentially isolates the contaminant from the affected ecosystem.

In terms of microbial remediation (bioremediation), which involves the breakdown of pollutants by microorganisms, aerobic processes are considered the most efficient and generally applicable—aerobic degradation is dependent on the presence of molecular oxygen and is catalyzed by enzymes that have evolved for the catabolism of natural substrates and exhibit low specificities. Depending upon the type of enzyme catalyzing the reaction, either one (monooxygenase) or two (dioxygenase) oxygen atoms are inserted into the molecule via an electrophilic attack on an unsubstituted carbon atom. Anaerobic degradation proceeds via reductive dehalogenation, wherein an electron transfer to the compound results in hydrogenation.

Petroleum hydrocarbons are widespread common environmental pollutants (Megharaj et al., 2000). The search for an effective remediation for petroleum-contaminated soil is a huge challenge to environmental researchers. The requirement of bioremediation is highlighted after the case of Van Daze Oil Spill in 1989 (Bragg et al., 1994). Bioremediation research has recently attracted widespread attention (Atlas, 1991; Grishchenkov et al., 2000; Chen et al., 2008). It has been reported that suitable microbes are available and can be used to effectively remediate petroleum contamination, even if at low environmental temperature (Rike et al., 2003; Sanscartier et al., 2009).

Although regulations are strictly enforced in developed countries like the United States and most of the European countries to meet the challenges of petroleum-related contamination, these regulations often remain unenforced in most of the developing countries. Cleaning up such sites is often not only technically challenging but also very expensive. Considerable pressure encourages the adoption of waste management alternatives to burial, the traditional means of disposing of solid and liquid wastes.

Approaches such as air-stripping (to remove volatile compounds) and incineration have been used (Speight, 1996, 2005; Speight and Lee, 2000). However, where the contaminants infect a large area but are in low (albeit significant) concentration, such methods are either very costly or simply not feasible (Blackburn and Hafker, 1993; Singh et al., 2001). In such schemes, microorganisms can provide an effective alternative through the biodegradation of the contaminants.

Since most of the contaminants of concern in crude oil are readily biodegradable under the appropriate conditions, the success of oil-spill bioremediation depends mainly on the ability to establish these conditions in the contaminated environment using the above new developing technologies to optimize the microorganisms' total efficiency. The technologies used at various polluted sites depend on the limiting factor present at the location. For example, where there is insufficient dissolved oxygen, bioventing or sparging is applied, biostimulation or bioaugmentation is suitable for instances where the biological count is low.

Over the past three decades, opportunities for applying bioremediation to a much broader set of contaminants have been identified. Indigenous and enhanced organisms have been shown to degrade industrial solvents, polychlorinated biphenyls (PCBs), explosives, and many different agricultural chemicals. Pilot demonstration and full-scale applications of bioremediation have been carried out on a continuing basis. However, the full benefits of bioremediation have not been realized because processes and organisms that are effective in controlled laboratory tests are not always equally effective in full-scale applications. The failure to perform optimally in the field setting stems from a lack of predictability due, in part, to inadequacies in the fundamental scientific understanding of how and why these processes work.

Thus, if bioremediation is to be effective, the microorganisms must enzymatically attack the pollutants and convert them to noncontaminating products—some microbes may produce products that are not only toxic to themselves but also the ecosystem. Parameters that affect the bioremediation process include temperature, nutrients (fertilizers), and the amount of oxygen present in the soil and/or the affected water system. These conditions allow the microbes to grow and multiply and consume more of the contaminant. When conditions are adverse, microbes grow too slowly or die, or they can create more harmful chemicals. In addition, the application of any technology is dependent not only on the availability of the technology but also on the reliability of the technology, as well as on the suitability of the technology for the specific site conditions and whether the technology is readily available (i.e., emerging, developing, or proven).

Microorganisms excel at using organic substances, natural or synthetic, as sources of nutrients and energy. Indeed, the diversity of petroleum-related constituents for growth had led to the discovery of enzymes capable of transforming many unrelated natural organic compounds by many different catalytic mechanisms (Butler and Mason, 1997; Ellis, 2000).

However, depending on behavior in the environment, organic compounds are often classified as biodegradable, persistent, or recalcitrant. A biodegradable organic compound is one that undergoes a biological transformation as long as one or more requirements are fulfilled for the reaction to proceed towards completion (Table 8.4) (Blackburn and Hafker, 1993; Liu and Suflita, 1993). A persistent organic chemical does not undergo biodegradation in

TABLE 8.4 Essential Factors for Biotransformation

Factor	Optimal Conditions
Microbial population	Suitable kinds of organisms that can biodegrade all of the contaminants
Oxygen	Enough to support aerobic biodegradation (about 2% oxygen in the gas phase or 0.4 mg/L in the soil water)
Water	Soil moisture should be from 50% to 70% of the water holding capacity of the soil
Nutrients	Nitrogen, phosphorus, sulfur, and other nutrients to support good microbial growth
Temperature	Appropriate temperatures for microbial growth (0 −40°C, 32−104°F)
pH	Optimal range: 6.5−7.5

certain environments; and a recalcitrant compound resists biodegradation in a wide variety of environments. While partial biodegradation is usually an alteration by a single reaction, primary biodegradation involves a more extensive chemical change. The term *mineralization* is a parallel term to biodegradation and refers to complete degradation to the endproducts of carbon dioxide, water, and other inorganic products.

Biodegradation and its application in bioremediation of organic pollutants have benefited from the biochemical and molecular studies of microbial processes (Lal et al., 1986; Fewson, 1988; Sangodkar et al., 1989; Chaudhary and Chapalamadugu, 1991; Bollag and Bollag, 1992; Van der Meer et al., 1992; Dickel et al., 1993; Deo et al., 1994; Johri et al., 1996, 1999; Kumar et al., 1996; Faison, 2001; Janssen et al., 2001). Indeed, the biotransformation of organic chemical contaminants in the natural environment has been extensively studied to understand microbial ecology, physiology, and evolution for their potential in bioremediation (Bouwer and Zehnder, 1993; Chen et al., 1999; Johan et al., 2001; Mishra et al., 2001a,b; Watanabe, 2001).

As a result, there is a strong demand to increase the adoption of bioremediation as an effective technique for risk reduction on hydrocarbon impacted soils. However, the biodegradation effectiveness diminishes with the time extension and the inhibiting effect may become dominant with time. The key solution to bioremediation is to speed up the restoration process and eliminate or delay the inhibitory effect, such as through the selection of specifically targeted strains or microorganisms (Marijke and Vlerken, 1998), or through the alteration of microbial community structure (Antizar-Ladislao et al., 2008) during the treatment.

Like bioremediation, phytoremediation has recently been developed as a remedial strategy for organic chemical contaminants. It is believed that phytoremediation is much less disruptive to the environment and may have a high probability of public acceptance as a low-cost alternative (Alkorta and Garbisu, 2001). Furthermore, rhizosphere microbes can become contaminant degraders under stress condition (Wenzel, 2009; Gerhart et al., 2009; Korade and Fulekar, 2009).

The recent successful implementation of phytoremediation by a number of researchers (Escalante-Espinosa et al., 2005; Merkl et al., 2005; Erute et al., 2009) has indicated that such an approach is quite promising and can be a viable alternative to the conventional bioremediation. However, the use of phytoremediation is constrained by climate and geological conditions of the sites to be cleaned such as temperature, altitude, soil type, and the accessibility by agricultural equipment (Macek et al., 2000). Plants growing in petroleum-contaminated soil have to cope with the nutrient deficiency and hydrocarbon toxicity; the complicated soil characteristics such as the heterogeneous distribution of the petroleum, soil structure, nutrient shortage, and transplanting may cause some unexpected consequences. Furthermore, the selection of proper plant species (Salt et al., 1998) is also crucial for the experiment. In general, phytoremediation can be a commercial strategy for low TPH contaminated soil, which could improve soil characteristics with minimal risks.

Pollutants sometimes cannot be removed completely by a single remediation process, especially when using biological methods. Therefore, multiprocess bioremediation provides a promising and environmentally friendly solution, which is cost-effective and pollution-free during the active cleaning process. In addition, multiprocess bioremediation may become an effective strategy for rapid biodegradation by altering microbial community structure. It is thus a challenging and rewarding research to search for an innovative solution to speed up TPH reduction for effective environmental cleaning.

The natural processes that drive bioremediation can be enhanced to increase the effectiveness and to reduce time required to meet cleanup objectives. Enhanced bioremediation involves the addition of microorganisms (e.g., fungi, bacteria, and other microbes) or nutrients (e.g., oxygen, nitrates) to the subsurface environment to accelerate the natural biodegradation process—a process in which indigenous microorganisms degrade (metabolize) organic chemical contaminants found in soil and/or ground water and convert them to innocuous end products. The process relies on general availability of naturally occurring microbes to consume contaminants as a food source (petroleum hydrocarbons in aerobic processes) or as an electron acceptor (chlorinated solvents). In addition to microbes being present, in order to be successful, these processes require nutrients (carbon:nitrogen:phosphorus).

The potential of microorganisms in the remediation of some of the compounds hitherto known to be nonbiodegradable has been widely acknowledged globally. With advances in biotechnology, bioremediation has become a rapidly growing area and has been commercially applied for the treatment of hazardous wastes and contaminated sites. Although a wide range of new microorganisms have been discovered that are able to degrade highly stable, toxic organic xenobiotic chemicals, many pollutants still persist in the various environmental ecosystems.

Briefly, a xenobiotic is a chemical which is found in an organism, such as a bacterium, but which is not normally produced or expected to be present in the organism. The word is very often used in the context of a pollutant and is a substance that is not indigenous to an ecosystem or a biological system and which did not exist in nature before human intervention.

A number of reasons have been identified as challenges posed to the microorganisms working in contaminated sites. Such potential limitations to biological treatments include: poor bioavailability of chemicals, presence of other toxic compounds, inadequate supply of nutrients, and insufficient biochemical potential for effective biodegradation. A wide range of bioremediation strategies have been developed for the treatment of contaminated soils using natural and modified microorganisms.

Selecting the most appropriate strategy to treat a specific site can be guided by considering three basic principles: the amenability of the pollutant to biological transformation to less toxic products, the bioavailability of the contaminant to microorganisms, and the opportunity for bioprocess optimization. With the help of advances in bioinformatics, biotechnology holds a bright future for developing bioprocesses for environmental applications.

Biotechnological processes for the bioremediation of petroleum-related pollutants offer the possibility of in situ treatments and are mostly based on the natural activities of microorganisms. Biotechnological processes to destroy contaminants of the type found in petroleum and petroleum products offer many advantages over physicochemical processes. When successfully operated, biotechnological processes may achieve complete destruction of petroleum-related pollutants. However, an important factor limiting the bioremediation of sites contaminated with such contaminants is the slow rate of degradation (Iwamoto and Nasu, 2001), which may limit the practicality of using microorganisms in remediating contaminated sites. This is an area where genetic engineering can make a marked improvement. Molecular techniques can be used to increase the level of a particular protein or enzyme or series of enzymes in bacteria which can lead to an increase in the reaction rate (Chakrabarty, 1986).

Biosurfactants are surface-active microbial products that have numerous industrial applications (Desai and Banat, 1997; Sullivan, 1998; Sekelsky and Shreve, 1999). Many microorganisms, especially bacteria, produce biosurfactants when grown on water-immiscible substrates. Most common biosurfactants are glycolipids in which carbohydrates are attached to a long-chain aliphatic acid, while others such as lipopeptides, lipoproteins, and heteropolysaccharides are more complex. The most promising applications of biosurfactants are in the cleaning of oil-contaminated tankers; oil-spill management; transportation of heavy crude; enhanced oil recovery; recovery of oil from sludge; and bioremediation of sites contaminated with hydrocarbons, heavy metals, and other petroleum-related pollutants.

In addition, heavy metals occur naturally in the soil environment from the pedogenetic processes of weathering of parent materials at levels that are regarded as *trace amounts* (<1000 mg/kg) and rarely toxic (Pierzynski et al., 2000; Kabata-Pendias and Pendias, 2001). Due to the disturbance and acceleration of nature's slowly occurring geochemical cycle of metals by man, most soils of rural and urban environments may accumulate one or more of the heavy metals above defined background values high enough to cause risks to human health, plants, animals, ecosystems, or other media (D'Amore et al., 2005). The heavy metals essentially become contaminants in the soil environments because (1) their rates of generation via anthropogenic processes and cycles are more rapid relative to natural processes, and once (2) they become transferred from mines to random environmental locations where higher potentials of direct exposure occur, (3) the concentrations of the metals in discarded products are relatively high compared to those in the receiving environment, and (4) the chemical form (species) in which a metal is found in the receiving environmental system may render it more bioavailable (D'Amore et al., 2005; Wuana and Okieimen, 2011).

Furthermore, the bioremediation of polynuclear aromatic hydrocarbons was designed by the addition of surfactants; and mathematical models were constructed to explain the effect of surfactants on biodegradation. With the increasing awareness of the applicability of biosurfactants, the focus is now on the utilization of biosurfactants for the bioremediation of NAPLs.

Advances in genetic and protein engineering techniques have opened up new avenues to move toward the goal of *genetically engineered microorganisms* to function as biocatalysts, in which certain desirable biodegradation pathways or enzymes from different organisms are brought together in a single host with the aim of performing specific reactions (Masai et al., 1995; Hauschild et al., 1996; Timmis and Piper, 1999; Sayler and Ripp, 2000). A strategy has also been suggested (Timmis and Piper, 1999) for designing organisms with novel pathways and the creation of a bank of genetic modules encoding broad-specificity enzymes or pathway segments that can be combined at will to generate new or improved activities.

Petroleum and petroleum products, as target pollutants, are difficult to identify because of the complexity of the petroleum system (Speight, 2014). The presence of metals and metalloorganic constituents, which are not destroyed biologically but are only transformed from one oxidation state to another, interfere with the bioremediation processes. Genetic engineering allows transferring the heavy metal–resistance genes to identifiable microorganism hosts that can then serve as an excellent base from which to construct recombinant strains to overcome the challenge of the metal constituents of petroleum.

Methods for the rapid and specific identification of microorganisms within their natural environments continue to be developed. Classic methods are time-consuming and only work for a limited number of microorganisms (Amann et al., 2001). An increasing need to develop new methods for characterization of microorganisms able to degrade the various types of petroleum-related pollutants has led to the use of molecular probes to identify, enumerate, and isolate microorganisms with degradative potential.

Through the genetic engineering of metabolic pathways, it is possible to extend the range of substrates that an organism can utilize. Aromatic hydrocarbon dioxygenases have broad substrate specificity and catalyze enantiospecific reactions with a wide range of substrates.

The biodegradation of many components of petroleum hydrocarbons has been reported in a variety of terrestrial and marine cold ecosystems, extreme environments such as alpine soil (Margesin, 2000), Arctic soil (Braddock et al., 1997), Arctic seawater (Siron et al., 1995), Antarctic soil (Aislabie et al., 1998, 2004), as well as Antarctic seawater and sediments (Delille et al., 1998; Delille and Delille, 2000; Al-Darbi et al., 2005).

Antarctic exploration and research have led to some significant, although localized, impacts on the environment. Human impacts occur around current or past scientific research stations, typically located on ice-free areas that are predominantly soils. Fuel spills, the most common occurrence, have the potential to cause the greatest environmental impact in the Antarctic through accumulation of aliphatic and aromatic compounds. Effective management of hydrocarbon spills is dependent on understanding how they impact soil properties such as moisture, hydrophobicity, soil temperature, and microbial activity. Number of hydrocarbon-degrading bacteria, typically *Rhodococcus*, *Sphingomonas*, and *Pseudomonas* species, for example, may become elevated in contaminated soils, but overall microbial diversity declines. Alternative management practices to the current approach of *dig it up and ship it out* are required but must be based on sound information (Aislabie et al., 2004).

Cold-tolerant bacteria, isolated from oil-contaminated soils in Antarctica, were able to degrade *n*-alkanes (C_6 to C_{20}) typical of the hydrocarbon contaminants that persist in Antarctic soil (Bej et al., 2000). Representative isolates were identified as *Rhodococcus* species; they retained metabolic activity at subzero temperatures of −2°C. A psychrotrophic *Rhodococcus* sp. from Arctic soil (Whyte et al., 1998) utilized a broad range of aliphatic compounds (C_{10} to C_{21} alkanes, branched alkanes, and a substituted cyclohexane) present in diesel oil at 5°C (41°F). The strain mineralized the short-chain alkanes (C_{10} and C_{16}) to a significantly greater extent than the long-chain alkanes (C_{28} and C_{32}) at 0 and 5°C (32 and 41°F). The decreased bioavailability of the long-chain alkanes at low temperature (many form crystals at 0°C, 32°F) may be responsible for their increased recalcitrance, which affects in situ bioremediation in cold climates.

In addition, the physical environment is also important for hydrocarbon biodegradation. This has been demonstrated in sub-Antarctic intertidal beaches (Delille and Delille, 2000) and in Arctic soils (Mohn and Stewart, 2000). Soil nitrogen and concentrations of TPH together accounted for 73% of the variability of the induction period (lag time) for dodecane mineralization at 7°C (45°F) in Arctic soils. High total carbon concentrations were associated with high mineralization rates; high sand content resulted in longer half-times for mineralization. Dodecane mineralization was limited by both N and P; mineralization kinetics varied greatly among different soils (Mohn and Stewart, 2000).

Cold habitats possess sufficient indigenous microorganisms, psychrotrophic bacteria (bacteria capable of surviving or even thriving in a cold environment) being predominant. They adapt rapidly to the presence of the contaminants, as demonstrated by significantly increased numbers of oil degraders shortly after a pollution event, even in the most northerly areas of the world (Whyte et al., 1999). However, the temperature threshold for significant petroleum biodegradation is approximately 0°C (32°F) (Siron et al., 1995).

However, the bulk of information on hydrocarbon degradation borders on activities of mesophiles, although significant biodegradation of hydrocarbons have been reported in psychrophilic environments in temperate regions (Yumoto et al., 2002; Pelletier et al., 2004). Full-scale in situ remediation of petroleum contaminated soils has not yet been used in Antarctica, for example, partly because it has long been assumed that air and soil temperatures are too low for an effective biodegradation. Such omissions in research programs need to be corrected.

The technical feasibility of in situ bioremediation of hydrocarbons in cold groundwater systems has been demonstrated (Bradley and Chapelle, 1995). Rapid aerobic toluene mineralization was demonstrated in sediments from a cold (mean groundwater temperature 5°C/41°F) petroleum-contaminated aquifer in Alaska. The mineralization rate obtained at 5°C in this aquifer was comparable to that measured in sediments from a temperate aquifer (mean temperature 20°C). Rates of overall microbial metabolism in the two sediments were comparable at their respective in situ temperatures.

Bioaugmentation of contaminated cold sites with hydrocarbon-degrading bacteria has been tested as a bioremediation strategy. For the most part, indigenous microbial populations degrade hydrocarbons more efficiently than the introduced microbial strains. However, bioaugmentation (through the use of nonindigenous microbial species) may biodegrade through the onset of a shorter hydrocarbon acclimation period (Mohammed et al., 2007).

Inoculation of contaminated Arctic soils with consortia (Whyte et al., 1999) or with alkane-degrading *Rhodococcus* sp. (Whyte et al., 1998) decreased the lag time and increased the rate of C_{16} mineralization at 5°C (41°F). However, nitrogen-phosphorus-potassium fertilization alone had a comparable effect on hydrocarbon loss like fertilization plus bioaugmentation; this has been shown both in chronically oil-polluted Arctic soil (Whyte et al., 1999) and in artificially diesel oil–contaminated alpine soil (Margesin and Schinner, 1997a,b; Vieira et al., 2009).

Finally, the application of cold-active solubilizing agents could be useful for enhancing hydrocarbon bioavailability. Two hydrocarbon-degrading Antarctic marine bacteria, identified as *Rhodococcus fascians*, produced bioemulsifiers when grown with *n*-alkanes as the sole carbon source. The strains utilized hexadecane and biphenyl as sole carbon sources at temperatures ranging from 4 to 35°C (39–95°F), the optimum temperature was 15–20°C (59–68°F) (Yakimov et al., 1999; Chugunov et al., 2000).

In summary, hydrocarbon biodegradation rates in cold groundwater systems are not necessarily lower than in temperate systems but activity measurements should be performed at the prevailing in situ temperature in order to obtain a realistic estimate of the naturally occurring biodegradation.

In conclusion, biodegradation (with subsequent bioremediation) has shown great promise for cleanup of spills of petroleum-related contaminant. Caution is advised in extolling its virtues as the panacea for cleanup of all petroleum-related spills. It is, nevertheless, one of several viable options that are available as a single or piggyback method for environmental cleanup, but considerations such as (1) site parameters, (2) the properties of the spilled material, (3) prevailing climatic conditions, and (4) the nature of the microorganisms that are available must be considered.

However, with advances in biotechnology, bioremediation has become one of the most rapidly developing fields of environmental restoration, utilizing microorganisms to reduce the concentration and toxicity of petroleum-related hydrocarbons, polycyclic aromatic hydrocarbons, and other constituents such as metals and metalloorganic compounds. However, although simple aromatic compounds are biodegradable by a variety of degradative pathways, their halogenated counterparts are more resistant to bacterial attacks and often necessitate the evolution of novel pathways (Chakrabarty, 1982; Engasser et al., 1990).

A number of bioremediation strategies have been developed to treat petroleum-contaminated wastes and sites. However, selecting the most appropriate strategy to treat a specific site can be guided by considering three basic principles: the amenability of the pollutant to biological transformation to less toxic products (biochemistry), the accessibility of the contaminant to microorganisms (bioavailability), and the opportunity for optimization of biological activity (bioactivity).

Bioremediation is a preferred method for the long-term restoration of petroleum hydrocarbon polluted systems, with the added advantage of cost efficiency and environmental friendliness. Although exhaustive investigations have been done in relation to biodegradation of petroleum and petroleum products, these studies must continue. The identification of active microbial strains is not always ascertained to a sufficient degree, and misidentifications or incomplete identifications are sometimes reported. There is much to be done in terms of the optimization of the process conditions for more efficient application of biological degradation of petroleum and petroleum products under different climatic conditions.

Although a wide range of new microorganisms have been discovered that are able to degrade highly stable, toxic organic chemicals, many pollutants continue to persist in the environment. A number of reasons have been identified as challenges posed to the microorganisms working in contaminated sites, such as: (1) poor bioavailability of chemicals, (2) presence of nonpetroleum compounds, (3) inadequate supply of nutrients, and (4) insufficient biochemical potential for effective biodegradation.

Thus, selecting the most appropriate strategy to treat a specific site can be guided by considering three basic principles: (1) the amenability of the pollutant to biological transformation to less toxic products, (2) the bioavailability of the contaminant to microorganisms, and (3) the opportunity for bioprocess optimization.

In summary, the potential and success of microorganisms in the remediation of spills of petroleum and petroleum-related products has been widely acknowledged. With advances in biotechnology, bioremediation has become a rapidly growing area and has been commercially applied for the treatment of hazardous wastes and contaminated sites.

REFERENCES

Aislabie, J.M., McLeod, M., Fraser, R., 1998. Potential for biodegradation of hydrocarbons in soil from the Ross dependency, Antarctica. Applied Microbiology and Biotechnology 49, 210–214.

Aislabie, J.M., Balks, M.R., Foght, J.M., Waterhouse, E.J., 2004. Hydrocarbon spills on Antarctic soils: effects and management. Environ. Sci. Technol. 38 (5), 1265–1274.

Al-Darbi, M.M., Saeed, N.O., Islam, M.R., Lee, K., 2005. Biodegradation of natural oils in seawater. Energy Sources 27, 19–34.

Alkorta, I., Garbisu, C., 2001. Phytoremediation of organic chemical contaminants in soils. Bioresource Technology 79, 273–276.

Amann, R., Fuchs, B.M., Behrens, S., 2001. The identification of microorganisms by fluorescence in situ hybridization. Current Opinion in Biotechnology 12, 231–236.

Antizar-Ladislao, B., Spanova, K., Beck, A.J., Russell, N.J., 2008. Microbial community structure changes during bioremediation of PAHs in an aged coal tar-contaminated soil by in-vessel composting. International Biodeterioration & Biodegradation 61, 357–364.

Atlas, R.M., 1977. Stimulated crude oil biodegradation. Critical Reviews in Microbiology 5, 371–386.

Atlas, R.M., 1991. Bioremediation: using nature's helpers-microbes and enzymes to remedy mankind's pollutants. In: Lyons, T.P., Jacques, K.A. (Eds.), Proceedings of the Biotechnology in the Feed Industry. Alltech's Thirteenth Annual Symposium. Alltech Technical Publications, Nicholasville, Kentucky, pp. 255–264.

Azadi, H.P., 2010. Genetically modified and organic crops in developing countries: a review of options for food security. Biotechnology Advances 28, 160–168.

Bej, A.K., Saul, D., Aislabie, J., 2000. Cold-tolerant alkane-degrading *Rhodococcus* species from Antarctica. Polar Biology 23, 100–105.

Blackburn, J.W., Hafker, W.R., 1993. The impact of biochemistry, bioavailability and bioactivity on the selection of bioremediation techniques. Trends in Biotechnology 11, 328–333.

Bollag, W.B., Bollag, J.M., 1992. Biodegradation. Encyclopedia of Microbiology 1, 269–280.

Borch, T., Fendorf, S., 2008. Phosphate interactions with iron (hydr)oxides: mineralization pathways and phosphorus retention upon bioreduction. In: Barnett, M.O., Kent, D.B. (Eds.), Adsorption of Metals by Geomedia II: Variables, Mechanisms, and Model Applications, first ed., vol. 7. Elsevier, Amsterdam, Netherlands, pp. 321–348.

Borch, T., Kretzschmar, R., Kappler, A., Van Cappellen, P., Ginder-Vogel, M., Campbell, K., 2010. Biogeochemical redox processes and their impact on contaminant dynamics. Environ. Sci. Technol. 44, 15–23.

Bouwer, E.J., Zehnder, A.J.B., 1993. Bioremediation of organic compounds: microbial metabolism to work. Trends in Biotechnology 11, 360–367.

Braddock, J.F., Ruth, M.L., Walworth, J.L., McCarthy, K.A., 1997. Enhancement and inhibition of microbial activity in hydrocarbon-contaminated arctic soils: implications for utrientamended bioremediation. Environmental Science and Technology 31, 2078–2084.

Bradley, P.M., Chapelle, F.H., 1995. Rapid toluene mineralization by aquifer microorganisms at Adak, Alaska: implications for intrinsic bioremediation in cold environments. Environmental Science and Technology 29, 2778–2781.

Bragg, J.R., Prince, R.C., Harner, E.J., Atlas, R.M., 1994. Effectiveness of bioremediation for the Exxon Valdez oil spill. Nature 368, 413–418.

Butler, C.S., Mason, J.R., 1997. Structure, function analysis of the bacterial aromatic ring hydroxylating dioxygenases. Advances in Microbial Physiology 38, 47–84.

Chakrabarty, A.M., 1982. Genetic mechanisms in the dissimilation of chlorinated compounds. In: Chakrabarty, A.M. (Ed.), Biodegradation and Detoxification of Environmental Pollutants. CRC Press, Boca Raton, Florida, pp. 127–139.

Chakrabarty, A.M., 1986. Genetic engineering and problems of environmental pollution. Biotechnology 8, 515–530.

Chaudhary, G.R., Chapalamadugu, S., 1991. Biodegradation of halogenated organic compounds. Microbiological Reviews 55, 59–78.

Chen, W., Bruhlmann, F., Richnis, R.D., Mulchandani, A., 1999. Engineering of improved microbes and enzymes for bioremediation. Current Opinion in Biotechnology 10, 137–141.

Chen, Y.D., Barker, J.F., Gui, L., 2008. A strategy for aromatic hydrocarbon bioremediation under anaerobic condition and the impacts of ethanol: a microcosm study. Journal of Contaminant Hydrology 96, 17–31.

Chhatre, S., Purohit, H., Shanker, R., Khanna, P., 1996. Bacterial consortia for crude oil spill remediation. Water Science and Technology 34, 187–193.

Chugunov, V.A., Ermolenko, Z.M., Martovetskaya, I.I., Mironava, R.I., Zhirkova, N.A., Kholodenko, V.P., Urakov, N.N., 2000. Development and application of a liquid preparation with oil-oxidizing bacteria. Applied Biochemistry and Microbiology 36, 577–581.

Compeau, G.C., Mahaffey, W.D., Patras, L., 1991. Full-scale bioremediation of a contaminated soil and water site. In: Sayler, G.S., Fox, R., Blackburn, J.W. (Eds.), Environmental Biotechnology for Waste Treatment. Plenum Press, New York, pp. 91–110.

Cunningham, J.A., Hopkins, G.D., Lebron, C.A., Reinhard, M., 2000. Enhanced anaerobic bioremediation of groundwater contaminated by fuel hydrocarbons at Seal Beach, California. Biodegradation 11, 159–170.

D'Amore, J.J., Al-Abed, S.R., Scheckel, K.G., Ryan, J.A., 2005. Methods for speciation of metals in soils: a review. Journal of Environmental Quality 34 (5), 1707–1745.

Delille, D., Delille, B., 2000. Field observations on the variability of crude oil impact in indigenous hydrocarbon-degrading bacteria from sub-Antarctic intertidal sediments. Marine Environmental Research 49, 403–417.

Delille, D., Bassères, A., Dessommess, A., 1998. Effectiveness of bioremediation for oil-polluted Antarctic seawater. Polar Biology 19, 237–241.

Delille, D., Pelletier, E., Rodriguez-Blanco, A., Ghiglione, J., 2009. Effects of nutrient and temperature on degradation of petroleum hydrocarbons in sub-Antarctic coastal seawater. Polar Biology 32, 1521–1528.

Deo, P.G., Karanth, N.G., Karanth, N.G.K., 1994. Biodegradation of hexachlorocyclohexane isomers in soil and food environment. Critical Reviews in Microbiology 20, 57–78.

Desai, J.D., Banat, I.M., 1997. Microbial production of surfactants and their commercial potential. Microbiology and Molecular Biology Reviews 61, 47–64.

Dickel, O., Haug, W., Knackmus, H.-J., 1993. Biodegradation of nitrobenzene by a sequential anaerobic–aerobic process. Biodegradation 4, 187–194.

Dowling, D.N., 2009. Improving phytoremediation through biotechnology. Current Opinion in Biotechnology 20, 204–209.

Dua, M., Singh, A., Sethunathan, N., Johri, A.K., 2002. Biotechnology and bioremediation: successes and limitations. Applied Microbiology and Biotechnology 59, 143–152.

Ellis, B.M.L., 2000. Environmental biotechnology informatics. Current Opinion in Biotechnology 11, 232–235.

Engasser, K.H., Auling, G., Busse, J., Knackmus, H.-J., 1990. 3-Fluorobenzoate enriched bacterial strain FLB 300 degrades benzoate and all three isomeric monofluoro-benzoates. Archives of Microbiology 153, 193–199.

Erute, O., Ufuoma, A., Gloria, O., 2009. Screening of four common Nigerian weeds for use in phytoremediation of soil contaminated with spent lubricating oil. African Journal of Plant Science 3 (5), 102–106.

Escalante-Espinosa, E., Gallegos-Martinez, M.E., Favela-Torres, E., Gutierrez-Rojas, M., 2005. Improvement of the hydrocarbon phytoremediation rate by *Cyperus laxus* Lam. inoculated with a microbial consortium in a model system. Chemosphere 59, 405–413.

Faison, B.D., 2001. Hazardous waste treatment. SIMULA Newsletter 51, 193–208.

Fewson, C.A., 1988. Biodegradation of xenobiotic and other persistent compounds: the causes of recalcitrance. Trends in Biotechnology 6, 148–153.

Gerhart, K.E., Huang, X., Glick, B.R., Greenberg, B.M., 2009. Phytoremediation and rhizoremediation of organic soil contaminants: potential and challenges. Plant Science 176, 20–30.

Ginder-Vogel, M., Borch, T., Mayes, M.A., Jardine, P.M., Fendorf, S., 2005. Chromate reduction and retention processes within arid subsurface environments. Environ. Sci. Technol. 39, 7833–7839.

Grishchenkov, V.G., Townsend, R.T., McDonald, T.J., Autenrieth, R.L., Bonner, J.S., Boronin, A.M., 2000. Degradation of petroleum hydrocarbons by facultative anaerobic bacteria under aerobic and anaerobic conditions. Process Biochemistry 35, 889–896.

Harms, H., Smith, K.E.C., Wick, L.Y., 2010. Problems of hydrophobicity/bioavailability. In: Timmis, K.N. (Ed.), Handbook of Hydrocarbon and Lipid Microbiology. Springer, Berlin, pp. 1439–1450 (Chapter 42).

Harms, H., Schlosser, D., Wick, L.Y., 2011. Untapped potential: exploiting fungi in bioremediation of hazardous chemicals. Nature Reviews Microbiology 9 (3), 177–192.

Hatti-Kaul, R., Törnvall, U., Gustafsson, L., 2007. Industrial biotechnology for the production of bio-based chemicals – a cradle-to-grave perspective. Trends in Biotechnology 25, 119–124.

Hauschild, J.E., Masai, E., Sugiyama, K., Hatta, T., Kimbara, K., Fukuda, M., Yano, K., 1996. Identification of an alternative 2,3-dihydroxybiphenyl 1,2-dioxygenase in *Rhodococcus* sp. Strain RHA1 and cloning of the gene. Applied and Environmental Microbiology 62, 2940–2946.

Hayman, M., Dupont, R.R., 2001. Groundwater and Soil Remediation: Process Design and Cost Estimating of Proven Technologies. ASCE Press, Reston, Virginia.

Iwamoto, T., Nasu, M., 2001. Current bioremediation practice and perspective. Journal of Bioscience and Bioengineering 92, 1–8.

Janssen, D.B., Oppentocht, J.E., Poelarends, G., 2001. Microbial dehalogenation. Current Opinion in Biotechnology 12, 254–258.

Johan, E.T., Vlieg, V.H., Janssen, D.B., 2001. Formation and detoxification of reactive intermediates in the metabolites of chlorinated ethenes. Journal of Biotechnology 85, 81–102.

Johri, A.K., Dua, M., Tuteja, D., Saxena, R., Saxena, D.M., Lal, R., 1996. Genetic manipulations of microorganisms for the degradation of hexachlorocyclohexane. FEMS Microbiology Reviews 19, 69–84.

Johri, A.K., Dua, M., Singh, A., Sethunathan, N., Legge, R.L., 1999. Characterization and regulation of catabolic genes. Critical Reviews in Microbiology 25, 245–273.

Kabata-Pendias, A., Pendias, H., 2001. Trace Metals in Soils and Plants, second ed. CRC Press, Taylor & Francis Group, Boca Raton, Florida.

Kappler, A., Haderlein, S.B., 2003. Natural organic matter as reductant for chlorinated aliphatic pollutants. Environ. Sci. Technol. 37, 2714–2719.

Karigar, C.S., Rao, S.S., 2011. Role of microbial enzymes in the bioremediation of pollutants: a review. SAGE-Hindawi Access to Research Enzyme Research Volume 2011, 805187. https://doi.org/10.4061/2011/805187. https://www.hindawi.com/journals/er/2011/805187/.

Koenigsberg, S., Hazen, T., Peacock, A., 2005. Environmental biotechnology: a bioremediation perspective. Remediation Journal 15, 5–25.

Korade, D.L., Fulekar, M.H., 2009. Development and evaluation of Mycorrhiza for rhizosphere bioremediation. Journal of Applied Biosciences 17, 922–929.

Kumar, S., Mukerjim, K.G., Lal, R., 1996. Molecular aspects of pesticide degradation by microorganisms. Critical Reviews in Microbiology 22, 1–26.

Lal, R., Lal, S., Shivaji, S., 1986. Use of microbes for detoxification of pesticides. Critical Reviews in Microbiology 3, 1–14.

Leavitt, M.E., Brown, K.L., 1994. Bioremediation versus bioaugmentation – three case studies. In: Hinchee, R.E., Alleman, B.C., Hoeppel, R.E., Miller, R.N. (Eds.), Hydrocarbon Bioremediation. CRC Press, Inc., Boca Raton, Florida, pp. 72–79.

Liu, S., Suflita, J.M., 1993. Ecology and evolution of microbial populations for bioremediation. Trends in Biotechnology 11, 344–352.

Macek, T., Mackova, M., Kas, J., 2000. Exploitation of plants for the removal of organics in environmental remediation. Biotechnology Advances 18, 23–34.

Margesin, R., 2000. Potential of cold-adapted microorganisms for bioremediation of oil-polluted alpine soils. International Biodeterioration & Biodegradation 46, 3–10.

Margesin, R., Schinner, F., 1997a. Efficiency of indigenous and inoculated cold-adapted soil microorganisms for biodegradation of diesel oil in alpine soils. Applied and Environmental Microbiology 63, 2660–2664.

Margesin, R., Schinner, F., 1997b. Bioremediation of diesel-oil-contaminated alpine soils at low temperatures. Applied Microbiology and Biotechnology 47, 462–468.

Marijke, M.A., van Vlerken, F., 1998. Chances for biological techniques in sediment remediation. Water Science and Technology 37, 345–353.

Masai, E., Yamada, A., Healy, J.M., Hatta, T., Kimbara, K., Fukuda, M., Yano, K., 1995. Characterization of biphenyl catabolic genes of gram positive polychlorinated biphenyl degrader *Rhodococcus* sp. RH1. Applied and Environmental Microbiology 61, 2079–2085.

Megharaj, M., Singleton, I., McCluture, N.C., Naidu, R., 2000. Influence of petroleum hydrocarbon contamination on microalgae and microbial activities in a long–term contaminated soil. Archives of Environmental Contamination and Toxicology 38, 439–445.

Merkl, N., Schultze-Kraft, R., Infante, C., 2005. Phytoremediation in the tropics-influence of heavy crude oil on root morphological characteristics of graminoids. Environmental Pollution 138, 86–91.

Mishra, S., Jyot, J., Kuhad, R.C., Lal, B., 2001a. In situ bioremediation potential of an oily sludge-degrading bacterial consortium. Current Microbiology 43, 328–335.

Mishra, V., Lal, R., Srinivasan, S., 2001b. Enzymes and operons mediating xenobiotic degradation in bacteria. Critical Reviews in Microbiology 27, 133–166.

Moberly, J., Borch, T., Sani, R., Spycher, N., Sengör, S., Ginn, T., Peyton, B., 2009. Heavy Metal-Mineral Associations in Coeur d'Alene river sediments: a synchrotron-based analysis. Water, Air, and Soil Pollution 201, 195–208.

Mohammed, D., Ramsubhag, A.S., Beckles, D.M., 2007. An assessment of the biodegradation of petroleum hydrocarbons in contaminated soil using non-indigenous, commercial microbes. Water, Air, and Soil Pollution 182, 349–356.

Mohn, W.W., Stewart, G.R., 2000. Limiting factors for hydrocarbon biodegradation at low temperature in arctic soils. Soil Biology and Biochemistry 32, 1161–1172.

Osswald, P., Baveye, P., Block, J.C., 1996. Bacterial influence on partitioning rate during the biodegradation of styrene in a biphasic aqueous-organic system. Biodegradation 7, 297–302.

Pelletier, E., Delille, D., Delille, B., 2004. Crude oil bioremediation in sub-Antarctic intertidal sediments: chemistry and toxicity of oil residues. Marine Environmental Research 57, 311–327.

Pierzynski, G.M., Sims, J.T., Vance, G.F., 2000. Soils and Environmental Quality, second ed. CRC Press, Taylor & Francis Group, Boca Raton, Florida.

Polizzotto, M.L., Kocar, B.D., Benner, S.G., Sampson, M., Fendorf, S., 2008. Near-surface wetland sediments as a source of arsenic release to ground water in Asia. Nature 454, 505–508.

Rike, A.G., Haugen, K.B., Børresen, M., Engene, B., Kolstad, P., 2003. In situ biodegradation of petroleum hydrocarbons in frozen arctic soils. Cold Regions Science and Technology 37, 97–120.

Salt, D.E., Smith, R.D., Raskin, I., 1998. Phytoremediation. Annual Review of Plant Physiology and Plant Molecular Biology 49, 643–668.

Sangodkar, U.M.X., Aldrich, T.L., Haugland, R.A., Johnson, J., Rothmel, R.K., Chapman, P.J., Chakrabarty, A.M., 1989. Molecular basis of biodegradation of chloroaromatic compounds. Acta Biotechnologica 9, 301–316.

Sanscartier, D., Zeeb, B., Koch, I., Reimer, K., 2009. Bioremediation of diesel-contaminated soil by heated and humidified biopile system in cold climates. Cold Regions Science and Technology 55, 167–173.

Sayler, G.S., Ripp, S., 2000. Field application of genetically engineered microorganisms for bioremediation processes. Current Opinion in Biotechnology 11, 286–289.

Sekelsky, A.M., Shreve, G.S., 1999. Kinetic model of biosurfactant-enhanced hexadecane biodegradation by *Pseudomonas aeruginosa*. Biotechnology and Bioengineering 63, 401–409.

Semple, K.T., Reid, B.J., Fermor, T.R., 2001. Impact of composting strategies on the treatment of soils contaminated with organic pollutants. Environmental Pollution 112, 269–283.

Senesi, N., Xing, B., Huang, P.M., 2009. In: Biophysico-chemical Processes Involving Natural Nonliving Organic Matter in Environmental Systems. John Wiley & Sons Inc., Hoboken, New Jersey.

Singh, A., Mullin, B., Ward, O.P., 2001. Reactor-based process for the biological treatment of petroleum wastes. In: Proceedings. Middle East Petrotech 2001 Conference. Bahrain, pp. 1–13.

Siron, R., Pelletier, E., Brochu, C., 1995. Environmental factors influencing the biodegradation of petroleum hydrocarbons in cold seawater. Archives of Environmental Contamination and Toxicology 28, 406–416.

Speight, J.G., 1996. Environmental Technology Handbook. Taylor & Francis, Washington, DC.

Speight, J.G., 2005. Environmental Analysis and Technology for the Refining Industry. John Wiley & Sons Inc., Hoboken, New Jersey.

Speight, J.G., 2007. Natural Gas: A Basic Handbook. GPC Books, Gulf Publishing Company, Houston, Texas.

Speight, J.G., 2014. The Chemistry and Technology of Petroleum, fifth ed. CRC Press, Taylor & Francis Group, Boca Raton, Florida.

Speight, J.G., 2017a. Environmental Organic Chemistry for Engineers. Butterworth-Heinemann, Elsevier, Oxford, United Kingdom.

Speight, J.G., 2017b. Environmental Inorganic Chemistry for Engineers. Butterworth-Heinemann, Elsevier, Oxford, United Kingdom.

Speight, J.G., Arjoon, K.K., 2012. Bioremediation of Crude Oil and Crude Oil Products. Scrivener Publishing, Salem, Massachusetts.

Speight, J.G., Lee, S., 2000. Environmental Technology Handbook, second ed. Taylor & Francis, New York.

Sullivan, E.R., 1998. Molecular genetics of biosurfactant production. Current Opinion in Biotechnology 9, 263–269.

Timmis, K.N., Piper, D.H., 1999. Bacteria designed for bioremediation. Trends in Biotechnology 17, 201–204.

Van der Meer, J.R., de Vos, W.M., Harayama, S., Zehnder, A.Z.B., 1992. Molecular mechanisms of genetic adaptation to xenobiotic compounds. Microbiological Reviews 56, 677–694.

Vasudevan, N., Rajaram, P., 2001. Bioremediation of oil sludge-contaminated soil. Environment International 26, 409–411.

Vidali, M., 2001. Bioremediation: an overview. Pure and Applied Chemistry 73 (7), 1163–1172.

Vieira, P.A., Faria, S.R., Vieira, B., De Franca, F.P., Cardoso, V.L., 2009. Statistical analysis and optimization of nitrogen, phosphorus, and inoculum concentrations for the biodegradation of petroleum hydrocarbons by response surface methodology. Journal of Microbiology and Biotechnology 25, 427–438.

Watanabe, K., 2001. Microorganisms relevant to bioremediation. Current Opinion in Biotechnology 12, 237–241.

Wenzel, W., 2009. Rhizosphere processes and management in plant-assisted bioremediation (phytoremediation) of soils. Plant and Soil 321 (1–2), 385–408.

Whyte, L.G., Hawari, J., Zhou, E., Bourbonnière, L., Inniss, W.E., Greer, C.W., 1998. Biodegradation of variable-chain-length alkanes at low temperatures by a psychrotrophic *Rhodococcus* sp. Applied and Environmental Microbiology 64, 2578–2584.

Whyte, L.G., Bourbonnière, L., Bellerose, C., Greer, C.W., 1999. Bioremediation assessment of hydrocarbon-contaminated soils from the high arctic. Bioremediation Journal 3, 69–79.

Wuana, R., Okieimen, F.E., 2011. Heavy metals in contaminated soils: a review of sources, chemistry, risks and best available strategies for remediation. International Scholarly Research Network. ISRN Ecology Volume 2011, 402647. https://doi.org/10.5402/2011/402647. www.scirp.org/reference/ReferencesPapers.aspx?ReferenceID=1367955.

Yakimov, M.M., Giuliano, L., Bruni, V., Scarfi, S., Golyshin, P.N., 1999. Characterization of Antarctic hydrocarbon-degrading bacteria capable of producing bioemulsifiers. Microbiologica 22, 249–256.

Yumoto, I., Nakamura, A., Iwata, H., Kojima, K., Kusumuto, K., Nodasaka, Y., Matsuyama, H., 2002. *Dietzia psychralcaliphila* sp. nov., a novel facultatively psychrophilic alkaliphile that grows on hydrocarbons. International Journal of Systematic and Evolutionary Microbiology 52, 85–90.

Zhu, X., Venosa, A.D., Suidan, M.T., 2004. Literature Review on the Use of Commercial Bioremediation Agents for Clean-up of Oil Contaminated Estuarine Environments. Report No. EPA/600/R-04/075. National Risk Management Research Laboratory, Environmental Protection Agency, Cincinnati, Ohio.

FURTHER READING

Aislabie, J., Saul, D.J., Foght, J.M., 2006. Bioremediation of hydrocarbon-contaminated polar soils. Extremophiles 10, 171–179.

Atlas, R.M., Bartha, R., 1998. Fundamentals and applications. In: Microbial Ecology, fourth ed. Benjamin/Cummings Publishing Company Inc., California, p. 523.

Nester, E.W., Anderson, D.G., Roberts Jr., C.E., Pearsall, N.N., Nester, M.T., 2001. Microbiology: A Human Perspective, third ed. McGraw-Hill, New York.

Nugroho, A., Effendi, E., Karonta, Y., 2010. Crude oil degradation in soil by thermophilic bacteria with biopile reactor. Makara Journal of Technology 14 (1), 43–46.

US EPA, 2006. In Situ and Ex Situ Biodegradation Technologies for Remediation of Contaminated Sites. Report No. EPA/625/R-06/015. Office of Research and Development National Risk Management Research Laboratory, United States Environmental Protection Agency, Cincinnati, Ohio.

Van Eyk, J., 1994. Venting and bioventing for the in-situ removal of crude oil from soil. In: Hinchee, R.E., Alleman, B.C., Hoeppel, R.E., Miller, R.N. (Eds.), Hydrocarbon Bioremediation. CRC Press, Boca Raton, Florida, pp. 234–251.

Chapter 9

Molecular Interactions, Partitioning, and Thermodynamics

1. INTRODUCTION

When the use of chemicals results in discharge into the environment with ensuing contamination, it is necessary to set standards for acceptable concentrations of contaminants in water, air, soil, and biota (Chapter 2) (Dean, 1998; Crompton, 2012). Monitoring of these concentrations in the environment, and resultant mobility of the chemical, as well as any biological effects—such as the effects of indigent microbial species on the chemical (Chapter 8)—must be undertaken to ensure that the standards as set in any regulation are realistic and provide protection of the environment from any adverse effects (Thibodeaux, 1996). Furthermore, considerable attention continues to be focused on regulation of the use of all chemicals and a primary aspect involves the prediction of the behavior and effects of a chemical from its properties. With this concept, the characteristics of the molecule govern the physicochemical properties of the chemical, which in turn influences transformation and distribution in the environment and the biological effects. This suggests that the transformation and distribution of the chemical in the environment, as well as any effects it has on the floral and faunal species, can be predicted from the physicochemical properties of that chemical. However, the prediction of biological effects is the most complex of the set of predictions (Speight and Lee, 2004; Tinsley, 2004).

Industrial chemical manufacturers use and generate both large numbers and quantities of chemicals. In the past, prior to the institution of environmental regulations, the chemical industry discharged chemicals into all types of environmental ecosystems including air (through both fugitive nd direct emissions) water (direct discharge and runoff), and land. The types of pollutants a single facility will release depend on (1) the type of process, (2) the process feedstocks, (3) the equipment in use, such as the reactor, and (4) the equipment and process maintenance practices, which can vary over short periods of time (such as from hour to hour) and also with the part of the

Reaction Mechanisms in Environmental Engineering. https://doi.org/10.1016/B978-0-12-804422-3.00009-2

process that is underway. For example, for batch reactions in a closed vessel, the chemicals are more likely to be emitted at the beginning and end of a reaction step (that are associated with reactor or treatment vessel loading and product transfer operations) than during the reaction.

The chemical pollutants that are most likely to present ecological risks are those that are (1) highly bioaccumulative, building up to high levels in floral and faunal tissues even when concentrations in the ecosystem remain relatively low, and (2) highly toxic, so that they cause harm at comparatively low doses. In addition, atmosphere—water interactions that control the input and outgassing of persistent pollutants in environmental systems are critically important in determining the life cycle and residence times of these chemicals and the extent of contamination. Although the effects of various types of chemical pollutants are usually evaluated independently, many ecosystems are subject to multiple pollutants, and their fate and impacts are intertwined (Mackay et al., 2006). For example, the effects of nutrient deposition in an ecosystem can alter the methods by which the chemical contaminants are assimilated and bioaccumulated, as well as the means by which the organisms in the ecosystem are affected.

Of all the chemical pollutants released into the environment by anthropogenic activity, persistent pollutants are among the most dangerous and need extreme measures for removal (Chapters 1, 2, and 10). Persistent pollutants (PPs) are chemical substances that persist in the environment, bioaccumulate through the food web, and pose a risk of causing adverse effects to human health and the environment. PPs can be transported across international boundaries far from their sources, even to regions where they have never been used or produced. Consequently, PPs pose a threat to the environment and to human health all on a global scale.

Releases into the environment of chemicals that persist in the ecosystem (rather than undergo some form of biodegradation) lead to an exposure level that is not only subject to the length of time the chemical remains in circulation (in the environment) but also on the number of times that the chemical is recirculated before it is ultimately removed from the ecosystem. Typically, PPs are chemicals or unwanted byproducts of industrial processes that have been used and disposed for decades (prior to the inception of the various regulations and often without due regard for the environment) but have more recently been found to share a number of significant characteristics that need consideration before disposal is planned, including: (1) persistence in the environment insofar as these chemicals resist degradation in air, water, and sediments, (2) bioaccumulation insofar as these chemicals accumulate in living tissues at concentrations higher than those in the surrounding environment, and (3) long-range transport insofar as these chemicals can travel great distances from the source of release through air, water, and the internal organs of migratory animals, any of which can result in the contamination of ecosystems thousands of significant distances (up to thousands of miles) away from the source of the chemicals.

Briefly and by way of explanation, bioaccumulation is a process by which persistent environmental pollution leads to the uptake and accumulation of one or more contaminants, by organisms in an ecosystem. The amount of a pollutant available for exposure depends on its persistence and the potential for its bioaccumulation. Any chemical is considered to be capable of bioaccumulation if the chemical has a degradation half-life in excess of 30 days or if the chemical has a bioconcentration factor (BCF) greater than 1000 or if the log of the octanol–water partition coefficient (K_{ow}) is greater than 4.2:

$$BCF = (\text{Concentration in biota})/(\text{Concentration in ecosystem})$$

The BCF indicates the degree to which a chemical may accumulate in fish (and other aquatic animals, such as mussels, etc.) by transport across the gills or other membranes, excluding feeding. Bioconcentration is distinct from food-chain transport, bioaccumulation, or biomagnification. The BCF is a constant of proportionality between the chemical concentration in flora or fauna in an ecosystem. It is possible to estimate the BCF from the K_{ow} for many chemicals (Bergen et al., 1993). Thus:

$$\text{Log}(BCF) = m\log K_{ow} + b$$

In terms of actual numbers, for many lipophilic chemicals, the BCF can be calculated using the regression equation:

$$\log BCF = -2.3 + 0.76 \times (\log K_{ow}).$$

Furthermore, empirical relationships between K_{ow} and the BCF can be developed on a chemical-by-chemical basis.

On this note, it is worth defining the source of chemical contaminant insofar as chemical contaminants can originate from a point source or a nonpoint source (NPS). Point source of pollution is a single identifiable source of pollution which may have negligible extent, distinguishing it from other pollution source geometries. On the other hand, NPS pollution generally results from land runoff, precipitation, atmospheric deposition, drainage, seepage, or hydrologic modification. NPS pollution, unlike pollution from industrial and sewage treatment plants, comes from many diffuse sources and is often caused by rainfall or snowmelt moving over and through the ground. As the runoff moves, it picks up and carries away natural and humanmade pollutants, finally depositing them into lakes, rivers, wetlands, coastal waters, and groundwater.

Typically, PPs are highly toxic and long-lasting (hence the name *persistent*) and cause a wide range of adverse effects to environmental flora and fauna. Moreover, PPs do not respect international borders and the serious environmental hazards created by these chemicals affect not only developing countries, where systems and technology for monitoring, tracking, and disposing them of can be weak or nonexistent, but also affect developed countries. As long as the chemical can be transported by air, water, and land, no country is immune from the effects of these chemicals.

In the last four decades it has become increasingly clear that the chemical and allied industries, such as the pharmaceutical industry, have been facing serious environmental problems. Many industrial processes have broad scope but often generate extreme amounts of chemical waste. Indeed, an analysis of the amount of waste formed in processes for the manufacture of a range of fine chemicals and pharmaceutical intermediates has revealed that the generation of tens of kilograms of waste per kilogram of desired product was not exceptional in the chemical industry. This led to the introduction of the E (environmental) factor (kilograms of waste per kilogram of product) as a measure of the environmental footprint of manufacturing processes in various segments of the chemical industry. Thus:

$$\text{E} = \text{Kilograms of waste/Kilogram of product}$$

The E-factor can be conveniently calculated from knowledge of the number of tons of raw materials purchased and the number of tons of product sold, the calculation being for a particular product or a production site or even a whole company. A higher E-factor means more waste and, consequently, a larger environmental footprint; thus, the ideal E-factor for any process is zero.

However, in the context of environmental protection and to be all inclusive, the E-factor is the total mass of raw materials plus ancillary process requirements minus the total mass of product, all divided by the total mass of product. Thus, the E-factor should represent the *actual amount* of waste produced in the process, defined as everything but the desired product and takes the chemical yield into account and includes reagents, solvent losses, process aids, and (in principle) even the fuel necessary for the process. Water has been generally excluded from the E-factor as the inclusion of all process water could lead to exceptionally high E-factors in many cases and make meaningful comparisons of the technical factors (E-factors excluding water use) for processes difficult. However, in the modern environmentally conscious era, there is no reason for water requirement to be omitted since the disposal of process water is an environmental issue. Moreover, use of the E-factor has been widely adopted by many of the chemical industries, and the pharmaceutical industries in particular. Thus, a major aspect of process development recognized by process chemists and process engineers is the need for determining an E-factor—whether or not it is called by that name (i.e., E-factor). But the need to know the chemical balance, that is, *chemicals discharged* versus *chemicals remediated* has become a major factor for consideration in many industries that manufacture chemicals or use chemicals to manufacture products.

It is clear that the E-factor increases substantially on going from bulk chemicals to fine chemicals and then to pharmaceuticals. This is partly a reflection of the increasing complexity of the products, necessitating not only processes which use multistep syntheses, but is also a result of the widespread use of stoichiometric amounts of the reagents, that is, the required amounts of the reagents (some observers would advocate the stoichiometric amounts of the reagents plus 10%) to accomplish conversion of the starting chemical to the product(s). A reduction in the number of steps of a process for the synthesis of chemicals will, in most cases (but not always), lead to a reduction in the amounts of reagents and solvents used and hence a reduction in the amount of waste generated. This has led to the introduction of the concepts of step economy and function-oriented synthesis (FOS) of pharmaceuticals. The main issues behind the concept of FOS is that the structure of an active chemical, which may be a natural product, can be reduced to simpler structures designed for ease of synthesis while retaining or enhancing the biological activity. This approach can provide practical access to new (designed) structures with novel activities while at the same time allowing for a relatively straightforward synthesis.

As noted above, knowledge of the stoichiometric equation allows the process chemist or process engineer to predict the theoretical minimum amount of waste that can be expected. This has led to the concept of *atom economy* or *atom utilization* to quickly assess the environmental acceptability of alternatives to a particular product before any experiment is performed. It is a theoretical number, that is, it assumes a chemical yield of 100% and exact stoichiometric amounts and disregards substances which do not appear in the stoichiometric equation. In short, the key to minimizing waste is precision or *selectivity* in synthesis which is a measure of how efficiently a synthesis is performed. The standard definition of selectivity is the yield of product divided by the amount of substrate converted, expressed as a percentage.

Chemists distinguish between different categories of selectivity: (1) chemoselectivity, which relates to competition between different functional groups, each of which can interact differently with the constituents of an ecosystem (Table 9.1), and (2) regioselectivity, which is the selective formation of one regioisomer, for example, *ortho* or *para* substitution in

TABLE 9.1 Potential Reactions of Organic Functional Groups in the Environment

Functional Group	Interaction
Carboxylic acid, $-COOH$	ion exchange, complexation
Alcohol, phenol, $-OH$	hydrogen bonding, complexation
Carbonyl, $>C=O$	reduction–oxidation
Hydrocarbon, $[-CH_2-]_n$	hydrophobic

aromatic ring systems. However, one category of selectivity was, traditionally, largely ignored by chemists: the *atom selectivity* or *atom utilization* or *atom economy*. The virtually complete disregard for this important parameter by chemists and engineers has been a major cause of the waste problem in the manufacture of chemicals. Quantification of the waste generated in chemicals manufacturing, by way of E-factors, served to illustrate the omissions related to the production of chemical waste and focussed the attention of fine chemical companies, the pharmaceutical companies, and the petrochemicals companies on the need for a paradigm shift from a concept of process efficiency, which was exclusively based on chemical yield, to a need that more focused on the elimination of waste chemicals and maximization of raw material utilization.

Many of the global environmental changes forced by human activities are mediated through the chemistry of the environment (Andrews et al., 1996; Schwarzenbach et al., 2003; Manahan, 2010; Spellman, 2016). Important changes include the global spread of air pollution, groundwater pollution, pollution of the oceans, increases in the concentration of tropospheric oxidants (including ozone), stratospheric ozone depletion, and global warming (the so-called greenhouse effect). Since the onset and establishment of the agricultural revolution and the Industrial Revolution, the delicate balance between physical, chemical, and biological processes within the earth system has been perturbed, as a result. Example of the causes of perturbation of the earth systems include (1) the exponential growth in the world population, (2) the use of increasing amounts of fossil fuel, (3) fossil fuel—related emissions of carbon to the atmosphere, and (4) the intensification of agricultural practices including the more frequent use of fertilizers. The observed increase in the atmospheric abundance of carbon dioxide (CO_2), which has been ascribed (correctly or incorrectly) mainly to fossil fuel burning (Speight and Islam, 2016), although biomass destruction is an important secondary source of carbon dioxide emissions. Atmospheric concentrations are additionally influenced by exchanges of carbon with the ocean and the continental biosphere (Firor, 1990; WMO, 1992; Calvert, 1994; Goody, 1995).

The progressive modification and fertilization of the terrestrial biosphere are believed to have caused the observed increase in atmospheric nitrous oxide (N_2O), a tropospheric greenhouse gas and a source of reactive species in the stratosphere. Methane (CH_4), which also contributes to *greenhouse forcing* (the technical term for the influence of the greenhouse effect that causes a shift in the climate—this can occur due to changes in the level of gasses that share two properties: they are transparent to visible light, but absorb the infrared, which we typically perceive as heat) and also plays an important role in the photochemistry of the troposphere and the stratosphere, is produced by biosphere-related processes (wetlands, livestock, landfills, biomass burning), as well as by leakage from gas distribution systems in various countries (Calvert, 1994). The global atmospheric concentration of methane has also grown in the past.

Observed increases in the abundance of tropospheric ozone (O_3), which contribute to deteriorating air quality, result from complex photochemical processes involving industrial and biological emissions of nitrogen oxides, hydrocarbons, and certain other chemicals. Ozone is a strong absorber of solar ultraviolet radiation and also contributes to greenhouse forcing. Anthropogenic emissions of sulfur resulting from combustion of sulfur-containing fuels and also from coal combustion (without the necessary end-of-pipe gas cleaning protocols) (Speight, 2013, 2014), coal burning in highly populated and industrialized regions of the Northern Hemisphere, and the related increase in the aerosol load of the troposphere, have contributed to regional pollution and have probably produced a cooling of the surface in these regions by backscattering a fraction of the incoming solar energy.

The bad news is that the rapid increase in the abundance of industrially manufactured chlorofluorocarbons in the atmosphere produced an observed depletion in stratospheric ozone and the formation each spring (since the late 1970s) of an *ozone hole* over Antarctica. By way of explanation, the ozone hole is not technically a *hole* where no ozone is present, but it is actually a region of exceptionally depleted ozone in the stratosphere over the Antarctic that happens at the beginning of Southern Hemisphere spring (August–October). The good news is that, as a result of environmental caution and a reduction in terms of the release of contaminants, the ozone hole has diminished in size (i.e., the depletion of ozone has stopped and may even be on a turnaround to an increase in the amount of ozone in the area).

Although toxic and hazardous chemicals are produced by the various chemical industries, frequent reference is made here to the chemical products produced by the crude oil refining industry, without any attempt to point to this industry and the major polluter. The chemicals produced in the various crude oil–derived product mixtures offer a wide range of properties and behavior that make the crude oil–derived products (Speight, 2014, 2015) suitable for this text.

In summary, many of the specific chemicals in petroleum are hazardous because of their chemical reactivity, fire hazard, toxicity, and other properties. In fact, a simple definition of a hazardous chemical (or hazardous waste) is that it is a chemical substance (or chemical waste) that has been inadvertently released, discarded, abandoned, neglected, or designated as a waste material and has the potential to be detrimental to the environment. Alternatively, a hazardous chemical may be a chemical that may interact with other (chemical) substances to give a product that is hazardous to the environment.

2. MOLECULAR INTERACTIONS

In the typical approach to environmental science and technology, there is often lesser interest in the concept of molecular interactions of the discharged chemicals with the environment compared to the interaction between the discharged chemicals. However, the variety of structures of various chemicals actually is the result of weak ordering because of noncovalent interactions. Indeed, for self-assembly to be possible in soft materials, it is evident that forces between molecules must be much weaker than covalent bonds between the atoms of a molecule. The weak intermolecular interactions responsible for molecular ordering include, for example, ionic and dipolar interactions, van der Waals forces, and hydrogen bonds (Table 9.2). Recent evolutions in environmental science provide convincing arguments to support the opportunity and importance of noncovalent bonds since the fundamental and applicative aspects related to molecular interactions are of considerable interest in all forms of interaction of discharged chemicals with the environment.

The forces between molecules (i.e., the intermolecular interactions) govern a wide range of chemical properties and physical properties of chemicals and the interactions of chemicals with various environmental systems. The structures of solids, liquids, and gases and their behavior in various ecosystems are examples of fields where this topic is of central importance.

TABLE 9.2 Different Types of Bond Arrangements

Covalent bond: a bond in which one or more electrons (often a pair of electrons) are drawn into the space between the two atomic nuclei. These bonds exist between two identifiable atoms and have a direction in space, allowing them to be shown as single connecting lines between atoms in drawings, or modeled as sticks between spheres in models.
Ionic bond: Occurs between ionized functional groups such as carboxylic acids and amines.
Hydrogen bond: Occurs between derivatives of alcohol, carboxylic acid, amide, amine, and phenol. Hydrogen bonding involves the interaction of the partially positive hydrogen ion one molecule and the partially negative heteroatom on another molecule. Hydrogen bonding is also possible with elements other than nitrogen or oxygen and can occur intermolecularly or intramolecularly.
Dipole–dipole interaction: Possible between molecules having polarizable bonds, in particular the carbonyl group (CO) which has a dipole moment and molecules can align themselves such that their dipole moments are parallel and in opposite directions. Ketones and aldehydes are capable of interacting through dipole–dipole interactions.
Van der Waals interaction: Weak intermolecular bonds between regions of different molecules bearing transient positive and negative charges which are caused by the movement of electrons. Alkanes, alkenes, alkynes, and aromatic rings interact through van der Waals interactions.
Intermolecular bond: Occurs between different molecules and can take the form of ionic bonding, hydrogen bonding, dipole–dipole interactions, and van der Waals interactions.
Intramolecular bond: Occurs within a molecule and can take the form of ionic bonding, hydrogen bonding, dipole–dipole interactions, and van der Waals interactions.

By definition, molecular interactions are the attractive or repulsive forces *between molecules* and between nonbonded atoms and are also known as noncovalent interactions or intermolecular interactions. In this sense, molecular interactions are not bonds (which break and form during chemical reactions) but constitute a series of interactions that are important in diverse fields of environmental technology.

By way of explanation for the nonchemist, bonds hold atoms together *within molecules* —a molecule is a set of atoms that associates and does not dissociate or lose the structure when the molecule interacts with an ecosystem. Understanding intermolecular interactions helps scientists and engineers to understand the function and behavior of a molecule, as well as assist in predicting the outcome of environmental processes such as the interactions of discharged chemical contaminants with the soil, particularly the adsorption/desorption of discharged chemicals on to clay minerals (Chapter 5).

Moreover, a basic understanding of the kinetics and mechanisms of important reactions and processes (e.g., sorption/desorption, precipitation/dissolution, and redox) in natural systems such as soils and sediments are necessary to accurately determine the speciation, mobility, and bioavailability of contaminants in the environment. Ideally, one should make investigations over a range of spatial and temporal scales and environmental conditions. The rates of these processes can vary over a range of temporal scales ranging from milliseconds to years. In many cases, kinetic studies of metals, metalloids, nutrients, radionuclides, and organic chemicals have shown that reaction rates are initially rapid followed by a slow approach to a steady state. The rapid reaction has been ascribed to chemical reactions and film diffusion, while the slow reactions have been attributed to interparticle and intraparticle diffusion, retention on sites of lower reactivity, and surface precipitation. With many metals and organic chemicals, desorption is typically much slower than adsorption and that the longer the contact time (residence time or aging time) between the sorbent and sorbate, the more difficult it is to release the contaminant (Chapter 5). This is most likely due to the physical entrapment within the pore systems of the sorbent, as well as to other diffusion phenomena and surface precipitation (sometimes referred to as sedimentation).

Intermolecular interactions (also called intermolecular forces) are the interactions which mediate interaction between molecules, including forces of attraction or repulsion which act between molecules and other types of neighboring particles, such as atoms or ions. Intermolecular forces are weak relative to intramolecular interactions (also called intramolecular forces) which are the interactions (forces) that hold a molecule together. For example, the covalent bond (which involves sharing electron pairs between atoms) is much stronger than the forces present between neighboring molecules.

The importance of intermolecular interactions starts from macroscopic observations which indicate the existence and action of forces at a molecular level. These observations include nonideal gas thermodynamic behavior reflected by properties such as vapor pressure, viscosity, surface tension, and adsorption (Chapters 3 and 5). Examples of the various interactions to be considered in this text are (1) covalent bonds, (2) van der Waals interactions, (3) electrostatic interactions, and (4) hydrogen bonds.

Covalent bonds remain intact and unchanged when a physical change occurs in a molecule (the simple example is ice melting—this is not a chemical reaction). The enthalpy of a given molecular interaction, between a pair of nonbonded atoms, is 1−10 kcal/mol, and in the upper limit is significantly less than a covalent bond. Even though the bonds are weak individually, cumulatively the energies of molecular interactions are significant. Moreover, the phrase *van der Waals interaction* (also called *van der Waals force*) has come to mean cohesive (attraction between like), adhesive (attraction between unlike), and/or repulsive forces between molecules. Put simply, these interactions arise from interaction between uncharged atoms or molecules, leading not only to such phenomena as the cohesion of condensed phases and physical adsorption of gases, but also to a universal force of attraction between macroscopic bodies.

Electrostatic interactions are interactions that occur between and among cations (are ions with a net positive charge) and anions (are ions with a net negative charge) and can be either attractive or repulsive, depending on the signs of the charges. Like charges repel. Unlike charges attract. A *hydrogen bond* is a favorable interaction between an atom with a basic lone pair of electrons (a Lewis base) and a hydrogen atom that has been partially stripped of the electrons because it is covalently bound to an electronegative atom (N, O, or S). In a hydrogen bond, the Lewis base is the hydrogen bond acceptor (A) and the partially exposed proton is bound to the hydrogen bond donor (H-D).

Thus, a *hydrogen bond* is the interaction between the lone pair of electrons of an electronegative atom and a hydrogen atom that is bonded to an oxygen atom. Thus:

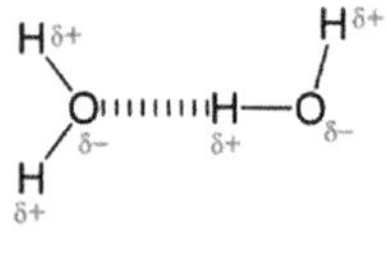

$$H^{\delta+}{-}O^{\delta-}({-}H^{\delta+})\cdots H^{\delta+}{-}O^{\delta-}{-}H^{\delta+}$$

Other atoms such a nitrogen and fluorine may also play a role in a hydrogen bonding interaction.

The hydrogen bond is often described as a strong electrostatic dipole−dipole interaction. However, it also has some features of covalent bonding, such as (1) the hydrogen bond is directional, (2) the hydrogen bond is stronger than a van der Waals interaction, and (3) the hydrogen bond usually involves a limited number of interaction partners. Intermolecular hydrogen bonding is responsible for the high boiling point of water (100°C, 212°F) compared to the

TABLE 9.3 Melting Point and Boiling Point Temperatures (at Normal Pressure) for Hydrides of the Groups 15–17 Elements in the Periodic Table That Are Analogous to Water

Melting Points (°C)		
NH_3: −78	H_2O: 0	HF: −83
PH_3: −133	H_2S: −86	HCl: −115
AsH_3: −116	H_2Se: −60	HBr: −89
SbH_3 -88	H_2Te: −49	HI: −51
Boiling Points (°C)		
NH_3: −33	H_2O: 100	HF: 20
PH_3: −88	H_2S: −61	HCl: −85
AsH_3: −55	H_2Se: −42	HBr −67
SbH_3: −17	H_2Te: −2	HI: −35

other Group hydrides of the periodic table (Table 9.3, Fig. 9.1), which have no hydrogen bonds. Hydrogen−oxygen bonding can also play a role in the adsorption of a chemical on to a mineral (Chapter 5). These types of reactions are necessary considerations of the type of interactions that can occur when a chemical is discharged into an ecosystem. Similar forces play a role in the adsorption/desorption of a chemical on to, for example, soil, which will play a role in the selection of the appropriate remediation process (Chapters 5 and 10).

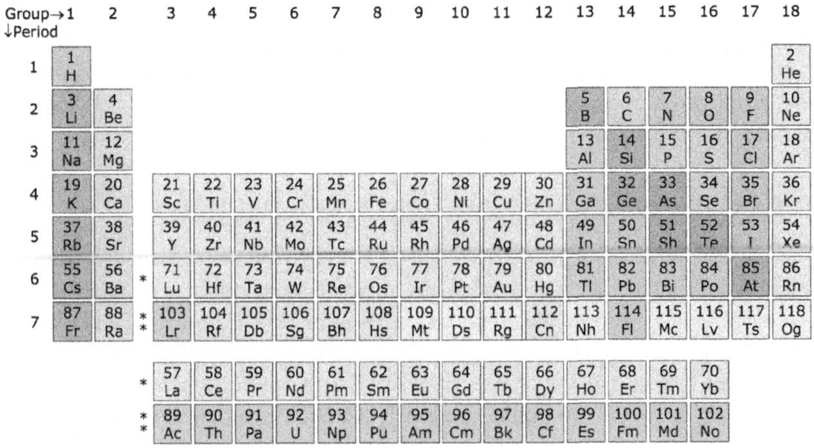

Group→1 ↓Period	1	2		3	4	5	6	7	8	9	10	11	12	13	14	15	16	17	18
1	1 H																		2 He
2	3 Li	4 Be												5 B	6 C	7 N	8 O	9 F	10 Ne
3	11 Na	12 Mg												13 Al	14 Si	15 P	16 S	17 Cl	18 Ar
4	19 K	20 Ca		21 Sc	22 Ti	23 V	24 Cr	25 Mn	26 Fe	27 Co	28 Ni	29 Cu	30 Zn	31 Ga	32 Ge	33 As	34 Se	35 Br	36 Kr
5	37 Rb	38 Sr		39 Y	40 Zr	41 Nb	42 Mo	43 Tc	44 Ru	45 Rh	46 Pd	47 Ag	48 Cd	49 In	50 Sn	51 Sb	52 Te	53 I	54 Xe
6	55 Cs	56 Ba	*	71 Lu	72 Hf	73 Ta	74 W	75 Re	76 Os	77 Ir	78 Pt	79 Au	80 Hg	81 Tl	82 Pb	83 Bi	84 Po	85 At	86 Rn
7	87 Fr	88 Ra	**	103 Lr	104 Rf	105 Db	106 Sg	107 Bh	108 Hs	109 Mt	110 Ds	111 Rg	112 Cn	113 Nh	114 Fl	115 Mc	116 Lv	117 Ts	118 Og
			*	57 La	58 Ce	59 Pr	60 Nd	61 Pm	62 Sm	63 Eu	64 Gd	65 Tb	66 Dy	67 Ho	68 Er	69 Tm	70 Yb		
			**	89 Ac	90 Th	91 Pa	92 U	93 Np	94 Pu	95 Am	96 Cm	97 Bk	98 Cf	99 Es	100 Fm	101 Md	102 No		

FIGURE 9.1 The periodic table of the elements showing the groups and periods including the Lanthanide elements and the Actinide elements.

Since an extremely high number of different pollutants are usually simultaneously present in a certain environmental area, it must be expected that interactions between these pollutants may occur frequently. Nevertheless, not very much is known about the combined impact of several pollutants and the environmental policy ignores these effects in defining pollution standards and limits just for single pollutants. Moreover, it is not at all clear how these effects should precisely be described.

Toxic interaction refers to the qualitative and/or quantitative modification of the toxicity of one chemical by another, the process principally occurring within the organism after the exposure phase. Interactions can either result in greater-than-additive or less than-additive toxic response. Over a 1000 studies published to date report the occurrence of supra- or infraadditive toxicity from combined exposure to two chemicals. The interactive toxicity resulting from combined exposure to chemicals is a consequence of the alteration of the toxicity kinetics and/or toxicity dynamics of one chemical by another. Interference at the kinetic level would imply a modulation of absorption, distribution, metabolism, and/or excretion of one chemical by another. Interference at the toxicodynamic level might involve a competition between two chemicals for binding to the target site or an alteration by one chemical of the susceptibility of the target cells to the effects of another agent. Despite the report of the occurrence of several significant interactions between environmental pollutants, their relevance for humans and relative importance for regulation remain unclear and ill-defined. This situation is a consequence of the lack of chronic exposure studies with chemical mixtures and the lack of understanding of the mechanistic basis of interactions at a quantitative level. Until these issues are resolved, we probably would not be able to confidently use animal data on interactions to make quantitative predictions for humans. However, there is some direct and/or epidemiological evidence for the occurrence of several supraadditive and infraadditive chemical interactions in humans. In this article, we corroborate laboratory observations with human experience as they relate to toxic interactions among environmental pollutants and propose an approach to consider data on toxic interactions for human health risk assessment.

3. PARTITIONING AND PARTITION COEFFICIENTS

The increasing awareness of chemicals in the environment, their disposition, and their ultimate fate has created the need to find reliable mechanisms to assess the environmental behavior and effects of new chemicals. Whether or not a chemical will pose a hazard to the environment will depend on the concentration levels it will reach in various ecosystems, and whether or not those concentrations are toxic to the floral and faunal species within the ecosystem. It is, therefore, important to determine expected environmental distribution patterns of chemicals in order to identify which areas will be of primary environmental concern (McCall et al., 1983).

When a small amount of a chemical is added to two immiscible phases and then shaken, the phases will eventually separate, and the chemical will partition between the two phases according to its solubility in each phase. In fact, the simplest and most common method of estimating contaminant retardation, which is the inverse of the relative transport rate of a contaminant compared to that of water, is based on the value of the partition coefficient (also called the distribution coefficient), $K_{d,2}$ values, which is the concentration ratio at equilibrium:

$$C_1/C_2 = K_{12}$$

In the laboratory, K_{12} is typically determined from the slope of C_1 versus C_2 over a range of concentrations. Partition coefficients can be measured for essentially any two-phase system, such as: air–water, octanol–water, lipid–water, and particle–water. In situ partition coefficients also can be measured where site-specific environmental conditions might influence the equilibrium phase distribution. The partition coefficient is a direct measure of the partitioning of a contaminant between the solid and aqueous phases. It is an empirical metric that attempts to account for various chemical and physical retardation mechanisms that are influenced by a myriad of variables. Ideally, site-specific values for the partition coefficient would be available for the range of aqueous and geological conditions in the system to be modeled.

Values for the partition coefficient vary greatly between contaminants, but also vary as a function of aqueous and solid-phase chemistry. Another approach to describing the partitioning of contaminants between the aqueous and solid phases is the recognition and insertion of the appropriate data, that the value for the partition coefficient also varies according to the chemistry and mineralogy of the system under study.

Thus, any mechanistic model must accommodate the dependency of partition coefficient values on (1) the contaminant concentration, (2) any competing ion concentration, (3) the variable surface charge on the absorbent, and (4) the distribution of the species in the solution.

3.1 Single Chemicals

The term *partitioning* refers to the distribution of a solute, S, between two immiscible solvents (such as aqueous phase and organic phase) and is an equilibrium state according to the following equation:

$$\mathrm{S}(aq) \leftrightarrow \mathrm{S}(org)$$

The equilibrium constant for this equilibrium state is:

$$\mathrm{K} = [\mathrm{S}(org)]/[\mathrm{S}(aq)]$$

S(*org*) and S(*aq*) are the concentration of the solute in the organic phase and the aqueous phase, respectively, and K is the equilibrium constant, which is also called the partition coefficient (P) or the distribution coefficient (D).

More specifically, the focus of partitioning studies is on the equilibrium distribution of a chemical that is established between the phases and the associated problem of calculating the distribution of a chemical between the different phases (*partitioning equilibrium*). There are many situations in which it is correct to assume that phase transfer processes are fast compared to the other processes (such as chemical transformation of the contaminant to other [benign or hazardous] chemicals that play a role in determining the fate of a contaminant). In such cases, it is appropriate to describe the distribution of the chemical as a change of phase but using the equilibrium approach using the valid assumption that an equilibrium condition will be reached with the chemical passing to one phase or remaining distributed between different passes.

By way of explanation, an example of phase equilibrium is the partitioning of a contaminant (such a carboxylic acid or a carboxylic acid derivative—that is, the sodium salt of the acid) between the pore water and other water in the bed of a sediment and solids in sediment beds. Thus, from the equilibrium partition coefficient it is possible to calculate the rate of transfer of a chemical contaminant across interfaces and the rate at which such transfer can be anticipated to occur.

However, before too much excitement is generated at the thought of the answers that a study of partitioning will provide, it must be recognized that there are many situations where an equilibrium of the chemical between the immiscible phases is not reached. However, some observers (justifiably) find the information useful insofar as the data are used (correctly or incorrectly) to characterize what the equilibrium distribution of the chemical would be if sufficient time (often difficult to define) is allowed. However, in such cases a quantitative description of the potential partitioning equilibrium can be employed to estimate the possible direction of the transport of the chemical contaminant from one environmental ecosystem to another. If this is the case, it may be possible to evaluate whether or not the chemical component(s) of a contaminant (such as a solvent or naphtha) will continue to dissolve in the groundwater and/or volatilize into the overlying soil. Both options are viable, but (hopefully) the data will provide the most likely option to occur, given the position of the chemical within the underground chemical and geological system. Thus, the goal of any partitioning study of the ability of a contaminant is to gain insight into the role played by the chemical (and physical) structure of a contaminant in determining its fate in the environment (McCall et al., 1983).

Furthermore, when dealing with phase partitioning parameters, it is necessary to develop an understanding of the intramolecular and intermolecular interactions that are a result of various chemical and physical properties (Chapter 3) between the chemical(s) and the specific molecular environment in which the chemical is spilled or disposed. Thus, there is the necessity to understand the means by which structural groups function within the contaminants and the means by which the functional groups are related to chemical and physical behavior (Chapters 2 and 3).

Therefore, there must be an effort to understand (1) the interactions arising from contacts of functional groups within a molecule, that is, the intramolecular interactions, (2) the influence of these intramolecular interactions on the existence of the molecule in the environment, (3) the interactions arising from contacts of functional groups with the functional groups of another molecule, that is, the intermolecular interactions, and (4) the influence of these intermolecular interactions on the existence of the molecule in the environment. However, a partition coefficient is not the only means of estimating the transfer of a chemical contaminant between phases. It is also valuable to correlate the partition coefficients on the basis of solubility of the chemical(s) which are convenient and readily understood measurable expressions of single-phase activity coefficients (Cole and Mackay, 2000).

Thus, this section presents an examination of the PPs of the chemical and any relevant environmental factors that are needed for quantifying such partitioning.

3.1.1 Acid–Base Partitioning

Acid–base extraction is a procedure using sequential liquid–liquid extractions to separate acid from bases or the use of an acidic solution or a basic solution to separate acid and base mixtures based on their chemical properties. In either case, the product is largely free of neutral and acidic or basic impurities. It is not possible to separate chemically similar acids or bases using this simple method. In the process, the addition of an acid to, for example, a mixture of an organic base and acid will result in the acid remaining uncharged while the base will be protonated to form a salt. Conversely, the addition of a base to a mixture of an organic acid and base will result in the base remaining uncharged while the acid is deprotonated to give the corresponding salt.

Thus, acid–base partitioning (also called *pH partitioning*) is the tendency for acids to accumulate in basic fluid compartments and bases to accumulate in acidic regions. The reason for this phenomenon is that acids become negatively charged in basic fluids, since they donate a proton. On the other hand, bases become positively charged in acidic liquids because the base receives a proton (H^+). Since electric charge decreases the membrane permeability of substances, once an acid enters a basic fluid and becomes electrically charged, it cannot escape that compartment with ease and therefore accumulates, and vice versa with bases. Thus, by manipulating the pH of the aqueous layer, the partitioning of a solute can be changed.

3.1.2 Air–Water Partitioning

The transfer between the atmosphere and the aquasphere (the various bodies of water) is one of the key processes affecting the transport of many chemicals in the environment (Schwarzenbach et al., 2004). For neutral chemicals (such as nonpolar chemicals), at dilute solution concentrations in pure water, the air–water distribution ratio is referred to as the Henry's Law constant K_H (or K_{aw}). For real aqueous solutions (i.e., solutions that contain many other chemical species), the term *air–water distribution ratio* may also be used which is, for practical purposes, approximated by the Henry's Law constant. The Henry's Law constant K_H can be approximated as the ratio of the abundance of the chemical in the gas phase to that in the aqueous phase at equilibrium:

$$X(aq) = X(g)$$

$$K_H = P_i/C_w \text{ atm L/mol}$$

In this equation, P_i is the partial pressure of the gas phase of the chemical and C_w is its molar concentration in water. The Henry's Law constant has been defined in other terms, which can introduce confusion and caution and should be used to check the units. K_H (K_{aw}) and is also defined in terms of a unitless ratio of concentrations. Nevertheless, the air–water partition coefficient (K_{aw}) is the constant of proportionality between the concentration of a chemical in air and its concentration in water at low partial pressures and below its saturation limits in either air or water. The coefficient can be estimated from Henry's law constant and is sometimes referred to as dimensionless.

In a homologous series, the partitioning may decrease with molecular size; the water solubility is more sensitive to size increases due to the greater energy costs associated with cavity formation in water. For a series of structurally related chemicals, the magnitude of K_{aw} increases with size (i.e., molar volume of the chemical) because molecules that have similar functional groups will experience the same type of intermolecular interactions. Increasing the size within the same functional group class will result in a systematic increase in the dispersive forces. For example, in considering organic chemicals as example, alcohol derivatives have a much lower K_{aw} than alkane derivatives due to the more favorable solute–solvent hydrogen bonding interactions. For molecules of similar size, alkenes are more polarizable than alkanes, which results in more favorable solute solvent–induced dipole dispersive interactions. Monopolar molecules such as ether, ester, ketone, and aldehyde derivatives experience dipole–dipole interactions and H-accepting interactions with the solvent, water. Although, the presence of salts in the aqueous phase will decrease the water solubility of solutes, there will be little or no effect on the vapor pressure of the pure liquid (Chapter 5). Hence, K_H values will increase with increasing salinity.

Whatever the system, it is essential to have reliable data for the air–water partition coefficient or Henry's law constant for the chemicals under investigation for elucidating the environmental dynamics of many natural and anthropogenic chemicals. When a chemical (here referred to as the solute) is introduced into the environment, it tends to diffuse from phase to phase in the direction towards establishing equilibrium between all phases. Frequently, the physical–chemical properties of the solute dictate that it will partition predominantly into a different phase from the one into which it is normally emitted.

For example, benzene emitted in waste water will tend to partition or transfer from that water into the atmosphere where it becomes subject to atmospheric photolytic degradation processes. Knowledge of the air–water partition characteristics of a solute is thus important for elucidating where the solute will tend to accumulate and also in calculating the rates of transfer between the phases. Conventionally these rates are expressed as the product of a kinetic constant such as a mass transfer coefficient (or diffusivity divided by a diffusion path length) and the degree of departure from equilibrium which exists between the two phases. Elucidating the direction and rate of transfer of such solutes thus requires accurate values for the Henry's law.

The transfer between the atmosphere and bodies of water is one of the key processes affecting the transport of many chemicals in the environment. For neutral (nonpolar) chemicals, at dilute concentrations in pure water, the air–water distribution ratio is referred to as the Henry's Law constant (K_H or K_{aw}). For real aqueous solutions (i.e., solutions that contain many other chemical species), the term *air–water distribution ratio* is often used which, for practical purposes, is approximated by the Henry's Law constant. The Henry's Law constant K_H can be approximated as the ratio of the abundance of a chemical in the gas phase to that in the aqueous phase at equilibrium.

$$X(aq) = X(g)$$

$$K = P_i/C_w \text{ atm L/mol}$$

In this equation, P_i is the partial pressure of the gas phase of the chemical and C_w is the molar concentration of the chemical in water.

Knowledge of the air–water partitioning behavior of volatile chemicals and greenhouse gases is important in a number of environmental applications. For example, volatile chemicals emitted from open process streams in a paper mill can promote ground level ozone and lead to respiratory problems in humans. Therefore, it is important to have reliable estimates of the amount of volatile chemicals in the atmosphere in contact with open process streams. In global climate models, the partitioning of carbon dioxide and methane between the atmosphere and ocean water is of current interest since oceans represent a substantial storage reservoir for these gases.

3.1.3 Molecular Partitioning

Typically, in a two-layer system of water (bottom layer) and a solvent (top layer), a chemical will be (predominantly) in one or the other of the layers. After stirring and allowing time for the layers to settle, the chemical could well be in both phases, albeit to a different extent (concentration, C) in each phase. The partition coefficient (K) is:

$$K = C_{organic}/C_{water}$$

From this equation, a high value of K suggests that the chemical is not very soluble in water but is more soluble in the solvent, that is, the chemical is hydrophobic (sometimes referred to as lipophilic, which is the ability of the chemical to dissolve in nonpolar solvents such as hexane or toluene).

Molecular partitioning occurs when a chemical dissolves in each of the two immiscible solvent phases and is measured by the *partition coefficient* or *distribution coefficient,* which is the ratio of concentrations of the chemical in a mixture of the two immiscible phases at equilibrium. This ratio is therefore a measure of the difference in solubility of the chemical in these two phases. Most commonly, one of the solvents is water while the second is hydrophobic such as 1-octanol. Hence the partition coefficient is a measure of the degree of hydrophilic or hydrophobic character of the chemical.

In the environment, the hydrophobic character of the chemical can give an indication of the relative ease that a chemical might be taken up in groundwater and pollute the aquasphere. The partition coefficient can also be used to predict the mobility of a chemical in groundwater and, in the field of hydrogeology (the area of geology that deals with the distribution and movement of groundwater in the soil and rocks of the crust of the earth, commonly in aquifers), the K_{ow} is used to predict and model the migration of dissolved hydrophobic chemicals in the soil and in the groundwater.

3.1.4 Octanol–Water Partitioning

The distribution of nonpolar chemicals between water and natural solids (e.g., soils, sediments, and suspended particles) or organisms, can be viewed in many cases as a partitioning process between the aqueous phase and the bulk matter present in natural solids or biota. More recently, environmental chemists have found similar correlations with soil humus and other naturally occurring phases. These correlations arise because the same molecular forces that control the distribution of chemicals between water-immiscible solvents and water also control (or determine) the environmental partitioning from water into natural phases.

The K_{ow} is a key parameter in understanding and predicting the environmental fate and transport behavior of chemicals (Lyman et al., 1990; Boethling and Mackay, 2000). K_{ow} is often used as a surrogate for the lipophilicity of a chemical and its tendency to concentrate in phases such as within plant lipids or fish from the aqueous solution. Chemicals with relatively low values of K_{ow} are considered relatively hydrophilic and will tend to have high water solubility and low BCFs (Lyman et al., 1990). The K_{ow} is also used in the prediction of other parameters including water solubility and the carbon–water partition coefficient.

For a series of neutral (nonpolar) chemicals partitioning between octanol and water, the K_{ow} value is determined largely by the magnitude of the aqueous activity coefficient (a measure of the dissimilarity between the solute and the aqueous solvent). Thus, the major property that determines the magnitude of the partition constant of a nonpolar or moderately polar chemical between a solvent and water is the incompatibility of the chemical with water. In many cases, the nature of the solvent is generally of secondary importance relative to the properties of the chemical.

3.1.5 Particle–Water Partitioning

The persistence of discharged chemical pollutants in soil, their migration to groundwater, and the evaluation of the degree of contamination expected in a groundwater system after an accidental spill or as consequence of the presence of a waste disposal site are problems of particular environmental concern. To resolve these issues, knowledge of the sorption characteristics of the pollutants as well as the knowledge of the type of soil and its characteristics are required. Furthermore, sorption of a chemical also affects (1) the volatility of the chemical, (2) the bioavailability and bioactivity of the chemical, (3) phytotoxicity of the chemical that is the effect of the chemical on plant growth, (4) the effect of the chemical on faunal communities, and (5) the chemical or microbial transformations of the chemical (Delle Site, 2001).

In fact, many chemicals will preferentially associate with soil and sediment particles rather than the aqueous phase (sometime referred to as adsorption partitioning). The particle–water partition coefficient (K_P) describing this phenomenon is:

$$K_P = C_S/C_W$$

C_S is the concentration of chemical in the soil or sediment (mg/kg dry weight) and C_W is the concentration in water (mg/L). In this form, K_P has units of L/kg or reciprocal density. Dimensionless partition coefficients are sometimes used where K_P is multiplied by the particle density (in kg/L). It has also been observed that nonionic organic chemicals were primarily associated with the organic carbon phase(s) of particles. A plot of K_P versus the mass fraction of organic carbon in the soil (f_{OC}, g/g) is linear with a near-zero intercept yielding the relationship:

$$K_P = f_{OC} K_{OC}$$

where K_{OC} is the organic carbon–water partition coefficient (L/kg) which is useful when the fraction of organic carbon is about 0.5% or greater. At lower organic carbon fractions, interaction with the mineral phase becomes relatively more important (though highly variable) resulting in a small positive intercept of K_P versus f_{OC}. The strongest interaction between organic chemicals and mineral phases appears to be with dry clay minerals. Thus, K_P will likely change substantially as a function of water content in clay–mineral soil having a low organic carbon content.

3.2 Mixtures of Chemicals

The sorption of a chemical (especially chemical mixtures) on a naturally occurring solid (such as a nonclay mineral or a clay mineral) is a complicated process which involves many sorbent properties, besides the physicochemical properties of the chemical (or chemical constituents of the mixture). These properties are especially the relative amount of the mineral and material in soil/sediment and their respective composition with associated physical characteristics. Also, different regions of a soil or sediment matrix may contain different types, amounts, and distributions of surfaces and of soil material, even at the smaller size of many of the particles in soil (Booth et al., 2006; Mao et al., 2009; Nabi et al., 2014).

In most cases, the tendency of a chemical to be adsorbed or desorbed can be expressed in terms of the carbon partition coefficient (K_{OC}) which is largely independent of soil or sediment properties (Lyman et al., 1982). Thus, the carbon partition coefficient is a chemical specific adsorption parameter and may be determined as the ratio of the amount of chemical adsorbed per unit weight of carbon in the soil or sediment to the concentration of the chemical in solution at equilibrium:

$$K_{OC} = \text{(Microgram adsorbed per gram of carbon)/(Microgram per mL of solution)}$$

Factors which affect measured values of the carbon partition coefficient include (1) temperature, (2) acidity or alkalinity, measured as the pH, (3) salinity, which is the salt content, (4) the concentration of dissolved oxygen, (5) suspended particulates, and (6) the solids-to-solution ratio.

In the case of a spill of crude oil or a crude-oil product, the higher molecular weight constituents of the spilled material will influence the transport and distribution of the constituents among different environmental compartments, including air, water, natural organic matter, and other natural phases. Furthermore, the extent to which a chemical (or mixture) is partitioned between the solid phase and the solution phase is determined by several physical and chemical properties of both the chemical (or the mixture) and the soil or sediment aqueous solutions (Booth et al., 2006). Some mixtures (such as crude oil) contain hundreds or thousands of distinct organic chemicals that may arise in environmental samples and remain incompletely characterized (Speight, 2014), and it remains unclear how to assess bioaccumulation potential or to parametrize transport models describing interphase transfers for these mixtures. In fact, the so-called average structure concept for estimating the structure of the higher molecular (higher-boiling constituents) fraction of crude oil is totally inadequate because it does not allow for the different functional groups in these molecules (Chapter 1) which can and will affect the sorption properties and the relative reactivity of the constituents of crude oil (Speight, 2014, 2015).

The functional group of a chemical is a molecular moiety that typically dictates the behavior (reactivity) of the organic chemical in the environment and the reactivity of that functional group is assumed to be the same in a variety of molecules, within some limits, and if steric effects (that arise from the three-dimensional structure of the molecule) do not interfere (Chapter 1). Thus, most organic functional groups feature heteroatoms (atoms other than carbon and hydrogen, such as: nitrogen, oxygen, and sulfur). The concept of functional groups is a major concept in organic chemistry, both to classify the structure of organic chemicals and to predict the physical (sorption) and chemical properties, especially as these properties are exhibited in the environment (Speight, 2014). In fact, it is the functional groups in crude oil that are used to separate the various constituents using a variety of adsorbents (e.g., silica, alumina, and clay).

Thus, the presence of a variety of chemicals in a complex mixture could alter the activity coefficient of the chemical in the water, the pH of the water, or the solubility of the chemical in water, and, consequently, the sorption of the chemical to soils and sediments. The degree of adsorption will not only affect chemical mobility but will also affect volatilization, photolysis, hydrolysis, and biodegradation. Adsorption of chemicals will also occur on minerals free of matter. It may be significant under certain conditions such as: (1) the presence of clay minerals with a very large surface area, (2) those situations where cation exchange occurs, such as for dissociated bases, (3) situations where clay–colloid–induced polymerization occurs, and/or (4) situations where chemisorption is a factor (Lyman et al., 1982).

In all cases, the distribution of a solute between sorbent and solvent phases results from its relative affinity for each phase, which in turn relates to the nature of forces which exist between molecules of the solute and those of the solvent and sorbent phases. The type of interaction depends on the nature of the sorbent, as well as the physicochemical features of the sorbate that is hydrophobic or polar at various degrees.

Moreover, the physical sorption processes involve interactions between dipole moments—permanent or induced—of the sorbate and sorbent molecules. The relatively weak bonding forces associated with physical sorption are often amplified in the case of hydrophobic molecules by substantial thermodynamic gradients for repulsion from the solution in which they are dissolved. Chemical interactions involve covalent bond and hydrogen bond.

Finally, electrostatic interactions involve ion–ion and ion–dipole forces. In a more detailed way, the type of interactions and the approximate values of energy associated are: (1) van der Waals interactions, (2) hydrophobic bonding, (3) hydrogen bonding, (4) charge transfer association, (5) ligand-exchange and ion bonding, (6) direct and induced ion-dipole and dipole–dipole interactions, and (7) chemisorption by the formation of a covalent bond.

The sorption of pollutants sometimes can be explained with the simultaneous contribution of two of more of these mechanisms, especially when the nonpolar or polar character of the chemicals is not well defined.

This concept also deals with the energy content (through the properties and behavior of chemicals and chemical mixtures). This moves on to the concept of thermodynamics, which deals with energy levels and energy transfers between states of matter, fundamental to all branches of science.

3.3 Partition Coefficients

A partition coefficient is used to describe how a solute is distributed between two immiscible solvents. They are used in drug design as a measure of a solute's hydrophobicity and a proxy for its membrane permeability.

Partition coefficients (sometimes referred to as partition ratios) are widely used in environmental science to relate the concentration of a chemical solute in one phase to that in a second phase between which equilibrium applies or is approached. The solutes include organic and inorganic substances and the phases of interest include air, water, soils, sediments, and aerosols. The availability of reliable partition coefficients for contaminants is essential for regulatory and scientific purposes, the general aim being to understand and predict the distribution of the substances in multimedia environmental systems.

Typical physical and chemical properties play a role in assessing the behavior of a chemical in the environment, but a property that is not often considered in the partitioning of the chemical in an ecosystem is a function of the chemical and physical structure of the chemical. This leads to a determination of the way in which the chemical or the mixture of chemicals is distributed among the different environmental phases (McCall et al., 1983). These phases may include air, water, matter, mineral solids, and even organisms.

In the chemical sciences, both phases usually are solvents and, most commonly, one of the solvents is water while the second is a hydrophobic (nonmiscible) solvent such as 1-octanol (1-hydroxy octane, $C_8H_{17}OH$, $CH_3CH_2CH_2CH_2CH_2CH_2CH_2CH_2OH$). In this case, the partition coefficient is a measure of the hydrophilic character (literally: water-loving) or hydrophobic character (literally: water-fearing) of the chemical. Despite formal recommendation to the contrary, the term *partition coefficient* remains the predominantly used term in the scientific literature. On the other hand, the International Union of Pure and Applied Chemistry (IUPAC) has recommended that the title term no longer be used, rather, that it be replaced with more specific terms. For example, *partition constant*, defined as:

$$(K_D)_A = [A]_{org}/[A]_{aq}$$

In this equation, K_D is the process equilibrium constant, [A] represents the concentration of solute A that is being tested, and *org* and *aq* refer to the organic and aqueous phases, respectively.

The IUPAC has also recommended the use of the term *partition ratio* for cases where transfer activity coefficients can be determined and use of the term *distribution ratio* for the ratio of total analytical concentrations of a solute between phases, regardless of chemical form. However, the term *partition coefficient* and the term *distribution coefficient* (as defined above) are still preferred by many scientists and engineers.

In the environmental context, the transfer between the atmosphere and bodies of water is one of the key processes affecting the transport of many organic chemicals in the environment. For neutral chemicals, at dilute solution concentrations in pure water, the air–water distribution ratio is referred to as the Henry's Law constant K_H (or K_{aw}). For aqueous solutions (i.e., solutions that contain many other chemical species), the term *air–water distribution ratio* is used which, for practical purposes, we approximate by the Henry's Law constant K_H. K_H can be approximated as the ratio of the abundance of a chemical in the gas phase to the abundance of a chemical in the aqueous phase at equilibrium.

Partition coefficients are useful in estimating the chemicals within any system. In the environment, a hydrophobic chemical with a high octanol–water partition coefficient is mainly distributed to hydrophobic areas of an ecosystem. Conversely, a hydrophilic chemical (having low octanol–water partition coefficient) will be found primarily in aqueous regions, such as swamps (water-logged areas), ponds, rivers, and lakes. If one of the solvents is a gas and the other a liquid, a gas/liquid partition coefficient can be determined.

The most common agents studied in partitioning have been related to the discharge of single chemicals as related to discharge of these chemicals into the environment. Thus, the partition-coefficient or the distribution-coefficient is the ratio of the respective equilibrium concentrations of a chemical in a mixture of two immiscible phases and is also a measure of the difference in solubility of the chemical in the two phases. A point of differentiation between the partition coefficient and the distribution coefficient is that the partition-coefficient typically refers to the concentration ratio of nonionized species of a chemical whereas the distribution coefficient typically refers to the concentration ratio of all species of the chemical (that is, the ionized and the nonionized species).

In general, a large K_{ow} indicates that the chemical tends to prefer an organic (nonpolar) environment and not water (polar) and, thus, the chemical will have a low solubility in water. Most pesticides are less polar than water and tend to accumulate in soil or living organisms which contain organic matter. It is, therefore, predictable from the K_{ow} value where a chemical will be distributed in the environment (Linde, 1994).

4. THERMODYNAMICS

The environment is the external surroundings of a system and is a complex assembly of coupled components showing a common behavior. Thermodynamic concepts have been utilized by practitioners in a variety of disciplines with interests in environmental sustainability, including ecology, economics, and engineering. Widespread concern about environmental degradation is common to them all. It has been argued that these consequences of human development are reflected in thermodynamic parameters and methods of analysis that can illustrate the various chemical and physical transformations within an ecosystem.

A thermodynamic system is an aggregation of physical bodies that may interact with each other and may exchange energy and mass with other bodies. Thermodynamic systems are the objects of study of thermodynamics. A thermodynamic process occurs each time one form of energy is converted into another, which includes every material process, including mental activity. As an example, a common industrial thermodynamic process involves the conversion of the chemical energy of a fuel into the heat energy of combustion, and this heat is in turn converted to mechanical or electrical work.

In classical thermodynamics, the environment is the surroundings that have an influence on a given system. In environmental thermodynamics the emphasis is more on the outside, and the systems are so varied (one person, a community, one industry, an industrial sector, an ecosystem, humanity as a whole) that it is often not explicitly mentioned. Although there may be far-reaching influences (from the core of the earth, from the moon, and from other celestial bodies), there are only two extensions for the environment: (1) the whole ecosphere of the earth, and (2) the immediate local environment in contact with the flora and fauna.

While all chemical and physical aspects of the earth are important, thermodynamics is crucial because it embraces (1) all kind of energies—physical, chemical, and biological, (2) it governs all kind of equilibriums, and (3) it sets restrictions on all kind of processes. All major environmental concerns are directly related to thermodynamic processes including the increase in the greenhouse effect that causes climate changes, as well as acid rain and other forms of chemical pollution, as well as the decrease in the protective ozone layer thickness due to release of gases.

In the context of this book, it is appropriate to consider the thermodynamics of an ecosystem which is an open system that supports structure and functioning due to external energy input. Usually an ecosystem consumes solar energy in the form of relatively short-wave radiation (visible light), although some ecosystems (e.g., at great depth in ocean) use chemical energy. Nevertheless, the general rule is the reception of solar energy by green plants via photosynthesis. They assimilate approximately 0.01%–3% of energy of falling radiation and, using this energy, create organic matter (*primary production*) from inorganic compounds (water, carbon dioxide, nitrates, phosphates, and a lot of minor substances). The byproduct of photosynthesis is oxygen and energy, which is stored in organic matter and is used and dissipated during the processes of plant respiration, growth, and reproduction in the form of heat. The remaining energy, accumulated in plant biomass is used by animals to support their structure and functioning. These processes are balanced at global level, as well as in healthy mature ecosystems. The remains of dead plants and animals and organic waste produced by animals are decomposed and reduced to primary inorganic compounds, available for the new cycle of production/destruction by bacteria and other microorganisms. So, an ecosystem consumes high quality energy of solar radiation, uses part of it to support itself, and dissipates the rest in the form of heat, increasing the total entropy of whole system, sun—ecosystem—environment. Thus, an ecosystem functions according to both the first and the second laws of thermodynamics.

Following this, the most important contribution of thermodynamics is in determining whether a given conceivable process is possible or not. Thermodynamics answers this question by making use of its famous first and second laws. The first is the law of conservation of energy, while the second implies the impossibility of constructing a perpetual motion machine or the spontaneous transfer of heat from a low to a high temperature reservoir by a cyclic process. The latter also involves the concept of entropy, which is a measure of the unavailable energy in a closed thermodynamic system. Although energy is a quite well-understood concept, that of entropy is less so. A simple way to understand entropy is to consider it as an index of disorder or the absence of order. Disordered or high entropy states occur with the highest probability because there are relatively more ways to achieve them. Thus, there are many more ways to disorder (shuffle) a deck of cards than to place it in an ordered array.

A fundamental concept of thermodynamics is that of a *system* versus the *surroundings*. For example, in a particular system consisting of a chemical fuel, oxygen to burn it, and the combustion products, carbon dioxide and water, the total behavior of this system cannot be understood without reference to the surroundings, with which it may exchange both energy and matter. This concept is particularly important to the current context of this book because the term *surroundings* is synonymous with the ecosystem (or the environment).

A thermodynamic system consists of such a large number of particles that the system's state may be characterized by macroscopic parameters, for example, density, pressure, and the concentrations of the various substances forming the system. If the parameters of a system do not vary with time and if there are no steady fluxes in the system, the system is in equilibrium. In fact, the concept of temperature as a state parameter having the same value for all macroscopic parts of the system is introduced for systems in equilibrium.

The properties of thermodynamic systems in thermodynamic equilibrium are studied by equilibrium thermodynamics, or thermostatics; the properties of nonequilibrium systems are studied by nonequilibrium thermodynamics. Thermodynamics deals with closed, open, adiabatic, and isolated systems. Closed systems do not exchange mass with other systems. Open systems may exchange mass and energy with other systems. Adiabatic systems do not exchange heat with other systems. Isolated systems exchange neither energy nor mass with other systems.

The natural laws which govern the environment and, therefore, which are of interest to the scientist and engineer are the first two laws of thermodynamics. The third law of thermodynamics relates to the properties of systems in equilibrium at absolute zero temperature (−273.15°C, −459.67°F). Thus, the entropy of a perfect crystal at absolute zero is exactly equal to zero. As a result, this law is not taken into consideration when dealing with the thermodynamics of environmental systems.

4.1 First Law of Thermodynamics

The first law states that whenever energy is converted in form, its total quantity remains unchanged. In other words, energy (or matter) can be neither created nor destroyed. An example that is pertinent to the current text is of a coal-fired electricity generating plant. The coal is heated, which produces electricity. A byproduct of this process is waste heat that is transported away as cooling water or gases. In addition, various waste gases are emitted into the atmosphere, which cause pollution, such as acid rain.

The first law of thermodynamics expresses the fact that energy is always conserved or that any process must yield as much energy, albeit in different form, as was put into it in the first place. However, this simple relation is often ignored, particularly on a simple chemical discharged versus chemical remediated basis by failure to consider *all* significant energy inputs into a particular technology and the result may be the application of a less-than-appropriate technology.

4.2 Second Law of Thermodynamics

This law states that in a closed system, entropy (a measure of the disorder of energy) does not decrease. As an example, ordered energy is useful whereas disordered energy is not useful. Entropy is a thermodynamic property of matter and is related to the amount of energy that can be transferred from one system to another in the form of work. For an ecosystem with a fixed amount of energy, the value of the entropy ranges from zero to a maximum. If the entropy is at its maximum, the amount of work that can be transferred is equal to zero, and if the entropy is at zero, the amount of work that can be transferred is equal to the energy of the system. During an irreversible process the entropy of a system always increases. Thus, because of these natural laws: (1) increased extraction of minerals by the production process leads to an increase in wastes, (2) there is a limit on the substitutability of inputs, and (3) since production and consumption lead to the dissipation of matter, scarce energy is needed for recycling. The importance of these two laws relates to the use, reuse, and recycling of the environment after interactions with discharged chemicals.

By contrast to the first law of thermodynamics, the second law of thermodynamics expresses the spontaneous creation of entropy or disorder that drives the remediation process. Here the key word is *spontaneous*, since any independent process that is possible must be thermodynamically spontaneous which, in this sense, is interpreted as any process that does not require help from an outside energy source. The second law of thermodynamics states that in such a process the total entropy (of system plus surroundings) always *increases*. If the total entropy change for any conceivable process is negative, the process requires energy to be used in the process that is brought from another (or outside) source.

Put simply, the second law of thermodynamics states that heat can be transferred from a hotter zone to a cooler zone without input of work, but that work is necessary to move heat from a cooler zone to a hotter zone. Application of the second law of thermodynamics helps explain the various ways in which engines transform heat into mechanical work, as for instance, in a steam turbine.

Efficiency measures based on the second law of thermodynamics take into account the quality of energy—unlike efficiencies based on the first law of thermodynamics which take into account only the amount of energy. The first law of thermodynamics states that energy is conserved even when its form is changed, as for instance from mechanical energy to heat. By contrast, the second law of thermodynamics allows an explanation of the degree of efficiency of performance of an energy system in terms of the quality of the energy.

The second law of thermodynamics can also be stated in terms of entropy, which is a measure of disorder in a system. Since it takes work to increase the orderliness inside a closed system, an increase in order corresponds to a decrease of entropy. Hence, the second law of thermodynamics can also be expressed in terms of entropy: a decrease in the entropy of a system requires an input of work into that system. Zero thermal energy—in which there is no random motion of atoms or molecules—is achieved only at a temperature known as absolute zero, which is equal to −273°C (about −460°F). An absolute zero temperature cannot actually be reached by any practical device.

Thus, a fundamental concept of thermodynamics and the environment is that of a system (say, an ecosystem) and the surroundings. The total behavior of this system cannot be understood without reference to the surroundings, with which it may exchange both energy and matter. The most important contribution of thermodynamics is in determining whether a given conceivable process is possible. Thermodynamics gives an understanding of the possibility of a process by making use of the first and second laws of thermodynamics. The first is the law of the conservation of energy, while the second implies the impossibility of constructing a perpetual motion machine or the spontaneous transfer of heat from a low to a high temperature reservoir by a cyclic process. The latter also involves the concept of entropy and a simple way to understand the concept of entropy as an index of disorder or the absence of order. Disordered or high entropy states occur with the highest probability because there are relatively more ways to achieve them. Thus, environmental thermodynamics is concerned with energy conversions and flows and must also take into consideration the application of any remediation process to remove and transform one or more chemical contaminants.

4.3 Free Energy

There is another concept of thermodynamics known as *free energy*, which is energy only in a formal sense, since it simultaneously incorporates both the energy and entropy of the system. It is directly and simply related to the total entropy and similar to entrophy as an index of the possibility or impossibility of any conceptual process. Unlike total entropy, however, free energy *decreases* during a spontaneous process. Free energy, in thermodynamics, is an energy-like property or state function of a system in thermodynamic equilibrium. Free energy has the dimensions of energy and is used to determine how systems change and how much work they can produce, and it is an extensive property, meaning that its magnitude depends on the amount of a substance in a given thermodynamic state.

The changes in free energy are useful in determining the direction of spontaneous change and evaluating the maximum work that can be obtained from thermodynamic processes involving chemical or other types of reactions. Changes in free energy can be used to judge whether changes of state can occur spontaneously. Under constant temperature and volume, the transformation will happen spontaneously, either slowly or rapidly, if the Helmholtz free energy is smaller in the final state than in the initial state—that is, if the difference between the final state and the initial state is negative.

Thus, the concept of free energy is a convenient way to characterize pollutants of all kinds, since these always have more of free energy than the degradation products. For example, under surface conditions of the earth there is a marked decrease in free energy when the pollutant carbon monoxide is converted to carbon dioxide and subsequently, by reaction with calcium oxide, to harmless carbonate minerals (limestone, $CaCO_3$).

REFERENCES

Andrews, J.E., Brimblecombe, P., Jickells, T.D., Liss, P.S., 1996. An Introduction to Environmental Chemistry. Blackwell Science Publications, Oxford.

Bergen, B.J., Nelson, W.G., Pruell, R.J., 1993. Bioaccumulation of PCB congeners by blue mussels (*Mytilus edulis*) deployed in New Bedford Harbor, Massachusetts. Environmental Toxicology & Chemistry 12, 1671–1681.

Boethling, R.S., Mackay, D., 2000. Handbook of Property Estimation Methods. Lewis Publishers, Boca Raton, Florida.

Booth, A.M., Sutton, P.A., Lewis, C.A., Lewis, A.C., Scarlett, A., Chau, W., Widdows, J., Rowland, S.J., 2006. Unresolved complex mixtures of aromatic hydrocarbons: thousands of overlooked persistent, bioaccumulative, and toxic contaminants in Mussels. Environmental Science and Technology 41 (2), 457–464.

Calvert, J., 1994. The Chemistry of the Atmosphere: Its Impact on Global Change. Blackwell Scientific Publications, Oxford, United Kingdom.

Cole, J.G., Mackay, D., 2000. Correlating environmental partitioning properties of compounds: the three solubility approach. Environmental Toxicology & Chemistry 19 (2), 265–270.

Crompton, T.R., 2012. Compounds in Soils, Sediments and Sludge: Analysis and Determination. CRC Press, Taylor & Francis Group, Boca Raton, Florida.

Dean, J.R., 1998. Extraction Methods for Environmental Analysis. John Wiley & Sons, Inc., New York.

Delle Site, A., 2001. Factors affecting sorption of compounds in natural sorbent/water systems and sorption coefficients for selected pollutants. A Review. Journal of Physical and Chemical Reference Data 30 (1), 187–439.

Firor, J., 1990. The Changing Atmosphere. Yale University Press, New Haven, Connecticut.

Goody, R., 1995. Principles of Atmospheric Physics and Chemistry. Oxford University Press, Oxford, United Kingdom.

Linde, C.D., 1994. Physico-chemical Properties and Environmental Fate of Pesticides. Report EH 94–03. Environmental Hazards Assessment Program, State of California, Environmental Protection Agency, Department of Pesticide Regulation, Environmental Monitoring and Pest Management Branch, Sacramento, California. http://www.cdpr.ca.gov/docs/emon/pubs/ehapreps/eh9403.pdf.

Lyman, W.J., Reehl, W.F., Rosenblatt, D.H., 1982. Handbook of Chemical Property Estimation Methods. McGraw-Hill, New York.

Lyman, W.J., Reehl, W.F., Rosenblatt, D.H., 1990. Handbook of Chemical Property Estimation Methods. American Chemical Society, Washington, DC.

McCall, P.J., Laskowski, D.A., Swann, R.L., Dishburger, H.J., 1983. Estimation of environmental partitioning of organic chemicals in model ecosystems. In: Gunther, F.A., Gunther, J.D. (Eds.), Residue Reviews, vol. 85. Springer, New York, pp. 231–244.

Mackay, D., Shiu, W.-Y., Ma, K.-C., Lee, S., 2006. Handbook of Physical-chemical Properties and Environmental Fate of Chemicals, second ed. CRC Press, Taylor & Francis Group, Boca Raton, Florida.

Manahan, S.E., 2010. Environmental Chemistry, ninth ed. CRC Press, Taylor & Francis Group, Boca Raton, Florida.

Mao, D., Lookman, R., Weghe, H.V.D., Weltens, R., Vanermen, G., Brucker, N.D., Diels, L., 2009. Combining HPLC-GCxGC, GCxGC/TOF-MS, and selected ecotoxicity assays for detailed monitoring of petroleum hydrocarbon degradation in soil and leaching water. Environmental Science and Technology 43 (20), 7651–7657.

Nabi, D., Gros, J., Dimitriou-Christidis, P., Arey, J.S., 2014. Mapping environmental partitioning properties of nonpolar complex mixtures by use of GC × GC. Environmental Science and Technology 48, 6814–6826.

Schwarzenbach, R.P., Gschwend, P.M., Imboden, D.M., 2004. Environmental Organic Chemistry, second ed. John Wiley & Sons Inc., Hoboken, New Jersey.

Speight, J.G., Lee, S., 2000. Environmental Technology Handbook, second ed. Taylor & Francis, New York. Also, CRC Press, Taylor and Francis Group, Boca Raton, Florida.

Speight, J.G., 2013. The Chemistry and Technology of Coal, third ed. CRC Press, Taylor & Francis Group, Boca Raton, Florida.

Speight, J.G., 2014. The Chemistry and Technology of Petroleum, fifth ed. CRC Press, Taylor & Francis Group, Boca Raton, Florida.

Speight, J.G., 2015. Handbook of Petroleum Products Analysis, second ed. John Wiley & Sons Inc., Hoboken, New Jersey.

Speight, J.G., Islam, M.R., 2016. Peak Energy – Myth or Reality. Scrivener Publishing, Salem, Massachusetts.

Spellman, F.R., 2016. Handbook of Environmental Engineering. CRC Press, Taylor & Francis Group, Boca Raton, Florida.

Thibodeaux, L.J., 1996. Environmental Chemodynamics: Movement of Chemicals in Air, Water, and Soil. John Wiley & Sons Inc., Hoboken, New Jersey.

Tinsley, I.J., 2004. Chemical Concepts in Pollutant Behavior, second ed. John Wiley & Sons Inc., Hoboken, New Jersey.

WMO, 1992. Scientific Assessment of Stratospheric Ozone 1991. WMO Global Ozone Research and Monitoring Project, Report No. 25. World Meteorological Organization, Geneva, Switzerland.

Chapter 10

Mechanisms of Transformation

1. INTRODUCTION

Chemicals can be emitted directly into the environment or formed through chemical reactions of so-called precursor species. The pollutants can undergo chemical reactions converting highly toxic substances to less toxic or inert products, as well as convert nontoxic chemicals or chemicals of low-toxicity to chemicals of high toxicity. Thus, exposure to hazardous chemicals can occur on time scales that range from minutes to days or even weeks and a variety of chemistry and physical processes can play an important role in defining the effect of the exposure and the fate of the discharged chemical (Speight and Lee, 2000; Tinsley, 2004).

As used here, the term *fate* refers to the ultimate disposition of the chemical in the ecosystem, either by chemical or biological transformation to a new form which (hopefully) is nontoxic (degradation) or, in the case of ultimately persistent pollutants, by conversion to less offensive chemicals or even by sequestration in a sediment or other location which is expected to remain undisturbed. This latter option—the sequestration in a sediment or other location—is not a viable option as for safety reasons the chemical must be dealt with at some stage of its environmental life cycle. Using the old adage *bad pennies always turn up* can also be applied to a hidden chemical and it is likely to manifest its presence (usually by an adverse action) at some future date. In summary, hiding the chemical away on paper (a note in a file giving the written but unproven or hypothetical reason why the chemical is considered to be of limited danger) is not an effective way of protecting the environment. However, for chemicals that are effectively degraded in nature, whether by hydrolysis, photolysis, microbial degradation, or other chemical transformation in the ecosystem, it is necessary (even essential) to collect, tabulate, and store (for ready retrieval) any information related to the chemical reaction parameters, which can serve as indicators of the processes and the rates at which transformation (that is, degradation) would occur.

The discharge of chemicals into the environment and the fate of the chemical can take many forms. For example, there can be adsorption, dilution, dissolution (Chapter 5), hydrolysis (Chapter 6), oxidation–reduction (Chapter 7), and biological transformation (Chapter 8). In addition, acid–base reactions can

Reaction Mechanisms in Environmental Engineering. https://doi.org/10.1016/B978-0-12-804422-3.00010-9

cause partitioning of the chemical (especially when the discharge is a mixture), as well as neutralization of acids (by bases) or bases (by acids) (Chapter 9). In addition, decomposition reactions involve the decomposition or cleavage of one molecule to one or more product molecules and *displacement reactions* involve displacement of one or more cation or anion between two molecules. All of these reactions are subject to the laws of chemistry (Chapter 9) and the products can take many forms.

It is not surprising, therefore, that there is no one remediation process that can be claimed as the remedy for application to all forms of chemical discharge. Given this conclusion, it must also be concluded that the remediation must meet and defeat the chemistry of the pollution and be able to sever any chemical bonds or physical arrangements that might exist between the chemical and the relevant parts of an ecosystem. This is where the knowledge of the interaction(s) between the discharged chemical and the ecosystem will be invaluable.

Thus, the successful treatment of an ecosystem in which a chemical has (or chemicals have) been discharged requires a detailed understanding of the nature and distribution of contaminant mass, as well as an aggressive application approach that will maximize oxidant–contaminant contact and deliver a sufficient amount of oxidant to treat the dissolved, sorbed, and separate-phase contaminants. The remediation technology must offer a safe and effective treatment technique that can maximize treatment longevity, ensure efficient oxidation, and promote safe use in the field. Furthermore, the remediation process applied to any site that is contaminated by chemicals requires a comprehensive understanding of the fate of the chemical(s) through an understanding of the structure, properties, and behavior of the chemical(s) (Borda and Sparks, 2008; Chorover and Brusseau, 2009; Thompson and Goyne, 2012), all of which discuss the kinetics of sorption with a treatment of surface precipitation. Finally, we emphasize that in soils, sorption occurs concurrently with biotic and abiotic transformation reactions in an open hydrodynamic system.

From the standpoint of environmental chemistry, it is essential to know the important groups of potential pollutants: (1) bulk chemicals, which include a wide range of inorganic and organic chemicals, (2) agrochemicals including inorganic and organic herbicides and pesticides, (3) chemicals used in plastics, mining, metal-working, wood preservation, paints, textiles, paints and pigments, flame retardants, and household products, (4), pharmaceutical chemicals, and (5) refined petrochemicals, which include gasoline, diesel fuel, fuel oil, and lubricating oil, especially those products that contain monocyclic aromatic hydrocarbons, as well as a variety of inorganic chemicals (Speight, 2005, 2011; Speight and Arjoon, 2012).

Within the major groups of chemical pollutants, the principal focus to understand the reactions for the degradation or transformation of these chemicals requires (1) the knowledge of the type of pollutants and (2) the reactivity of the pollutants so that the design of the relevant method to remove the pollutants from the environment can be achieved. Although much emphasis has been, and continues to be, placed on biotic reactions carried out by the various biota (such as bacteria), important transformation reactions of the pollutants in the environment must not be omitted. Some of these chemical reactions will be beneficial—in terms of reduced toxicity and pollutant removal—while other chemical reactions may have adverse effects on pollutant removal by an opposite effect in which the pollutant is converted to a product that is more capable of remaining in the environment and may even prove to be a persistent chemical. Thus, emphasis must be placed on the partial degradation or full degradation of the chemical and the role of any intermediate products that are toxic to floral and faunal organisms, inhibit further degradation, or have adverse effects on the environment.

The transformation of a chemical in the environment is an issue that needs to be given serious consideration because of the chemical and physical changes (often nonbenign) that can occur to the chemical. It would be unusual if the chemical transformation did not show some effect on the properties of the discharged chemical. Thus, the chemical transformations of chemicals released into the environment are, in the context of this book, considered to be the transformation of the released chemical into a product that is still of environmental concern in terms of toxicity, but which may require a different mechanism for the chemical to be released from the environment compared to the original chemical. Furthermore, knowledge of the relative amounts of each species present is critical because of the potential for differences in behavior and toxicity (including the possibility of enhanced toxicity) which are of concern when examining the potential fate of such chemicals.

Transformation in the current context is conversion of a discharged chemical (to a more benign product) by means of physical, chemical, or biological processes (Chapters 5–8). When no transformation occurs, the system is classed as a *conservative system*, and this characteristic is represented mathematically using the conservation of mass equation:

$$dMi\ dt = 0$$

where Mi is the total mass of species i.

When transformation does occur, the system is said to be classed as *reactive system* and, for a given chemical, the system is no longer a *conservative system*. Thus:

$$dMi\ dt = Si$$

where Si is the source of the chemical.

Furthermore, transformation reactions are broadly categorized as either (1) homogeneous or (2) heterogeneous. Homogeneous reactions occur everywhere within the area of interest insofar as the chemical is distributed throughout the affected area. On the other hand, heterogeneous reactions typically occur only at fluid boundaries and the chemical is not distributed throughout the affected area. However, some reactions have properties of both homogeneous and heterogeneous reactions. As an example, if a reaction occurs on the surface of suspended sediment particles since the reaction occurs only at the sediment/water interface, the reaction is heterogeneous. But, because the sediment is suspended throughout the water column, the effect of the reaction is homogeneous in nature. Models that represent the reaction through boundary conditions (i.e., they treat the reaction as heterogeneous) are sometimes called two-phase, or multiphase, models. Models that simplify the reaction to treat it as a homogeneous reaction are called single-phase, or mixture, models. To obtain analytical solutions, the single-phase approach is the approach that is often used.

Moreover, a discharged chemical is subject to two processes that determine the fate of the chemical in the environment: (1) the potential for transportation of the chemical and (2) the chemical changes that can occur once the chemical has been released to the environment and which depend upon a variety of physical and chemical properties (Table 10.1), some of which may lead to corrosion (Table 10.2). Thus, release into the environment of a persistent pollutant leads to an exposure level which ultimately depends on the length of time the chemical remains in circulation, and how many times it is recirculated, in some sense, before ultimate termination of the environmental life cycle of the chemical—the same rationale applies to a product formed from the pollutant by any form of chemical transformation. In addition, the potential for transportation and chemical change (either before or after transportation) raises the potential for the chemical to behave in an unpredictable manner.

Many processes arising in chemical science and technology involve physical transformations in addition to, or even without, chemical reactions, such as a phase change (Fig. 10.1) and/or adsorption on to a solid surface (Table 10.3, Fig. 10.1). Most mechanistic work has focused on chemical reactions in solution or extremely simple processes in the gas phase (Delle Site, 2001). There is increasing interest in reactions in solids or on solid surfaces, such as the surfaces of solid catalysts in contact with reacting gases (Mokhatab et al., 2006; Speight, 2007). Some such catalysts act inside pores of defined size, such as those in zeolites.

In these cases, only certain molecules can penetrate the pores to get to the reactive surface, and they are held in defined positions when they react. Perhaps, the saving grace is that in the environment there is the potential that physical changes can be reversed with the substance changed back without extraordinary means, if at all, unlike in a chemical change. However, reversibility is not always a certain criterion for classification between chemical and physical changes, but other issues must be given attention for the process to run efficiently (Table 10.4).

TABLE 10.1 Various Physical and Chemical Properties That Influence the Behavior of a Chemical in the Environment

Chemical Transport Processes	Chemical Fate Processes
• Erosion	• Biological Processes
• Leaching	• Sorption
• Movement in water systems streams, rivers, or groundwater	• Transport
• Runoff	• Transformation/Degradation
• Wind	• Volatilization
Transformation and Degradation Processes	
• Biological transformations due to microorganism (1) aerobic processes (2) anaerobic processes	
Physical Properties	**Chemical Properties**
Boiling point	Ability to act as an oxidizing agent
Color	Ability to act as a reducing agent
Density	Absorption into a liquid or porous solid
Electrical conductivity	Adsorption to a surface
Hardness	Can cause corrosion
Melting point	Decomposition into lower molecular–weight chemicals
Solubility	Reaction with acids
Sublimation point	Reaction with alkalis (bases)
State (gas, liquid, solid)	Reaction with other chemicals
Vapor pressure	Reaction with oxygen (oxidation)
	Reaction with oxygen (combustion)
	Decomposition into lower molecular–weight chemicals

In fact, to predict the persistency of a chemical in the environment, the physical–chemical properties and reactivity of the inorganic chemical in the environment need to be known or at least estimated. The chemicals that react in the environment are not likely to persist in the environment, while those that did not show any observable reactivity may persist for a very long time.

TABLE 10.2 Typical Corrosive Chemicals

Acidic Corrosives:
• Acids
• Hydrochloric acid (HCl)
• Nitric acid (HNO_3)
• Sulfuric acid (H_2SO_4)
Alkaline, or Basic, Corrosives:
• Sodium hydroxide ($NaOH$)
• Potassium hydroxide (KOH)
Corrosive Dehydrating Agents:
• Phosphorous pentoxide (P_2O_5)
• Calcium oxide (CaO)
Corrosive Oxidizing Agents:
• Halogen gases (F_2, Cl_2, Br_2)
• Hydrogen peroxide (H_2O_2, concentrated)
• Perchloric acid ($HClO_4$)

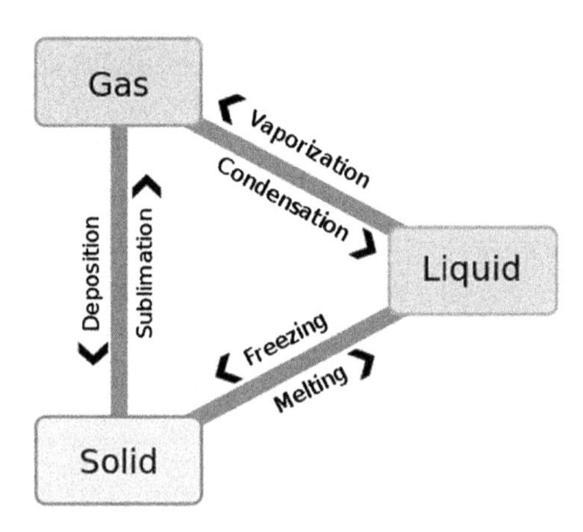

FIGURE 10.1 Examples of phase changes in the environment.

Thus, the focus in this chapter is to present the chemical aspects of the various remediation processes and how they might be applied to environmental cleanup. Thus, as examples:

TABLE 10.3 Phase Transformations From Gas to Liquid to Solid and the Reverse

	To:	Gas	Liquid	Solid
From:	Gas	N/A	Condensation	Deposition
	Liquid	Evaporation*	N/A	Freezing
	Solid	Sublimation	Melting	Transformation**

*Also, boiling.**For example, change in crystal structure.

TABLE 10.4 Factors Affecting the Adsorption Process. Factors Which Affect the Extent of Adsorption

Nature of the adsorbate (gas) and adsorbent (solid)
In general, easily liquefiable gases such as CO_2, NH_3, Cl_2, and SO_2 are adsorbed to a greater extent than the elemental gases such as H_2, O_2, N_2, and He (while chemisorption is specific in nature).
Porous and finely powdered solid (such as clay minerals) adsorb more as compared to the harder and nonporous materials.
Surface area of the solid adsorbent
The extent of adsorption depends directly upon the surface area of the adsorbent, i.e., larger the surface area of the adsorbent, greater is the extent of adsorption.
Surface area of a powdered solid adsorbent depends upon its particle size. Smaller the particle size, greater is its surface area.
Effect of pressure on the adsorbate gas
An increase in the pressure of the adsorbate gas increases the extent of adsorption.
At low temperature, the extent of adsorption increases rapidly with pressure.
At small range of pressure, the extent of adsorption is found to be directly proportional to the pressure.
At high pressure (closer to the saturation vapor pressure of the gas), the adsorption tends to achieve a limiting value.
Effect of temperature
As adsorption is accompanied by evolution of heat, the magnitude of adsorption should decrease with rise in temperature (Le-Chatelier's principle).

Process	Minimum Knowledge Required
Thermal desorption	Adsorption bonding
	Absorption bonding
Pump and treat	Adsorption of chemical
	Solubility in water
	Solubility in surfactant-enhanced water
Solidification and stabilization	Formation of precipitate
	Formation of sediment
In situ oxidation	Oxidation chemistry
	Properties and behavior of products
In situ hydrogenation	Hydrogenation chemistry
	Properties and behavior of products
Soil vapor extraction	Volatility
	Vapor pressure
	Tendency to steam distil
Bioremediation	Response to microbial attack
	Properties and behavior of products

While this list is by no means comprehensive (because of the variations is the properties and behavior of the discharged chemical, as well as the properties and behavior of the ecosystem into which the chemical is discharged), the list does give an indication of the background knowledge that would be required to select an appropriate remediation process.

2. TRANSFORMATION IN THE ENVIRONMENT

Reliable estimates of the behavior of chemicals in the environment are not a simple task and may only experience rare success. While acknowledging that the use of E-factors (i.e., the mass efficiency in terms of mass of the reactants) is a step in the right direction, it appears that there are too many unknowns and the unreliability of the environmental reaction parameters needs to be finely tuned to provide a much-needed tool for estimating the behavior of chemicals under such variable conditions. In view of the lack of any estimation of satisfactory error limits for the produced data, an arbitrary outer difference limit may not be sufficiently accurate to define whether or not the data are satisfactory and of any use.

In the chemical sciences, a transformation is the chemical or physical conversion of a chemical (the substrate) to another chemical (the product). A typical chemical process generates products and wastes from raw materials such as substrates and excess reagents. If most of the reagents and the solvent can be recycled, the mass flow looks quite different and the prevention of waste (to be disposed into the environment) can be achieved if most of the reagents can be recycled. Furthermore, there has been the suggestion that an efficiency factor (E-factor) of a chemical process can be used to assess the transformation of a chemical in the environment (Chapter 4):

E-factor = (Mass of original waste)/(Mass of transformed product)

Alternatively:

E-factor = (Mass of raw feedstock – mass of product)/Mass of product

Typically, E-factors used in the manufacturing processes for bulk and fine chemicals give reliable data, but in those processes the conditions are controlled and any examination of amounts and properties can be achieved relatively conventionally (Chapter 3). The examination of chemical properties and chemical transformation in an ecosystem is much more difficult and the E-factors may, more than likely, not be as meaningful. In addition, any such E-factor and any related factors do not account for any type of toxicity of the chemical waste. Efforts (to determine an E-factor) are at a very preliminary stage and the parameters for the calculation of these factors need much more consideration and development before being applied to chemical wastes in the environment (Zhou and Haynes, 2010).

Finally, it must be recognized that there are certain limitations of chemical equations. While the phenomena described in the following phrases may be identified in the laboratory, they are difficult to identify as occurring in the environment. For example, (1) the chemical equation does not usually clarify the state of the substances and (s) for solid, (l) for liquid, (g) for gas and (vapor) may have to be added, (2) the reaction may or may not be concluded but the equation does not reveal it, (3) the chemical equation does not give any information about the rate (speed) of the reaction, (4) the chemical equation does not give the concentration of the substances and in some cases, the terms like diluted and concentrated are used, (5) the chemical equation does not give the general parameter of the reaction such as temperature, pressure, and whether or not a catalyst is necessary, (6) the chemical equation will not give any idea about color changes during the exchange, which has to be mentioned separately, and (7) the chemical equation will never give any indication regarding the production or absorption of heat. This is mentioned separately.

2.1 Chemical Transformation

Chemical transformation refers to the broad range of chemical and physical reactions that do not involve transformations at the atomic level. Thus, the periodic table of the elements contains all the building blocks of chemical transformations. A classic example from aqueous phase chemistry that can lead to a transformation of the reactant (carbon dioxide) is the dissolution of carbon dioxide (CO_2) in water (H_2O), given by the equilibrium equation:

$$CO_2 + H_2O \leftrightarrow HCO_3^- + H^+$$

A chemical transformation involves changing a chemical (or chemicals) from a beginning chemical (the reactants or reactants) mass to a different resulting (the product or products) chemical substance. The transformation produces new chemicals which have different physical and chemical properties when compared to the physical and chemical properties of the starting material(s). Such changes are usually irreversible in environment because the newly formed chemical(s) cannot easily change back into the original chemical(s). In the laboratory, chemical changes can easily be identified with the help of change in color, odor, energy level, and physical state. Any chemical change can easily be represented with the help of chemical equations which is a simple representation of a chemical reaction and which involves the molecular or atomic formulas of reactants and products. Based on cleavage and formation of chemical bonds, most chemical reactions can be classified into four broad categories: (1) combination reactions, (2) decomposition reactions, (3) single displacement reactions, and (4) double displacement reactions. Thus:

Combination reaction:

$$A + B \rightarrow AB$$

$$2Na(s) + Cl_2(g) \rightarrow 2NaCl(s)$$

$$8Fe + S8 \rightarrow 8FeS$$

$$S + O_2 \rightarrow SO_2$$

Decomposition reaction:

$$AB \rightarrow A + B$$

$$2HgO \rightarrow 2Hg + O_2$$

Single displacement reaction:

$$A + BC \rightarrow AC + B$$

$$Mg + 2H_2O \rightarrow Mg(OH)_2 + H_2$$

$$Cu(s) + 2AgNO_3(aq) \rightarrow 2Ag(s) + Cu(NO_3)_2(aq)$$

$$Zn(s) + CuSO_4(aq) \rightarrow Cu(s) + ZnSO_4(aq)$$

Double displacement reaction:

$$AB + CD \rightarrow AD + CB$$

$$Pb(NO_3)_2 + 2KI \rightarrow PbI_2 + 2KNO_3$$

$$CaCl_2(aq) + 2AgNO_3(aq) \rightarrow Ca(NO_3)_2(aq) + 2AgCl(s)$$

Double displacement reaction–acid–base reaction:

$$HA + BOH \rightarrow H_2O + BA$$

$$HBr(aq) + NaOH(aq) \rightarrow NaBr(aq) + H_2O(l)$$

$$HCl(aq) + NaOH(aq) \rightarrow NaCl(aq) + H_2O(l)$$

These reactions are equally likely to occur in the environment as in the laboratory and can occur on a regular basis once a chemical is released into an ecosystem. However, the rate of the reaction does depend upon the chemical and the conditions that exist in the ecosystem.

One reaction that is often omitted from the above is the *combustion reaction*. While this type of reaction is typically assigned plied to the effect of chemicals on the environment, the combustion reaction does (e.g., butane) produce gases that are of serious environmental concern, such as carbon dioxide (complete combustion) and carbon monoxide, with some particulate matter such as soot or carbon (incomplete combustion). Thus:

$$2C_4H_{10} + 13O_2 \rightarrow 8CO_2 + 10H_2O \text{ (complete combustion)}$$

$$C_4H_{10} + 3O_2 \rightarrow CO + 3C + 5H_2O \text{ (incomplete combustion)}$$

In the environment, a chemical transformation is typically based on the same principle as in the laboratory or in any of the chemical process industries—the transformation of a substrate to a product—but whether or not the product is benign and less likely to harm the environment (relative to the substrate) or is more detrimental by exerting a greater impact on the environment depends upon the origin, properties, and reactivity pathways of the starting substrate. Thus, a chemical transformation requires a chemical reaction to lead to the transformation of one chemical substance to another (Habashi, 1994). Typically, chemical reactions encompass changes that only involve the positions of electrons in the forming and breaking of chemical bonds between atoms, with no change to the nuclei (no change to the elements present), and can often be described by a relatively simple chemical equation; thus:

$$A + B \rightarrow C$$

However, although the various chemical transformation processes that occur and which influence the presence and the analysis of chemicals at a particular site are often represented by simple (and convenient) chemical equations, the chemical transformation reactions can be much more complex than the equation indicates.

The chemical (substrate, reactant) initially involved in a chemical reaction (the reactant) in a chemical reaction is usually characterized by a chemical change, and yields one or more products, which usually have properties different from the original reactant. Reactions often consist of a sequence of individual (and often complex) substeps and the information on the precise course of action as part of the reaction mechanism is not always clear. Chemical reactions typically occur under a specific set of parameters (temperature, pressure, chemical concentration, time, ratios of reactants) and (under these parameters) at a characteristic reaction rate. Typically, reaction rates increase with increasing temperature because there is more thermal energy available to reach the activation energy necessary for breaking the bonds between the constituent atoms. The general rule of thumb is that for every 10°C (18°F) increase in temperature, the rate of a chemical reaction is doubled.

In addition, a chemical reaction may proceed in the forward direction and processed to completion, as well as in the reverse direction, until the reactants and products reach an equilibrium state:

$$A + B \rightarrow C + D$$

$$C + D \rightarrow A + B$$

Thus,

$$A + B \leftrightarrow C + D$$

Reactions that proceed in the forward direction to approach equilibrium are often described as spontaneous, requiring no input of free energy to go forward. On the other hand, nonspontaneous reactions require input of free energy to go forward (e.g., application of heat for the reaction to proceed). In chemical synthesis, different chemical reactions are used in combinations during the reaction in order to obtain a desired product. Also, in chemistry, a consecutive series of chemical reactions (where the product of one reaction is the reactant of the next reaction) is often encouraged to proceed by a variety of catalysts (Centi et al., 2002; Grassian, 2005). These catalysts increase the rates of biochemical reactions without themselves (the catalysts) being consumed in the reaction, so that syntheses and decompositions impossible under ordinary conditions can occur at the temperatures, pressures, and reactant concentrations present within a reactor and, by inference, in the current content within the environment.

$$A + B \rightarrow C$$

$$C \rightarrow D + E$$

This simplified equation does not truly illustrate the potential complexity of a chemical reaction, but such complexity must be anticipated when a chemical is transformed in an environmental ecosystem. Mother Nature, a tricky Old Lady, can often encourage or instigate complexity!

The energy requirements of a reaction fall within the realm of *thermodynamics* (Chapter 9) but in the present context, the rate of the reaction falls within the realm of *kinetics*. It is essential that the rate of the reaction can be controlled not only for commercial but also for safety reasons. For example, if a reaction takes too long to reach equilibrium or completion, the rate at which a product is manufactured would not be viable. On the other hand, a runaway reaction—a reaction which progresses too fast and becomes uncontrollable—can result in inherent dangers such as fires and/or explosions.

The rate at which a reaction takes place can be affected by the concentration of reactants, pressure, temperature, the size of particles of solid reactants, or the presence of catalysts (i.e., additives which alter the speed of reactions without being consumed during the reaction), as well as impurities. Catalysts tend to be specific to a particular reaction or family of reactions. For example, nickel is used to facilitate hydrogenation reactions (e.g., add hydrogen to carbon–carbon double bonds in a refinery) whereas platinum is used to catalyze specific oxidation reactions. In most cases, chemical reactions require that the reactants be pure since impurities can act as unwanted catalysts; alternatively, catalysts can be deactivated by poisoning—the deposition of metals or coke-like products on to the catalyst during the reaction.

For reactions which progress slowly at room temperature, it may be necessary to heat the mixture or add a catalyst for the reaction to occur at an economically viable rate. For very rapid reactions, the reactant mixture may need to be cooled or solvent added to dilute the reactants and hence reduce the speed of reaction to a manageable (a controllable and nondangerous) rate. In general, the rate of a chemical reaction: (1) doubles for every 10°C (18°F) rise in temperature, (2) is proportional to the concentration of reactants in solution, (3) increases with decreased particle size for reactions involving a solid, and (4) increases with pressure for gas phase reactions.

2.1.1 Hydrolysis Reactions

Hydrolysis is part of the larger class of chemical reactions called nucleophilic displacement reactions in which a nucleophile (electron-rich species with an unshared pair of electrons) attacks an electrophile (electron-deficient), cleaving one covalent bond to form a new one. Hydrolysis is usually associated with surface waters but also takes place in the atmosphere (fogs and clouds), groundwater, at the particle–water interface of soils and sediments, and in living organisms.

More specifically, hydrolysis is a double decomposition reaction (Chapter 6) with water as one of the reactants. Put simply, if a chemical is represented by the formula *AB* in which *A* and *B* are atoms or groups and water is represented by the formula HOH, the hydrolysis reaction may be represented by the reversible chemical equation:

$$AB + \mathrm{HOH} \rightleftharpoons A\mathrm{H} + B\mathrm{OH}$$

The reactants other than water, and the products of hydrolysis, may be neutral molecules (as in most hydrolysis reactions involving chemicals) or ionic molecules (charged species), as in hydrolysis reactions of salts, acids, and bases.

In the current context, hydrolysis typically means the cleavage of chemical bonds by the addition of water or a base that supplies the hydroxyl ion ($-OH^-$) in which a chemical bond is cleaved, and two new bonds are formed, each one having either the hydrogen component (H) or the hydroxyl component (OH) of the water molecule. Many chemicals can be altered by a direct reaction of the chemical with water. The rate of hydrolysis reactions are typically expressed in terms of the acid-catalyzed, neutral-catalyzed, and base-catalyzed hydrolysis rate constants.

Typically, the hydroxyl replaces another chemical group on the molecule and hydrolysis reactions are usually catalyzed by hydrogen ions or hydroxyl ions. This produces the strong dependence on the acidity or alkalinity (pH) of the solution often observed but, in some cases, hydrolysis can occur in a neutral (pH = 7) environment. In chemistry, a common type of hydrolysis occurs when a salt of a weak acid or weak base (or both) is dissolved in water. The water spontaneously ionizes into hydronium cations (H_3O^+, for simplicity usually represented as H^+) and hydroxide anions ($-OH^-$). The salt (for example), using sodium acetate (CH_3COONa) as the example, dissociates into the constituent cations (Na^+) and anions (CH_3COO^-). The sodium ions tend to remain in the ionic form (Na^+) and react very little with the hydroxide ions (OH^-), whereas the acetate ions combine with hydronium ions to produce acetic acid (CH_3COOH). In this case the net result is a relative excess of hydroxide ions, and the solution has basic properties. On the other hand, strong acids also undergo hydrolysis as when dissolving sulfuric acid (H_2SO_4) in (mixed with) water— the dissolution is accompanied by hydrolysis to produce hydronium ion (H_3O^+) and a bisulfate ion $\left(HSO_4{}^-\right)$, which is the conjugate base of sulfuric acid.

There are four possible general rules involving the reactions of salts in the aqueous environment: (1) if the salt is formed from a *strong* base and *strong* acid, then the salt solution is neutral, indicating that the bonds in the salt solution will not break apart (indicating no hydrolysis occurred) and is basic, (2) if the salt is formed from a *strong* acid and *weak* base, the bonds in the salt solution will break apart and becomes acidic, (3) if the salt is formed from a *strong* base and *weak* acid, the salt solution is basic and hydrolyzes, and (4) if the salt is formed from a weak base and weak acid, it will hydrolyze but the acidity or basicity depends on the equilibrium constants of K_a and K_b. If the K_a value is greater than the K_b value, the resulting solution will be acidic and if the K_b value is greater than the K_a value, the resulting solution will be basic (Petrucci et al., 2010).

The rate of hydrolysis and the relative abundance of reaction products is often a function of the acidity–alkalinity (pH) of the reaction medium because alternative reaction pathways are preferred at different ranges of acidity or alkalinity. In the case of halogenated hydrocarbons, base-catalyzed hydrolysis will result in elimination reactions while neutral hydrolysis will take place via nucleophilic displacement reactions.

2.1.2 Photolysis Reactions

Photolysis (also called photodissociation and photodecomposition) is a chemical reaction in which a chemical (or a chemical) is broken down by photons and is the interaction of one or more photons with one target molecule. Photolysis should not be confused with photosynthesis which is a two-part process in which natural chemicals (typically chemicals) are synthesized by a living organism as part of the life cycle (life chemistry) of the organism.

The photolysis reaction is not limited to the effects of visible light but any photon with sufficient energy can cause the chemical transformation of the bonds of a chemical. Since the energy of a photon is inversely proportional to the wavelength, electromagnetic waves with the energy of visible light or higher, such as ultraviolet light, X-rays, and gamma rays can also initiate photolysis reactions.

The photolysis of a chemical can proceed either by direct absorption of light (direct photolysis) or by reaction with another chemical species that has been produced or excited by light (indirect photolysis). In either case photochemical transformations such as bond cleavage, isomerization, intramolecular rearrangement, and various intermolecular reactions can result. Photolysis can take place wherever sufficient light energy exists, including the atmosphere (in the gas phase and in aerosols and fog/cloud droplets), surface waters (in the dissolved phase or at the particle–water interface), and in the terrestrial environment (on plant and soil/mineral surfaces).

Photolysis dominates the fate of many chemicals in the atmosphere because of the high solar irradiation (George et al., 2015). Near to the surface of the earth, chromophores such as nitrogen oxides, carbonyls, and aromatic hydrocarbons play a large role in contaminant fate in urban areas. In the stratosphere, light is absorbed by ozone, oxygen, organohalogen derivatives, and hydrocarbon derivatives with environmental implications. The rate of photolysis in surface waters depends on light intensity at the air–water interface, the transmittance through this interface, and the attenuation through the water column. Open ocean waters can transmit blue light to depths of 150 m while highly eutrophic or turbid waters might absorb all light within 1 cm of the surface.

The primary step of a ?

$$X + h\nu \rightarrow X^*$$

X*: an electronically excited state of molecule X and subsequently undergo either physical or chemical processes:

Physical processes:
Fluorescence:

$$X^* \rightarrow X + h\nu$$

Collisional deactivation:

$$X^* + M \rightarrow X + M$$

Chemical processes:
Dissociation:

$$X^* \rightarrow Y + Z$$

Isomerization:

$$X^* \rightarrow X^{\wedge\prime}$$

Direct reaction:

$$X^* + Y \rightarrow Z_1 + Z_2$$

Intramolecular rearrangement:

$$X^* \rightarrow Y \text{ Ionization } X^* \rightarrow X + e$$

The general form of photolysis reactions:

$$X + h\nu \rightarrow Y + Z$$

The rate of reaction is:

$$-d/dt\,[X] = d/dt[Y] = d/dt[Z] = k[X]$$

k is the photolysis rate constant for this reaction in units of s^{-1}.

Thus, photolysis is a chemical process by which chemical bonds are broken as the result of transfer of light energy (direct photolysis) or radiant energy (indirect photolysis) to these bonds. The rate of photolysis depends upon numerous chemical and environmental factors including the light absorption properties, reactivity of the chemical, and the intensity of solar radiation. In the process, the photochemical mechanism of photolysis is divided into three stages: (1) the absorption of light which excites electrons in the molecule, (2) the primary photochemical processes which transform or deexcite the excited molecule, and (3) the secondary (*dark*) thermal reactions which transform the intermediates produced in the previous step (step 2).

Indirect photolysis or sensitized photolysis occurs when the light energy captured (absorbed) by one molecule is transferred to the molecule of concern. The donor species (the sensitizer) undergoes no net reaction in the process but has an essentially catalytic effect. Moreover, the probability of a sensitized molecule donating its energy to an acceptor molecule is proportional to the concentration of both chemical species. Thus, complex mixtures may, in some cases, produce enhancement of photolysis rates of individual constituents through sensitized reactions.

Photolysis occurs in the atmosphere as part of a series of reactions by which primary pollutant nitrogen oxides react to form secondary pollutants such as peroxyacyl nitrate derivatives. The two most important photolysis reactions in the troposphere are:

$$O_3 + h\nu \rightarrow O_2 + O^*$$

The excited oxygen atom (O^*) can react with water to give the hydroxyl radical:

$$O^* + H_2O \rightarrow 2OH\bullet$$

The hydroxyl radical is central to atmospheric chemistry because it can initiate the oxidation of hydrocarbons in the atmosphere.

Another reaction involves the photolysis of nitrogen dioxide to produce nitric oxide and an oxygen radical that is a key reaction in the formation of tropospheric ozone:

$$NO_2 + h\nu \rightarrow NO + O$$

The formation of the ozone layer is also caused by photolysis. Ozone in the stratosphere is created by ultraviolet light striking oxygen molecules (O_2) and splitting the oxygen molecules into individual oxygen atoms (atomic oxygen). The atomic oxygen then combines with an unbroken oxygen molecule to create ozone (O_3):

$$O_2 \rightarrow 2O$$

$$O_2 + O \rightarrow O_3$$

In addition, the absorption of radiation in the atmosphere can cause photodissociation of nitrogen (as one of several possible reactions) that can lead to the formation of nitric oxide (NO) and nitrogen dioxide (NO_2) that can act as a catalyst to destroy ozone:

$$N_2 \rightarrow 2N$$

$$O_2 \rightarrow 2O$$

$$CO_2 \rightarrow C + 2O$$

$$H_2O \rightarrow 2H + O$$

$$2NH_3 \rightarrow 3H_2 + N_2$$

$$N + 2O \rightarrow NO_2$$

As shown in the above equations, most photochemical transformations occur through a series of simple steps (Wayne and Wayne, 2005).

However, the application of photochemistry or photosensitized processes at interfaces of atmospheric relevance still represents a largely unknown area of remediation technologies. These processes have the potential to alter not only the chemical composition and properties of the irradiated surfaces, but also in some cases of the air surrounding them. For an understanding of the reaction perspectives, there is a clear requirement to bridge the physical properties of an interface, especially the air/water interface, with chemical features. For example, while it has been known for decades that organic molecules can alter surface energy (or surface tension), the consequence of this in terms of photochemical processing remains to be understood. For example, there is the need and space for basic photochemical investigation of such processes. Moreover, a wide range of environmental conditions and excitation wavelengths must be investigated for processes that have the potential to have a significant effect on the chemical composition and properties of the discharged chemical.

2.1.3 Radioactive Decay

Radionuclides are inorganic chemicals that emit radiation because their atoms are unstable, and they disintegrate or decay as they release energy in the form of radioactive particles or waves (Table 10.5). Unlike other elements or chemicals, radionuclides degrade naturally and require very little in the way of chemical or physical stimulation in the environment, but the degradation process itself is hazardous to living things. The risk posed by radionuclides is a function of the type and amount of radiation, as well as exposure pathways, and the half-life (the time required to reduce the concentration of a chemical to 50% of the initial concentration) of the radioactive element.

The elements that undergo radioactive decay are the radionuclides which are chemicals that emit radiation because of the instability of the atoms which

TABLE 10.5 Common Radionuclides

Americium-241	Iodine-129, 131	Ruthenium-103, 106
Barium-140	Krypton-85	Silver-110
Carbon-14	Molybdenum-99	Strontium-89, 90
Cerium-144	Neptunium-237	Technetium-99
Cesium-134, -137	Plutonium-238, -239,-241	Tellurium-132
Cobalt-60	Polonium-210	Thorium-228, 230, 232
Curium-242, 244	Radium-224, -226	Tritium
Europium-152, 154, 155	Radon-222	Uranium-234, 235, 238

causes the decay (disintegration) as they release energy in the form of radioactive particles or waves. There are many different types of radioactive decay (Table 10.6). A decay, or loss of energy from the nucleus, results when an atom with an initial type of nucleus (the *parent radionuclide* or *parent radioisotope*) transforms into a *daughter nuclide*. The transformation produces an atom in a different state (a nucleus containing a different number of protons and neutrons). In some decay processes, the parent and the daughter nuclides are different chemical elements, and thus the decay process results in the creation of an atom of a different element (nuclear transmutation). Another type of radioactive decay results in products that are not defined but appear in a range of molecular fragments of the original nucleus (spontaneous fission) which occurs when a large unstable nucleus spontaneously splits into two (and occasionally three) smaller daughter nuclei, and generally leads to the emission of gamma rays, neutrons, or other particles from those products. Chemicals such as radionuclides are not readily digested and destroyed by the flora and fauna in an ecosystem and exhibit variable and destructive behavior in the environment (Summers and Silver, 1978).

TABLE 10.6 Common Forms of Radioactive Decay (See Also Fig. 10.1)

Mode of Decay	Participating Particles
Decay With Emission of Nucleons:	
Alpha decay	An alpha particle emitted from the nucleus
Proton emission	A proton ejected from the nucleus
Neutron emission	A neutron ejected from the nucleus
Spontaneous fission	Nucleus disintegrates into two or more smaller nuclei and other particles
Different Modes of Beta Decay:	
β^- decay	A nucleus emits an electron and an electron antineutrino
β^+decay	A nucleus emits a positron and an electron neutrino
Electron capture	A nucleus captures an orbiting electron and emits a neutrino
Transitions Between States of the Same Nucleus:	
Isomeric transition	Excited nucleus releases a high-energy photon (gamma ray)
Internal conversion	Excited nucleus transfers energy to an orbital electron, which is subsequently ejected from the atom

Radioactive decay is an example of physical transformations which result from processes governed by the laws of physics. It is the process by which an atomic nucleus emits particles or electromagnetic radiation to become either a different isotope of the same element or an atom of a different element. The three radioactive decay paths are alpha decay (α-decay, the emission of a helium nucleus), beta decay (β-decay, the emission of an electron or positron), and gamma decay (γ-decay, the emission of a photon). Gamma decay alone does not result in transformation, but it is generally accompanied by beta emission, which does result in transformation. For example, radon decays to polonium by alpha decay. Thus:

$$^{222}\mathrm{Rn} \rightarrow {}^{218}\mathrm{Po} + \alpha \mathrm{In}$$

In this equation, α represents the ejected helium nucleus. This single-step reaction is first-order, and the concentration of radon decreases exponentially with time. The time it takes for half the initial mass of radon to be transformed is called the half-life.

In the environment, radioactive decay is a natural process that happens spontaneously and, as a radionuclide decays, radioactive isotopes transform into other, often less radioactive, isotopes of the same element, or even sometimes into isotopes of other elements. The identity and chemical properties of an element are determined by the number of protons in its nucleus: the atomic number. Many elements, however, occur as different isotopes, which are defined by the number of neutrons in the nucleus. For example, the common isotope of hydrogen has one proton and no neutrons in its nucleus. Tritium, the radioactive isotope of hydrogen, has one proton and two neutrons. Uranium-235 and Uranium-238, two isotopes of the element Uranium, both have 92 protons, but they have 143 and 146 neutrons respectively (Majumder and Wall, 2017).

If the flora and fauna of an ecosystem are isolated from the radionuclide, the risk of contamination is low. However, active remediation does not affect the decay rate but is designed to stabilize, prevent the migration of, or isolate the substance. In comparing the natural attenuation of radionuclides to active remediation, sorption and dilution are often considered as the mechanisms for isolating the material and decreasing the concentration. Some radioactive isotopes decay to form other more hazardous radionuclides. As with the degradation of contaminants, therefore, it is essential to consider the hazard posed by the daughter products, as well as radiation from the original hazard, when evaluating the effects of natural radioactive decay. Radioactive chemicals exhibit a wide range in half-lives—for example, Uranium^{238} (also written as ^{238}Uranium, Uranium-238) has a half-life of 4.5 billion years while tritium (^{3}H) has a half-life of 12.33 years.

Though nonradioactive metals do not break down in the environment, sometimes chemical changes in the subsurface environment reduce the behavior in the environment—such as the reaction with water—although changes can lead to positive or negative issues as they relate to remediation processes (Table 10.7). Copper in certain forms is considered so safe that we use it to pipe drinking water yet trace concentrations can be detrimental to aquatic life. Chromium in one form (hexavalent, or Cr^{6+}) forms chemicals which are highly toxic and very soluble, yet under certain conditions it transforms into trivalent chromium (Cr^{3+}), which is less toxic, as well as insoluble. The potential for such a change differs for each substance, but it can be estimated by conducting a chemical and biological analysis of the subsurface environment.

Finally, the radioactive half-life for a given radioisotope is a measure of the tendency of the nucleus to decay (disintegrate) and is based upon that probability. The half-life is independent of the physical state (solid, liquid, gas), temperature, pressure, the chemical in which the nucleus exists, and essentially any other outside influence. The half-life is also independent of the chemistry of the atomic surface and of the ordinary physical factors of the environment in which the chemical exists.

2.1.4 Rearrangement Reactions

A rearrangement reaction falls into a broad class of reactions where a molecule is rearranged to produce a structural isomer of the original molecule. Thus, a rearrangement reaction generates an isomer, a compound with the same formula but a different arrangement of atoms in the molecule and different properties. In a rearrangement reaction, the number of bonds normally does not change, and rearrangement reactions are more common in organic chemistry than in inorganic chemistry.

An isomerization reaction is the process by which one molecule is transformed into another molecule which has exactly the same number of atoms, but the atoms have a different arrangement. In some molecules and under some conditions, isomerization occurs spontaneously. When the isomerization occurs intramolecularly (that is, within the molecule), it is considered a rearrangement reaction. For example:

$$CH_3CH_2CH_2CH_2CH_3 \rightarrow CH_3CH_2CH(CH_3)CH_3 \rightarrow CH_3C(CH_3)_2CH_3$$

	n-pentane	2-methylbutane	2,2-dimethylpropane
aka		iso-pentane	neopentane

Inorganic reactions typically yield products that are in accordance with the generally accepted mechanism of the reactions. However, in some instances,

TABLE 10.7 Reactions of Metals With Water That Can Occur in an Ecosystem

Metal	Reaction With Air	Reaction With Water	Reaction With Steam	Reaction With Dilute Acid	Solubility in Water
Sodium	Burns readily to form oxide	Violent reaction	Violent reaction	Violent reaction	Readily soluble
Potassium	Burns readily to form oxide	Violent reaction	Violent reaction	Violent reaction	Readily soluble
Magnesium	Burns readily to form oxide	No reaction	Strong reaction	Ready reaction	Sparingly soluble
Calcium	Burns readily to form oxide	Slow reaction	Violent reaction	Violent reaction	Slightly soluble
Iron	Forms oxide when heated	No reaction	Reversible reaction	Reaction	Insoluble
Copper	Forms oxide when heated	No reaction	No reaction	No reaction	Insoluble
Silver	No reaction	No reaction	No reaction	No reaction	Insoluble
Gold	No reaction	No reaction	No reaction	No reaction	Insoluble
Zinc	Forms oxide when heated	No reaction	Reaction	Reaction	Insoluble
Mercury	Reversible reaction	No reaction	No reaction	No reaction	Insoluble
Aluminium	Burns readily to form oxide	No reaction	Reaction	Reaction	Insoluble
Tin	Forms oxide when heated	No reaction	No reaction	Weak reaction	Insoluble
Lead	Forms oxide when heated	No reaction	No reaction	Weak reaction	Insoluble

reactions do not give exclusively and solely the anticipated products but may lead to other products that arise from an unexpected and mechanistically different reaction path. This type of unexpected product is often referred to as a rearranged product and, while such a product may not be the expected product it may be the major product of the reaction. Thus, the reaction has involved a rearrangement of the expected product to an unexpected product—a rearrangement reaction has occurred.

More than likely, this may have resulted from a plausible rearrangement occurring during the mechanistic course of the reaction to fulfill the principle of the minimum energy state of the whole system, that is, of the transition state which assumed another configuration to maintain a minimum energy balance to the system. In many cases, the rearrangement affords products of an isomerization, coupled with some stereochemical changes. An energetic requirement is also observed in order for a rearrangement to take place; that is, the rearrangement usually involves an evolution of energy (typically in the form of heat, i.e., the reaction is overall an exothermic reaction) to be able to yield a more stable chemical.

2.1.5 Redox Reactions

Although many oxidation–reduction (redox) reactions are reversible, they are included here because many of the redox reactions that influence the fate of toxicants are irreversible on the temporal and spatial scales that are important to toxicity.

Oxidation is the loss of electrons—oxidizing agents are electrophiles and thus gain electrons upon reaction. An oxidation reaction can result in the increase in the oxidation state of the chemical as in the oxidation of metals or oxidation can incorporate oxygen into the molecule. Typical organic chemical oxidative reactions include dealkylation, epoxidation, aromatic ring cleavage, and hydroxylation.

The term autooxidation, or weathering, is commonly used to describe the general oxidative degradation of a chemical (or chemical mixture, e.g., petroleum) upon exposure to air. Chemicals can react abiotically in both water and air with oxygen, ozone, peroxides, free radicals, and singlet oxygen. The last two are common intermediate reactants in indirect photolysis.

Mineral surfaces are known to catalyze many oxidative reactions. Clay minerals and the mineral that consists of the oxides of silicon, aluminum, iron, and manganese can provide surface active sites that increase rates of oxidation. There are a variety of complex mechanisms associated with this catalysis, so it is difficult to predict the catalytic activity of soils and sediment in nature.

On the other hand, reduction of a chemical species takes place when an electron donor (reductant) transfers electrons to an electron acceptor (oxidant). Organic chemicals typically act as the oxidant, while abiotic reductants including sulfide minerals, reduce metals or sulfur compounds, and natural organic matter. There are also extracellular biochemical reducing agents such as porphyrins, corrinoids, and metal-containing coenzymes. Most of these reducing agents are present only in anaerobic environments where anaerobic bacteria are themselves busy reducing chemicals. Thus, it is usually very difficult to distinguish biotic and abiotic reductive processes in nature. Many abiotic reductive transformations could be important in the environment, including dehalogenation, dealkylation, and the reduction of quinone, nitrosamine, azoaromatic, nitroaromatic, and sulfoxide derivatives. Functional groups (Tables 10.8 and 10.9) that are resistant to reduction (and therefore, to reductive remediation processes) include aldehyde, ketone, carboxylic acid (and ester derivatives), amide, alkene, and aromatic hydrocarbon derivatives. Each type of functional group chemical will have a specific reaction in the environment and require a specific remediation process for cleanup.

Thus, redox reactions (reduction–oxidation reactions) are reactions in which one of the reactants is reduced and another reactant is oxidized. Therefore, the oxidation state of the species involved must change. The word *reduction* originally referred to the loss in weight upon heating a metallic ore such as a metal oxide to extract the metal—the ore was *reduced* to the metal. However, the meaning of *reduction* has become generalized to include all processes involving gain of electrons. Thus, in redox reactions, one species is oxidized while another is reduced by the net transfer of electron from one to the other. As may be expected, the change in the oxidation states of the oxidized species must be balanced by any changes in the reduced species. For example, the production of iron from the iron oxide ore:

$$Fe_2O_3 + 3CO \rightarrow 2Fe + 3CO_2$$

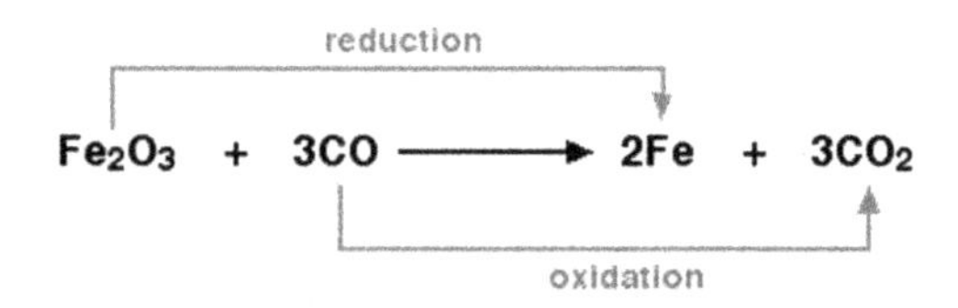

To complicate matters even further, the oxidizing and reducing agents can be the same element or chemical, as is the case when disproportionation of the reactive species occurs. For example:

$$2A \rightarrow (A + n) + (A - n)$$

In this equation, n is the number of electrons transferred. Disproportionation reactions do not need to commence with a neutral molecule and can involve more than two species with differing oxidation states.

Within redox reactions, the pair of reactions must always occur, that is, a reduction reaction must be accompanied by an oxidation process, as electrons

TABLE 10.8 Functional Groups in Organic Chemicals

Functional Group	Type	Compound	Example	IUPAC Name	Common Name
C=C	Double bond	alkene	H_2CCH_2	Ethene	Ethylene
C≡C	Triple bond	alkyne	HC≡CH	Ethyne	Acetylene
–OH	Hydroxyl	alcohol	CH_3OH	Methanol	Methyl alcohol
–O–	Oxy	ether	H_3COCH_3	Methoxymethane	Methyl ether
>CO	Carbonyl	aldehyde	H_2CO	Methanal	Formaldehyde
>CO	Carbonyl	ketone	CH_3COCH_3	Propanone	Acetone
$-CO_2^-$	Carboxyl	carboxylic acid	HCOOH	Methanoic acid	Formic acid
$-CO_2^-$	Carboxyl	ester	$HCOOCH_2CH_3$	Ethyl methanoate	Ethyl formate
$-NH_2$	Amino	amine	CH_3NH_2	Aminomethane	Methylamine
–CN	Cyano	nitrile	CH_3CN	Ethanenitrile	Acetonitrile
–X	Halogen	haloalkane	CH_3Cl	Chloromethane	Methyl chloride

TABLE 10.9 Potential Reactions of Organic Functional Groups in the Environment

Functional Group	Interaction
Carboxylic acid, –COOH	Ion exchange, complexation
Alcohol, phenol, –OH	Hydrogen bonding, complexation
Carbonyl, >CO	Reduction–oxidation
Hydrocarbon, $[-CH_2-]_n$	Hydrophobic

are transferred from one species to another. Each of the singular reactions in this pair is called a half-reaction, in which the electrons lost or gained are included explicitly, allowing electron balance to be accounted as well. The two sides of the reaction, given by the half-reactions, should be balanced accordingly. The additional terminology comes from the definition that within redox processes, a reductant transfers one electron or more electrons to an oxidant; hence, the reductant (reducing agent) loses electrons, and is oxidized, while the oxidant (oxidizing agent) gains electrons, and is reduced.

Redox reactions are important for a number of applications, including energy storage devices (batteries), photographic processing, and energy production and utilization in living systems including humans. For example, a *reduction reaction* is a reaction in which an atom gains an electron and therefore decreases (or reduces its oxidation number). The result is that the positive character of the species is reduced. On the other hand, an oxidation reaction is a reaction in which an atom loses an electron and therefore increases its oxidation number. The result is that the positive character of the species is increased.

Although oxidation reactions are commonly associated with the formation of oxides from oxygen molecules, these are only specific examples of a more general concept of reactions involving electron transfer. Redox reactions are a matched set, that is, there cannot be an oxidation reaction without a reduction reaction happening simultaneously. The oxidation reaction and the reduction reaction always occur together to form a whole reaction. Although oxidation and reduction properly refer to *a change in the oxidation state*, the actual transfer of electrons may never occur. The oxidation state of an atom is the fictitious charge that an atom would have if all bonds between atoms of different elements were 100% ionic. Thus, oxidation is best defined as an *increase in oxidation state*, and reduction as a *decrease in oxidation state*. In practice, the transfer of electrons will always cause a change in oxidation state, but there are many reactions that are classified as redox reactions even though no electron transfer occurs (such as those involving covalent bonds).

The key to identifying oxidation–reduction reactions is recognizing when a chemical reaction leads to a change in the oxidation number of one or more atoms.

2.1.6 Complexation

Complexation is an important process that will determine in some cases if mineral solubility limits are reached, the amount of adsorption that occurs, and the redox state that exists in the water. Chemicals can also form stable, soluble complexes with ligands. Ligands in leachates could include synthetic chelating agents (such as ethylenediamine), partially oxidized biodegradation products (such as acids), or natural humic materials. In fact, the movement of metals in the subsurface is strongly influenced by the concentration and chemical properties of ligands. However, the types and concentrations of ligands in most waste leachates are largely unknown.

2.2 Physical Transformation

A physical change involves a change in physical properties such as melting point or melting range, transition to another phase, change in crystal form, as well as changes in color, volume, and density. Many physical changes also involve the rearrangement of atoms most noticeably in the formation of crystals. Although chemical changes may be recognized by an indication such as odor, color change, or production of a gas, every one of these indicators can result from physical change.

However, other natural processes, such as chemical transformation, dilution, sorption, and radioactive decay, are responsible for natural attenuation, as broadly defined by United States Environment Protection Agency (www.epa.gov). The health and environmental risks posed by those contaminants may be reduced by changing the amount of exposure, the exposure pathway, or the toxicity of the chemical (Rahm et al., 2005).

A physical conversion that is often ignored (or not recognized as such) is the change in composition that occurs when precipitation and dissolution reactions exert a major effect on the concentrations of ions in solution. The precipitation of a solid phase in environmental systems rarely results in a pure mineral phase. Minor elements such as cadmium can coprecipitate with major elements such as calcium to form a solid solution. In addition, the equilibrium activity of the component of a solid solution does not correspond to the solubility calculated from the solubility products of pure minerals. Furthermore, it is often impossible to distinguish chemically between a minor component coprecipitated in a solid, and the adsorption of that minor component onto the solid surface.

Physical transformations result from processes governed by the laws of physics and the classical example, which comes from the field of nuclear physics, is radioactive decay which is the process by which an atomic nucleus emits particles or electromagnetic radiation to become either a different isotope of the same element or an atom of a different element.

Physical changes are changes affecting the form of a chemical but not always the chemical composition. Physical changes are used to separate mixtures into their component chemicals, such as the use of liquid–solid chromatography, but cannot usually be used to separate chemicals into chemical elements or simpler chemicals. Thus, physical changes occur when objects or substances undergo a change that does not change their chemical composition. This contrasts with the concept of chemical change in which the composition of a substance changes or one or more substances combine or break up to form new substances.

Physical transformations between states of matter, not necessarily involving chemical reaction, can occur on changes in temperature or pressure or application of external forces or fields. Such phase transitions have been central to quantitative research in chemical sciences. Suitable choices or changes in temperature, pressure, and other controllable properties can produce abrupt changes (first-order transitions) in the state of a substance: for example, boiling or freezing, or formation of systems having two phases coexisting with distinct phase boundaries.

The importance of physical transformations in causing changes to spills or chemical disposal implies that behavior of the spilled material can change and requires an understanding of the behavior of the various constituents of the mixture. For example, absorption of part of the spillage into water or adsorption on to a mineral (clay) surface can cause part of the spilled material to be retained in the water or on the soil. Control over such forces is difficult and the selective absorption or adsorption of any part of a spilled chemical mixture can cause changes in the effect of the nonabsorbed or nonadsorbed chemical on the environment. The converse is also true insofar as the selective absorption or adsorption of any part of a spilled chemical mixture can cause changes in the effect of absorbed or adsorbed chemical on the environment. And in either cases, the environmental effects can be considerable.

A common example that is treated as a physical transformation is the settling of suspended sediment particles. Although settling does not actually transform the sediment into something else, it does remove sediment from our control volume by depositing it on the river bed. This process can be expressed mathematically by heterogeneous transformation equations at the river bed; hence, we will discuss it as a transformation.

Another example of a physical change is radioactive decay, which is the process by which an atomic nucleus emits particles or electromagnetic radiation to become either a different isotope of the same element or an atom of a different element.

2.2.1 Sorption

Sorption is a physical process with chemical tendencies which is typically considered to be a physical process insofar as a chemical may be adsorbed on to, say, a mineral surface without any chemical bonding between the chemical and the mineral. If one substance (the chemical) becomes attached to the other (the mineral), the process may be referred to as a physicochemical process. Some examples of the sorption phenomenon are: (1) adsorption, which is, in the current context, the physical adherence or bonding of a molecule or an ion on to the surface of another phase such as the surface of a mineral—especially a clay mineral, (2) absorption, which is the incorporation of a substance in one state into another of a different state (such as a solid or a gas being absorbed by a liquid), (3) desorption, which is the reversal of adsorption and absorption in which the adsorbate (the chemical adsorbed) or the absorbate (the chemical absorbed) is returned to the free state. There is also a fourth phenomenon—not typically recognized in environmental chemistry but which has a real effect—and this is ion-exchange in which there is an exchange of ions between two electrolytes or between an electrolyte solution and a complex (Chapter 1).

By way of definition, an electrolyte is a chemical that produces an electrically conducting solution when it is dissolved in a polar solvent, such as water. The dissolved electrolyte separates into cations (positively charged species) and anions (negatively charged species) which disperse uniformly through the solvent.

Thus, the phenomenon known as sorption covers three types of physical interactions that are the focus of the current text and which can lead to the transformation of chemical in the environment: (1) adsorption, (2) absorption, and (3) desorption. *Adsorption*, in the context of this text (Chapter 5), refers to the adhesion of ions or molecules from a gas, a liquid, or a dissolved solid on the surface of a mineral (of which a clay mineral is an example). The process creates a film of the *adsorbate* on the surface of the adsorbent. Adsorbed chemicals are those that are resistant to washing with the same solvent medium in the case of adsorption from a solution. Adsorption on to a mineral sediment (such as a clay sediment that has strong adsorptive powers) reduces the mobility and (quite often) the reactivity of the adsorbed chemical, although there is always the possibility that the mineral sediment can cause catalyzed chemical transformation reactions. On the other hand, *absorption* is the process by which a gas or a liquid (the *absorbate*) is dissolved in a liquid or permeates the pore system of a solid (the *absorbent*) (Chapter 5). In the process, the gaseous or liquid absorbate diffuses into liquid or solid materials. Thus, while the adsorption phenomenon is a surface-based process, the absorption process involves the whole volume of the material.

Desorption (Chapter 5) is the process whereby a, adsorbed or absorbed chemical is released from a surface or through a surface. In chemistry, using chromatography as an example, it is possible to explain the ability of the chemical to move with the mobile phase. After adsorption, the adsorbed chemical will remain on the substrate nearly indefinitely, provided the temperature remains low. However, if there is a rise in temperature, the likelihood of desorption increases according to the equation:

$$R = rN^x$$

In this equation, R is the rate of desorption, r is the rate constant for desorption, N is concentration of the adsorbed chemical, and *x* is the kinetic order of the desorption process. Typically, the kinetic order of the process can be predicted by the number of elementary steps involved—for example, simple molecular desorption will typically be a first-order process.

In terms of adsorption of chemicals on to soil (Thompson and Goyne, 2012), the mechanism dictates that the chemical adsorbents fall into three categories (1) anionic chemicals that are negatively charged because they have more electrons than protons, such as the nutrient orthophosphate anion, PO_4^{3-}, (2) cationic chemicals that are positively charged because they have fewer electrons than protons, such as the divalent cations of calcium (Ca^{2+}) and lead (Pb^{2+}), and (3) uncharged organic chemicals that exhibit a range of polarities—varying from nonpolar to polar—based on the distribution of electrons across the molecule, such as the aromatic ring in benzene, C_6H_6, that has a uniform electron distribution and hence is nonpolar, whereas sucrose, $C_{12}H_{22}O_{11}$, has a nonuniform distribution of electrons and is highly polar. These categories can also be applied to portions of large adsorbates that contain a diversity of functional groups (Table 10.10).

Similarly, the reactive surface sites on adsorbents can be categorized as (1) positively charged, (2) negatively charged, or (3) uncharged, that is nonpolar. The major solid phase materials (sorbents) in soils are layer silicate clay minerals, metal-(oxyhydr)oxides, and soil organic matter (SOM). Layer silicate clay minerals are primarily negatively charged because their stacks of aluminum–oxygen and silicon–oxygen sheets are often chemically substituted by ions of lower valance. In many soils, these clay minerals represent the largest source of negative charge. On the other hand, the metal-(oxyhydr)oxides are variably charged because their surfaces become hydroxylated when exposed to water (Liu et al., 1998) and assume anionic, neutral, or cationic forms based on the degree of protonation ($\equiv$M-O, $\equiv$MOH, or $\equiv MOH_2^+$, where $\equiv$M represents a metal bound at the edge of a crystal structure), which varies as a function of solution pH. These variably charged minerals adopt a net positive surface charge at low pH and a net negative surface charge at high pH (Qafoku et al., 2004).

Also, an extra consideration in the field is the presence of chemicals that are nonpolar and cannot bond effectively with the polar water molecule, which minimizes the interfacial area between water and the nonpolar sorptive, and is energetically favorable. This leads to the common segregation of nonpolar

TABLE 10.10 Functional Groups in Organic Chemicals

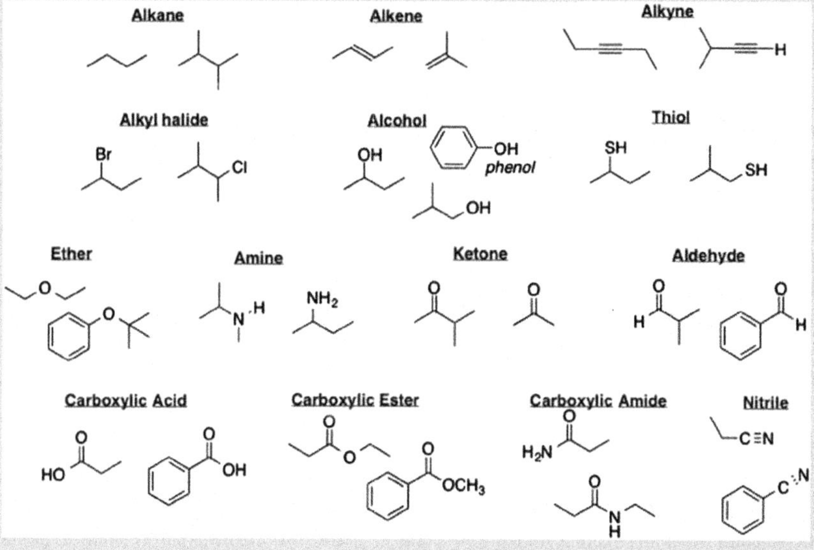

Alkane
Alkene
Alkyne
H
Alkyl halide
Br
Cl
Alcohol
OH
OH
phenol
OH
Thiol
SH
SH
Ether
O
O
Amine
N
H
NH_2
Ketone
O
O
Aldehyde
O
H
O
H
Carboxylic Acid
O
HO
O
OH
Carboxylic Ester
O
O
O
OCH_3
Carboxylic Amide
O
H_2N
O
N
H
Nitrile
C≡N
C
N

compounds from water commonly (the *hydrophobic effect*). When the relevant sites of a nonpolar compound come in contact with a nonpolar sorbent, they are no longer exposed to water and thus the overall nonpolar/polar interfacial area is reduced, which in turn influences the efficiency of the remediation process.

SOM includes living and partially decayed (nonliving) material, as well as assemblages of biomolecules and transformation products of organic residue decay (often referred to as *humic substances*). SOM contains a multitude of reactive sites such as potentially: (1) anionic hydroxyl derivatives (ROH), (2) carboxylic group derivatives (R-COOH), (3) cationic sulfhydryl derivatives (R-SH), (4) amino group derivatives (R-NH_2), as well as (5) aromatic (-Ar-) and aliphatic ($[-CH_2-]_n$) moieties that are the principal noncharged, nonpolar regions of the soil solid phase. Variations in the abundance, surface area, and chemical composition of these three sorbents significantly influence the sorption characteristics of a given soil (Johnston and Tombácz, 2002) and, hence, the ability of a discharged chemical to be freed during the application of a remediation process.

The pH of the solution used in the remediation process can have a pronounced effect on sorption by influencing both the magnitude and sign of sorptive and sorbent charge. As the solution pH increases, sorbent hydroxyl and carboxyl functional groups of metal-(oxyhydr)oxides and the SOM lose protons leaving an increased negative charge density on the sorbent, which will facilitate cation adsorption but have an adverse effect on anion adsorption (Thompson and Goyne, 2012).

When applying a remediation process to adsorbed chemicals, it must be recognized that despite heterogeneity in the chemical composition of sorbents (solids), sorptive concentration has perhaps the most significant influence on the accumulation of a sorbate on or within a sorbent. The standard (or common) method for assessing sorption characteristics of a solid (such as soil) is to measure the relationship (or distribution) between the equilibrium concentration of the sorptive chemical (C_{eq} in mol/L) and the sorbate (γ_{ads} in mol/kg) across a range of sorptive concentrations while holding temperature and other parameters constant, which gives the distribution coefficient (K_d, L/kg). Thus:

$$K_d = \gamma_{ads}/C_{eq}$$

The outcome is an *adsorption isotherm* which can be used in a predictive fashion to describe the adsorption behavior. For many organic compounds and ions in solution at low concentration (Chiou, 2002), this relationship is linear and summarized by a distribution coefficient (K_d). However, at higher ion concentrations the relationship between γ_{ads} and the equilibrium concentration of the sorptive chemical (C_{eq}) and the sorbate (C_{eq}) can exhibit significant nonlinearity, often approaching a maximum sorption plateau after which subsequent increases in sorptive concentration do not result in additional sorption.

In addition, the adsorption isotherm can (from the shape of the graphical line) give an indication of whether the chemical is adsorbed at one point (i.e., lying at right angles to the surface of the adsorbent) or is adsorbed through several points of contact (especially a multifunctional chemical) with the adsorbent (i.e., lying parallel to the surface of the adsorbent). In either case, the efficiency of the remediation process in terms of recovery of the adsorbed chemical can be assessed.

2.2.2 Dilution

Dilution is the process of decreasing the concentration, for example, of a solute in a solution, usually by mixing the solution with more solvent without the addition of more solute (Chapter 5). The dilution capacity of a water-based ecosystem is the effective volume of receiving water available for the dilution of the effluent. The effective volume can vary according to tidal cycles and transient physical phenomena such as stratification. In estuaries, in particular, the effective volume is much greater at high spring tides than at low neap tides. It is important to consider concentrations of substances in worst-case scenarios (usually low neap tides except, for example, when pollutants might be carried further into a sensitive location by spring tides) when calculating appropriate discharge consent conditions. Stratification can reduce the effective volume of the receiving water by reducing vertical mixing and constraining the effluent to either the upper or lower layers of the water column.

Dilution has often been used to solve pollution issues, but not all chemical discharges can be mitigated by dilution of the chemical. Thus, when chemicals are discharged into an ecosystem, such as a water system, they are subject to a number of physical processes which result in their dilution in the receiving water. In some cases, dilution is more important for reducing the concentration of conservative substances (those that do not undergo rapid degradation, e.g., metals) than for nonconservative substances (those that do undergo rapid degradation).

In fact, dilution is one of the main processes for reducing the concentration of substances away from the discharge point. Using dilution of a chemical in a water-based ecosystem as the example, the resulting solution is thoroughly mixed so as to ensure that all parts of the solution are identical. The same relationship applies to gases (and vapors) that are diluted in air, although thorough mixing of gases and vapors may not be as easily accomplished. The relationship can be represented mathematically, thus:

$$C_1 \times V_1 = C_2 \times V_2$$

C_1 is = initial concentration of the solute, V_1 is the initial volume of the solvent, C_2 is the final concentration of the solute, and V_2 is the final volume of the solvent.

However, dilution is not always the answer to mitigating the effect of a chemical discharge. In many cases an ecosystem can be overwhelmed by the amount of discharged chemical and dilution cannot always render the chemical harmless. In addition, dilution is not effective for chemicals that bioaccumulate, that is, which persist through the food chain and increase in any of the organisms that make up the flora and/or the faunal inhabitants of an ecosystem.

2.3 Catalytic Transformation

Catalytic transformation of a discharged chemical can occur in the environment because of the presence of catalysts. Catalysts are substances that increase the rate of a reaction by providing a low energy alternate route from reactants to products. In some cases, reactions occur so slowly that without a catalyst, they are of little value.

The study of environmental interfaces and environmental catalysis is central to finding more effective solutions to air pollution and in understanding how pollution impacts the natural environment. Surface catalysis of airborne particles—including ice, trace atmospheric gases, aerosolized soot nanoparticles, and mineral dust surfaces—as well as particles in contact with ground water and their role in surface adsorption, surface catalysis, hydrolysis, dissolution, precipitation, oxidation, and ozone decomposition must be continuously investigated. With increasing ground water pollution, and increasing particulates in the atmosphere, there is an increasing need to remove pollutants from industrial and automotive sources.

In catalysis reactions, the reaction does not proceed directly, but through reaction with a third substance (the catalyst) and, although the catalyst takes part in the reaction, it is (in theory) returned to its original state by the end of the reaction and so is not consumed. However, the catalyst is not immune to being inhibited, deactivated, or destroyed by secondary processes.

Industrial catalysts are often metals, as most metals have many electrons which help out in reactions before resuming the normal electronic configuration when the reaction is over. Examples are iron-based catalysts used for making ammonia (the Haber–Bosch process):

$$N_2 + 3H_2 \rightarrow 2NH_3$$

In the absence of a catalyst a high temperature on the order of 3000°C (5430°F) would be required to cause the reactant molecules to collide with enough force as to break chemical bonds.

Catalysts can be used in a different phase (heterogeneous catalysis) or in the same phase (homogeneous catalysis) as the reactants. In heterogeneous catalysis, typical secondary processes include coking (coke production from inorganic starting materials) where the catalyst becomes covered by ill-defined high molecular–weight byproducts. Heterogeneous catalysis is used in automobile exhaust systems to decrease nitrogen oxide, carbon monoxide, and unburned hydrocarbon emissions. The exhaust gas is vented through a high-surface area chamber lined with platinum, palladium, and rhodium. For example, the carbon monoxide is catalytically converted to carbon dioxide by reaction with oxygen.

Additionally, heterogeneous catalysts can dissolve into the solution in a solid–liquid system or evaporate in a solid–gas system. Catalysts can only speed up the reaction—chemicals that slow down the reaction are called inhibitors and there are chemicals that increase the activity of catalysts (catalyst promoters), as well as chemicals that deactivate catalysts (catalytic poisons). With a catalyst, a reaction which is kinetically inhibited by a high activation energy can take place by circumvention of this activation energy. Heterogeneous catalysts are usually solids, powdered in order to maximize their surface area. Of particular importance in heterogeneous catalysis are the platinum metals and other transition metals, which are used in crude oil refining processes such as hydrogenation and catalytic reforming.

Homogeneous catalysis involves a reaction in which the soluble catalyst is in solution—the catalyst is in the same phase as the reactants. Although the term is used almost exclusively to describe reactions (and catalysts) in solution, it often implies catalysis by organometallic compounds but can also apply to gas phase reactions and solid phase reactions. In heterogeneous catalysis, the catalyst is in a different phase than the reactants. The advantage of homogeneous catalysts is the ease of mixing them with the reactants, but they may also be difficult to separate from the products. Therefore, heterogeneous catalysts are preferred in many industrial processes for the production and transformation (conversion) of the starting compound(s). However, heterogeneous catalysis offers the advantage that products are readily separated from the catalyst, and heterogeneous catalysts are often more stable and degrade much slower than homogeneous catalysts. However, heterogeneous catalysts are difficult to study, so their reaction mechanisms are often unknown.

Environmental catalysis has continuously grown in importance over the latter half of the 20th century and the development of innovative catalysts for protection of the environment is a crucial factor towards the objective of developing a new sustainable industrial chemistry. In the last decade, considerable expansion of the traditional area of environmental catalysis has occurred, mainly removal of nitrogen oxides from stationary and mobile sources and conversion of volatile organic compounds (VOCs). New areas include: (1) catalytic technologies for liquid or solid waste reduction or purification; (2) use of catalysts in energy-efficient technologies and processes; (3) reduction of the environmental impact in the use or disposal of catalysts; (4) new ecocompatible refinery, chemical or nonchemical catalytic processes; (5) catalysis for greenhouse gas control; (6) use of catalysts for user-friendly technologies and reduction of indoor pollution; (7) catalytic processes for sustainable chemistry; (8) reduction of the environmental impact of transport (Centi et al., 2002; Grassian, 2005).

3. BIOLOGICAL TRANSFORMATION

Biological transformation (Chapter 8) refers to chemical reactions that are mediated by living organisms through the processes of photosynthesis and respiration (Shea, 2004). These reactions involve the consumption of a nutritive substance to produce biomass and are accompanied by an input or output of energy. The classical photosynthesis equation shows the production of glucose ($C_6H_{12}O_6$) by the process known as photosynthesis. Thus:

$$6CO_2 + 6H_2O \rightarrow C_6H_{12}O_6 + 6O_2$$

Photosynthesis and respiration (particularly in the form of biodegradation) are of particular interest in environmental engineering because they affect the concentration of oxygen, a component essential for most aquatic life.

Biotransformation reactions follow a complex series of chemical reactions (and often physical reactions) that are enzymatically mediated and are usually irreversible reactions that are energetically favorable. Thus, the potential for the biotransformation of a chemical depends on the reduction in free energy that results from reacting the chemical with other chemicals in its environment (e.g., oxygen). As with inorganic catalysts, microbes simply use enzymes to lower the activation energy of the reaction and increase the rate of the transformation. Each successive chemical reaction further degrades the chemical, eventually mineralizing it to inorganic compounds (CO_2, H_2O, inorganic salts).

Usually microbial growth is stimulated because the microbes capture the energy released from the biotransformation reaction. As the microbial population expands, overall biotransformation rates increase, even though the rate for each individual microbe may be constant or even decrease. This complicates the modeling and prediction of biotransformation rates in nature. When the toxicant concentrations (and the potential energy) are small relative to other substrates or when the microbes cannot efficiently capture the energy from the biotransformation, microbial growth is not stimulated but biotransformation often still proceeds inadvertently through cometabolism (Shea, 2004).

3.1 Biodegradation

Biodegradation (biotic degradation, biotic decomposition) is the chemical degradation of contaminants by bacteria or other biological means and is often referred to as the natural method of degrading discharged chemicals (Chapter 8). Most biodegradation systems operate under aerobic conditions, but a system under anaerobic conditions may permit microbial organisms to degrade chemical species that are otherwise nonresponsive to aerobic treatment and vice versa. Thus, biodegradation is a natural process (or a series of processes) by which spilled chemicals or other waste material can be broken down (degraded) into nutrients that can be used by other organisms. The ability of a chemical to be biodegraded is an indispensable element in the understanding of the risk posed by that chemical on the environment.

Biodegradation is a key process in the natural attenuation (reduction or disposal) of chemicals at hazardous waste sites. The contaminants of concern must be amenable to degradation under appropriate conditions, but the success of the process depends on the ability to determine these conditions and establish them in the contaminated environment. Important site factors required for success include (1) the presence of metabolically capable and sustainable microbial populations, (2) suitable environmental growth conditions, such as the presence of oxygen, (3) temperature, which is an important variable—keeping a substance frozen or below the optimal operating temperature for microbial species, can prevent biodegradation—most biodegradation occurs at temperatures between 10 and 35°C (50 and 95°F), (4) the presence of water, (5) appropriate levels of nutrients and contaminants, and (6) favorable acidity or alkalinity (Table 10.11). In regard to the last parameter, soil pH is extremely important because most microbial species can survive only within a certain pH range—for example, the biodegradation of chemicals might be optimal at a pH 7 (neutral) but the

TABLE 10.11 Essential Factors for Microbial Bioremediation

Factor	Optimal Conditions
Microbial population	Suitable kinds of organisms that can biodegrade all of the contaminants
Oxygen	Enough to support aerobic biodegradation (about 2% oxygen in the gas phase or 0.4 mg/L in the soil water)
Water	Soil moisture should be from 50% to 70% of the water holding capacity of the soil
Nutrients	Nitrogen, phosphorus, sulfur, and other nutrients to support good microbial growth
Temperature	Appropriate temperatures for microbial growth (0–40°C, 32–104°F)
pH	Optimal range: 6.5 to 7.5

acceptable (or optimal) pH range might be in the order of 6–8. Furthermore, soil (or water) pH can affect availability of nutrients. Thus, through biodegradation processes, living microorganisms (primarily bacteria, but also yeasts, molds, and filamentous fungi) can alter and/or metabolize various classes of chemicals.

Temperature influences rate of biodegradation by controlling rate of enzymatic reactions within microorganisms. Generally, the rate of an enzymatic reaction approximately doubles for each 10°C (18°F) rise in temperature (Nester et al., 2001). However, there is an upper limit to the temperature that microorganisms can withstand. Most bacteria found in soil, including many bacteria that degrade petroleum hydrocarbons, are mesophile organisms which have an optimum working temperature range in the order of 25–45°C (77–113°F) (Nester et al., 2001). Thermophilic bacteria (those which survive and thrive at relatively high temperatures) which are normally found in hot springs and compost heaps exist indigenously in cool soil environments and can be activated to degrade chemicals with an increase in temperature to 60°C (140°F). This indicates the potential for natural attenuation in cool soils through thermally enhanced bioremediation techniques.

In the absence of oxygen, some microorganisms obtain energy from fermentation and anaerobic oxidation of carbon. Many anaerobes use nitrate, sulfate, and salts of iron (Fe^{3+}) as practical alternates to oxygen acceptor. The anaerobic reduction process of nitrates, sulfates, and salts of iron is an example:

$$2NO_3^- + 10e^- + 12H^+ \rightarrow N_2 + 6H_2O$$

$$SO_4^{2-} + 8e^- + 10H^+ \rightarrow H_2S + 4H_2O$$

$$Fe(OH)_3 + e^- + 3H^+ \rightarrow Fe^{2+} + 3H_2O$$

Anaerobic biodegradation is a multistep process performed by different bacterial groups. It involves hydrolysis of polymeric substances like proteins or carbohydrates to monomers and the subsequent decomposition to soluble acids, alcohols, molecular hydrogen, and carbon dioxide. Depending on the prevailing environmental conditions, the final steps of ultimate anaerobic biodegradation are performed by denitrifying, sulfate-reducing, or methanogenic bacteria.

In contrast to the strictly anaerobic sulfate-reducing and methanogenic bacteria, the nitrate-reducing microorganisms, as well as many other decomposing bacteria, are mostly facultative anaerobic insofar as these microorganisms are able to grow and degrade substances (under aerobic as well as anaerobic conditions). Thus, aerobic and anaerobic environments represent the two extremes of a continuous spectrum of environmental habitats which are populated by a wide variety of microorganisms with specific biodegradation abilities.

3.2 Bioremediation

Bioremediation is the use of microbial species to clean up soil and groundwater that has been contaminated by discharged chemicals (Chapter 8). The bioremediation process stimulates the growth of specific microbes that use the discharged chemical contaminants as a source of food and energy. In a nonpolluted environment, bacteria, fungi, protists, and other microorganisms are constantly at work breaking down organic matter. Bioremediation works by providing these pollution-eating organisms with fertilizer, oxygen, and other conditions that encourage their rapid growth. These organisms would then be able to break down the organic pollutant at a correspondingly faster rate. In fact, bioremediation is often used to help clean up crude oil spills. Bioremediation of a contaminated site typically works in one of two ways: (1) methods are found to enhance the growth of whatever pollution-eating microbes might already be living at the contaminated site, and (2) specialized microbes are added to degrade the contaminants.

Some types of microbes eat and digest contaminants, usually changing them into small amounts of water and relatively harmless gases such as carbon dioxide (CO_2) and ethylene (CH_2CH_2). If the contaminated ecosystem (such as soil and groundwater) does not have enough of the right microbes, they can be added (bioaugmentation). However, like any chemical reaction, for the bioremediation process to be effective, the site must be at the appropriate temperature and nutrients must also be present, otherwise the microbes grow too slowly or die. The site conditions may be improved by adding amendments which can range from chemical mixtures such as molasses and vegetable oil to air and chemicals that produce oxygen.

The conditions necessary for bioremediation in soil cannot always be achieved in situ, however. At some sites, the climate may be too cold for the microbes to be active, or the soil might be too dense to allow amendments to spread evenly underground. At such sites, the soil may be dug up for ex-situ (above ground) cleaning operation on a pad or in tanks. The soil may then be heated, stirred, or mixed with amendments to improve conditions. Moreover, mixing soil can cause contaminants to evaporate before the microbes can attack. To prevent the vapors from contaminating the air, the soil can be mixed inside a special tank or building where vapors from chemicals that evaporate may be collected and treated.

In order to clean up contaminated groundwater in situ, wells are drilled to pump some of the groundwater into above ground tanks in which the water is mixed with amendments before it is pumped back into the ground. The groundwater enriched with amendments promotes the microbes to attack the contaminating chemicals. As a part of the ex situ treatment process, the groundwater also can be pumped into a bioreactor which is a tank in which groundwater is mixed with microbes and amendments for treatment. Depending on the site and the extent of the chemical contamination, the treated water may be pumped back to the ground or discharged to surface water or to a municipal wastewater system.

Bioremediation provides a good cleanup strategy for some types of pollution, but as you might expect, it will not work for all. For example, bioremediation may not provide a feasible strategy at sites with high concentrations of chemicals that are toxic to most microorganisms. These chemicals include metals such as cadmium or lead, and salts such as sodium chloride. Nonetheless, bioremediation provides a technique for cleaning up pollution by enhancing the same biodegradation processes that occur in nature. Depending on the site and its contaminants, bioremediation may be safer and less expensive than alternative solutions such as incineration or landfilling of the contaminated materials. The process also has the advantage of treating the contamination in place so that large quantities of soil, sediment, or water do not have to be dug up or pumped out of the ground for treatment.

4. TRANSPORT OF CHEMICALS

Transport in the subsurface environment is relatively slow compared with the other environmental media. Many chemicals that would rapidly volatilize into the atmosphere from surface waters may reside in groundwater for decades or longer. The subsurface environment also shares many processes with surface waters. Microbially mediated redox reactions and biodegradation processes are significant in each medium. The presence of particles, and their potential to absorb chemicals, also is common to both surface waters and groundwater; when modeling surface transport, the high ratio of solid material to water requires special recognition of even moderate sorbing tendencies.

Chemical transport in the atmosphere has many parallels with transport in water; both advection and Fickian processes are important (Warneck, 2000).

However, the velocities and the Fickian transport coefficients tend to be larger in the atmosphere, and the distances over which pollutant sources have influence can be greater. Atmospheric chemical processes are dominated by the photochemical biota, which compete with photochemistry, as agents of chemical transformation in surface waters are few in the air. Also, light intensities in the atmosphere are higher than those in surface waters, and the presence of ultraviolet light of shorter wavelengths contributes to an expanded suite of photochemical reactions that may occur.

Thus, a particular question which needs to be addressed more often for persistent pollutants relates to the fraction that remains in circulation (until the end of the life cycle) and the means by which the environmental existence of the chemical can be terminated as expeditiously as possible and without further harm to the environment. The findings may not always be positive but must be given serious consideration in terms of as near-as-possible efficient remediation (that is, as near as possible complete removal) of the chemical and any products of chemical transformation.

When a chemical (or a mixture of chemicals) is released into the environment, the issues that need to be considered: (1) the toxicity of the chemical, (2) the concentration of the released chemical, (3) if a mixture is released, the concentration of the toxic chemical in the mixture, (4) the potential of the chemical to migrate to other sites, (5) the potential of the chemical to produce a toxic degradation product, (6) whether or not the toxicity is lower or higher than the toxicity of the released chemical, (7) the potential of the toxic degradation product or products to migrate to other sites, (8) the persistence of the chemical in an ecosystem, (9) the persistence of any toxic degradation product in an ecosystem (10) the potential for the toxic degradation product to degrade even further into harmful or nonharmful constituents and the rate of degradation, and (11) the degree to which the chemical or any degradation product of the chemical can accumulate in an ecosystem. Other factors that may be appropriate because of site specificity may also be considered—this list is not meant to be complete but does serve to indicate the types of issues that must be given serious consideration preferably before a spill or discharge of a chemical into the environment.

Thus, in order to complete such a list and monitor the behavior and effects of chemicals in an ecosystem, an understanding of chemical transformation processes in which a disposed or discharged chemical might particulate is valuable. Typically, chemical transformation processes change the chemical composition and structure of the discharged chemical which can change the properties (and possibly the toxicity) of the chemical and influence behavior and life cycle of the chemical in the environment.

As an example of chemical transformation of chemicals in the environment where weathering processes (such as oxidative processes and physical-change processes) are ever present and include such phenomena as (1) evaporation and sublimation, in which part of the chemical or part of a mixture can vaporize into the atmosphere, (2) leaching, in which part of the chemical or part of a mixture is transferred to the aqueous phase through dissolution, (3) entrainment, in which part of the chemical or part of a mixture is subject to physical transport along with the aqueous phase, (4) chemical oxidation, in which part of the chemical or part of a mixture is oxidized to a derivative or to a completely different chemical, and (5) microbial degradation, in which part of the chemical or part of a mixture is converted by bacteria to a derivative or to a completely different chemical. Furthermore, the rate of transformation of the chemical is highly dependent on environmental conditions. Unfortunately, the database on such transformations and that available on the composition of spilled chemicals that have been transformed in the environment are limited.

However, the various chemical transformation processes, which influence the presence and the analysis of chemicals at a particular site, although often represented by simple (and convenient) chemical equations, can be very complex and the true nature of the chemical transformation process is difficult to elucidate. The extent of transformation is dependent on many factors including the (1) the properties of the chemical, (2) the geology of the site, (3) the climatic conditions, such as temperature, oxygen levels, and moisture, (4) the type of microorganisms present at the site, and (5) any other environmental conditions that can influence the life cycle of the chemical. In fact, the primary factor controlling the extent of chemical transformation is the molecular composition of the chemical contaminant.

However, it must be reemphasized that a chemical deposited into the environment has a potential to undergo transformation (unless the chemical is a persistent pollutant) to another chemical form which is still of concern in terms of toxicity. Moreover, when released into the environment, the fate of chemicals depends on the physical and chemical properties of the chemical(s) and the ability of these chemicals to undergo transformation to products—that is, the reactivity of the chemical. In addition, it is not only the structure of the chemical deposited into the environment but also the chemical forms that can result in chemical transformation, which is the result of the chemicals undergoing weathering (oxidation) and other environmental effects,. Thus, chemicals that are not directly toxic to an environmental floral species and faunal species (including humans) at current environmental concentrations can become capable of causing environmental damage after chemical transformation has occurred.

Knowledge of the principles underlying the fate and transport of chemicals in the environment allows problems ranging from local to global scales to be defined and analyzed. Most physical transport of chemicals in the environment occurs in the fluids, air and water. There are primarily two kinds of physical processes by which chemicals are transported in these fluids: bulk movement of fluids from one location to another, and random (or seemingly random) mixing processes within the fluids. Another common example that can be treated as a physical transformation is the settling of suspended sediment particles. Although settling does not actually transform the sediment into something else, it does remove sediment from our control volume by depositing it on the river bed. This process can be expressed mathematically by heterogeneous transformation equations at the river bed; hence, the consideration of sedimentation as a transformation (Chapter 5).

4.1 Advection

The first type of process, *advection*, is due to bulk, large-scale movement of air or water, as seen in blowing wind and flowing streams. This bulk advective movement, resulting in chemical transport, passively carries a chemical present in air or water. In the second type of transport process, a chemical is transported from one location in the air or water where its concentration is relatively high to another location where its concentration is lower, due to random motion of the chemical molecules (*molecular diffusion*), random motion of the air or water that carries the chemical (*turbulent diffusion*), or a combination of the two. Transport by such random motions, also called *diffusive* transport, is often termed as being *Fickian*. Sometimes the motions of the fluid are not entirely random; they have a discernible pattern, but it is too complex to characterize. In this situation, the mass transport process is called *dispersion*, and it is also commonly treated as a Fickian transport. In a given amount of time, the distances over which mass is carried by Fickian transport (molecular diffusion, turbulent diffusion, and dispersion) are usually not as great as those covered by advection.

Thus, advection is the passive movement of a chemical in bulk transport media either within the same medium (intraphase or homogeneous transport) or between different media (interphase or heterogeneous transport). Examples of homogeneous advection include transport of a chemical in air on a windy day or a chemical dissolved in water moving in a flowing stream, in surface runoff (nonpoint source), or in a discharge effluent (point source). Examples of heterogeneous advection include the deposition of a discharged chemical sorbed to a suspended particle that settles to the bottom, atmospheric deposition to soil or water, and even ingestion of contaminated particles or food by an organism (i.e., bioaccumulation). Advection is not influenced by diffusion and can transport a chemical either in the same or opposite direction as diffusion and is often called nondiffusive transport.

As with source emissions, advection of air and water can vary substantially with time and space within a given environmental compartment. Advection in a stream might be several orders of magnitude higher during a large rain event compared to a prolonged dry period, while at one point in time, advection within a stagnant pool might be several orders of magnitude lower than a connected stream. In surface waters advective currents often dominate the transport of toxicants, and they can be estimated from hydrodynamic models or current measurements. In many cases advective flow can be approximated by the volume of water exchanged per unit time by assuming conservation of mass and measuring flow into or out of the system.

In water bodies that experience density stratification (i.e., thermocline) separate advective models or residence times can be used for each water layer. In air, advection also dominates the transport of chemicals, with air currents being driven by pressure gradients. The direction and magnitude of air velocities are recorded continuously in many areas, and daily, seasonal, or annual means can be used to estimate advective air flow.

Heterogeneous advective transport involves a secondary phase within the bulk advective phase, such as when a particle in air or water acts as a carrier of a chemical. In many cases, heterogeneous advection is treated in the same manner as homogeneous advection, as long as the flow rate of the secondary phase and the concentration of chemical in the secondary phase are known. Although the flow rate of particles is much lower than that of water, the concentration of the chemical is much higher in the suspended particles than dissolved in the water. In soil and sedimentary systems, colloidal particles (often macromolecular detritus) can play a very important role in heterogeneous advective transport because they have greater mobility than larger particles, and they often have greater capacity to sorb many toxicants because of the higher organic carbon content and higher surface area/mass ratio. In highly contaminated sites, organic cosolvents can be present in the water (usually groundwater) and act as a high-capacity and high-efficiency carrier of toxicants through heterogeneous advection in the water.

In addition to advection of particles with flowing water, aqueous phase heterogeneous transport includes particle settling, resuspension, burial in bottom sediments, and mixing of bottom sediments. Particle settling can be an important mechanism for transporting hydrophobic toxicants from the water to the bottom sediments. Modeling this process can be as simple as using an overall mass transfer coefficient or can include rigorous modeling of particles with different size, density, and organic carbon content. Resuspension of bottom sediments occurs when sufficient energy is transferred to the sediment bed from advecting water, internal waves, boats, dredging, fishing, and the movement of sediment dwelling organisms (i.e., bioturbation).

Heterogeneous advective transport in air occurs primarily through the absorption of chemicals into falling water droplets (wet deposition) or the sorption of chemicals into solid particles that fall to earth's surface (dry deposition). Under certain conditions both processes can be treated as simple first-order advective transport using a flow rate and concentration in the advecting medium. For example, wet deposition is usually characterized by a washout coefficient that is proportional to rainfall intensity.

4.2 Diffusion

Diffusion is the transport of a chemical by random motion due to a state of disequilibrium. For example, diffusion causes the movement of a chemical within a phase (e.g., water) from a location of relatively high concentration to a place of lower concentration until the chemical is homogeneously distributed throughout the phase. Also, transport of a chemical by diffusion will drive a chemical between media (e.g., water and air) until the equilibrium concentrations are reached and, thus, the chemical potential or fugacity are equal in each phase.

Transport of a chemical by diffusion within a phase can result from random (thermal) motion of the chemical (molecular diffusion), the random turbulent mixing of the transport medium (turbulent diffusion), or a combination of both. Turbulent diffusion usually dominates the diffusive (but not necessarily the advective) chemical transport in air and water due to the turbulent motions or eddies that are common in nature. In porous media (sediment and soil) the water velocities are typically too low to create eddies, but random mixing still occurs as water tortuously flows around particles (sometime referred to as dispersion).

Although different physical mechanisms can cause diffusive mixing, they all cause a net transport of a chemical from areas of higher concentration to areas of lower concentration. All diffusive processes are also referred to as *Fickian* transport because they all can be described mathematically by Fick's first law, which states that the flow (or flux) of a chemical (N, g/h) is proportional to its concentration gradient (dC/dx):

$$N = -DA(dC/dx)$$

In this equation, D is the diffusivity or mass transfer coefficient (m^2/h), A is the area through which the chemical is passing (m^2), C is the concentration of the diffusing chemical (g/m^3), and x is the distance being considered (m). The negative sign is simply the convention that the direction of diffusion is from high to low concentration (diffusion is positive when dC/dx is negative). Note that many scientists and texts define diffusion as an area-specific process with units of g/m^2 h and thus the area term (A) is not included in the diffusion equation. This is simply an alternative designation that describes transport as a flux density (g/m^2 h) rather than as a flow (g/h). In either case the diffusion equation can be integrated numerically and even expressed in three dimensions using vector notation. However, for most environmental situations, there may be no accurate estimate of D or dx, so we combine the two into a one-dimensional mass transfer coefficient (k_M) with units of velocity (m/h). Thus, the chemical flux is then the product of this velocity, area, and concentration:

$$N = -k_M AC.$$

Mass transfer coefficients can be derived separately for molecular diffusion, turbulent diffusion, and dispersion in porous media, and all three terms can be added to the chemical flux equation.

The transport of a chemical between phases is sometimes treated as a third category of transport processes or even as a transformation reaction. Interphase or intermedia transport is not a transformation reaction because the chemical is moving only between phases; it is not reacting with anything or changing its chemical structure. Instead, intermedia transport is simply driven by diffusion between two phases. When a chemical reaches an interface such as air–water, particle–water, or (biological) membrane–water, two diffusive regions are created at either side of the interface, such as when a chemical passes through two stagnant boundary layers by molecular diffusion, while the two bulk phases are assumed to be homogeneously mixed. This can be represented by the use of a first-order function of the concentration gradient in the two phases, where the mass transfer coefficient will depend only on the molecular diffusivity of the chemical in each phase and the thickness of the boundary layers.

Diffusive transport between phases can be described mathematically as the product of the departure from equilibrium and a kinetic term:

$$N = kA(C_1 - C_2K_{12})$$

In this equation, N is the transport rate (g/h), k is a transport rate coefficient (m/h), A is the interfacial area (m^2), C_1 and C_2 are the concentrations in the two phases, and K_{12} is the equilibrium partition coefficient. At equilibrium K_{12} is equal to C_1/C_2, the term describing the departure from equilibrium ($C_1 - C_2K_{12}$) becomes zero, and thus the net rate of transfer also is zero. The partition coefficients are readily obtained from thermodynamic data and equilibrium partitioning experiments. The transport rate coefficients are usually estimated from the transport rate equation itself by measuring intermedia transport rates (N) under controlled laboratory conditions (temperature, wind, and water velocities) at known values of A, C_1, C_2, and K_{12}. The knowledge that many interfacial regions have reached or are near equilibrium leads to the assumption that equilibrium exists at the interface and, thus, the net transport rate is zero and the phase distribution of a chemical is simply described by its equilibrium partition coefficient.

4.3 Concentration

Another important parameter in environment fate and transport of a discharged chemical is chemical *concentration* (C). The concentration of a chemical is a measure of the amount of that chemical in a specific volume or mass of air, water, soil, or other material. Not only is concentration a key quantity in fate and transport equations; a chemical's concentration in an environmental medium also in part determines the magnitude of its biological effect. Most laboratory analysis methods measure concentration. The choice of units for concentration depends in part on the medium and in part on the process that is being measured or described.

No matter which units are used, however, concentration is the relevant measure for predictions of the effect of a chemical on an organism or the environment. Concentration is also critical in one of the most important concepts of environmental fate and transport: the bookkeeping of chemical mass in the environment.

Three possible outcomes exist for a chemical present at a specific location in the environment at a particular time: the chemical can remain in that location, can be carried elsewhere by a transport process, or it can be eliminated through transformation into another chemical. This very simple observation is known as *mass balance*.

Mass balance is a concept around which an analysis of the fate and transport of any environmental chemical can be organized. Mass balance also serves as a check on the completeness of knowledge of a chemical's behavior. The key elements in a mass balance are: (1) a defined control volume, (2) a knowledge of input and output which cross the boundary of the control volume, (3) a knowledge of the transport characteristics within the control volume and across its boundaries, and (4) a knowledge of the reaction kinetics within the control volume.

A control volume can be as small as an infinitesimal thin slice of water in a swiftly flowing stream or as large as the entire body of oceans on the planet earth. The important point is that the boundaries are clearly defined with respect to their location (element 1) so that the volume is known and mass fluxes across the boundaries can be determined (element 2).

Within the control volume, the transport characteristics (degree of mixing) must be known either by measurement or an estimate based on the hydrodynamics of the system. Likewise, the transport in adjacent or surrounding control volumes may contribute mass to the control volume (much as smoke can travel from another room to your room within a house), so transport across the boundaries of the control volume must be known or estimated (element 3).

Knowledge of the chemical, biological, and physical reactions that the substance can undergo within the control volume (element 4) is the subject of next part of this section. If there were no degradation reactions taking place in aquatic ecosystems, every pollutant which was ever released to the environment would still be here to haunt us. Fortunately, there are natural purification processes that serve to assimilate some wastes and to ameliorate aquatic impacts. We must understand these reactions from a quantitative viewpoint in order to assess the potential damage to the environment from pollutant discharges and to allocate allowable limits for these discharges.

A mass balance is simply an accounting of mass inputs, outputs, reactions, and accumulation as described by the following equation.

$$\text{Accumulation within the control volume} = \text{Mass inputs} - \text{Mass outputs} \pm \text{Reactions} \quad (10.1)$$

If a chemical is being formed within the control volume (such as the combination of two reactants to form a product, $A + B \rightarrow P$), then the algebraic sign in front of the "Reactions" term is positive when writing a mass balance for the product. If the chemical is being destroyed or degraded within the control volume, then the algebraic sign of the "Reactions" term is negative. If the chemical is conservative (i.e., nonreactive or inert), then the "Reactions" term is zero.

$$\text{Accumulation} = \text{Inputs} - \text{Outputs} \pm \text{Reactions} \tag{10.2}$$

If the system is at steady state (i.e., no change in concentration in the system and outputs are simply equal to inputs plus or minus reactions), Outputs = Inputs ± Reactions.

REFERENCES

Borda, M.J., Sparks, D.L., 2008. Kinetics and mechanisms of sorption-desorption in soils: a multiscale assessment. In: Violette, A., Huang, P.M., Gadd, G.M. (Eds.), Biophysical-chemical Processes of Heavy Metals and Metalloids in Soil Environments. John Wiley & Sons Inc., Hoboken, New Jersey, pp. 97–124.

Centi, G., Ciambelli, P., Perathoner, S., Russo, P., 2002. Environmental catalysis: trends and outlook. Catalysis Today 75 (1–4), 3015.

Chiou, C.T., 2002. Partition and Adsorption of Organic Contaminants in Environmental Systems. John Wiley & Sons Inc., Hoboken, New Jersey.

Chorover, J., Brusseau, M.L., 2009. Kinetics of sorption-desorption. In: Brantley, S.L., Kubicki, J.D., White, A.F. (Eds.), Kinetics of Water-rock Interaction. Springer, New York, pp. 109–149.

Delle Site, A., 2001. Factors affecting sorption of organic compounds in natural sorbent/water systems and sorption coefficients for selected pollutants. A review. Journal of Physical and Chemical Reference Data 30 (1), 187–439.

George, C., Ammann, M., D'Anna, B., Donaldson, D.J., Nizkorodov, S.A., 2015. Heterogeneous photochemistry in the atmosphere. Chemical Reviews 115, 4218–4258.

Grassian, V.H., 2005. Environmental Catalysis. CRC Press, Taylor & Francis Group, Boca Raton, Florida.

Habashi, F., 1994. Conversion reactions in chemistry. Journal of Chemical Education 71 (2), 130.

Johnston, C.T., Tombácz, E., 2002. Surface chemistry of soil minerals. In: Dixon, J.B., Schulze, D.G. (Eds.), Soil Mineralogy with Environmental Applications. Soil Science Society of America, Madison, WI, pp. 37–67.

Liu, P., Kendelewicz, T., Brown, G.E., Nelson, E.J., Chambers, 1998. Reaction of water vapor with α-Al_2O_3(0001) and α -Fe_2O_3(0001) surfaces: synchrotron X-ray photoemission studies and thermodynamic calculations. Surface Science 417, 53–65.

Majumder, E.L.-W., Wall, J.D., 2017. Uranium bio-transformations: chemical or biological processes? Open Journal of Inorganic Chemistry 7, 28–60. http://www.scirp.org/journal/PaperInformation.aspx?paperID=75871.

Mokhatab, S., Poe, W.A., Speight, J.G., 2006. Handbook of Natural Gas Transmission and Processing. Elsevier, Amsterdam, Netherlands.

Nester, E.W., Anderson, D.G., Roberts Jr., C.E., Pearsall, N.N., Nester, M.T., 2001. Microbiology: A Human Perspective, third ed. McGraw-Hill, New York.

Petrucci, R.H., Herning, G.E., Madura, J., Bissonnette, C., 2010. General Chemistry: Principles and Modern Application, eleventh ed. Prentice Hall, Upper Saddle River, New Jersey.

Qafoku, N.P., Ranst, E.V., Noble, A.D., Baert, G., 2004. Variable charge soils: their mineralogy, chemistry and management. In: Sparks, D.L. (Ed.), Advances in Agronomy, vol. 84. Academic Press, New York, pp. 159–215.

Rahm, S., Green, N., Norrgran, J., Bergman, Å., 2005. Hydrolysis of environmental contaminants as an experimental tool for indication of their persistency. Environmental Science and Technology 39 (9), 3128–3133.

Shea, D., 2004. Transport and fate of toxicants in the environment. In: Hodgson, E. (Ed.), A Textbook of Modern Toxicology 3rd Edition. John Wiley & Sons Inc., Hoboken, New Jersey.

Speight, J.G., 2005. Environmental Analysis and Technology for the Refining Industry. John Wiley & Sons Inc., Hoboken, New Jersey.

Speight, J.G., 2007. Natural Gas: A Basic Handbook. GPC Books. Gulf Publishing Company, Houston, Texas.

Speight, J.G., 2011. Handbook of Industrial Hydrocarbon Processes. Gulf Professional Publishing, Elsevier, Oxford, United Kingdom.

Speight, J.G., Arjoon, K.K., 2012. Bioremediation of Petroleum and Petroleum Products. Scrivener Publishing, Beverly, Massachusetts.

Speight, J.G., Lee, S., 2000. Environmental Technology Handbook, second ed. Taylor & Francis, New York (Also, CRC Press, Taylor and Francis Group, Boca Raton, Florida).

Summers, A.O., Silver, S., 1978. Microbial transformation of metals. Annual Review of Microbiology 32, 1–709.

Thompson, A., Goyne, K.W., 2012. Introduction to the sorption of chemical constituents in soils. Nature Education Knowledge 4 (4), 7.

Tinsley, I.J., 2004. Chemical Concepts in Pollutant Behavior, second ed. John Wiley & Sons Inc., Hoboken, New Jersey.

Warneck, P., 2000. Chemistry of the Natural Atmosphere, second ed. Academic Press Inc., New York.

Wayne, C.E., Wayne, R.P., 2005. Photochemistry. Oxford University Press, Oxford, United Kingdom.

Zhou, Y.-F., Haynes, R.J., 2010. Sorption of heavy metals by and components of solid wastes: significance to use of wastes as low-cost adsorbents and immobilizing agents. Critical Reviews in Environmental Science and Technology 40 (11), 909–977.

FURTHER READING

NRC, 2014. Physicochemical Properties and Environmental Fate: A Framework to Guide Selection of Chemical Alternatives. National Research Council, Washington, DC.

Part III

Conversion Tables and Glossary

Conversion Tables

1. AREA

1 square centimeter (1 cm^2) = 0.1550 square inches
1 square meter (1 m^2) = 1.1960 square yards
1 hectare = 2.4711 acres
1 square kilometer (1 km^2) = 0.3861 square miles
1 square inch (1 $inch^2$) = 6.4516 square centimeters
1 square foot (1 ft^2) = 0.0929 square meters
1 square yard (1 yd^2) = 0.8361 square meters
1 acre = 4046.9 square meters
1 square mile (1 mi^2) = 2.59 square kilometers

2. CONCENTRATION CONVERSIONS

1 part per million (1 ppm) = 1 microgram per liter (1 μg/L)
1 microgram per liter (1 μg/L) = 1 milligram per kilogram (1 mg/kg)
1 microgram per liter (μg/L) $\times 6.243 \times 10^8$ = 1 lb per cubic foot (1 lb/ft^3)
1 microgram per liter (1 μg/L) $\times 10^{-3}$ = 1 milligram per liter (1 mg/L)
1 milligram per liter (1 mg/L) $\times 6.243 \times 10^5$ = 1 pound per cubic foot (1 lb/ft^3)
1 gram mole per cubic meter (1 g mol/m^3) $\times 6.243 \times 10^5$ = 1 pound per cubic foot (1 lb/ft^3)
10,000 ppm = 1% w/w
1 ppm hydrocarbon in soil × 0.002 = 1 lb of hydrocarbons per ton of contaminated soil

3. NUTRIENT CONVERSION FACTOR

1 pound, phosphorus × 2.3 (1 lb P × 2.3) = 1 pound, phosphorous pentoxide (1 lb P_2O_5)
1 pound, potassium × 1.2 (1 lb K × 1.2 = 1 pound, potassium oxide (1 lb K_2O)

4. TEMPERATURE CONVERSIONS

°F = (°C × 1.8) + 32
°C = (°F − 32)/1.8
(°F − 32) × 0.555 = °C
Absolute zero = −273.15°C
Absolute zero = −459.67°F

5. SLUDGE CONVERSIONS

1700 lbs wet sludge = 1 yd^3 wet sludge
1 yd^3 sludge = wet tons/0.85
Wet tons sludge × 240 = gallons sludge
1 wet ton sludge × % dry solids/100 = 1 dry ton of sludge

6. VARIOUS CONSTANTS

Atomic mass	mu = $1.6605402 \times 10^{-27}$
Avogadro number	N = 6.0221367×10^{23} mol^{-1}
Boltzmann constant	k = 1.380658×10^{-23} J/K
Elementary charge	e = $1.60217733 \times 10^{-19}$ C
Faraday constant	F = 9.6485309 × 104 C/mol
Gas (molar) constant	R =∼ 8.314510 J/mol K
	= 0.08205783 L atm/mol K
Gravitational acceleration	g = 9.80665 m/s^2
Molar volume of an ideal gas at 1 atm and 25°C	$V_{ideal\ gas}$ = 24.465 L/mol
Planck constant	h = $6.6260755 \times 10^{-34}$ J s
Zero, Celsius scale	0°C = 273.15K

7. VOLUME CONVERSION

Barrels (petroleum, U S) to Cu feet multiply by 5.6146
Barrels (petroleum, U S) to Gallons (U S) multiply by 42
Barrels (petroleum, U S) to Liters multiply by 158.98
Barrels (US, liq.) to Cu feet multiply by 4.2109
Barrels (US, liq.) to Cu inches multiply by 7.2765×10^3
Barrels (US, liq.) to Cu meters multiply by 0.1192
Barrels (US, liq.) to Gallons multiply by (US, liq.) 31.5
Barrels (US, liq.) to Liters multiply by 119.24
Cubic centimeters to Cu feet multiply by 3.5315×10^{-5}
Cubic centimeters to Cu inches multiply by 0.06102

Cubic centimeters to Cu meters multiply by 1.0×10^{-6}
Cubic centimeters to Cu yards multiply by 1.308×10^{-6}
Cubic centimeters to Gallons (US liq.) multiply by 2.642×10^{-4}
Cubic centimeters to Quarts (US liq.) multiply by 1.0567×10^{-3}
Cubic feet to Cu centimeters multiply by 2.8317×10^{4}
Cubic feet to Cu meters multiply by 0.028317
Cubic feet to Gallons (US liq.) multiply by 7.4805
Cubic feet to Liters multiply by 28.317
Cubic inches to Cu cm multiply by 16.387
Cubic inches to Cu feet multiply by 5.787×10^{-4}
Cubic inches to Cu meters multiply by 1.6387×10^{-5}
Cubic inches to Cu yards multiply by 2.1433×10^{-5}
Cubic inches to Gallons (US liq.) multiply by 4.329×10^{-3}
Cubic inches to Liters multiply by 0.01639
Cubic inches to Quarts (US liq.) multiply by 0.01732
Cubic meters to Barrels (US liq.) multiply by 8.3864
Cubic meters to Cu cm multiply by 1.0×10^{6}
Cubic meters to Cu feet multiply by 35.315
Cubic meters to Cu inches multiply by 6.1024×10^{4}
Cubic meters to Cu yards multiply by 1.308
Cubic meters to Gallons (US liq.) multiply by 264.17
Cubic meters to Liters multiply by 1000
Cubic yards to Bushels (Brit.) multiply by 21.022
Cubic yards to Bushels (US) multiply by 21.696
Cubic yards to Cu cm multiply by 7.6455×10^{5}
Cubic yards to Cu feet multiply by 27
Cubic yards to Cu inches multiply by 4.6656×10^{4}
Cubic yards to Cu meters multiply by 0.76455
Cubic yards to Gallons multiply by 168.18
Cubic yards to Gallons multiply by 173.57
Cubic yards to Gallons multiply by 201.97
Cubic yards to Liters multiply by 764.55
Cubic yards to Quarts multiply by 672.71
Cubic yards to Quarts multiply by 694.28
Cubic yards to Quarts multiply by 807.90
Gallons (US liq.) to Barrels (US liq.) multiply by 0.03175
Gallons (US liq.) to Barrels (petroleum, US) multiply by 0.02381
Gallons (US liq.) to Bushels (US) multiply by 0.10742
Gallons (US liq.) to Cu centimeters multiply by 3.7854×10^{3}
Gallons (US liq.) to Cu feet multiply by 0.13368
Gallons (US liq.) to Cu inches multiply by 231
Gallons (US liq.) to Cu meters multiply by 3.7854×10^{-3}
Gallons (US liq.) to Cu yards multiply by 4.951×10^{-3}

Gallons (US liq.) to Gallons (wine) multiply by 1.0
Gallons (US liq.) to Liters multiply by 3.7854
Gallons (US liq.) to Ounces (US fluid) multiply by 128.0
Gallons (US liq.) to Pints (US liq.) multiply by 8.0
Gallons (US liq.) to Quarts (US liq.) multiply by 4.0
Liters to Cu centimeters multiply by 1000
Liters to Cu feet multiply by 0.035315
Liters to Cu inches multiply by 61.024
Liters to Cu meters multiply by 0.001
Liters to Gallons (US liq.) multiply by 0.2642
Liters to Ounces (US fluid) multiply by 33.814

8. WEIGHT CONVERSION

1 ounce (1 ounce) = 28.3495 g (18.2495 g)
1 pound (1 lb) = 0.454 kg
1 pound (1 lb) = 454 g (454 g)
1 kilogram (1 kg) = 2.20462 pounds (2.20462 lb)
1 stone (English) = 14 pounds (14 lb)
1 ton (US; 1 short ton) = 2000 lbs
1 ton (English; 1 long ton) = 2240 lbs
1 metric ton = 2204.62262 pounds
1 tonne = 2204.62262 pounds

9. OTHER APPROXIMATIONS

14.7 pounds per square inch (14.7 psi) = 1 atmosphere (1 atm)
1 kilopascal (kPa) $\times 9.8692 \times 10^{-3}$ = 14.7 pounds per square inch (14.7 psi)
1 yd^3 = 27 ft^3
1 US gallon of water = 8.34 lbs
1 imperial gallon of water = 10 lbs
1 ft^3 = 7.5 gallon = 1728 cubic inches = 62.5 lbs
1 yd^3 = 0.765 m^3
1 acre-inch of liquid = 27,150 gallons = 3.630 ft^3
1-foot depth in 1 acre (in situ) = 1613 × (20%−25% excavation factor) = ~2000 yd^3
1 yd^3 (clayey soils-excavated) = 1.1−1.2 tons (US)
1 yd^3 (sandy soils-excavated) = 1.2−1.3 tons (US)
Pressure of a column of water in psi = height of the column in feet by 0.434.

Glossary

Abiotic: Not associated with living organisms; synonymous with *abiological.*

Abiotic transformation: The process in which a substance in the environment is modified by nonbiological mechanisms.

Absorption: The penetration of atoms, ions, or molecules into the bulk mass of a substance.

Abyssal zone: The portion of the ocean floor below 3281–6561 ft where light does not penetrate and where temperatures are cold, and pressures are intense; this zone lies seaward of the continental slope and covers approximately 75% of the ocean floor; the temperature does not rise above 4°C (39°F); since oxygen is present, a diverse community of invertebrates and fishes do exist, and some have adapted to harsh environments such as hydrothermal vents of volcanic creation.

Acceleration: A measure of how fast velocity is changing; so we can think of it as the change in velocity over change in time. The most common use of acceleration is acceleration due to gravity, which can also appear as the gravitational constant (9.8 m/s^2).

Acetic acid (CH_3CO_2H): Trivial name for ethanoic acid, formed by the oxidation of ethanol with potassium permanganate.

Acetone (CH_3COCH_3): Trivial name for propanone, formed by the oxidation of 2-propanol with potassium permanganate.

Achiral molecule: A molecule that does not contain a stereogenic carbon; an achiral molecule has a plane of symmetry and is superimposable on its mirror image.

Acid: A chemical containing the carboxyl group and capable of donating a positively charged hydrogen atom (proton, H^+) or capable of forming a covalent bond with an electron pair; an acid increases the hydrogen ion concentration in a solution, and it can react with certain metals.

Acid anhydride: An organic compound that reacts with water to form an acid.

Acid–base partitioning: The tendency for acids to accumulate in basic fluid compartments and bases to accumulate in acidic regions; also called *pH partitioning*.

Acid/base reaction: A reaction in which an acidic hydrogen atom is transferred from one molecule to another.

Acidic: A solution with a high concentration of H^+ ions.

Acidity: The capacity of the water to neutralize OH^-.

Acidophiles: Metabolically active in highly acidic environments, and often have a high heavy metal resistance.

Acids, bases, and salts: Many inorganic compounds are available as acids, bases, or salts.

Acyclic: A compound with straight or branched carbon–carbon linkages but without cyclic (ring) structures.

Addition reaction: A reaction where a reagent is added across a double or triple bond in an organic compound to produce the corresponding saturated compound.

Additivity: The effect of the combination equals the sum of individual effects.

Adhesion: The degree to which oil will coat a surface, expressed as the mass of oil adhering per unit area. A test has been developed for a standard surface that gives a semi-quantitative measure of this property.

Adsorbent (sorbent): The solid phase or substrate onto which the sorbate adsorbs.

Adsorption: The retention of atoms, ions, or molecules on to the surface of another substance; the two-dimensional accumulation of an adsorbate at a solid surface. In the case of surface precipitation; also used when there is diffusion of the sorbate into the solid phase.

Advection: A process due to the bulk, large-scale movement of air or water, as seen in blowing wind and flowing streams.

Aerobe: An organism that needs oxygen for respiration and hence for growth.

Aerobic: In the presence of, or requiring, oxygen; an environment or process that sustains biological life and growth, or occurs only when free (molecular) oxygen is present.

Aerobic bacteria: Any bacteria requiring free oxygen for growth and cell division.

Aerobic conditions: Conditions for growth or metabolism in which the organism is sufficiently supplied with oxygen.

Aerobic respiration: The process whereby microorganisms use oxygen as an electron acceptor.

Aerosol: A colloidal-sized atmospheric particle.

Alcohol: An organic compound with a carbon bound to a hydroxyl (−OH) group; a hydroxyl group attached to an aromatic ring is called a phenol rather than an alcohol; a compound in which a hydroxyl group (−OH) is attached to a saturated carbon atom (e.g., ethyl alcohol, C_2H_5OH).

Aldehyde: An organic compound with a carbon bound to a −(CO)−H group; a compound in which a carbonyl group is bonded to one hydrogen atom and to one alkyl group [RC(O)H].

Algae: Microscopic organisms that subsist on inorganic nutrients and produce organic matter from carbon dioxide by photosynthesis.

Aliphatic compound: Any organic compound of hydrogen and carbon characterized by a linear chain or branched chain of carbon atoms; three subgroups of such compounds are alkanes, alkenes, and alkynes.

Alkali metal: A metal in Group IA in the periodic table; an active metal which may be used to react with an alcohol to produce the corresponding metal alkoxide and hydrogen gas.

Alkalinity: The capacity of water to accept H^+ ions (protons).

Alkaliphiles: Organisms that have their optimum growth rate at least 2 pH units above neutrality.

Alkalitolerants: Organisms that are able to grow or survive at pH values above 9, but their optimum growth rate is around neutrality or less.

Alkane (paraffin): A group of *hydrocarbons* composed of only carbon and hydrogen with no double bonds or aromaticity. They are said to be "saturated" with hydrogen. They may by straight-chain (normal), branched, or cyclic. The smallest alkane is methane (CH_4), the next, ethane (CH_3CH_3), then propane ($CH_3CH_2CH_3$), and so on.

Alkanes: The homologous group of linear (acyclic) aliphatic hydrocarbons having the general formula C_nH_{2n+2}; alkanes can be straight chains (linear), branched chains, or ring structures; often referred to as paraffins.

Alkene (olefin): An unsaturated *hydrocarbon*, containing only hydrogen and carbon with one or more double bonds, but having no aromaticity. *Alkenes* are not typically found in crude oils, but can occur as a result of heating.

Alkenes: Acyclic branched or unbranched hydrocarbons having one carbon−carbon double bond (−CC−) and the general formula C_nH_{2n}; often referred to as olefins.

Alkoxide: An ionic compound formed by removal of hydrogen ions from the hydroxyl group in an alcohol using a reactive metal such as sodium or potassium.

Alkoxy group (RO^-): A substituent containing an alkyl group linked to an oxygen.

Alkyl benzene (C_6H_5-R): A benzene ring that has one alkyl group attached; the alkyl group (except quaternary alkyl groups) is susceptible to oxidation with hot $Kmno_4$ to yield benzoic acid ($C_6H_5CO_2H$).

Alkyl groups: A hydrocarbon functional group (C_nH_{2n+1}) obtained by dropping one hydrogen from fully saturated compound; e.g., methyl ($-CH_3$), ethyl ($-CH_2CH_3$), propyl ($-CH_2CH_2CH_3$), or isopropyl [$(CH_3)_2CH-$].

Alkyl radicals: Carbon-centered radicals derived formally by removal of one hydrogen atom from an alkane, for example, the ethyl radical (CH_3CH_2).

Alkynes: The group of acyclic branched or unbranched hydrocarbons having a carbon–carbon triple bond ($-C{\equiv}C-$).

Ambient: The surrounding environment and prevailing conditions.

Amide: An organic compound that contains a carbonyl group bound to nitrogen; the simplest amides are formamide ($HCONH_2$) and acetamide (CH_3CONH_2).

Amine: An organic compound that contains a nitrogen atom bound only to carbon and possibly hydrogen atoms; examples are methylamine, CH_3NH_2; dimethylamine, CH_3NHCH_3; and trimethylamine, $(CH_3)_3N$.

Amino acid: A molecule that contains at least one amine group ($-NH_2$) and at least one carboxylic acid group ($-COOH$); when these groups are both attached to the same carbon, the acid is an α-amino acid—α-amino acids are the basic building blocks of proteins.

Amorphous solid: A noncrystalline solid having no well-defined ordered structure.

Amphoteric molecule: A molecule that behaves both as an acid and a base, such as hydroxy pyridine:

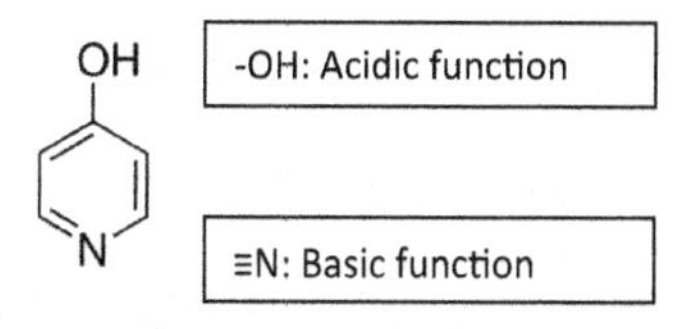

Anaerobe: An organism that does not need free form of oxygen for growth. Many anaerobes are even sensitive to free oxygen.

Anaerobic: A biologically mediated process or condition not requiring molecular or free oxygen; relating to a process that occurs with little or no oxygen present.

Anaerobic bacteria: Any bacteria that can grow and divide in the partial or complete absence of oxygen.

Anaerobic respiration: The process whereby microorganisms use a chemical other than oxygen as an electron acceptor; common substitutes for oxygen are nitrate, sulfate, and iron.

Analyte: The component of a system to be analyzed—for example, chemical elements or ions in groundwater sample.

Anion: An atom or molecule that has a negative charge; a negatively charged ion.

Anode: The electrode where electrons are lost (oxidized) in redox reactions.

Anoxic: An environment without oxygen.

Antagonism: The effect of the combination is less than the sum of individual effects.

Aphotic zone: The deeper part of the ocean beneath the photic zone, where light does not penetrate sufficiently for photosynthesis to occur.

API Gravity: An American Petroleum Institute measure of *density* for petroleum: API Gravity = [141.5/(specific gravity at 15.6°C) − 131.5]; fresh water has a gravity of 10°API. The scale is commercially important for ranking oil quality; heavy oils are typically <20°API; medium oils are 20−35°API; light oils are 35−45°API.

Aquasphere: The water areas of the earth; also called the hydrosphere.

Aquatic chemistry: The branch of environmental chemistry that deals with chemical phenomena in water.

Aquifer: A water-bearing layer of soil, sand, gravel, rock, or other geologic formation that will yield usable quantities of water to a well under normal hydraulic gradients or by pumping.

Arene: A hydrocarbon that contains at least one aromatic ring.

Aromatic: An organic cyclic compound that contains one or more benzene rings; these can be monocyclic, bicyclic, or polycyclic hydrocarbons and their substituted derivatives. In aromatic ring structures, every ring carbon atom possesses one double bond.

Aromatic ring: An exceptionally stable planar ring of atoms with resonance structures that consist of alternating double and single bonds, such as benzene:

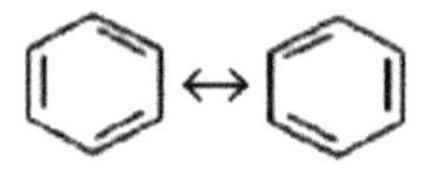

Aromatic compound: A compound containing an aromatic ring; aromatic compounds have strong, characteristic odors.

Aryl: A molecular fragment or group attached to a molecule by an atom that is on an aromatic ring.

Asphaltene fraction: A complex mixture of heavy organic compounds precipitated from crude oil and *bitumen* by natural processes or in laboratory by addition of excess *n*-pentane or *n*-heptane; after precipitation of the *asphaltene fraction*, the remaining oil or *bitumen* consists of *saturates*, *aromatics*, and *resins*.

Assay: Qualitative or (more usually) quantitative determination of the components of a material or system.

Association colloids: Colloids which consist of special aggregates of ions and molecules (micelles).

Asymmetric carbon: A carbon atom covalently bonded to four different atoms or groups of atoms.

Atmosphere: The thin layer of gases that cover surface of the earth; composed of two major components: nitrogen 78.08% and oxygen 20.955 with smaller amounts of argon 0.934%, carbon dioxide 0.035%, neon 1.818×10^{-3}%, krypton 1.14×10^{-4}%, helium 5.24×10^{-4}%, and xenon 8.7×10^{-6}%; may also contain 0.1%−5% water by volume, with a normal range of 1%−3%; the reservoir of gases, moderates the temperature of the earth, absorbs energy and damaging ultraviolet radiation from the sun, transports energy away from equatorial regions and serves as a pathway for vapor-phase movement of water in the hydrologic cycle.

Atomic number: The atomic number is equal to the number of positively charged protons in the nucleus of an atom which determines the identity of the element.

Atomic radius: The relative size of an atom; among the main group of elements, atomic radii mostly decrease from left to right across rows in the periodic table; metal ions are smaller than their neutral atoms, and nonmetallic anions are larger than the atoms from which they are formed; atomic radii are expressed in angstrom units of length (Å).

ATSDR: Agency for Toxic Substances and Disease Registry.

Attenuation: The set of human-made or natural processes that either reduce or appear to reduce the amount of a chemical compound as it migrates away or is disposed from one specific point towards another point in space or time; for example, the apparent reduction in the amount of a chemical in a groundwater plume as it migrates away from its source; degradation, dilution, dispersion, sorption, or volatilization are common processes of attenuation.

Autoignition Temperature (AIT): A fixed temperature above which a flammable mixture is capable of extracting sufficient energy from the environment to self-ignite.

Autotrophs: Organisms or chemicals that use carbon dioxide and ionic carbonates for the C that they require.

Avogadro's number: The number of molecules (6.023×10^{23}) in 1 g-mole of a substance.

Bacteria: Single-celled prokaryotic microorganisms that may be shaped as rods (bacillus), spheres (coccus), or spirals (vibrios, spirilla, spirochetes).

Base: A substance which gives off hydroxide ions (OH^-) in solution.

Basic: Having the characteristics of a base.

Benthic zone: The ecological region at the lowest level of a body of water such as an ocean or a lake, including the sediment surface and some subsurface layers; organisms living in this zone (benthos or benthic organisms) generally live in close relationship with the substrate bottom; many such organisms are permanently attached to the bottom; because light does not penetrate very deep ocean-water, the energy source for the benthic ecosystem is often organic matter from higher up in the water column which sinks to the depths.

Benzene: A colorless liquid formed from both anthropogenic activities and natural processes; widely used in the United States and ranks in the top 20 chemicals used; a natural part of crude oil, gasoline, and cigarette smoke; one of the major components of JP-8 fuel.

Benzoic acid ($C_6H_5CO_2H$): The simplest aromatic carboxylic acid, formed by the vigorous oxidation of alkyl benzene, benzyl alcohol, and benzaldehyde.

Bimolecular reaction: The collision and combination of two reactants involved in the rate-limiting step.

Bioaccumulation: The accumulation of substances, such as pesticides, or other chemicals in an organism; occurs when an organism absorbs a chemical—possibly a toxic chemical—at a rate faster than that at which the substance is lost by catabolism and excretion; the longer the biological half-life of a toxic substance, the greater the risk of chronic poisoning, even if environmental levels of the toxin are not very high; see Biomagnification.

Bioaugmentation: A process in which acclimated microorganisms are added to soil and groundwater to increase biological activity. Spray irrigation is typically used for shallow contaminated soils, and injection wells are used for deeper contaminated soils.

Biochemical oxygen demand (BOD): An important water-quality parameter; refers to the amount of oxygen utilized when the organic matter in a given volume of water is degraded biologically.

Biocide: A chemical substance or microorganism intended to destroy, deter, render harmless, or exert a controlling effect on any harmful organism by chemical or biological means.

Biodegradation: The natural process whereby bacteria or other microorganisms chemically alter and break down organic molecules; the breakdown or transformation of a chemical substance or substances by microorganisms using the substance as a carbon and/or energy source.

Biogeochemical cycle: The pathway by which a chemical moves through biotic (biosphere) and abiotic (atmosphere, aquasphere, lithosphere) compartments of the earth.

Bioinorganic compounds: Natural and synthetic compounds that include metallic elements bonded to proteins and other biological chemistries.

Biological marker (biomarker): Complex organic compounds composed of carbon, hydrogen, and other elements which are found in oil, *bitumen*, rocks, and sediments and which have undergone little or no change in structure from their parent organic molecules in living organisms; typically, biomarkers are isoprenoids, composed of isoprene subunits; biomarkers include compounds such as pristane, phytane, triterpane derivatives, sterane derivatives, and porphyrin derivatives.

Biomagnification: The increase in the concentration of heavy metals (i.e., mercury) or organic contaminants such as chlorinated hydrocarbons, in organisms as a result of their consumption within a food chain/web; an example is the process by which contaminants such as polychlorobiphenyl derivatives (PCBs) accumulate or magnify as they move up the food chain—PCBs concentrate in tissue and internal organs, and as big fish eat little fish, they accumulate all the PCBs that have been eaten by everyone below them in the food chain; can occur as a result of: (1) persistence, in which the chemical cannot be broken down by environmental processes, (2) food chain energetics, in which the concentration of the chemical increases progressively as it moves up a food chain, and (3) a low or nonexistent rate of internal degradation or excretion of the substance that is often due to water-insolubility.

Bioremediation: A treatment technology that uses biological activity to reduce the concentration or toxicity of contaminants: materials are added to contaminated environments to accelerate natural biodegradation.

Biosphere: A term representing all of the living entities on the earth.

Biota: Living organisms that constitute the plant and animal life of a region (Arctic region, temperate region, subtropical region, or tropical region).

Bitumen: A complex mixture of *hydrocarbonaceous constituents* of natural or pyrogenous origin or a combination of both.

Boiling liquid expanding vapor explosion (BLEVE): An event which occurs when a vessel ruptures which contains a liquid at a temperature above its atmospheric pressure boiling point; the explosive vaporization of a large fraction of the vessel contents; possibly followed by the combustion or explosion of the vaporized cloud if it is combustible (similar to a rocket).

Boiling point: The temperature at which a liquid begins to boil—that is, it is the temperature at which the vapor pressure of a liquid is equal to the atmospheric or external pressure. The boiling point distributions of crude oils and petroleum products may be in a range from 30 to in excess of 700°C (86–1290°F).

Breakdown product: A compound derived by chemical, biological, or physical action on a chemical compound; the breakdown is a process which may result in a more toxic or a less toxic compound and a more persistent or less persistent compound than the original compound.

BTEX: The collective name given to benzene, toluene, ethylbenzene, and the xylene isomers (*p*-, *m*-, and *o*-xylene); a group of volatile organic compounds (VOCs) found in petroleum hydrocarbons, such as gasoline, and other common environmental contaminants.

BTX: The collective name given to benzene, toluene, and the xylene isomers (*p*-, *m*-, and *o*-xylene); a group of volatile organic compounds (VOCs) found in petroleum hydrocarbons, such as gasoline, and other common environmental contaminants.

CH_3

benzene toluene

CH_3 CH_3 CH_3

CH_3

CH_3

CH_3

ortho-xylene *meta*-xylene *para*-xylene

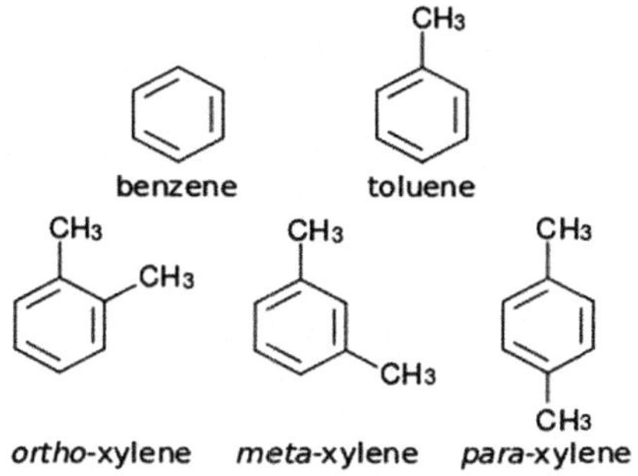

Buffer solution: A solution that resists change in the pH, even when small amounts of acid or base are added.

Carbenium ion: A generic name for carbocation that has at least one important contributing structure containing a tervalent carbon atom with a vacant *p* orbital.

Carbanion: The generic name for anions containing an even number of electrons and having an unshared pair of electrons on a carbon atom (e.g., Cl_3C^-).

Carbon: Element number 6 in the periodic table of elements.

Carbon preference index (CPI): The ratio of odd to even *n*-alkanes; odd/even CPI *alkanes* are equally abundant in petroleum but not in biological material—a CPI near 1 is an indication of petroleum.

Carbon tetrachloride: A manufactured compound that does not occur naturally; produced in large quantities to make refrigeration fluid and propellants for aerosol cans; in the past, carbon tetrachloride was widely used as a cleaning fluid, in industry and dry cleaning businesses, and in the household; also used in fire extinguishers and as a fumigant to kill insects in grain—these uses were stopped in the mid-1960s.

Carbonyl group: A divalent group consisting of a carbon atom with a double bond to oxygen; for example, acetone (CH_3–(CO)–CH_3) is a carbonyl group linking two methyl groups.

Carboxy group (–CO_2H or –COOH): A carbonyl group to which a hydroxyl group is attached; carboxylic acids have this functional group.

Carboxylic acid: An organic molecule with a –CO_2H group; hydrogen atom on the –CO_2H group ionizes in water; the simplest carboxylic acids are formic acid (H–COOH) and acetic acid (CH_3–COOH).

Catabolism: The breakdown of complex molecules into simpler ones through the oxidation of organic substrates to *provide* biologically available energy—ATP (adenosine triphosphate) is an example of such a molecule.

Catalysis: The process where a catalyst increases the rate of a chemical reaction without modifying the overall standard Gibbs energy change in the reaction.

Catalyst: A substance that alters the rate of a chemical reaction and may be recovered essentially unaltered in form or amount at the end of the reaction.

Cathode: The electrode where electrons are gained (reduction) in redox reactions.

Cation exchange: The interchange between a cation in solution and another cation in the boundary layer between the solution and surface of negatively charged material such as clay or organic matter.

Cation exchange capacity (CEC): The sum of the exchangeable bases plus total soil acidity at a specific pH, usually 7.0 or 8.0. When acidity is expressed as salt extractable acidity, the cation exchange capacity is called the effective cation exchange capacity (ECEC), because this is considered to be the CEC of the exchanger at the native pH value; usually expressed in centimoles of charge per kilogram of exchanger (cmol/kg) or millimoles of charge per kilogram of exchanger.

Cellulose: A polysaccharide, polymer of glucose, that is found in the cell walls of plants; a fiber that is used in many commercial products, notably paper.

CERCLA: Comprehensive Environmental Response, Compensation, and Liability Act. This law created a tax on the chemical and petroleum industries and provided broad federal authority to respond directly to releases or threatened releases of hazardous substances that may endanger public health or the environment.

Chain reaction: A reaction in which one or more reactive reaction intermediates (frequently radicals) are continuously regenerated, usually through a repetitive cycle of elementary steps (the *propagation step*); for example, in the chlorination of methane by a radical mechanism, Cl is continuously regenerated in the chain propagation steps:

$$Cl\bullet + CH_4 \rightarrow HCl + H_3C\bullet$$

$$H_3C\bullet + Cl_2 \rightarrow CH_3Cl + Cl\bullet$$

In chain polymerization reactions, reactive intermediates of the same types generated in successive steps or cycles of steps differ in relative molecular mass.

Check standard: An analyte with a well-characterized property of interest, e.g., concentration, density, and other properties that is used to verify method, instrument, and operator performance during regular operation; *check standards* may be obtained from a certified supplier, may be a pure substance with properties obtained from the literature, or may be developed inhouse.

Chemical bond: The forces acting among two atoms or groups of atoms that lead to the formation of an aggregate with sufficient stability to be considered as an independent molecular species.

Chemical change: A processes or events that alter the fundamental structure of a chemical.

Chemical dispersion: In relation to oil spills, this term refers to the creation of oil-in-water *emulsions* by the use of chemical dispersants made for this purpose.

Chemical induction (coupling): When one reaction accelerates another in a chemical system, there is said to be chemical induction or coupling. Coupling is caused by an intermediate or byproduct of the inducing reaction that participates in a second reaction; chemical induction is often *observed* in oxidation–reduction reactions.

Chemical reaction: A process that results in the interconversion of chemical species.

Chemical species: An ensemble of chemically identical molecular entities that can explore the same set of molecular energy levels on the time scale of the experiment; the term is applied equally to a set of chemically identical atomic or molecular structural units in a solid array.

Chemical waste: Any solid, liquid, or gaseous waste material that, if improperly managed or disposed of, may pose substantial hazards to human health and the environment.

Chemical weight: The weight of a molar sample as determined by the weight of the molecules (the molecular weight); calculated from the weights of the atoms in the molecule.

Chemistry: The science that studies matter and all of the possible transformations of matter.

Chemotrophs: Organisms or chemicals that use chemical energy derived from oxidation–reduction reactions for their energy needs.

Chirality: The ability of an object or a compound to exist in right- and left-handed forms; a chiral compound will rotate the plane of polarized light.

Chlorinated solvent: A volatile organic compound containing chlorine; common solvents are trichloroethylene, tetrachloroethylene, and carbon tetrachloride.

Chlorofluorocarbon: Gases formed of chlorine, fluorine, and carbon whose molecules normally do not react with other substances; formerly used as spray-can propellants, they are known to destroy the protective ozone layer of the earth.

Chromatography: A method of chemical analysis where compounds are separated by passing a mixture in a suitable carrier over an absorbent material; compounds with different absorption coefficients move at different rates and are separated.

***Cis–trans* isomers:** The difference in the positions of atoms (or groups of atoms) relative to a reference plane in an organic molecule; in a *cis*-isomer, the atoms are on the same side of the molecule, but are on opposite sides in the *trans*-isomer; sometimes called stereoisomers; these arrangements are common in alkenes and cycloalkanes.

Clay: A very fine-grained soil that is plastic when wet but hard when fired; typical clay minerals consist of silicate and aluminosilicate minerals that are the products of weathering reactions of other minerals; the term is also used to refer to any mineral of very small particle size.

Clean Water Act: The Clean Water Act establishes the basic structure for regulating discharges of pollutants into the waters of the United States. It gives EPA the authority to implement pollution control programs such as setting waste water standards for industry; also continued requirements to set water quality standards for all contaminants in surface waters and makes it unlawful for any person to discharge any pollutant from a point source into navigable waters, unless a permit was obtained under its provisions.

Cluster compounds: Ensembles of bound atoms; typically larger than a molecule yet more defined than a bulk solid.

Coefficient of linear thermal expansion: The ratio of the change in length per degree C to the length at 0°C.

Cofferdam (also called a *coffer*): A temporary enclosure built within, or in pairs across, a body of water and constructed to allow the enclosed area to be pumped out.

Coke: A hard, dry substance containing carbon that is produced by heating bituminous coal or other carbonaceous materials to a very high temperature in the absence of air; used as a fuel.

Colligative properties: The properties of a solution that depend only on the number of particles dissolved in it, not the properties of the particles themselves; the main colligative properties addressed are boiling point elevation and freezing point depression.

Colloidal particles: Particles which have some characteristics of both species in solution and larger particles in suspension, which range in diameter from about 0.001 μm to approximately 1 μm, and which scatter white light as a light blue hue observed at right angles to the incident light.

Combination reactions: Reactions where two substances combine to form a third substance; an example is two elements reacting to form a compound of the elements and is shown in the general form: $A + B \rightarrow AB$; examples include: $2Na(s) + Cl_2(g) \rightarrow 2NaCl(s)$ and $8Fe + S_8 \rightarrow 8FeS$

Cometabolism: The process by which compounds in petroleum may be enzymatically attacked by microorganisms without furnishing carbon for cell growth and division; a variation on biodegradation in which microbes transform a contaminant even though the contaminant cannot serve as the primary energy source for the organisms. To degrade the contaminant, the microbes require the presence of other compounds (primary substrates) that can support their growth.

Complex modulus: A measure of the overall resistance of a material to flow under an applied stress, in units of force per unit area. It combines *viscosity* and elasticity elements to provide a measure of "stiffness," or resistance to flow. The *complex modulus* is more useful than *viscosity* for assessing the physical behavior of very nonNewtonian materials such as *emulsions*.

Complex inorganic chemicals: Molecules that consist of different types of atoms (atoms of different chemical elements) which, in chemical reactions, are decomposed with the formation of several other chemicals.

Compound: The combination of two or more different elements, held together by chemical bonds; the elements in each compound are always combined in the same proportion by mass (law of definite proportion).

Concentration: Composition of a mixture characterized in terms of mass, amount, volume, or number concentration with respect to the volume of the mixture.

Condensation aerosol: Formed by condensation of vapors or reactions of gases.

Conjugate acid: A substance which can lose a H^+ ion to form a base.

Conjugate base: A substance which can gain a H^+ ion to form an acid.

Conservative constituent or compound: One that does not degrade, is unreactive, and its *movement* is not retarded within a given environment (aquifer, stream, contaminant plume).

Constituent: An essential part or component of a system or group (that is, an ingredient of a chemical mixture); for example, benzene is one constituent of gasoline.

Contaminant: A pollutant, unless it has some detrimental effect, can cause deviation from the normal composition of an environment. Contaminants are not classified as pollutants unless they have some detrimental effect.

Coordination compounds: Compounds where the central ion, typically a transition metal, is surrounded by a group of anions or molecules.

Corrosion: Oxidation of a metal in the presence of air and moisture.

Covalent bond: A region of *relatively* high electron density between atomic nuclei that results from sharing of electrons and that *gives* rise to an attractive force and a characteristic internuclear distance; carbon–hydrogen bonds are covalent bonds.

Cracking: The process in which large molecules are broken down (thermally decomposed) into smaller molecules; used especially in the petroleum refining industry.

Critical point: The combination of critical temperature and critical pressure; the temperature and pressure at which two phases of a substance in equilibrium become identical and form a single phase.

Critical pressure: the pressure required to liquefy a gas at its critical temperature; the minimum pressure required to condense gas to liquid at the critical temperature; a substance is still a fluid above the critical point, neither a gas nor a liquid, and is referred to as a supercritical fluid; expressed in atmosphere or psi.

Critical temperature: The temperature above which a gas cannot be liquefied, regardless of the amount of pressure applied; the temperature at the critical point (end of the vapor pressure curve in phase diagram); at temperatures above critical temperature, a substance cannot be liquefied, no matter how great the pressure; expressed in °C.

Culture: The growth of cells or microorganisms in a controlled artificial environment.

Cyclic compound: A molecule which has the two ends of the carbon chain connected to form a ring.

Cyclo: The prefix used to indicate the presence of a ring.

Cycloalkanes (naphthene, cycloparaffin): A saturated, cyclic compound containing only carbon and hydrogen. One of the simplest *cycloalkanes* is cyclohexane (C_6H_{12}); sterane derivatives and triterpane derivatives are branched naphthene derivatives consisting of multiple condensed five- or six-carbon rings.

Daughter product: A compound that results directly from the degradation of another chemical.

Decomposition reactions: Reactions in which a single compound reacts to give two or more products; an example of a decomposition reaction is the decomposition of mercury (II) oxide into mercury and oxygen when the compound is heated; a compound can also decompose into a compound and an element, or two compounds.

Deflagration: An explosion with a flame front moving in the unburned gas at a speed below the speed of sound (1250 ft/s).

Degradation: The breakdown or transformation of a compound into byproducts and/or end products.

Degree of completion: The percentage or fraction of the limiting reactant that has been converted to products.

Dehydration reaction (condensation reaction): A chemical reaction in which two organic molecules become linked to each other via covalent bonds with the removal of a molecule of water; common in synthesis reactions of organic chemicals.

Dehydrohalogenation: Removal of hydrogen and halide ions from an alkane resulting in the formation of an alkene.

Denitrification: Bacterial reduction of nitrate to nitrite to gaseous nitrogen or nitrous oxides under anaerobic conditions.'

Density: The mass per unit volume of a substance. *Density* is temperature-dependent, generally decreasing with temperature. The density of oil relative to water, its specific gravity, governs whether a particular oil will float on water. Most fresh crude oils and fuels will float on water. Bitumen and certain residual fuel oils, however, may have densities greater than water at some temperature ranges and may submerge in water. The density of a spilled oil will also increase with time as components are lost due to weathering.

Desorption: The release of ions or molecules from solids into solution.

Detection limit (in analysis): The minimum single result that, with a stated probability, can be distinguished from a representative blank value during the laboratory analysis of substances such as water, soil, air, rock, and biota.

Detonation: An explosion with a shock wave moving at a speed greater than the speed of sound in the unreacted medium.

1,4-Dichlorobenzene: A chemical used to control moths, molds, and mildew and to deodorize restrooms and waste containers; does not occur naturally but is produced by chemical companies to make products for home use and other chemicals such as resins; most of the 1,4-dichlorobenzene enters the environment as a result of its use in moth-repellant products and in toilet-deodorizer blocks. Because it changes from a solid to a gas easily (sublimes), almost all 1,4-dichlorobenzene produced is released into the air.

Dichloroelimination: Removal of two chlorine atoms from an alkane compound and the formation of an alkene compound within a reducing environment.

Dichloromethane (CH_2Cl_2): An organic solvent often used to extract organic substances from samples; toxic, but much less so than chloroform or carbon tetrachloride, which were previously used for this purpose.

Diene: A hydrocarbon with two double bonds.

Differential thermal analysis (DTA) and thermogravimetric analysis (TGA): Techniques that may be used to measure the water of crystallization of a salt and the thermal decomposition of hydrates.

Diffuse layer: The region of ion adsorption near a sorbent surface that is subject to diffusion with the bulk solution; diffuse layer ions are not immediately adjacent the surface, but rather are distributed between the inner, stern layer ions and the bulk solution by balance of electrostatic attraction to the sorbent and diffusion away from the sorbent; see Stern layer.

Dihaloelimination: Removal of two halide atoms from an alkane compound and the formation of an alkene compound within a reducing environment.

Dilution: The process of decreasing the concentration, for example, of a solute in a solution, usually by mixing the solution with more solvent without the addition of more solute.

Dilution capacity (of a water-based ecosystem): The effective volume of receiving water available for the dilution of the discharged chemical.

Diols: Chemical compounds that contain two hydroxy (−OH) groups, generally assumed to be, but not necessarily, alcoholic; aliphatic diols are also called glycols.

Dipole−dipole forces: Intermolecular forces that exist between polar molecules. Active only when the molecules are close together. The strengths of intermolecular attractions increase when polarity increases.

Dispersion: It is an intermolecular attraction force that exists between all molecules. These forces are the result of the movement of electrons which cause slight polar moments. Dispersion forces are generally very weak but as the molecular mass increases so does their strength.

Dispersion forces: (also called London dispersion forces)

Direct emissions: Emissions from sources that are owned or controlled by the reporting entity.

Dispersant (chemical dispersant): A chemical that reduces the surface tension between water and a hydrophobic substance such as oil. In the case of an oil spill, dispersants facilitate the breakup and dispersal of an oil slick throughout the water column in the form of an oil-in-water emulsion; chemical dispersants can only be used in areas where biological damage will not occur and must be approved for use by government regulatory agencies.

Dispersion aerosol: Formed by grinding of solids, atomization of liquids, or dispersion of dusts; a colloidal-sized particle in the atmosphere formed.

Dissolved oxygen (DO): The key substance in determining the extent and kinds of life in a body of water.

Double bond: A covalent bond resulting from the sharing of two pairs of electrons (four electrons) between two atoms.

Double displacement reactions: Reactions where the anions and cations of two different molecules switch places to form two entirely different compounds.

These reactions are in the general form:

$$AB + CD \rightarrow AD + CB$$

An example is the reaction of lead (II) nitrate with potassium iodide to form lead (II) iodide and potassium nitrate:

$$Pb(NO_3)_2 + 2\ KI \rightarrow PbI_2 + 2KNO_3$$

A special kind of double displacement reaction takes place when an acid and base react with each other; the hydrogen ion in the acid reacts with the hydroxyl ion in the base causing the formation of water. Generally, the product of this reaction is some ionic salt and water:

$$HA + BOH \rightarrow H_2O + BA$$

An example is the reaction of hydrobromic acid (HBr) with sodium hydroxide:

$$HBr + NaOH \rightarrow NaBr + H_2O$$

Downgradient: In the direction of decreasing static hydraulic head.

Drug: Any substance presented for treating, curing, or preventing disease in human beings or in animals; a drug may also be used for making a medical diagnosis, managing pain, or for restoring, correcting, or modifying physiological functions.

Ecology: The scientific study of the relationships between organisms and their environments.

Ecological chemistry: The study of the interactions between organisms and their environment that are mediated by naturally occurring chemicals.

Ecology: The study of environmental factors that affect organisms and how organisms interact with these factors and with each other.

Ecosystem: A community of organisms together with their physical environment which can be viewed as a system of interacting and interdependent relationships; this can also include processes such as the flow of energy through trophic levels as well as the cycling of chemical elements and compounds through living and nonliving components of the system; the trophic level of an organism is the position it occupies in a food chain; a term representing an assembly of mutually interacting organisms and their environment in which materials are interchanged in a largely cyclical manner.

Electron acceptor: The atom, molecule, or compound that receives electrons (and therefore is reduced) in the energy-producing oxidation–reduction reactions that are essential for the growth of microorganisms and bioremediation—common electron acceptors in bioremediation are oxygen, nitrate, sulfate, and iron.

Electron affinity: The electron affinity of an atom or molecule is the amount of energy released or spent when an electron is added to a neutral atom or molecule in the gaseous state to form a negative ion.

Electron configuration of an atom: The extranuclear structure; the arrangement of electrons in shells and subshells; chemical properties of elements (their valence states and reactivity) can be predicted from the electron configuration.

Electron donor: The atom, molecule, or compound that donates electrons (and therefore is oxidized); in bioremediation, the organic contaminant often serves as an electron donor.

Electronegativity: The tendency of an atom to attract electrons in a chemical bond; nonmetals have high electronegativity, fluorine being the most electronegative while alkali metals possess least electronegativity; the electronegativity difference indicates polarity in the molecule.

Elimination: A reaction where two groups such as chlorine and hydrogen are lost from adjacent carbon atoms and a double bond is formed in their place.

Empirical formula: The simplest whole-number ratio of atoms in a compound.

Emulsan: Is a polyanionic heteropolysaccharide bioemulsifier produced by *Acinetobacter calcoaceticus* RAG-1; used to stabilize oil-in-water emulsions.

Emulsion: A stable mixture of two immiscible liquids, consisting of a continuous phase and a dispersed phase. Oil and water can form both oil-in-water and water-in-oil emulsions. The former is termed a dispersion, while *emulsion* implies the latter. Water-in-oil emulsions formed from petroleum and brine can be grouped into four stability classes: stable, a formal emulsion that will persist indefinitely; meso-stable, which gradually degrade over time due to a lack of one or more stabilizing factors; entrained water, a mechanical mixture characterized by high viscosity of the petroleum component which impedes separation of the two phases; and unstable, which are mixtures that rapidly separate into immiscible layers.

Emulsion stability: Generally accompanied by a marked increase in *viscosity* and elasticity, over that of the parent oil which significantly changes behavior. Coupled with the increased volume due to the introduction of brine, emulsion formation has a large effect on the choice of countermeasures employed to combat a spill.

Emulsification: The process of *emulsion* formation, typically by mechanical mixing. In the environment, *emulsions* are most often formed as a result of wave action. Chemical agents can be used to prevent the formation of *emulsions* or to "break" the *emulsions* to their component oil and water phases.

Endergonic reaction: A chemical reaction that requires energy to proceed. A chemical reaction is endergonic when the change in free energy is positive.

Endothermic reaction: A chemical reaction in which heat is absorbed.

Engineered bioremediation: A type of remediation that increases the growth and degradative activity of microorganisms by using engineered systems that supply nutrients, electron acceptors, and/or other growth-stimulating materials.

Enhanced bioremediation: A process which involves the addition of microorganisms (e.g., fungi, bacteria, and other microbes) or nutrients (e.g., oxygen, nitrates) to the subsurface environment to accelerate the natural biodegradation process.

Entering group: An atom or group that forms a bond to what is considered to be the main part of the substrate during a reaction, for example, the attacking nucleophile in a bimolecular nucleophilic substitution reaction.

Enthalpy of formation (ΔH_f): The energy change or the heat of reaction in which a compound is formed from its elements; energy cannot be created or destroyed but is converted from one form to another; the enthalpy change (or heat of reaction) is: $\Delta H = H_2 - H_1$

H_1 is the enthalpy of reactants and H_2 the enthalpy of the products (or heat of reaction); when H_2 is less than H_1 the reaction is exothermic and $\triangle H$ is negative, i.e., temperature increases; when H_2 is greater than H_1 the reaction is endothermic and the temperature falls.

Entropy: A thermodynamic quantity that is a measure of disorder or randomness in a system; the total entropy of a system and its surroundings always increases for a spontaneous process; the total entropy of a system and its surroundings always increases for a spontaneous process; the standard entropies are entropy values for the standard states of substances.

Environment: The total living and nonliving conditions of an organism's internal and external surroundings that affect an organism's complete life span; the conditions that surround someone or something; the conditions and influences that affect the growth, health, progress, etc., of someone or something; the total living and nonliving conditions (internal and external surroundings) that are an influence on the existence and complete life span of the organism.

Environmental analytical chemistry: The application of analytical chemical techniques to the analysis of environmental samples—in a regulatory setting.

Environmental biochemistry: The discipline that deals specifically with the effects of environmental chemical species on life.

Environmental chemistry: The study of the sources, reactions, transport, effects, and fates of chemical species in water, soil, and air environments, and the effects of technology thereon.

Environmentalist: A person working to solve environmental problems, such as air and water pollution, the exhaustion of natural resources, and uncontrolled population growth.

Environmental pollution: The contamination of the physical and biological components of the earth system (atmosphere, aquasphere, and geosphere) to such an extent that normal environmental processes are adversely affected.

Environmental science: The study of the environment, its living and nonliving components, and the interactions of these components.

Environmental studies: The discipline dealing with the social, political, philosophical, and ethical issues concerning man's interactions with the environment.

Enzyme: A macromolecule, mostly proteins or conjugated proteins produced by living organisms, that facilitate the degradation of a chemical compound (catalyst); in general, an enzyme catalyzes only one reaction type (reaction specificity) and operates on only one type of substrate (substrate specificity); any of a group of catalytic proteins that are produced by cells and that mediate or promote the chemical processes of life without themselves being altered or destroyed.

Epoxidation: A reaction wherein an oxygen molecule is inserted in a carbon–carbon double bond and an epoxide is formed.

Epoxides: A subclass of epoxy compounds containing a saturated three-membered cyclic ether. See *Epoxy compounds*.

Epoxy compounds: Compounds in which an oxygen atom is directly attached to two adjacent or nonadjacent carbon atoms in a carbon chain or ring system; thus cyclic ethers.

Equilibrium: A state when the reactants and products are in a constant ratio. The forward reaction and the reverse reactions occur at the same rate when a system is in equilibrium.

Equilibrium constant: A value that expresses how far the reaction proceeds before reaching equilibrium. A small number means that the equilibrium is towards the reactants side while a large number means that the equilibrium is towards the products side.

Equilibrium expression: The expression giving the ratio between the products and reactants. The equilibrium expression is equal to the concentration of each product raised to its coefficient in a balanced chemical equation and multiplied together, divided by the concentration of the product of reactants to the power of their coefficients.

Equipment blank: A sample of analyte-free media which has been used to rinse the sampling equipment. It is collected after completion of decontamination and prior to sampling. This blank is useful in documenting and controlling the preparation of the sampling and laboratory equipment.

Ester: A compound formed from an acid and an alcohol; in esters of carboxylic acids, the −COOH group and the −OH group lose a molecular of water and form a −COO− bond (R_1 and R_2 represent organic groups):

$$R_1COOH + R_2OH \rightarrow R_1COOR_2 + H_2O$$

Ether: A compound with an oxygen atom attached to two hydrocarbon groups. Any carbon compound containing the functional group C−O−C, such as diethyl ether ($C_2H_5O \cdot C_2H_5$).

Ethoxy group (CH_3CH_2O-): A two-carbon alkoxy substituent.

Ethylbenzene: A colorless, flammable liquid found in natural products such as coal tar and crude oil; it is also found in manufactured products such as inks, insecticides, and paints; a minor component of JP-8 fuel.

Ethyl group (CH_3CH_2-): A two-carbon alkyl substituent.

Eukaryotes: Microorganisms that have well-defined cell nuclei enclosed by a nuclear membrane.

Eutrophication: The growth of algae may become quite high in very productive water, with the result that the concurrent decomposition of dead algae reduces oxygen levels in the water to very low values.

Excess reactant: The excess of a reactant over the stoichiometric amount, with the exception of the limiting reactant; the term may refer to more than one reactant.

Exergy: A combination property of a system and its environment because it depends on the state of both the system and environment; the maximum useful work possible during a process that brings the system into equilibrium with a heat reservoir; when the surroundings are the reservoir, exergy is the potential of a system to cause a change as it achieves equilibrium with its environment and after the system and surroundings reach equilibrium, the exergy is zero; determining exergy is a prime goal of thermodynamics.

Exothermic reaction: A reaction that produces heat.

Ex situ bioremediation: A process which involves removing the contaminated soil or water to another location before treatment.

Extent of reaction: The extent to which a reaction proceeds and the material actually reacting can be expressed by the extent of reaction in moles—conventionally relates the feed quantities to the amount of each component present in the product stream, after the reaction has proceeded to equilibrium, through the stoichiometry of the reaction.

Facultative anaerobes: The extent to which a reaction proceeds and the material actually reacting can be expressed by the extent of reaction in moles—conventionally relates the feed quantities to the amount of each component present in the product stream, after the reaction has proceeded to equilibrium, through the stoichiometry of the reaction. Facultative anaerobes:Microorganisms that use (and prefer) oxygen when it is available, but can also use alternate electron acceptors such as nitrate under anaerobic conditions when necessary.

Fate: The ultimate disposition of the inorganic chemical in the ecosystem, either by chemical or biological transformation to a new form which (hopefully) is nontoxic (degradation) or, in the case of an ultimately persistent inorganic pollutants, by conversion to less offensive chemicals or even by sequestration in a sediment or other location which is expected to remain undisturbed.

Fatty acids: Carboxylic acids with long hydrocarbon side chains; most natural fatty acids have hydrocarbon chains that don't branch; any double bonds occurring in the chain are *cis* isomers—the side chains are attached on the same side of the double bond.

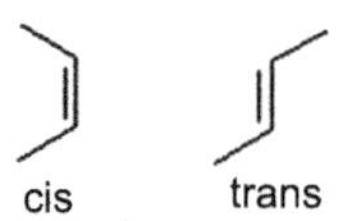

Fauna: All of the animal life of any particular region, ecosystem, or environment; generally, the naturally occurring or indigenous animal life (native animal life).

Fermentation: The process whereby microorganisms use an organic compound as both electron donor and electron acceptor, converting the compound to fermentation products such as organic acids, alcohols, hydrogen, and carbon dioxide; microbial metabolism in which a particular compound is used both as an electron donor and an electron acceptor resulting in the production of oxidized and reduced daughter products.

Field capacity or in situ (field water capacity): The water content, on a mass or volume basis, remaining in soil 2 or 3 days after having been wetted with water and after free drainage is negligible.

Fingerprint: A chromatographic signature of relative intensities used in oil–oil or oil–source rock correlations; mass chromatograms of sterane or terpane derivatives are examples of fingerprints that can be used for qualitative or quantitative comparison of crude oil.

Flammability limits: A gas mixture will not burn when the composition is lower than the lower flammable limit (LFL); the mixture is also not combustible when the composition is above the upper flammability limit (UFL).

Flammable chemical (flammable substance): A chemical or substance is usually termed flammable if the flash point of the chemical or substance is below 38°C (100°F).

Flash point: The temperature at which the vapor over a liquid will ignite when exposed to an ignition source. A liquid is considered to be flammable if its *flash point* is less than 60°C. *Flash point* is an extremely important factor in relation to the safety of spill cleanup operations. Gasoline and other light fuels can ignite under most ambient conditions and therefore are a serious hazard when spilled. Many freshly spilled crude oils also have low *flash points* until the lighter components have evaporated or dispersed.

Flora: The plant life occurring in a particular region or time; generally, the naturally occurring or indigenous plant life (native plant life).

Fluids: Liquids; also a generic term applied to all substances that flow freely, such as gases and liquids.

Foam: A colloidal suspension of a gas in a liquid.

Fog: A term denoting high level of water droplets.

Fraction: One of the portions of a chemical mixture separated by chemical or physical means from the remainder.

Free radical: A molecule with an odd number of electrons—they do not have a completed octet and often undergo vigorous redox reactions.

Fugacity (of a real gas): An effective partial pressure which replaces the mechanical partial pressure in an accurate computation of the chemical equilibrium constant.

Fugitive emissions: Emissions that include losses from equipment leaks, or evaporative losses from impoundments, spills, or leaks.

Functional group: An atom or a group of atoms attached to the base structure of a compound that has similar chemical properties irrespective of the compound to which it is a part; a means of defining the characteristic physical and chemical properties of families of organic compounds.

Functional isomers: Compounds which have the same molecular formula that possess different functional groups.

Fungi: Nonphotosynthetic organisms, larger than bacteria, aerobic, and can thrive in more acidic media than bacteria. Important function is the breakdown of cellulose in wood and other plant materials.

Gas: Matter that has no definite volume or definite shape and always fills any space given in which it exists.

Gas chromatography (GC): A separation technique involving passage of a gaseous moving phase through a column containing a fixed liquid phase; it is used principally as a quantitative analytical technique for compounds that are volatile or can be converted to volatile forms.

Gaseous nutrient injection: a process in which nutrients are fed to contaminated groundwater and soil via wells to encourage and feed naturally occurring microorganisms—the most common added gas is air in the presence of sufficient oxygen, microorganisms convert many organic contaminants to carbon dioxide, water, and microbial cell mass. In the absence of oxygen, organic contaminants are metabolized to methane, limited amounts of carbon dioxide, and trace amounts of hydrogen gas. Another gas that is added is methane. It enhances degradation by cometabolism in which as bacteria consume the methane, they produce enzymes that react with the organic contaminant and degrade it to harmless minerals.

GC–MS: Gas chromatography–mass spectrometry.

GC–TPH: GC detectable total petroleum hydrocarbons, that is the sum of all GC-resolved and unresolved hydrocarbons. The resolvable hydrocarbons appear as peaks and the unresolvable hydrocarbons appear as the area between the lower baseline and the curve defining the base of resolvable peaks.

Geological time: The span of time that has passed since the creation of the earth and its components; a scale use to measure geological events millions of years ago.

Geometric isomers: Stereoisomers which differ in the geometry around either a carbon–carbon double bond or ring.

Geosphere: A term representing the solid earth, including soil, which supports most plant life.

Gibb's free energy: The energy of a system that is available to do work at constant temperature and pressure.

Graham's law: The rate of diffusion of a gas is inversely proportional to the square root of the molar mass.

Glycerol: A small molecule with three alcohol groups ($HOCH_2CH(OH)CH_2OH$); basic building block of fats and oils.

$$\begin{array}{c} HOCH_2 \\ | \\ HOCH_2 \\ | \\ HOCH_2 \end{array}$$

Gram equivalent weight (nonredox reaction): The mass in grams of a substance equivalent to 1 g-atom of hydrogen, 0.5 g-atom of oxygen, or 1 g-ion of the hydroxyl ion; can be determined by dividing the molecular weight by the number of hydrogen atoms or hydroxyl ions (or their equivalent) supplied or required by the molecule in a reaction.

Gram equivalent weight (redox reaction): The molecular weight in grams divided by the change in oxidation state.

Gravimetric analysis: A technique of quantitative analytical chemistry in which a desired constituent is efficiently recovered and weighed.

Greenhouse effect: The warming of an atmosphere by its absorption of infrared radiation while shortwave radiation is allowed to pass through.

Greenhouse gases: Any of the gases whose absorption of solar radiation is responsible for the greenhouse effect, including carbon dioxide, ozone, methane, and the fluorocarbons.

Guest molecule (or ion): An organic or inorganic ion or molecule that occupies a cavity, cleft, or pocket within the molecular structure of a host molecular entity and forms a complex with the host entity or that is trapped in a cavity within the crystal structure of a host.

Half-life (abbreviated to $t_{1/2}$): The time required to reduce the concentration of a chemical to 50% of its initial concentration; units are typically in hours or days; the term is commonly used in nuclear physics to describe how quickly (radioactive decay) unstable atoms undergo radioactive decay, or how long stable atoms survive the potential for radioactive decay.

Halide: An element from the halogen group, which include fluorine, chlorine, bromine, iodine, and astatine.

Halogen: Group 17 in the periodic table of the elements; these elements are the reactive nonmetals and are electronegative.

Halogenation: The addition of a halogen molecule to an alkene to produce an alkyl dihalide or alkyne to produce an alkyl tetrahalide.

Halo group (X-): A substituent which is one of the four halogens; fluoro (F), chloro (Cl), bromo (Br), or iodo (I).

Hardness scale (Mohs scale): A measure of the ability of a substance to abrade or indent one another; the Mohs hardness is based on a scale from 1 to 10 units in which diamond, the hardest substance, is given a value of 10 Mohs and talc given a value of 0.5.

Hazardous waste: A potentially dangerous chemical substance that has been discarded, abandoned, neglected, released, or designated as a waste material, or one that may interact with other substances to pose a threat.

Haze: A term denoting decreased visibility due to the presence of particles.

Heat capacity ($C\rho$): The quantity of thermal energy needed to raise the temperature of an object by 1°C; the heat capacity is the product of mass of the object and its specific heat: $C\rho = \text{mass} \times \text{specific heat}$.

Heat of fusion (ΔH_{fus}): the amount of thermal energy required to melt 1 mol of the substance at the melting point; also termed as latent heat of fusion and expressed in kcal/mol or kJ/mol.

Heat of vaporization (ΔH_{vap}): the amount of thermal energy needed to convert 1 mol of a substance to vapor at boiling point; also known as latent heat of vaporization and expressed kcal/mol or kJ/mol.

Henry's law: The relation between the partial pressure of a compound and the equilibrium concentration in the liquid through a proportionality constant known as the Henry's Law constant.

Henry's law constant: The concentration ratio between a compound in air (or vapor) and the concentration of the com-pound in water under equilibrium conditions.

Herbicide: A chemical that controls or destroys unwanted plants, weeds, or grasses.

Heteroatoms: Elements other than carbon and hydrogen that are commonly found in organic molecules, such as nitrogen, oxygen, and the halogens.

Heterocyclic: An organic group or molecule containing rings with at least one noncarbon atom in the ring.

Heterogeneous: Varying in structure or composition at different locations in space.'

Heterotroph: An organism that cannot synthesize its own food and is dependent on complex organic substances for nutrition.

Heterotrophic bacteria: Bacteria that utilize organic carbon as a source of energy; organisms that derive carbon from organic matter for cell growth.

Heterotrophs: Organisms or chemicals that obtain their carbon from other organisms.

Hopane: A pentacyclic *hydrocarbon* of the *triterpane* group believed to be derived primarily from bacteriohopanoids in bacterial membranes.

Homogeneous: Having uniform structure or composition at all locations in space.

Homolog: A compound belonging to a series of compounds that differ by a repeating group; for example, propanol ($CH_3CH_2CH_2OH$), *n*-butanol ($CH_3CH_2CH_2CH_2OH$), and *n*-pentanol ($CH_3CH_2CH_2CH_2CH_2OH$) are homologs; they belong to the homologous series of alcohols: $CH_3(CH_2)_nOH$.

Homologous series: Compounds which differ only by the number of CH_2 units present.

Humic substances: Dark, complex, heterogeneous mixtures of organic materials that form in the geological systems of the earth from microbial transformations and chemical reactions that occur during the decay of organic biomolecules, polymers, and resides.

Hydration: The addition of a water molecule to a compound within an aerobic degradation pathway.

Hydration sphere: Shell of water molecules surrounding an ion in solution.

Hydrocarbon: One of a very large and diverse group of chemical compounds composed only of carbon and hydrogen; the largest source of hydrocarbons is petroleum crude oil; the principal constituents of crude oils and refined petroleum products.

Hydrogen bond: A form of association between an electronegative atom and a hydrogen atom attached to a second, relatively electronegative atom; best considered as an electrostatic interaction, heightened by the small size of hydrogen, which permits close proximity of the interacting dipoles or charges.

Hydrogenation: A reaction where hydrogen is added across a double or triple bond, usually with the assistance of a catalyst; a process whereby an enzyme in certain microorganisms catalyzes the hydrolysis or reduction of a substrate by molecular hydrogen.

Hydrogenolysis: A reductive reaction in which a carbon–halogen bond is broken, and hydrogen replaces the halogen substituent.

Hydrology: The scientific study of water.

Hydrolysis: A chemical transformation process in which a chemical reacts with water; in the process, a new carbon–oxygen bond is formed with oxygen derived from the water molecule, and a bond is cleaved within the chemical between carbon and some functional group.

Hydrophilic: Water loving; the capacity of a molecular entity or of a substituent to interact with polar solvents, in particular with water, or with other polar groups; hydrophilic molecules dissolve easily in water, but not in fats or oils.

Hydrophilic colloids: Generally, macromolecules, such as proteins and synthetic polymers, that are characterized by strong interaction with water resulting in spontaneous formation of colloids when they are placed in water.

Hydrophilicity: The tendency of a molecule to be solvated by water.

Hydrophobic: Fear of water; the tendency to repel water.

Hydrophobic colloids: Colloids that interact to a lesser extent with water and are stable because of their positive or negative electrical charges.

Hydrophobic effect: The attraction of nonionic, nonpolar compounds to surfaces that occurs due to the thermodynamic drive of these molecules to minimize interactions with water molecules.

Hydrophobic interaction: The tendency of hydrocarbons (or of lipophilic hydrocarbon-like groups in solutes) to form intermolecular aggregates in an aqueous medium, and analogous intramolecular interactions.

Hydrosphere: The water areas of the earth; also called the aquasphere.

Hydroxylation: Addition of a hydroxyl group to a chlorinated aliphatic hydrocarbon.

Hydroxyl group: A functional group that has a hydrogen atom joined to an oxygen atom by a polar covalent bond (−OH).

Hydroxyl ion: One atom each of oxygen and hydrogen bonded into an ion (OH^-) that carries a negative charge.

Hydroxyl radical: A radical consisting of one hydrogen atom and one oxygen atom; normally does not exist in a stable form.

Ideal gas law: A law which describes the relationship between pressure (P), temperature (T), volume (V), and moles of gas (n). This equation expresses behavior approached by real gases at low pressure and high temperature.

$$PV = nRT$$

Indirect emissions: Emissions that are a consequence of the activities of the reporting entity, but occur at sources owned or controlled by another entity.

Infiltration rate: The time required for water at a given depth to soak into the ground.

Inhibition: The decrease in rate of reaction brought about by the addition of a substance (inhibitor), by virtue of its effect on the concentration of a reactant, catalyst, or reaction intermediate; a component having no effect reduces the effect of another component.

Inner-sphere adsorption complex: Sorption of an ion or molecule to a solid surface where waters of hydration are distorted during the sorption process and no water molecules remain interposed between the sorbate and sorbent.

Inoculum: A small amount of material (either liquid or solid) containing bacteria removed from a culture in order to start a new culture.

Inorganic: Pertaining to, or composed of, chemical compounds that are not organic, that is, contain no carbon–hydrogen bonds; examples include chemicals with no carbon and those with carbon in nonhydrogen-linked forms.

Inorganic acid: An inorganic compound that elevates the hydrogen concentration in an aqueous solution;alphabetically, examples are:

Carbonic acid (HCO_3)**:**An inorganic acid.
Hydrochloric acid (HCl): A highly corrosive, strong inorganic acid with many uses.
Hydrofluoric acid (HF): An inorganic acid that is highly reactive with silicate, glass, metals, and semi-metals.
Nitric acid (HNO_3): A highly corrosive and toxic strong inorganic acid.
Phosphoric acid: Not considered a strong inorganic acid; found in solid form as a mineral and has many industrial uses.
Sulfuric acid: A highly corrosive inorganic acid. It is soluble in water and widely used.

Inorganic base: An inorganic compound that elevates the hydroxide concentration in an aqueous solution; alphabetically, examples are:

Ammonium hydroxide (ammonia water): a solution of ammonia in water.
Calcium hydroxide (lime water): a weak base with many industrial uses.
Magnesium hydroxide: referred to as brucite when found in its solid mineral form.
Sodium bicarbonate (baking soda): a mild alkali.
Sodium hydroxide (caustic soda): a strong inorganic base; used widely used in industrial and laboratory environments.

Inorganic chemistry: The study of inorganic compounds, specifically the structure, reactions, catalysis, and mechanism of action.

Inorganic compound: A compound that consists of an ionic component (an element from the periodic table) and an anionic component; a compound that does not contain carbon chemically bound to hydrogen; carbonates, bicarbonates, carbides, and carbon oxides are considered as inorganic compounds, even though they contain carbon; a large number of compounds occur naturally while others may be synthesized; in all cases, charge neutrality of the compound is key to the structure and properties of the compound.

Inorganic reaction chemistry: Inorganic chemical reactions fall into four broad categories: combination reactions, decomposition reactions, single displacement reactions, and double displacement reactions.

Inorganic salts: Inorganic salts are neutral, ionically bound molecules and do not affect the concentration of hydrogen in an aqueous solution.

Inorganic synthesis: The process of synthesizing inorganic chemical compounds used to produce many basic inorganic chemical compounds.

In situ: In its original place; unmoved; unexcavated; remaining in the subsurface.

In situ bioremediation: A process which treats the contaminated water or soil where it was found.

Interfacial Tension: The net energy per unit area at the interface of two substances, such as oil and water or oil and air. The air/liquid interfacial tension is often referred to as surface tension; the SI units for *interfacial tension* are milli-Newtons per meter (mN/m). The higher the *interfacial tension*, the less attractive the two surfaces are to each other and the more size of the interface will be minimized. Low surface tensions can drive the spreading of one fluid on another. The surface tension of an oil, together with its viscosity, affects the rate at which spilled oil will spread over a water surface or into the ground.

Intermolecular forces: Forces of attraction that exist between particles (atoms, molecules, ions) in a compound.

Internal Standard (IS): A pure analyte added to a sample extract in a known amount, which is used to measure the relative responses of other analytes and surrogates that are components of the same solution. The *internal standard* must be an analyte that is not a sample component.

Intramolecular: (1) Descriptive of any process that involves a transfer (of atoms, groups, electrons, etc.) or interactions (such as forces) between different parts of the same molecular entity; (2) relating to a comparison between atoms or groups within the same molecular entity.

Intrinsic bioremediation: A type of bioremediation that manages the innate capabilities of naturally occurring microbes to degrade contaminants without taking any engineering steps to enhance the process.

Inversions: Conditions characterized by high atmospheric stability which limit the vertical circulation of air, resulting in air stagnation and the trapping of air pollutants in localized areas.

Ionic bond: A chemical bond or link between two atoms due to an attraction between oppositely charged (positive–negative) ions.

Ionic bonding: Chemical bonding that results when one or more electrons from one atom or a group of atoms is transferred to another. Ionic bonding occurs between charged particles.

Ionic compounds: Compounds where two or more ions are held next to each other by electrical attraction.

Ionic liquids: An ionic liquid is a salt in the liquid state or a salt with a melting point lower than 100°C (212°F); variously called liquid electrolytes, ionic melts, ionic fluids, fused salts, liquid salts, or ionic glasses; powerful solvents and electrically conducting fluids (electrolytes).

Ionic radius: A measure of ion size in a crystal lattice for a given coordination number (CN); metal ions are smaller than their neutral atoms, and nonmetallic anions are larger than the atoms from which they are formed; ionic radii depend on the element, its charge, and its coordination number in the crystal lattice; ionic radii are expressed in angstrom units of length (Å).

Ionization energy: The ionization energy is the energy required to remove an electron completely from its atom, molecule, or radical.

Ionization potential: The energy required to remove a given electron from its atomic orbital; the values are given in electron volts (eV).

Irreversible reaction: A reaction in which the reactant(s) proceed to product(s), but there is no significant backward reaction:

$$nA + mB \rightarrow \text{Products}$$

In this reaction, the products do not recombine or change to form reactants in any appreciable amount.

Isomers: Compounds that have the same number and types of atoms—the same molecular formula—but differ in the structural formula, i.e., the manner in which the atoms are combined with each other.

Isotope: a variant of a chemical element which differs in the number of neutrons in the atom of the element; all isotopes of a given element have the same number of protons in each atom and different isotopes of a single element occupy the same position on the periodic table of the elements.

IUPAC: International Union of Pure and Applied Chemistry; the organization that establishes the system of nomenclature for organic and inorganic compounds using prefixes and suffixes, developed in the late 19th century.

Kelvin: The SI unit of temperature. It is the temperature in degrees Celsius plus 273.15.

Ketone: An organic compound that contains a carbonyl group (R_1COR_2).

Lag phase: The growth interval (adaption phase) between microbial inoculation and the start of the exponential growth phase during which there is little or no microbial growth.

Latex: A polymer of *cis*-1-4 isoprene; milky sap from the rubber tree *Hevea brasiliensis.*

Law: A system of rules that are enforced through social institutions to govern behavior; can be made by a collective legislature or by a single legislator, resulting in statutes, by the executive through decrees and regulations, or by judges through binding precedent; the formation of laws themselves may be influenced by a constitution (written or tacit) and the rights encoded therein; the law shapes politics, economics, history, and society in various ways and serves as a mediator of relations between people. See also Regulation.

Layer silicate clay: Clay minerals composed of planes of aluminum (Al^{3+}) or magnesium (Mg^{2+}) in octahedral coordination with oxygen and planes of silica (Si^{4+}) in tetrahedral coordination to oxygen. Substitution of Al^{3+} for Si^{4+} in the tetrahedral plane or substitution of Mg^{2+} or Fe^{2+} for Al^{3+} in the octahedral plane (isomorphic substitution) results in a permanent charge imbalance (i.e., structural charge) that must be satisfied through cation adsorption.

Leaving group: An atom or group (charged or uncharged) that becomes detached from an atom in what is considered to be the residual or main part of the substrate in a specified reaction.

Le Chatelier's principle: The principle that states that a system at equilibrium will oppose any change in the equilibrium conditions.

Lignin: A complex amorphous polymer in the secondary cell wall (middle lamella) of woody plant cells that cements or naturally binds cell walls to help make them rigid; highly resistant to decomposition by chemical or enzymatic action; also acts as support for cellulose fibers.

Limiting reactant: The reactant that is present in the smallest stoichiometric amount and which determines the maximum extent to which a reaction can proceed; if the reaction is 100% complete then all of the limiting reactant is consumed and the reaction can proceed no further.

Limnology: The branch of science dealing with characteristics of freshwater, including biological properties, as well as chemical and physical properties.

Lipophilic: Fat-loving; applied to molecular entities (or parts of molecular entities) having a tendency to dissolve in fat-like (e.g., hydrocarbon) solvents.

Lipophilicity: The affinity of a molecule or a moiety (portion of a molecular structure) for a lipophilic (fat soluble) environment. Iis commonly measured by its distribution behavior in a biphasic system, like liquid-liquid (e.g., partition coefficient in octanol/water).

Lithosphere: The part of the geosphere consisting of the outer mantle and the crust that is directly involved with environmental processes through contact with the atmosphere, the hydrosphere, and living things; varies from (approximately) 40–60 miles in thickness; also called the terrestrial biosphere.

Loading rate: The amount of a chemical that can be absorbed on soil on a per volume of soil basis.

LTU: Land Treatment Unit; a physically delimited area where contaminated land is treated to remove/minimize contaminants and where parameters such as moisture, pH, salinity, temperature, and nutrient content can be controlled.

Macromolecule: A large molecule of high molecular mass composed of more than 100 repeated monomers (single chemical units of lower relative mass); a large complex molecule formed from many simpler molecules.

Masking: Occurs when two components have opposite, cancelling effects such that no effect is observed from the combination.

Mass number: The number of protons plus the number of neutrons in the nucleus of an atom.

Matter: Any substance that has inertia and occupies physical space; can exist as solid, liquid, gas, plasma, or foam.

Measurement: A description of a property of a system by means of a set of specified rules, that maps the property on to a scale of specified values, by direct or mathematical comparison with specified references.

Mechanical explosion: An explosion due to the sudden failure of a vessel containing a nonreactive gas at a high pressure.

Melting point: The temperature when matter is converted from solid to liquid.

Mesosphere: The portion of the atmosphere of the earth where molecules exist as charged ions caused by interaction of gas molecules with intense ultraviolet (UV) light.

Metabolic byproduct: A product of the reaction between an electron donor and an electron acceptor; metabolic byproducts include volatile fatty acids, daughter products of chlorinated aliphatic hydrocarbons, methane, and chloride.

Metabolism: The physical and chemical processes by which foodstuffs are synthesized into complex elements, complex substances are transformed into simple ones, and energy is made available for use by an organism; thus all biochemical reactions of a cell or tissue, both synthetic and degradative, are included; the sum of all of the enzyme-catalyzed reactions in living cells that transform organic molecules into simpler compounds used in biosynthesis of cellular components or in extraction of energy used in cellular processes.

Metabolite: A product of metabolism.

Metal (oxyhydr)oxide: Minerals composed of various structural arrangements of metal cations—principally Al^{3+}, Fe^{3+}, and Mn^{4+}—in octahedral coordination with oxygen or hydroxide anions. These minerals are dissolution byproducts of mineral weathering and they are often found as coatings on layer silicates and other soil particles.

Methanogens: Strictly anaerobic archaebacteria, able to use only a very limited spectrum of substrates (for example, molecular hydrogen, formate, methanol, methylamine, carbon monoxide, or acetate) as electron donors for the reduction of carbon dioxide to methane.

Methanogenic: The formation of methane by certain anaerobic bacteria (methanogens) during the process of anaerobic fermentation.

Methyl: A group ($-CH_3$) derived from methane; for example, CH_3Cl is methyl chloride (systematic name: chloromethane) and CH_3OH is methyl alcohol (systematic name: methanol).

Micelles: A spherical cluster formed by the aggregation of soap molecules in water.

Microclimate: A highly localized climatic condition; the climate that organisms and objects on the surface are exposed to close to ground, under rocks, and surrounded by vegetation and is often quite different from the surrounding macroclimate.

Microcosm: A diminutive, representative system analogous to a larger system in composition, development, or configuration.

Microorganism: An organism of microscopic size that is capable of growth and reproduction through biodegradation of food sources, which can include hazardous contaminants; microscopic organisms including bacteria, yeasts, filamentous fungi, algae, and protozoa; a living organism too small to be seen with the naked eye; includes bacteria, fungi, protozoans, microscopic algae, and viruses.

Microbe: The shortened term for microorganism.

Mineralization: The biological process of complete breakdown of organic compounds, whereby organic materials are converted to inorganic products (e.g., the conversion of hydrocarbons to carbon dioxide and water); the release of inorganic chemicals from organic matter in the process of aerobic or anaerobic decay.

Mist: Liquid particles.

Mixed waste: Any combination of waste types with different properties or any waste that contains both hazardous waste and source, special nuclear, or byproduct material; as defined by the US EPA, mixed waste contains both hazardous waste (as defined by RCRA and its amendments) and radioactive waste (as defined by AEA and its amendments).

Modulus of elasticity: The stress required to produce unit strain to cause a change of length (Young's modulus), or a twist or shear (shear modulus), or a change of volume (bulk modulus); expressed as $dynes/cm^2$.

Moiety: A term generally used to signify part of a molecule, e.g., in an ester R^1COOR^2, the alcohol moiety is R^2O.

Molality (m): The gram moles of solute divided by kilograms of solvent.

Molar: A term expressing molarity, the number of moles of solute per liter of solution.

Molarity (M): The gram moles of solute divided by the liters of solution.

Mole: A collection of 6.022×10^{23} number of objects. Usually used to mean molecules.

Mole fraction: The number of moles of a particular substance expressed as a fraction of the total number of moles.

Molecular weight: The mass of 1 mol of molecules of a substance.

Molecule: The smallest unit in a chemical element or compound that contains the chemical properties of the element or compound.

Mole fraction: The number of moles of a component of a mixture divided by the total number of moles in the mixture.

Monoaromatic: Aromatic hydrocarbons containing a single benzene ring.

Monosaccharide: A simple sugar such as fructose or glucose that cannot be decomposed by hydrolysis; colorless crystalline substances with a sweet taste that have the same general formula $C_nH_{2n}O_n$.

MTBE (Methyl Tertiary Butyl Ether): Is a fuel additive which has been used in the United States since 1979. Its use began as a replacement for lead in gasoline because of health hazards associated with lead. MTBE has distinctive physical properties that result in it being highly soluble, persistent in the environment, and able to migrate through the ground. Environmental regulations have required the monitoring and cleanup of MTBE at petroleum contaminated sites since February, 1990; the program continues to monitor studies focusing on the potential health effects of MTBE and other fuel additives.

Native fauna: The native and indigenous animals of an area.

Native flora: The native and indigenous plant life of an area.

Natural organic matter (NOM): An inherently complex mixture of polyfunctional organic molecules that occurs naturally in the environment and is typically derived from the decay of floral and faunal remains; although they do occur naturally, the fossil fuels (coal, crude oil, and natal gas) are usually not included in the term *natural organic matter.*

NCP: National Contingency Plan—also called the National Oil and Hazardous Substances Pollution Contingency Plan; provides a comprehensive system of accident reporting, spill containment, and cleanup, and established response headquarters (National Response Team and Regional Response Teams).

Nernst Equation: An equation that is used to account for the effect of different activities upon electrode potential:

$$E = E^0 + \frac{2.303RT}{nF}\log\frac{Reactants}{Products} = E^0 + \frac{0.0591}{n}\log\frac{Reactants}{Products}$$

Nitrate enhancement: A process in which a solution of nitrate is sometimes added to groundwater to enhance anaerobic biodegradation.

Nonpoint source pollution: Pollution that does not originate from a specific source. Examples of nonpoint sources of pollution include the following: (1) sediments from construction, forestry operations and agricultural lands; (2) bacteria and microorganisms from failing septic systems and pet wastes; (3) nutrients from fertilizers and yard debris; (4) pesticides from agricultural areas, golf courses, athletic fields and residential yards, oil, grease, antifreeze, and metals washed from roads, parking lots, and driveways; (5) toxic chemicals and cleaners that were not disposed of correctly; and (6) litter thrown onto streets, sidewalks and beaches, or directly into the water by individuals. See Point source pollution.

Normality (N): The gram equivalents of solute divided by the liters of solution.

Nucleophile: A chemical reagent that reacts by forming covalent bonds with electronegative atoms and compounds.

Nuclide: A nucleus rather than to an atom—isotope (the older term) is better known than the term nuclide, and is still sometimes used in contexts where the use of the term nuclide might be more appropriate; identical nuclei belong to one nuclide, for example, each nucleus of the carbon-13 nuclide is composed of 6 protons and 7 neutrons.

Nutrients: Major elements (for example, nitrogen and phosphorus) and trace elements (including sulfur, potassium, calcium, and magnesium) that are essential for the growth of organisms.

Oceanography: The science of the ocean and its physical and chemical characteristics.

Octane: A flammable liquid (C_8H_{18}) found in petroleum and natural gas; there are 18 different octane isomers which have different structural formulas but share the molecular formula C_8H_{18}; used as a fuel and as a raw material for building more complex organic molecules.

Octanol–water partition coefficient (K_{ow}): The equilibrium ratio of a chemical's concentration in octanol (an alcoholic compound) to its concentration in the aqueous phase of a two-phase octanol–water system, typically expressed in log units (log K_{ow}); K_{ow} provides an indication of a chemical's solubility in fats (lipophilicity), its tendency to bioconcentrate in aquatic organisms, or sorb to soil or sediment.

Oleophilic: Oil seeking or oil loving (e.g., nutrients that stick to or dissolve in oil).

Order of reaction: A chemical rate process occurring in systems for which concentration changes (and hence the rate of reaction) are not themselves measurable, provided it is possible to measure a chemical flux.

Organic: Compounds that contain carbon chemically bound to hydrogen; often contain other elements (particularly O, N, halogens, or S); chemical compounds based on carbon that also contain hydrogen, with or without oxygen, nitrogen, and other elements.

Organic carbon (soil) partition coefficient (K_{oc}): The proportion of a chemical sorbed to the solid phase, at equilibrium in a two-phase, water/soil or water/sediment system expressed on an organic carbon basis; chemicals with higher K_{oc} values are more strongly sorbed to organic carbon and, therefore, tend to be less mobile in the environment.

Organic chemistry: The study of compounds that contain carbon chemically bound to hydrogen, including synthesis, identification, modeling, and reactions of those compounds.

Organic liquid nutrient injection: An enhanced bioremediation process in which an organic liquid, which can be naturally degraded and fermented in the subsurface results in the generation of hydrogen. The most commonly added for enhanced anaerobic bioremediation include lactate, molasses, hydrogen release compounds (HRCs, and vegetable oils.

Organochlorine compounds (chlorinated hydrocarbons): Organic pesticides that contain chlorine, carbon, and hydrogen (such as DDT); these pesticides affect the central nervous system.

Organometallic compounds: Compounds that include carbon atoms directly bonded to a metal ion.

Organophosphorus compound: A compound containing phosphorus and carbon; many pesticides and most nerve agents are organophosphorus compounds, such as malathion.

Osmotic potential: Expressed as a negative value (or zero), indicates the ability of the soil to dissolve salts and organic molecules; the reduction of soil water osmotic potential is caused by the presence of dissolved solutes.

OPA: Oil Pollution Act of 1990; an act which addresses oil pollution and establishes liability for the discharge and substantial threat of a discharge of oil to navigable waters and shorelines of the United States.

Outer-sphere adsorption complex: Sorption of an ion or molecule to a solid surface where waters of hydration are interposed between the sorbate and sorbent.

Oven dry: The weight of a soil after all water has been removed by heating in an oven at a specified temperature (usually in excess of 100°C, 212°F) for water; temperatures will vary if other solvents have been used.

Oxidation: The transfer of electrons away from a compound, such as an organic contaminant; the coupling of oxidation to reduction (see below) usually supplies energy that microorganisms use for growth and reproduction. Often (but not always), oxidation results in the addition of an oxygen atom and/or the loss of a hydrogen atom.

Oxidation number: A number assigned to each atom to help keep track of the electrons during a redox reaction.

Oxidation reaction: A reaction where a substance loses electrons.

Oxidation–reduction reaction: A reaction involving the transfer of electrons.

Oxidize: The transfer of electrons away from a compound, such as an organic contaminant. The coupling of oxidation to reduction (see below) usually supplies energy that microorganisms use for growth and reproduction. Often (but not always), oxidation results in the addition of an oxygen atom and/or the loss of a hydrogen atom.

Oxygen enhancement with hydrogen peroxide: An alternative process to pumping oxygen gas into groundwater involves injecting a dilute solution of hydrogen peroxide. Its chemical formula is H_2O_2, and it easily releases the extra oxygen atom to form water and free oxygen. This circulates through the contaminated groundwater zone to enhance the rate of aerobic biodegradation of organic contaminants by naturally occurring microbes. A solid peroxide product [e.g., oxygen releasing compound (ORC)] can also be used to increase the rate of biodegradation.

Oxidation–reduction reactions (redox reactions): Reactions that involve oxidation of one reactant and reduction of another.

Ozone (O_3): A form of oxygen containing three atoms instead of the common two (O_2); formed by high-energy ultraviolet radiation reacting with oxygen.

PAHs: Polycyclic aromatic hydrocarbons. Alkylated *PAHs* are *alkyl group* derivatives of the parent *PAHs*. The five target alkylated *PAHs* referred to in this report are the alkylated naphthalene, phenanthrene, dibenzothiophene, fluorene, and chrysene series.

Paraffin: An alkane.

Partition coefficient: A partition coefficient is used describe how a solute is distributed between two immiscible solvents; used in environmental science as a measure of a hydrophobicity of a solute and a proxy for transportation of a chemical through an ecosystem.

Partitioning: The distribution of a solute, S, between two immiscible solvents (such as aqueous phase and organic phase); important aspect of the transportation of a chemical into, through, and out of an ecosystem.

Partitioning equilibrium: The equilibrium distribution of a chemical that is established between the phases; the distribution of a chemical between the different phases.

Pathogen: An organism that causes disease (e.g., some bacteria or viruses).

Percentage excess: The excess of a reactant above the amount required to react with the total quantity of limiting reactant.

Percent conversion: The percentage of any reactant that has been converted to products.

Perfluorocarbon (PFC): A derivative of hydrocarbons in which all of the hydrogens have been replaced by fluorine.

Periodic table: Grouping of the known elements by their number of protons; there are many other trends such as size of elements and electronegativity that are easily expressed in terms of the periodic table.

Permeability: The capability of the soil to allow water or air movement through it. The quality of the soil that enables water to move downward through the profile, measured as the number of inches per hour that water moves downward through the saturated soil.

Permeable reactive barrier (PRB): A subsurface emplacement of reactive materials through which a dissolved contaminant plume must move as it flows, typically under natural gradient and treated water exits the other side of the permeable reactive barrier.

Pesticide: A chemical that is designed and produced to control for pest control, including weed control.

pH: A measure of the acidity or basicity of a solution; the negative logarithm (base 10) of the hydrogen ion concentration in gram ions per liter; a number between 0 and 14 that describes the acidity of an aqueous solution; mathematically the pH is equal to the negative logarithm of the concentration of H_3O^+ in solution.

Phenol: A molecule containing a benzene ring that has a hydroxyl group substituted for a ring hydrogen.

Phenyl: A molecular group or fragment formed by abstracting or substituting one of the hydrogen atoms attached to a benzene ring.

Photic zone: The upper layer within bodies of water reaching down to about 200 m, where sunlight penetrates and promotes the process of photosynthesis; the richest and most diverse area of the ocean.

Photocatalysis: The acceleration of a photoreaction in the presence of a catalyst in which light is absorbed by a substrate that is typically adsorbed on a (solid) catalyst.

Photocatalyst: A material that can absorb light, producing electron-hole pairs that enable chemical transformations of the reaction participants and regenerate its chemical composition after each cycle of such interactions.

Phototrophs: Organisms or chemicals that utilize light energy from photosynthesis.

Physical change: Refers to the change that occurs when a material changes from one physical state to another without formation of intermediate substances of different composition in the process, such as the change from gas to liquid.

Phytodegradation: The process in which some plant species can metabolize VOC contaminants. The resulting metabolic products include trichloroethanol, trichloroacetic acid, and dichloracetic acid; mineralization products are probably incorporated into insoluble products such as components of plant cell walls.

Phytovolatilization: The process in which VOCs are taken up by plants and discharged into the atmosphere during transpiration.

PM_{10}: Particulate matter below 10 microns in diameter; this corresponds to the particles inhalable into the human respiratory system, and its measurement uses a size selective inlet.

$PM_{2.5}$: Particulate matter below 2.5 microns in diameter; this is closer to, but slightly finer than, the definitions of respirable dust that have been used for many years in industrial hygiene to identify dusts which will penetrate the lungs.

pOH: A measure of the basicity of a solution; the negative log of the concentration of the hydroxide ions.

Point emissions: Emissions that occur through confined air streams as found in stacks, ducts, or pipes.

Point source pollution: Any single identifiable source of pollution from which pollutants are discharged, such as a pipe. Examples of point sources include: (1) discharges from waste water treatment plants, (2) operational wastes from industries, and (3) combined sewer outfalls. See Nonpoint source pollution.

Polar compound: An organic compound with distinct regions of positive and negative charge. *Polar compounds* include alcohols, such as sterols, and some *aromatics*, such as monoaromatic steroids. Because of their polarity, these compounds are more soluble in polar solvents, including water, compared to nonpolar compounds of similar molecular structure.

Pollutant: Either (1) a nonindigenous chemical that is present in the environment or (2) an indigenous chemical that is present in the environment in greater than the natural concentration. Both types of pollutants are the result of human activity and have an overall detrimental effect upon the environment or upon something of value in that environment.

Polymer: A large molecule made by linking smaller molecules (monomers) together.

Positional isomers: Compounds which differ only in the position of a functional group; 2-pentanol and 3-pentanol are positional isomers.

Potentiation: A component having no effect increases the effect of another component.

Pour point: The lowest temperature at which an oil will appear to flow under ambient pressure over a period of 5 s. The *pour point* of crude oils generally varies from −60 to 30°C. Lighter oils with low *viscosities* generally have lower *pour points*.

Precipitation: Formation of an insoluble product that occurs via reactions between ions or molecules in solution.

Primary substrates: The electron donor and electron acceptor that are essential to ensure the growth of microorganism; these compounds can be viewed as analogous to the food and oxygen that are required for human growth and reproduction.

Producers: Organisms or chemicals that utilize light energy and store it as chemical energy.

Prokaryotes: Microorganisms that lack a nuclear membrane so that their nuclear genetic material is more diffuse in the cell.

Propagule: Any part of a plant (e.g., bud) that facilitates dispersal of the species and from which a new plant may form.

Propane: A colorless, odorless, flammable gas (C_3H_8) found in petroleum and natural gas; used as a fuel and as a raw material for building more complex organic molecules.

Protozoa: Microscopic animals consisting of single eukaryotic cells.

Radical (free radical): A molecular entity such as $CH_3^•$, $Cl^•$ possessing an unpaired electron.

Radioactive decay: The process by which the nucleus of an unstable atom loses energy by emitting radiation.

Rate: A derived quantity in which time is a denominator quantity so that the progress of a reaction is measured with time.

Rate constant, k: See Order of reaction.

Rate-controlling step (rate-limiting step, rate-determining step): The elementary reaction having the largest control factor exerts the strongest influence on the rate; a step having a control factor much larger than any other step is said to be rate-controlling.

Reactants: Substances initially present in a chemical reaction.

Reaction rate: The change in concentration of the starting chemical in a given time interval.

Reaction (irreversible): A reaction in which the reactant(s) proceed to product(s), but there is no significant backward reaction:

$$nA + mB \rightarrow \text{Products}$$

In this reaction, the products do not recombine or change to form reactants in any appreciable amount.

Reaction (reversible): A reaction in which the products can revert to the starting materials (A and B). Thus:

$$nA + MB \leftrightarrow \text{Products}$$

Recalcitrant: Unreactive, nondegradable, refractory.

Receptor: An object (animal, vegetable, or mineral) or a locale that is affected by the pollutant.

Redox (reduction–oxidation reactions): Oxidation and reduction occur simultaneously; in general, the oxidizing agent gains electrons in the process (and is reduced) while the reducing agent donates electrons (and is oxidized).

Reduce: The transfer of electrons to a compound, such as oxygen, that occurs when another compound is oxidized.

Reducers: Organisms or chemicals that break down chemical compounds to more simple species and thereby extract the energy needed for their growth and metabolism.

Reduction: The transfer of electrons to a compound, such as oxygen, that occurs when another compound is oxidized.

Reductive dehalogenation: A variation on biodegradation in which microbially catalyzed reactions cause the replacement of a halogen atom on an organic compound with a hydrogen atom. The reactions result in the net addition of two electrons to the organic compound.

Refractive index (index of refraction): The ratio of wavelength or phase velocity of an electromagnetic wave in a vacuum to that in the substance; a measure of the amount of refraction a ray of light undergoes as it passes through a refraction interface; a useful physical property to identify a pure compound.

Regulation: A concept of management of complex systems according to a set of rules (laws) and trends; can take many forms; legal restrictions promulgated by a government authority, contractual obligations (such as contracts between insurers and their insureds), social regulation, coregulation, third-party regulation, certification, accreditation, or market regulation. See Law.

Releases: Onsite discharge of a toxic chemical to the surrounding environment; includes emissions to the air, discharges to bodies of water, releases at the facility to land, as well as contained disposal into underground injection wells.

Releases (to air, point and fugitive air emissions): All air emissions from industry activity; point emissions occur through confined air streams as found in stacks, ducts, or pipes; fugitive emissions include losses from equipment leaks, or evaporative losses from impoundments, spills, or leaks.

Releases (to land): Disposal of toxic chemicals in waste to onsite landfills, land treated or incorporation into soil, surface impoundments, spills, leaks, or waste piles. These activities must occur within the boundaries of the facility for inclusion in this category.

Release (to underground injection): A contained release of a fluid into a subsurface well for the purpose of waste disposal.

Releases (to water, surface water discharges): Any releases going directly to streams, rivers, lakes, oceans, or other bodies of water; any estimates for storm water runoff and nonpoint losses must also be included.

Resins: The name given to a large group of *polar compounds* in oil. These include heterosubstituted *aromatics*, acids, ketones, alcohols, and monoaromatic steroids. Because of their polarity, these compounds are more soluble in *polar* solvents, including water, than the nonpolar compounds, such as *waxes* and *aromatics*, of similar molecular weight. They are largely responsible for oil *adhesion*.

Respiration: The process of coupling oxidation of organic compounds with the reduction of inorganic compounds such as oxygen, nitrate, iron (III), manganese (IV), and sulfate.

Reversible reaction: A reaction in which the products can revert to the starting materials (A and B). Thus:

$$nA + mB \leftrightarrow \text{Products}$$

Rhizodegradation: The process whereby plants modify the environment of the root zone soil by releasing root exudates and secondary plant metabolites. Root exudates are typically photosynthetic carbon, low molecular—weight molecules, and high molecular—weight organic acids. This complex mixture modifies and promotes the development of a microbial community in the rhizosphere. These secondary metabolites have a potential role in the development of naturally occurring contaminant-degrading enzymes.

Rhizosphere: The soil environment encompassing the root zone of the plant.

RRF: Relative response factor.

Saturated hydrocarbon: A saturated carbon–hydrogen compound with all carbon bonds filled; that is, there are no double or triple bonds, as in olefins or acetylenes.

Saturated solution: A solution in which no more solute will dissolve; a solution in equilibrium with the dissolved material.

Saturation: The maximum amount of solute that can be dissolved or absorbed under prescribed conditions.

Side chain: A chain of atoms which is attached to a longer chain of atoms; examples of side chains would be methyl, ethyl, propyl groups (among others).

SIM (Selecting Ion Monitoring): Mass spectrometric monitoring of a specific mass/charge (m/z) ratio. The *SIM* mode offers better sensitivity than can be obtained using the full scan mode.

Simple inorganic chemicals: Molecules that consist of one-type atoms (atoms of one element) which, in chemical reactions, cannot be decomposed to form other chemicals.

Single displacement reactions: Reactions where one element trades places with another element in a compound. These reactions come in the general form of:

$$A + BC \rightarrow AC + B$$

Examples include:

(1) magnesium replacing hydrogen in water to make magnesium hydroxide and hydrogen gas:

$$Mg + 2H_2O \rightarrow Mg(OH)_2 + H_2$$

(2) the production of silver crystals when a copper metal strip is dipped into silver nitrate:

$$Cu(s) + 2AgNO_3(aq) \rightarrow 2Ag(s) + Cu(NO_3)_2(aq)$$

Smoke: The particulate material assessed in terms of its blackness or reflectance when collected on a filter, as opposed to its mass; this is the historical method of measurement of particulate pollution; particles formed by incomplete combustion of fuel.

Soil organic matter: Living and partially decayed (nonliving) materials, as well as assemblages of biomolecules and transformation products of organic residue decay known as humic substances.

Solid state compounds: A diverse class of compounds that are solid at standard temperature and pressure, and exhibit unique properties as semiconductors, etc.

Solubility: The amount of a substance (solute) that dissolves in a given amount of another substance (solvent); a measure of the solubility of an inorganic chemical in a solvent, such as water; generally, ionic substances are soluble in water and other polar solvents while the nonpolar, covalent compounds are more soluble in the nonpolar solvents; in sparingly soluble, slightly soluble, or practically insoluble salts, degree of solubility in water and occurrence of any precipitation process may be determined from the solubility product, Ksp, of the salt—the smaller the Ksp value, the lower the solubility of the salt in water.

Soluble: Capable of being dissolved in a solvent.

Solute: Any dissolved substance in a solution.

Solution: Any liquid mixture of two or more substances that is homogeneous.

Solvolysis: Generally, a reaction with a solvent, involving the rupture of one or more bonds in the reacting solute; more specifically the term is used for substitution, elimination, or fragmentation reactions in which a solvent species is the nucleophile; hydrolysis, if the solvent is water or alcoholysis if the solvent is an alcohol.

Sorbate: Sometimes referred to as adsorbate; it is the solute that adsorbs on the solid phase.

Sorbent (adsorbent): The solid phase or substrate onto which the sorbate sorbs; the solid phase may be more specifically referred to as an absorbent or adsorbent if the mechanism of removal is known to be absorption or adsorption, respectively.

Sorption: A general term that describes removal of a solute from solution to a contiguous solid phase and is used when the specific removal mechanism is not known.

Sorption isotherm: Graphical representation of surface excess (i.e., the amount of substance sorbed to a solid) relative to sorptive concentration in solution after reaction at fixed temperature, pressure, ionic strength, pH, and solid-to-solution ratio.

Sorptive: Ions or molecules in solution that could potentially participate in a sorption reaction.

Specific heat: The amount of heat required to raise the temperature of 1 g of a substance by 1°C; the specific heat of water is 1 calorie or 4.184 J.

Stable: As applied to chemical species, the term expresses a thermodynamic property, which is quantitatively measured by relative molar standard Gibbs energies; a chemical species A is more stable than its isomer B under the same standard conditions.

Standard potential: Used to predict if a species will be oxidized or reduced in solution (under acidic or basic conditions) and whether any oxidation–reduction reaction will take place.

Starch: A polysaccharide containing glucose (long-chain polymer of amylose and amylopectin) that is the energy storage reserve in plants.

Stereochemistry: The branch of organic chemistry that deals with the three-dimensional structure of molecules.

Stereogenic carbon (asymmetric carbon): A carbon atom which is bonded to four different groups or atoms; a chiral molecule must contain a stereogenic carbon, and therefore has no plane of symmetry and is not superimposable on its mirror image.

Stereoisomers: Isomers which have the same bonding connectivity but have a different three-dimensional structure; examples would be *cis*-2-butene and *trans*-2-butene (geometric isomers), and the left- and right-handed forms of 2-butanol (enantiomers).

Stern layer: The layer of ions adsorbed immediately adjacent to a charged sorbent surface. Ions in the stern layer can be directly bonded to the sorbent through covalent and ionic bonds (inner-sphere complexes) or held adjacent to a sorbent through strictly electrostatic forces in outer-sphere complexes.

Stoichiometry: The calculation of the quantities of reactants and products (among elements and compounds) involved in a chemical reaction.

Stoke's Law:

$$v = \frac{gd^2(\rho_1 - \rho_2)}{18\eta}$$

Stratosphere: The portion of the atmosphere of the earth where ozone is formed by the reaction of ultraviolet light on dioxygen molecules.

Strong acid: An acid that releases H^+ ions easily—examples are hydrochloric acid and sulfuric acid.

Strong base: A basic chemical that accepts and holds proton tightly—an example is the hydroxide ion.

Structural formula: A convention used to represent the structures of organic molecules in which not all the valence electrons of the atoms are shown.

Structural isomerism: The relationship between two compounds which have the same molecular formula, but different structures; they may be further classified as functional, positional, or skeletal isomers. This relation is also called constitutional isomerism.

Styrene: A human-made chemical used mostly to make rubber and plastics; present in combustion products, such as cigarette smoke and automobile exhaust.

Sublimation: The direct vaporization or transition of a solid directly to a vapor without passing through the liquid state.

Substitution reaction: The process in which one group or atom in a molecule is replaced by another group or atom.

Substrate: A chemical species of particular interest, of which the reaction with some other chemical reagent is under observation (e.g., a compound that is transformed under the influence of a catalyst); also the component in a nutrient medium, supplying microorganisms with carbon (C-substrate), nitrogen (N-substrate) as food needed to grow.

Surface-active agent: A compound that reduces the surface tension of liquids, or reduces interfacial tension between two liquids or a liquid and a solid; also known as surfactant, wetting agent, or detergent.

Surface tension: Caused by molecular attractions between the molecules of two liquids at the surface of separation.

Sustainable development: Development and economic growth that meets the requirements of the present generation without compromising the ability of future generations to meet their needs; a strategy seeking a balance between development and conservation of natural resources.

Sustainable enhancement: An intervention action that continues until such time that the enhancement is no longer required to reduce contaminant concentrations or fluxes.

Steranes: A class of tetracyclic, saturated biomarkers constructed from six isoprene subunits ($\sim C_{30}$). *Steranes* are derived from sterols, which are important membrane and hormone components in eukaryotic organisms. Most commonly used *steranes* are in the range of C_{26} to $C_{30}$and are detected using m/z 217 mass chromatograms.

Surrogate analyte: A pure analyte that is extremely unlikely to be found in any sample, which is added to a sample aliquot in a known amount and is measured with the same procedures used to measure other components. The purpose of a *surrogate analyte* is to monitor the method performance with each sample.

Synergism: The effect of the combination is greater than the sum of individual effects.

Terminal electron acceptor (TEA): A compound or molecule that accepts an electron (is reduced) during metabolism (oxidation) of a carbon source; under aerobic conditions molecular oxygen is the terminal electron acceptor; under anaerobic conditions a variety of terminal electron acceptors may be used. In order of decreasing redox potential, these terminal electron acceptors include nitrate, manganese (Mn^{3+}, Mn^{6+}), iron (Fe^{3+}), sulfate, and carbon dioxide; microorganisms preferentially utilize electron acceptors that provide the maximum free energy during respiration; of the common terminal electron acceptors listed above, oxygen has the highest redox potential and provides the most free energy during electron transfer.

Terpanes: A class of branched, cyclic alkane biomarkers including *hopanes* and tricyclic compounds.

Terpenes: Hydrocarbon solvents, compounds composed of molecules of hydrogen and carbon; they form the primary constituents in the aromatic fractions of scented plants, e.g., pine oil, as well as turpentine and camphor oil.

Terrestrial biosphere: The part of the geosphere consisting of the outer mantle and the crust that is directly involved with environmental processes through contact with the atmosphere, the hydrosphere, and living things; varies from (approximately) 40–60 miles in thickness; also called the lithosphere.

Tetrachloroethylene (perchloroethylene): A human-made chemical that is widely used for dry cleaning of fabrics and for metal-degreasing operations; also used as a starting material (building block) for making other chemicals and is used in some consumer products such as water repellents, silicone lubricants, fabric finishers, spot removers, adhesives, and wood cleaners; can stay in the air for a long time before breaking down into other chemicals or coming back to the soil and water in rain; much of the tetrachloroethylene that gets into water and soil will evaporate; because tetrachloroethylene can travel easily through soils, it can get into underground drinking water supplies.

Thermal conductivity: A measure of the rate of transfer of heat by conduction through unit thickness, across unit area for unit difference of temperature; measured as calories per second per square centimeter for a thickness of 1 cm and a temperature difference of 1°C; units are cal/cm sec.°K or W/cm°K.

Thermodynamic equilibrium: The thermodynamic state that is characterized by absence of flow of matter or energy.

Thermodynamics: The study of the energy transfers or conversion of energy in physical and chemical processes defines the energy required to start a reaction or the energy given out during the process.

Thermogravimetric analysis (TGA) and differential thermal analysis (DTA): Techniques that may be used to measure the water of crystallization of a salt and the thermal decomposition of hydrates.

Toluene: A clear, colorless liquid that occurs naturally in crude oil and in the tolu tree; produced in the process of making gasoline and other fuels from crude oil; used in making paints, paint thinners, fingernail polish, lacquers, adhesives, and rubber, and in some printing and leather tanning processes; a major component of JP-8 fuel.

Total *n*-alkanes: The sum of all resolved *n-alkanes* (from C_8 to C_{40} plus pristane and phytane).

Total 5 alkylated PAH homologs: The sum of the 5 target PAHs (naphthalene, phenanthrene, dibenzothiophene, fluorene, chrysene) and their alkylated ($C_1$to C_4) homologs, as determined by GCMS. These 5 target alkylated PAH homologous series are oil-characteristic aromatic compounds.

Total aromatics: The sum of all resolved and unresolved aromatic hydrocarbons including the total of BTEX and other alkyl benzene compounds, total 5 target alkylated PAH homologs, and other EPA priority PAHs.

Total saturates: The sum of all resolved and unresolved aliphatic hydrocarbons including the total *n*-alkanes, branched alkanes, and cyclic saturates.

Total suspended particulate matter: The mass concentration determined by filter weighing, usually using a specified sampler which collects all particles up to approximately 20 microns depending on wind speed.

Toxicity: A measure of the toxic nature of a chemical; usually expressed quantitatively as LD_{50} (median lethal dose) or LC_{50} (median lethal concentration in air)—the latter refers to inhalation toxicity of gaseous substances in air; both terms refer to the calculated concentration of a chemical that can kill 50% of test animals when administered.

Toxicological chemistry: The chemistry of toxic substances with emphasis upon their interactions with biologic tissue and living organisms.

TPH: Total petroleum hydrocarbons; the total measurable amount of petroleum-based hydrocarbons present in a medium as determined by gravimetric or chromatographic means.

Transfers: A transfer of toxic (organic) chemicals in wastes to a facility that is geographically or physically separate from the facility reporting under the toxic release inventory; the quantities reported represent a movement of the chemical away from the reporting facility; except for offsite transfers for disposal, these quantities do not necessarily represent entry of the chemical into the environment.

Transfers (POTWs): Waste waters transferred through pipes or sewers to a publicly owned treatment works (POTW); treatment and chemical removal depend on the chemical's nature and treatment methods used; chemicals not treated or destroyed by the POTW are generally released to surface waters or land filled within the sludge.

Transfers (to disposal): Wastes that are taken to another facility for disposal generally as a release to land or as an injection underground.

Transfers (to energy recovery): Wastes combusted offsite in industrial furnaces for energy recovery; treatment of an organic chemical by incineration is not considered to be energy recovery.

Transfers (to recycling): Wastes that are sent offsite for the purposes of regenerating or recovering still valuable materials; once these chemicals have been recycled, they may be returned to the originating facility or sold commercially.

Transfers (to treatment): Wastes moved offsite for either neutralization, incineration, biological destruction, or physical separation; in some cases, the chemicals are not destroyed but prepared for further waste management.

1,1,1-Trichloroethane: Does not occur naturally in the environment; used in commercial products, mostly to dissolve other chemicals; beginning in 1996, 1,1,1-trichloroethane was no longer made in the United States because of its effects on the ozone layer; because of its tendency to evaporate easily, the vapor form is usually found in the environment; 1,1,1-trichloroethane also can be found in soil and water, particularly at hazardous waste sites.

Trichloroethylene: A colorless liquid that does not occur naturally; mainly used as a solvent to remove grease from metal parts and is found in some household products, including typewriter correction fluid, paint removers, adhesives, and spot removers.

Triglyceride: An ester of glycerol and three fatty acids; the fatty acids represented by 'R' can be the same or different:

$$\begin{array}{c} RCOOCH_2 \\ | \\ RCOOCH_2 \\ | \\ RCOOCH_2 \end{array}$$

Triterpanes: A class of cyclic saturated *biomarkers* constructed from six isoprene subunits; cyclic terpane compounds containing two, four, and six isoprene subunits are called monoterpane (C_{10}), diterpane (C_{20}), and *triterpane* (C_{30}), respectively.

Trophic: The trophic level of an organism is the position it occupies in a food chain.

Troposphere: The portion of the atmosphere of the earth that is closest to the surface.

Tyndall effect: The characteristic light-scattering phenomenon of colloids results from those being the same order of size as the wavelength of light.

Ultraviolet radiation (UV radiation): An electromagnetic radiation with a wavelength from 10 to 400 nm, shorter than the wavelength of visible light but longer than the wavelength of X-rays. UV radiation is present in sunlight constituting about 10% of the total light output of the sun.

UCM: Unresolved complex mixture of hydrocarbons on, for example, a gas chromatographic tracing; the UCM appears as the *envelope* or *hump area* between the solvent baseline and the curve defining the base of resolvable peaks.

Underground storage tank: A storage tank that is partially or completely buried in the earth.

Unsaturated compound: An organic compound with molecules containing one or more double bonds.

Unsaturated zone: The zone between land surface and the capillary fringe within which the moisture content is less than saturation and pressure is less than atmospheric; soil pore spaces also typically contain air or other gases; the capillary fringe is not included in the unsaturated zone (See Vadose zone).

Upgradient: In the direction of increasing potentiometric (piezometric) head. See also *Downgradient*.

US EPA: United States Environmental Protection Agency.

USGS: United States Geological Survey.

Vadose zone: The zone between land surface and the water table within which the moisture content is less than saturation (except in the capillary fringe) and pressure is less than atmospheric; soil pore spaces also typically contain air or other gases; the capillary fringe is included in the vadose zone.

Valence state of an atom: The power of an atom to combine to form compounds; determines the chemical properties.

Van der Waals forces: Intermolecular attractive forces that arise between nonionic, nonpolar molecules due to dipole–dipole interactions and instantaneous dipole interactions (London dispersion forces).

Van der Waals interaction: The cohesive interaction (attraction between like) or the adhesive interaction (attraction between unlike) and/or repulsive forces between molecules.

Vapor pressure: The pressure exerted by a solid or liquid in equilibrium with its own vapor; depends on temperature and is characteristic of each substance; the higher the vapor pressure at ambient temperature, the more volatile the substance.

Viscosity: The resistance of a fluid to shear, move, or flow; a function of the composition of a fluid; the viscosity of an ideal, noninteracting fluid does not change with shear rate—such fluids are called Newtonian; expressed as g/cm sec or Poise; 1 Poise = 100 cP.

Volatile: Readily dissipating by evaporation.

Volatile organic compounds (VOC): Organic compounds with high vapor pressures at normal temperatures. VOCs include light saturates and aromatics, such as pentane, hexane, BTEX, and other lighter substituted benzene compounds, which can make up to a few percent of the total mass of some crude oils.

Water solubility: The maximum amount of a chemical that can be dissolved in a given amount of pure water at standard conditions of temperature and pressure; typical units are milligrams per liter (mg/L), gallons per liter (gal/L), or pounds per gallon (lbs/gal).

Waxes: Waxes are predominately straight-chain *saturates* with melting points above 20°C (generally, the *n*-alkanes C_{18} and higher molecular weight).

Weak Acid: An acid that does not release H^+ ions easily—an example is acetic acid.

Weak base: A basic chemical that has little affinity for a proton—an example is the chloride ion.

Weathering: Processes related to the physical and chemical actions of air, water, and organisms after oil spill. The major weathering processes include evaporation, dissolution, dispersion, photochemical oxidation, water-in-oil *emulsification*, microbial degradation, adsorption onto suspended particulate materials, interaction with mineral fines, sinking, sedimentation, and formation of tar balls.

Wet Deposition: The term used to describe pollutants brought to ground either by rainfall or by snow; this mechanism can be further subdivided depending on the point at which the pollutant was absorbed into the water droplets.

Wilting point: The largest water content of a soil at which indicator plants, growing in that soil, wilt and fail to recover when placed in a humid chamber.

Xylenes: The term that refers to all three types of xylene isomers (meta-xylene, ortho-xylene, and para-xylene); produced from crude oil; used as a solvent and in the printing, rubber, and leather industries, as well as a cleaning agent and a thinner for paint and varnishes; a major component of JP-8 fuel.

Yield: The mass (or moles) of a chosen final product divided by the mass (or moles) of one of the initial reactants.

Zwitterion: A particle that contains both positively charged and negatively charged groups; for example, amino acids ($H_2NCHRCO_2H$) can form zwitterions ($^+H_3NCHRCOO^-$).

Index

'*Note*: Page numbers followed by "f" indicate figures, "t" indicate tables.'

A

Abiotic oxidation, 239
Absorption, 165, 166t, 178–179
  mechanism, 185–186
Acid–base partitioning, 321
Acid-catalyzed process, 211–212
Acid dissociation constants, 89, 90t, 96
Acid hydrolysis, 218–221
Acid rain, 28
Acids and bases
  acid compatibility, 87–89, 88t
  acid dissociation constants, 89, 90t, 96
  Antoine Lavoisier theory, 98
  aprotic solvents, 98–99
  in aqueous solution, 89
  Arrhenius theory, 95
  Brønsted–Lowry acids and bases, 100
  chemical properties of, 96–97
  electrolysis, 97
  hydrogen ions, 89
    concentration, 97
  hydronium ion equation, 94–95
  hydroxide compounds, 89
  inorganic acids and bases, 87–89, 88t
    binary acids, 91
    descaling process, 90–91
    hydronium ion, 91
    periodic table elements, 91–92, 92f
    properties, 90
    solvent leveling, 92
    water-soluble hydroxides, 91
  inorganic salts, 87–89, 88t
  ionic liquids, 95–96
  Lewis acids and bases, 100–101
  neutralization reaction, 99
  organic acids and bases
    acyl chlorides and anhydrides, 93
    aspirin, 94
    carboxylic acid derivatives, 92
    ester, 94
    ibuprofen, 94
    vinegar, 93
  oxy-acids, 96
  perchloric acid, 99
  pH scale, 89, 89f
  polyatomic ions, 95
  solvent system theory, 98–99
  strong acids/bases, 97
  weak acids/bases, 97
Acyl chlorides and anhydrides, 93
Adsorption, 149–150
  acidity and alkalinity, 180–181
  aluminosilicate mineral adsorbents, 167–168
  calcite, 167, 175
  characteristics of, 176, 177t
  chemical composition, 179–180
  chemical interactions, 174–175
  clay minerals, 168, 168t
    applications, 170
    classification, 169–170
    impurities, 175
  definition, 165, 166t
  desorption, 183
    process, 63
  ecosystem treatment, 178
  electrostatic surface charge, 172
  exchangeable/charge-balancing cations, 175
  heavy metal cations and anions, 176
  ion exchange, 173–174, 182–183
  mass transfer, 173
  materials for, 166–167
  mechanism, 183–185, 184t
  microbe surfaces reactivity, 172
  natural adsorbents, 177
  organic functional groups, 172, 172t
  partition coefficient, 181–182
  physical phenomenon, 167
  properties, 167
  silica-alumina ratio, 168–169
  sorbent, 179
  sorption rate, 181
  surface-based process, 174

Adsorption (*Continued*)
surface energy, 173
temperature, 181
unbalanced residual forces, 173
washing conditions, 176
Aerobic biodegradation, 283–284
Aggregation
aquasphere chemistry, 32
Agricultural activities, 38
Agricultural chemical, 67
Agricultural production, 138–139
Agrochemicals, 45–46
Airborne bioaccumulative chemicals, 118
Air pollution, 22, 115, 139–140
area sources, 49
biomass feedstock, 54
carbon monoxide (CO), 51–52
causes, 49
chemical and physical characteristics, 49
chlorine, 54
chlorofluorocarbons (CFCs), 54
classification, 49
fossil fuels, 51
herbaceous plant material ash, 55
invisible air pollution, 49
mobile sources, 49
natural sources, 49
nitrogen dioxide (NO_2), 52
ozone (O_3), 52–53
periodic table, 54, 55f
persistent toxic pollutants, 51
PM, 53–54
pollutants and effects, 49, 50t–51t
primary air pollutants, 51
renewable sources, 54
secondary air pollutants, 51
stationary sources, 49
sulfur dioxide (SO_2), 53
toxic chemicals, 49
troposphere/stratosphere boundary, 54
ultraviolet radiation, 54
visible air pollution, 49
volatile compounds (VOCs), 54
water leaching, 55
wood ash, 55
Air sparging, 197
Air–water partitioning, 321–323
Alkaline hydrolysis, 221–222
Anaerobic biodegradation, 284
Anaerobic sediments, 234
Anodic currents, 234–235
Anthropogenic chemicals, 43–44
Antoine Lavoisier theory, 98
Aprotic solvents, 98–99
Aquasphere chemistry
acid–base reactions, 33
aggregation, 32
biosedimentation, 31–32
colloidal stability, 32
dielectric constant, 34
dissolution, 31
electrolyte solutions, 34
electron density distribution, 29
emulsification, 31
factors, 30–31
functions, 29–30
inorganic chemical, 30
characteristics, 31
marine environment, 31
microbial degradation, 32
oxidation, 31
physical/chemical treatment, soils, 33, 33t
physical transport, 31
rivers and groundwater pollution, 30
salt components, 34
self-purification, 33
stabilizers/stabilizing agents, 32
Aquasphere pollution, 115
Aquatic toxicity, 154, 156t
Aromatization reactions, 69
Arrhenius theory, 95
Arsenic mobility, 239
Artificial pollution, 139–140
Aspirational codes of conduct, 20, 22
Aspirin, 94
Atmospheric chemistry
acid rain, 28
chemical and physical characteristics, 27
chemical reactions, 26–27
composition of, 26
definition, 26
fossil fuels, 27–28
hydrocarbons, 29
photochemical smog, 27–28
recombination reactions, 27
semivolatile organic compounds, 28
solar radiation, 27
trace gases, 26
water-soluble volatile organic compounds, 28
Atmospheric pollutants, 153
Atom economy/utilization, 125, 311–312

B

Base-catalyzed hydrolysis, 223–224
Bimetallic chemicals, 254–255
Bioaccumulation process, 122–123
Bioaccumulative substance, 118
Bioaugmentation, 280–281
Bioconcentration factor (BCF), 118, 122–123, 309
Biodegradation, 47, 372–374, 373t
Biodilution, 190
Biodiversity, 138–139
Biological transformation, 371–375
  anaerobic biodegradation, 374–375
  biodegradation, 372–374, 373t
  bioremediation, 374–375
  biotransformation, 371–372
  photosynthesis, 371
Biopiling, 279
Bioplastics, 12–17
Bioremediation, 374–375
Biosedimentation
  aquasphere chemistry, 31–32
Bioslurping, 198
Bioslurry reactor, 279
Biosparging, 196–197
Biostimulation, 280–281
Biosurfactants, 295
Biotransformation
  advantages, 276t, 287–290
  aerobic biodegradation, 283–284
  anaerobic biodegradation, 284
  anaerobic organisms, 271
  antarctic exploration, 297
  aquasphere, 277
  bioaugmentation, 280–281, 298
  bioavailability and transport of pollutants, 286
  biopiling, 279
  bioslurry reactor, 279
  biostimulation, 280–281
  biosurfactants, 295
  biotechnological processes, 295
  Burkholderia xenovorans LB400, 283–284
  catalytic action, 270
  chemical contaminants, 272
  cold-tolerant bacteria, 297
  cometabolism, 271
  composting, 279
  contaminant and preferred conditions, 273t–274t
  control and optimization, 277
  conventional biodegradation methods, 278
  crude oil biodegradation, 286
  definition, 269–282
  delivery system, 270
  disadvantages, 276t, 287–290
  emerging technologies, 287
  enhanced methods, 279–280
  ex situ biotransformation process, 281–282
  extracellular electron transfer, 285–286
  factors, 275t, 292t
  future, 290–300
  inorganic electron donors, 271
  isolation technology, 290
  land farming, 279
  mechanisms and methods, 282–287
  microbes, 271–272
  microbe selection, 284–285
  microbial biotransformation, 271
  microbial inocula, 281
  microorganisms, 270
  mineralization, 269
  molecular methods, 285
  natural methods, 278
  nonaqueous phase liquid (NAPL), 277
  petroleum and petroleum products, 296
  petroleum hydrocarbons, 291
  phytoremediation, 293
  pollutants, 294
  polyaromatic hydrocarbons (PAHs), 282
  polychlorinated biphenyls (PCBs), 282–283
  Rhodococcus sp., 283–284
  in situ biotransformation process, 281–282
  traditional methods, 278–279
  traditional molecular analyses, 285
  xenobiotic chemicals, 294
Bioventing, 198
Boiler slag, 70–71, 72t, 75–76
Bond arrangements, 314t
  types, 69, 70t
Bond cleavage reactions, 69
Bottom ash, 70–71, 72t, 74–75
Brønsted–Lowry acids and bases, 100
Burkholderia xenovorans LB400, 283–284

C

Carbon–carbon bond
  cleavage, 68–69
  formation, 68–69
Carbon–chlorine covalent bond, 210
Carbon–halogen covalent bonds, 210

Carbone–heteroatom bond
cleavage, 68–69
formation, 68–69
Carboxylic acid derivatives, 92
Catalytic transformation, 369–371
catalysis reactions, 370
environmental catalysis, 369–371
environmental interfaces, 369–370
heterogeneous catalysts, 370
homogeneous catalysis, 370–371
industrial catalysts, 370
surface catalysis, 369–370
Cement production, 118–119
Chemical contamination, 115–116
Chemical curare, 19
Chemical destruction, 4
Chemical distribution and process, 63, 65t–66t
Chemical management, 18
Chemical pollutants, 308
Chemical properties, 81, 82t–83t
categories of, 87
definition, 82
environmental ecosystems, 87
Chemical safety guidelines, 138, 139t
Chemicals types
bioaccumulation factors (BAFs), 136–137
bioconcentration, 137
hazard assessment criteria, 136
heavy metals, 138
metal contamination, 138
persistent, bioaccumulative, and toxic (PBT) criteria, 136
polynuclear aromatic hydrocarbons, 137
Chemicals use/misuse codes, 20, 22
Chemical synthesis process, 46
Chemical transformation, 345–363
complexation, 363
hydrolysis reactions, 349–351
hydroxyl radical, 353
indirect photolysis, 352–353
inorganic reactions, 357–359
isomerization reaction, 357
organic chemicals, 361t
photolysis reactions, 351–354
radioactive decay, 354–357, 354t
rearrangement reaction, 357
rearrangement reactions, 357–359
redox reactions, 359–363
Chemical transformations
alkali-soluble materials, 25
atmosphere (air), 25. *See also* Atmospheric chemistry
clay minerals, 24–25
ecosystem resilience, 23
ecosystem resistance, 23
humus, 25
hydrosphere (water), 25. *See also* Aquasphere chemistry
individual reactants properties, 23
inorganic chemical reaction, 23
lithosphere (land), 24–25. *See also* Lithosphere chemistry
melting point and boiling point temperatures, hydrides, 24, 24t
sand constituents, 24–25
silt, 24–25
soil structure, 24–25
surface-located chemicals, 25–26
Chemical transport process, 116
Chemical types
harmful chemical, 9–10
inorganic chemicals, 10–11, 11f
organic chemicals
bioplastics, 12–17
coal/wood, 12–17
ethane *vs.* propionic acid, 11–12
hydrocarbons, 11, 12t
nitrogen compounds, 11, 14t–15t
organic functional groups, 11, 17t
oxygen compounds, 11, 13t
petroleum/natural gas feedstocks, 12–17
soil characteristics, 11, 17t
sulfur compounds, 11, 16t
properties and behavior, 10
Chemical waste, 45, 138–139
characteristics, 4–5
definition, 47–48
disposal, 67
sources and types, 5–7, 6t
Chemoselectivity, 125, 311–312
Citric acid, 249
Clausius–Clapeyron equation, 102
Clay minerals, 24–25, 68, 168, 168t
applications, 170
classification, 169–170
impurities, 175
Cleanup methods, 47
Climate change, 22
Coal ash, 70–71
contaminants, 71
Coal-fired power plant, 70–71

Coal/wood, 12–17
Coarse particles, 53
Cold-tolerant bacteria, 297
Composting, 279
Contaminant, 47–48
  definition, 4–5
Conventional biodegradation methods, 278
Covalent bonds, 316
Crude oil biodegradation, 286
Cyclone-type furnaces, 70–71
Cyclosilicates, 68

D

Descaling process, 90–91
Dielectric constant, 34
Dilution
  biodilution, 190
  capacity, 187–188
  conservative substances, 187
  flushing time, 189
  mixing zone, 188–189
  sediment type, 189–190
  solute–solute interactions, 186
  solute–solvent interactions, 186
  solvent–solvent interactions, 186
  stratification, 188
Disilicates, 68
Dispersion process, 142–144
Dissolution
  acidic carboxylic acid, 191
  agglomeration and aggregation, 192
  aquasphere chemistry, 31
  chemical bonds/electrostatic interactions, 192
  equilibrium solubility, 194
  gaseous solutes, 195
  hydrogen gas, 191
  kinetics, 192, 194
  molecular level, 191
  nonvolatile solute molecules, 195
  parameters, 194
  particle dissolution kinetics, 194
  polar bonds, 191
  precipitation, 63, 165, 192
  pressure effect, 195
  process, 142, 144–145
  retention mechanisms, 193–194
  sedimentation, 193
  solubility product, 194
  water solubility, 195
Dithionite, 253–254
Dry deposition, 53

E

Earth system, physical and biological components, 43
Ecological damage assessment, 47
Ecological risk, 121
  assessment, 47
Ecosystem-related factors, 63
Electrolysis, 97
Electron transfer, 246
Electrostatic interactions, 316
Electrostatic surface charge, 172
Emulsification
  aquasphere chemistry, 31
  process, 142, 145–146
Endocrine disruptors, 9
Enforceable codes of conduct, 20, 22
Environmental chemistry, definition, 3
Environmental degradation, 121, 160–161
Environmental factor (E-factor), 123–124, 310
Environmental impact, humans, 7, 8t
Environmental legislation, 119–120
Environmental management, 21
Environmental process, 43, 44t
Ester, 94
Evaporation, definition, 102
Evaporation process, 142, 146–147
Ex situ biotransformation process, 281–282
Extracellular electron transfer, 285–286

F

Fenton reagent, 244–248
Fine particles, 53
First law of thermodynamics, 332
Fly ash, 70–71
  calcium-bearing minerals, 73
  Class C fly ash, 74
  Class F fly ash, 74
  crystalline silica (SiO_2), 74
  electrostatic precipitators, 71–73
  hydraulic cement/plaster, 73
  recycling process, 71
  refractory phases, 73
  self-cementing properties, 74
  silicosis, 74
  trace concentrations, 71–73
Fossil fuel–derived fuels, 53
Fossil fuels, 118–119
  atmospheric chemistry, 27–28
Free energy, 334

Fulvic acid, 25
Function-oriented synthesis (FOS), 124–125, 310–311

G

Global warming, 3
Greenhouse effect, 67, 312
Groundwater
  contaminants, 154, 156t–160t
  physical properties, 154, 155t

H

Habitat destruction, 4
Halogen atoms, 115
Hazardous waste, 47–48, 141
Henry's law, 225, 323
Heterogeneous photocatalysis, 264–265
Homogeneous photocatalysis, 263–264
Household chemicals, 20, 76–77
Human-induced chemicals, 154
Human-made sources, 49
Humic acid, 25
Humin, 25
Hydrocarbons, 11, 12t
  atmospheric chemistry, 29
Hydrogen abstraction, 246
Hydrogen bond, 316–317
Hydrogen peroxide, 241
Hydrolysis
  abiotic transformation process, 216
  acid hydrolysis, 218–221
  alkaline hydrolysis, 221–222
  biological reactions, 203–204
  chemical reactions, 204
  chemical transformation, 203
  freshwater and marine systems, 215–216
  nucleophilic substitution reactions
    acid-catalyzed process, 211–212
    carbon–chlorine covalent bond, 210
    carbon–halogen covalent bonds, 210
    chemical behavior, 206
    electron pair, 207
    fundamental events, 207
    inorganic and organic chemistry, 206
    kinetic order and steric effects, 212–214
    parameters, 207, 207t
    reactive nucleophiles, 209–210, 209f
    S_E1 and S_E2 reactions, 214
    S_N1 and S_N2 reactions, 207–208
    solvation, 210
    strong nucleophile thiocyanide (SCN), 210–211
    substitution and elimination reactions, 211, 211t
    substrate effect, 215
  octanol–water partitioning coefficient, 205
  parameters, 216
  pathways, 216
  physical-chemical properties, 205
  reaction profiles
    base-catalyzed hydrolysis, 223–224
    Henry's law constant (HLC), 225
    hydrolysis half-life, 225
    hydrolytic rate constants, 223
    organophosphate pesticides, 224
    pH-rate and temperature-rate profiles, 222–223
    polynuclear aromatic hydrocarbon derivatives, 223
    predictive methods, 223
    reaction rate, 223
    sulfur compounds, 224
    1,1,1-trichloroethane (TCA), 223–224
    vapor pressure, 225
  reaction rates, 204
  spontaneous reactions, 203
  surface-located chemicals, 204–205
  types, 216
  volatility, 204–205
  water-reactive inorganic chemicals, 215–216, 217t–218t
Hydrolytic rate constants, 223
Hydronium ion, 91
  equation, 94–95
Hydroxyl radicals, 245–246

I

Igneous rocks, 126
Illite group, 169–170
Indigenous chemical, 4
Indirect photolysis, 260–261
Industrial activities, 38
Industrial chemicals, 3, 69–70
  manufacturers, 307–308
Industrial processes, 118–119
  releases types, 119–120, 120t
Industrial revolution, 7, 18, 44–45
Inorganic acids and bases, 87–89, 88t
  binary acids, 91
  descaling process, 90–91
  hydronium ion, 91
  periodic table elements, 91–92, 92f

properties, 90
solvent leveling, 92
water-soluble hydroxides, 91
Inorganic chemical pollutants, 7
Inorganic electron donors, 271
Inorganic polyatomic ions, 81–82, 85t
Inosilicates, 68
In situ biotransformation process, 281–282
In situ chemical oxidation (ISCO), 240
In situ chemical reduction (ISCR), 251
Ion exchange, 36, 173–174, 182–183
Iron minerals, 253

K

Kaolinite group, 169–170
Kyoto Protocol, 21

L

Land-based floral and faunal organisms, 18–19
Land farming, 279
Land pollution
anthropogenic chemicals, 61–62
components of, 61
efficiency factor, 62–63
heavy metals, 62
metal-bearing solids, 62
pedogenesis/lithogenesis, 62
pore-water solution, 61
soil properties, 61
xenobiotic (manmade) chemicals, 61–62
Leaching process, 142, 147–148
Lewis acids and bases, 100–101
Lithosphere chemistry
agricultural activities, 38
biosphere benefits, 35
clay mineral groups, 35, 35t
colloidal clays and organic matter, 36
external factors, 36
industrial activities, 38
inorganic chemicals, 36
inorganic matrix, 38
internal factors, 36
ion exchange, 36
oxidation-reduction reactions, 38
soil pH, 37
sorption and precipitation, 38
waste disposal, 38
Lithosphere pollution, 115

M

Marine ecosystems, 18
Material property, 86
Mechanisms, transformation
adsorption process, 343t
biological transformation, 371–375
anaerobic biodegradation, 374–375
biodegradation, 372–374, 373t
bioremediation, 374–375
biotransformation, 371–372
photosynthesis, 371
biotic and abiotic transformation, 338
catalytic transformation, 369–371
catalysis reactions, 370
environmental catalysis, 369–371
environmental interfaces, 369–370
heterogeneous catalysts, 370
homogeneous catalysis, 370–371
industrial catalysts, 370
surface catalysis, 369–370
chemical pollutants, 338–339
chemical transformation, 345–363
complexation, 363
hydrolysis reactions, 349–351
hydroxyl radical, 353
indirect photolysis, 352–353
inorganic reactions, 357–359
isomerization reaction, 357
organic chemicals, 361t
photolysis reactions, 351–354
radioactive decay, 354–357, 354t
rearrangement reaction, 357
rearrangement reactions, 357–359
redox reactions, 359–363
chemical transport, 375–383
advection, 378–379
concentration, 381–383
diffusion, 379–381
heterogeneous advective transport, 379
combustion reaction, 347
corrosive chemicals, 342t
definition, 344–371
displacement reactions, 337–338
heterogeneous reactions, 339–340
homogeneous reactions, 339–340
nonspontaneous reactions, 348
phase changes, 342f
phase transformations, 343t
physical and chemical properties, 341t
physical–chemical properties, 341
physical transformation, 363–369
absorption, 365

Mechanisms, transformation (*Continued*)
adsorption, 365
desorption, 365
dilution, 369
electrolyte, 365
organic chemicals, 367t
sorption, 365–368
pollutants, 337
single-phase approach, 339–340
spontaneous reactions, 348
thermodynamics, 348–349
Metamorphic rocks, 126
Methane, 312
Microbial biotransformation, 271
Microbial degradation
aquasphere chemistry, 32
Microbial inocula, 281
Minerals, 68
characteristics, 126
chemical composition and structure, 126
chemical compounds, 126
classification, 126–127
earth crust elements, 126–127, 127t
elements and ores, 126–127, 128t–135t
igneous rocks, 126
metamorphic rocks, 126
names and chemical composition, 126–127, 127t
nonsilicate minerals, 126–127
phosphate minerals and salts, 126–127
sedimentary rocks, 126
silica-based minerals, 126–127
Mining operations, 118–119
Molecular interactions, 313–318
actinide elements, 317f
bond arrangements, 314t
covalent bonds, 316
electrostatic interactions, 316
hydrogen bond, 316–317
intermolecular interactions, 313–315
lanthanide elements, 317f
melting point and boiling point temperatures, 317t
toxic interaction, 318
Molecular partitioning, 323–324

N

Natural chemicals, 20, 154
Natural-occurring inorganic chemicals, 68
Natural organic matter (NOM), 67
Natural pollution, 139–140
Nitric oxide (NO), 262
Nitrogen compounds, 11, 14t–15t
Nonaqueous phase liquid (NAPL), 277
Nonindigenous chemical, 4
Nonpoint source (NPS) pollution, 48, 123, 309
Nonsilicate minerals, 126–127
Nonspontaneous reactions, 84–86
Nucleophilic substitution reactions
acid-catalyzed process, 211–212
carbon–chlorine covalent bond, 210
carbon–halogen covalent bonds, 210
chemical behavior, 206
electron pair, 207
fundamental events, 207
inorganic and organic chemistry, 206
kinetic order and steric effects, 212–214
parameters, 207, 207t
reactive nucleophiles, 209–210, 209f
S_E1 and S_E2 reactions, 214
S_N1 and S_N2 reactions, 207–208
solvation, 210
strong nucleophile thiocyanide (SCN), 210–211
substitution and elimination reactions, 211, 211t
substrate effect, 215

O

Octanol–water partition coefficient, 123, 205, 324–325
Organic acids and bases
acyl chlorides and anhydrides, 93
aspirin, 94
carboxylic acid derivatives, 92
ester, 94
ibuprofen, 94
vinegar, 93
Organic chemicals
pollutants, 7
reaction, 66
synthesis, 84–86
Organic functional groups, 11, 17t
Organic–inorganic interactions, 68–69
Organometallic complexes, 69
Organophosphate pesticides, 224
Orthosilicate minerals, 68
Oxidation
aquasphere chemistry, 31
reactions, 239–251
benzene, toluene, ethylbenzene, and the xylene isomers (BTEX), 240–241
chemical oxidation, 240

citric acid, 249
electron transfer, 246
Fenton reagent, 244–248
ferrous sulfate solution, 244
hydrogen abstraction, 246
hydrogen peroxide, 241, 245
hydroxyl radicals, 245–246
ozone, 250–251
perhydroxyl radical, 245
permanganate, 241–244
persulfate, 248–250
phenol-containing waste water, 246–247
pH reduction, 245–246
potassium permanganate, 241–242
radical interaction, 246
in situ chemical oxidation (ISCO), 240
sodium permanganate, 242
sulfate radicals, 249
trivalent chromium, 243–244
Oxygen compounds, 11, 13t
Ozone, 19–20, 250–251

P

Particle–water partitioning, 325
Partition coefficients, 328–329
Partitioning coefficients, 318–329
mixtures of chemicals, 326–328
single chemicals, 319–325
acid–base partitioning, 321
air–water partitioning, 321–323
Henry's law, 323
molecular partitioning, 323–324
octanol–water partitioning, 324–325
particle–water partitioning, 325
in situ partition coefficients, 319
Perhydroxyl radical, 245
Permanganate, 241–244
Persistent, bioaccumulative, and toxic (PBT) criteria, 136
Persistent pollutants (PPs), 45–46, 121, 141
characteristics, 122
human health impacts, 122
Persulfate, 248–250
Pesticides, 153–154
Petrochemicals, 12–17
Petroleum and products, 101
Petroleum/natural gas feedstocks, 12–17
Phenol-containing waste water, 246–247
Photocatalysis, 256
environment, 263–265
heterogeneous photocatalysis, 264–265
homogeneous photocatalysis, 263–264
Photochemical smog
atmospheric chemistry, 27–28
Photochemistry, 256–260
environment, 260–263
hydroxyl radical, 261
indirect photolysis, 260–261
nitric oxide (NO), 262
sensitized photolysis, 260–261
tropospheric ozone, 261
pH/oxidation–reduction (redox) gradients, 63
Phyllosilicates, 68
Physical chemistry, 39–40, 39f
Physical destruction, 4
Physical properties, 81, 82t–83t
definition, 86
environmental ecosystems, 87
extensive properties, 86
gases, 81, 84t
intensive properties, 86
liquids, 81, 84t
solids, 81, 84t
Physical transformation, 363–369
absorption, 365
adsorption, 365
desorption, 365
dilution, 369
electrolyte, 365
organic chemicals, 367t
sorption, 365–368
Phytoremediation, 293
Point sources, 48
Pollutants, 116–117, 117t, 139–140, 140t
types, 3, 4t
Pollution process, 48
Polyaromatic hydrocarbons (PAHs), 282
Polyatomic ions, 95
Polychlorinated biphenyls (PCBs), 282–283
Polyfunctional organic molecules, 67
Polynuclear aromatic hydrocarbons, 137, 223
Polysulfides, 253
Potassium permanganate, 241
Pulverized fuel ash. *See* Fly ash

R

Reactor system, 46
Receptor, definition, 47–48
Redox reactions
abiotic oxidation, 239
advantages, 237t
anaerobic sediments, 234

Redox reactions (*Continued*)
anodic currents, 234–235
arsenic mobility, 239
arsenic toxicity, 239
bacteria, 239
biogeochemical redox processes, 238
biomass production, 231–232
cathodic currents, 234–235
chemical species, 232
disadvantages, 237t
electron transfer, 232–233
exchange of electrons, 231
ferric (hydr)oxides, 236–237
ferrous iron, 237
fertilizer-impacted aquifers, 236
field-scale effects, 235
hexavalent uranium, 239
humic substances, 238
hydrogen peroxide, 233–234
inorganic sorbates, 237–238
iron, 233–234, 236
macronutrients, 231–232
manganese oxide minerals, 237
methane, 232
microbial reduction, 239
microbial respiration, 234
micronutrients, 231–232
microorganisms, 239
mineral pyrite, 236
natural organic matter (NOM), 237–238
near-neutral pH conditions, 237
nitrogen, 236
oxic–anoxic interfaces, 237
oxidation reactions, 239–251
benzene, toluene, ethylbenzene, and the xylene isomers (BTEX), 240–241
chemical oxidation, 240
citric acid, 249
electron transfer, 246
Fenton reagent, 244–248
Fenton's reagent, 241
ferrous sulfate solution, 244
hydrogen abstraction, 246
hydrogen peroxide, 241, 245
hydroxyl radicals, 245–246
ozone, 250–251
perhydroxyl radical, 245
permanganate, 241–244
persulfate, 248–250
phenol-containing waste water, 246–247
pH reduction, 245–246
potassium permanganate, 241–242
radical interaction, 246
in situ chemical oxidation (ISCO), 240
sodium permanganate, 242
sulfate radicals, 249
trivalent chromium, 243–244
ozone, 233–234
permanganate, 233–234
persulfate, 233–234
phosphate-rich groundwater, 236
photocatalysis, 256
environment, 263–265
heterogeneous photocatalysis, 264–265
homogeneous photocatalysis, 263–264
photochemical process, 255
photochemistry, 256–260
environment, 260–263
hydroxyl radical, 261
indirect photolysis, 260–261
nitric oxide (NO), 262
sensitized photolysis, 260–261
tropospheric ozone, 261
redox conditions, 234–235
redox-sensitive trace elements, 235
reduction reactions, 251–255
bimetallic chemicals, 254–255
dithionite, 253–254
iron minerals, 253
polysulfides, 253
in situ chemical reduction (ISCR), 251
surface reactions, 251–252
zero valent metals, 252–253
semiquinone and hydroquinone species, 239
sulfur dioxide, 233
waterlogged soils, 237–238
water-saturated environments, 232
Reduction reactions, 251–255
bimetallic chemicals, 254–255
dithionite, 253–254
iron minerals, 253
polysulfides, 253
in situ chemical reduction (ISCR), 251
surface reactions, 251–252
zero valent metals, 252–253
Regioselectivity, 125
Regression equation, 123
Remediation process, 116–117
Rhodococcus sp., 283–284

S

Safety protocols, 47
Sand constituents, 24–25
S_E1 and S_E2 reactions, 214
Second law of thermodynamics, 332–334
Sedimentary rocks, 126
Sedimentation process, 142, 148–150
Self-purification
  aquasphere chemistry, 33
Sensitized photolysis, 260–261
Separation techniques, 46
Silica-based minerals, 126–127
Silt, 24–25
Site cleanup operations, 47
Slag-type furnaces, 70–71, 72t
Smectite group, 169–170
Smog, 49
S_N1 and S_N2 reactions, 207–208
Sodium permanganate, 242
Soil
  characteristics, 151–152, 152t
  contamination, 64
  pH, 37
  pollution, 139–140
  structure, 24–25
  vapor extraction, 197
Solubility. *See* Dissolution
Solvent leveling, 92
Solvent system theory, 98–99
Sorption, 169
  absorption, 170, 178–179
  adhesion, 169
  adsorption, 170. *See also* Adsorption
  chemical degradation, 170
  chemical transformation reaction/
    stabilization, 169
  chromatographic techniques, 171
  definition, 171
  factors, 170
  irreversible reaction, 169
  irreversible sorption, 170
  natural attenuation technique, 171
Spreading process, 142, 150
Stoichiometric equation, 311
Stratospheric ozone, 3
Strong nucleophile thiocyanide (SCN),
    210–211
Sublimation, 150–151, 151f, 151t
Sulfate radicals, 249
Sulfur compounds, 11, 16t
Surface reactions, 251–252
Synthetic chemicals, 20

T

Tectosilicates, 68
Terrestrial ecosystems, 18
Thermal energy, 66
Thermodynamics, 330–334
  first law of thermodynamics, 332
  free energy, 334
  second law of thermodynamics, 332–334
Toxic interaction, 318
Toxic waste, 141
Traditional molecular analyses, 285
1,1,1-Trichloroethane (TCA), 223–224
Trivalent chromium, 243–244
Tropospheric ozone, 261, 313

U

United States Environmental Protection
    Agency, 49

V

Vapor pressure, 102–103, 196–198, 225
Vermiculite group, 169–170
Vinegar, 93
Volatility
  aromatic structures, 104
  environmental factors, 103
  high-boiling and nonvolatile chemicals,
    106–107
  inorganic compounds solubility, 103, 103t
  leachability measures, 104
  low-boiling chemicals, 104–106
  toxicological properties, 104

W

Waste disposal, 38
Water pollution, 139–140
  aggregation, 60
  biosedimentation, 60
  chemical/biochemical process, 55–56
  crude oil transformation, 59
  dissolution, 59
  emulsification, 59
  floral and faunal communities, 55
  in groundwater, 56
  heavy metals, 57
  industrial sources, 58
  in marine environment, 58
  methyl iodide chemistry, 61
  microbial degradation, 60
  nonporous aquifers, 57

Water pollution (*Continued*)
operative processes, 59
organic compounds, 56
oxidation, 59–60
photodegradation reactions, 61
physical factors, 58
physical recycling process, 56
physical transport, 59
self-purification, 60–61
by silt/sediments, 58
surface-derived water masses, 61
total organic-carbon approach, 56
transportation process, 57
Water-reactive inorganic chemicals, 215–216, 217t–218t
Water resources, 138–139
Water solubility
air–water partitioning coefficient, 107
applications, 108
aquatic ecotoxicological test, 108
concentration, 107
enthalpy and entropy, 109
of inorganic chemicals, 111–112
intermolecular forces, energy balance, 108–109
of organic chemicals, 112–113
precipitation reactions, 109–110
pressure effect, 110
quantification of, 111
rate of dissolution, 110–111
solubilization, definition, 109
supersaturated solution, 109
temperature effect, 110
Wet deposition, 53
Wind, 142

X

Xenobiotic chemicals, 294

Z

Zero valent metals, 252–253